Plant, Soil and Agricultural Sciences: Challenges and Concerns

Plant, Soil and Agricultural Sciences: Challenges and Concerns

Edited by **Wendel Mason**

SYRAWOOD
PUBLISHING HOUSE
New York

Published by Syrawood Publishing House,
750 Third Avenue, 9th Floor,
New York, NY 10017, USA
www.syrawoodpublishinghouse.com

Plant, Soil and Agricultural Sciences: Challenges and Concerns
Edited by Wendel Mason

International Standard Book Number: 978-1-68286-220-9 (Hardback)

Printed in China.

Contents

Preface

Plant-soil interactions are extremely crucial for the development of agricultural systems. This book aims to shed light on some of the challenges and concerns pertaining to agricultural sciences through the recent researches in this field. It comprises experimental and conceptual case studies revolving around environmental microbiology, biodiversity, ecosystem function, plant-soil interactions, etc. This book is appropriate for all those seeking detailed information in this area. Coherent flow of topics and extensive use of examples make this book an invaluable source of knowledge for students and researchers pursuing agricultural sciences and allied disciplines.

Various studies have approached the subject by analyzing it with a single perspective, but the present book provides diverse methodologies and techniques to address this field. This book contains theories and applications needed for understanding the subject from different perspectives. The aim is to keep the readers informed about the progress in the field; therefore, the contributions were carefully examined to compile novel researches by specialists from across the globe.

Indeed, the job of the editor is the most crucial and challenging in compiling all chapters into a single book. In the end, I would extend my sincere thanks to the chapter authors for their profound work. I am also thankful for the support provided by my family and colleagues during the compilation of this book.

Editor

1

Nitrogen control of ^{13}C enrichment in heterotrophic organs relative to leaves in a landscape-building desert plant species

J. Zhang[1,2], L. Gu[3], F. Bao[1], Y. Cao[1], Y. Hao[4], J. He[1], J. Li[1], Y. Li[1], Y. Ren[1], F. Wang[1], R. Wu[1], B. Yao[1], Y. Zhao[4], G. Lin[5], B. Wu[1], Q. Lu[1], and P. Meng[6]

[1]Institute of Desertification Studies, Chinese Academy of Forestry, Beijing, China
[2]Research Institute of Forestry, Chinese Academy of Forestry, Beijing, China
[3]Environmental Sciences Division and Climate Change Science Institute, Oak Ridge National Laboratory, Oak Ridge, TN 37831, USA
[4]The Experimental Center of Desert Forestry of the Chinese Academy of Forestry, Dengkou, Inner Mongolia Autonomous Region, China
[5]Center for Earth System Science, Tsinghua University, Beijing, China
[6]Headquarters, Chinese Academy of Forestry, Beijing, China

Correspondence to: L. Gu (lianhong-gu@ornl.gov)

Abstract. A longstanding puzzle in isotope studies of C_3 plant species is that heterotrophic plant organs (e.g., stems, roots, seeds, and fruits) tend to be enriched in ^{13}C compared to the autotrophic organ (leaves) that provides them with photosynthate. Our inability to explain this puzzle suggests key deficiencies in understanding post-photosynthetic metabolic processes. It also limits the effectiveness of applications of stable carbon isotope analyses in a variety of scientific disciplines ranging from plant physiology to global carbon cycle studies. To gain insight into this puzzle, we excavated whole plant architectures of *Nitraria tangutorum* Bobrov, a C_3 species that has an exceptional capability of fixing sands and building sand dunes, in two deserts in northwestern China. We systematically and simultaneously measured carbon isotope ratios and nitrogen and phosphorous contents of different parts of the excavated plants. We also determined the seasonal variations in leaf carbon isotope ratios on nearby intact plants of *N. tangutorum*. We found, for the first time, that higher nitrogen contents in heterotrophic organs were significantly correlated with increased heterotrophic ^{13}C enrichment compared to leaves. However, phosphorous contents had no effect on the enrichment. In addition, new leaves had carbon isotope ratios similar to roots but were progressively depleted in ^{13}C as they matured. We concluded that a nitrogen-mediated process, hypothesized to be the refixation of respiratory CO_2 by phosphoenolpyruvate (PEP) carboxylase, was responsible for the differences in ^{13}C enrichment among different heterotrophic organs, while processes such as fractionating foliar metabolism and preferentially loading into phloem of ^{13}C-enriched sugars may contribute to the overall autotrophic–heterotrophic difference in carbon isotope compositions.

1 Introduction

The natural abundance analysis of stable carbon isotopes in plants has become an essential tool for studying plant-environmental interactions, plant metabolism, carbon allocation, and biosphere–atmosphere exchanges of carbon fluxes (Dawson et al., 2002; Bowling et al., 2008; Tcherkez et al., 2011; Cernusak et al., 2013). Understanding processes and factors controlling carbon isotope compositions in different plant organs, which are not homogenous (Leavitt and Long, 1986), is crucial to the successful applications of this tool (Hobbie and Werner, 2004). The primary determinant of plant-carbon isotope compositions is the photosynthetic discrimination against the heavier carbon isotope ^{13}C. This primary discrimination process has been relatively well understood, and detailed theoretical models relating the discrimination to environmental forcing conditions, leaf physiology, and biochemistry have been developed (Farquhar et al., 1982; Farquhar and Cernusak, 2012; Gu and Sun, 2014). However, other processes must also influence plant-carbon isotope compositions as heterotrophic plant organs (e.g., stems, roots, seeds and fruits) in C_3 plant species have been found to be generally enriched in ^{13}C as compared to the autotrophic organ (leaves) that supplies them with carbohydrates (Craig, 1953; Leavitt and Long, 1982; Ehleringer et al., 1987; Hobbie and Werner, 2004; Badeck et al., 2005; Cernusak et al., 2009). In contrast to the relatively well-understood photosynthetic carbon isotope discrimination, processes controlling the observed heterotrophic ^{13}C enrichment in C_3 plant species remain unclear even though the phenomenon was first reported 60 years ago (Craig, 1953).

Cernusak et al. (2009) and Ghashghaie and Badeck (2014) summarized more than half a dozen of nonexclusive processes that may explain the heterotrophic ^{13}C enrichment in C_3 plant species. These processes generally belong to two broad groups. Group I processes involve the occurrence of contrasting biochemical and metabolic fractionations between autotrophic and heterotrophic organs, for example, ^{13}C-enriched autotrophic vs. ^{13}C-depleted heterotrophic mitochondrial respirations, low autotrophic vs. high heterotrophic CO_2 fixation by phosphoenolpyruvate (PEP) carboxylase, and low autotrophic vs. high heterotrophic loss rates of ^{13}C-depleted volatile organic compounds, surface waxes and other products from secondary plant metabolism. Group II processes involve the utilization of contrasting organ-building photoassimilates, which in turn may be a result of a number of processes, including preferential export of ^{13}C-enriched nighttime sucrose to heterotrophic organs, reduced photosynthetic discrimination against ^{13}C due to developmental shifts in exporting mature leaves, and asynchronous growth of autotrophic vs. heterotrophic organs in contrasting environmental conditions. Although the term post-photosynthetic discrimination or post-carboxylation discrimination has been often used to refer the processes included in both groups, some of the processes in Group II cannot be strictly considered as occurring post photosynthesis or carboxylation. Nearly all processes outlined above have supporting as well as opposing evidences from observational and experimental studies (Cernusak et al., 2009). Thus it remains a challenge to identify cause(s) for the 60-year-old puzzle of heterotrophic ^{13}C enrichment.

It is important to overcome this challenge as many fundamental issues in a variety of scientific disciplines ranging from plant physiology to global carbon cycle studies depend on a precise knowledge of plant-carbon isotope compositions. Towards this goal, we have identified two areas that require strengthening in the studies of heterotrophic ^{13}C enrichment. First, there is a need for systemic whole-plant studies. Although heterotrophic ^{13}C enrichment in C_3 plant species has been reported widely, most previous studies have been done by comparing heterotrophic organs independently and on a piecemeal basis with leaves. This lack of systemic, whole-plant studies is not conducive to understanding the mechanism of heterotrophic ^{13}C enrichment, because to achieve this understanding, one must first have a comprehensive picture of the enrichment (or depletion) across all organs of the same plant.

Second, whether and how nutrients affect heterotrophic ^{13}C enrichment needs to be investigated. Nutrients, particularly nitrogen (N) and phosphorous (P), control leaf photosynthetic capacity (Field and Mooney, 1986; Domingues et al., 2010), which in turn affects the drawdown of CO_2 along stomatal and mesophyll diffusional pathways. It has been shown that leaf N content is positively correlated with leaf $\delta^{13}C$ and negatively with carbon isotope discrimination (Sparks and Ehleringer, 1997; Livingston et al., 1999; Duursma and Marshall, 2006; Cernusak et al., 2007). This relationship is consistent with the expectation that higher leaf photosynthetic capacity associated with higher leaf N leads to a sharper drawdown of CO_2 along the diffusional pathways (Cernusak et al., 2007, 2013), resulting in an expected pattern according to the photosynthetic isotope discrimination equations (Farquhar et al., 1982; Farquhar and Cernusak, 2012; Gu and Sun, 2014). To our knowledge, hitherto there has been no effort to systematically investigate how plant nutrients might affect heterotrophic ^{13}C enrichment compared to leaves. A lack of such an effort is not justifiable because plant nutrients play important roles in many of the processes discussed in Cernusak et al. (2009) and Ghashghaie and Badeck (2014). Thus it would not be surprising if certain relationships exist between plant nutrients and heterotrophic ^{13}C enrichment. An identification of such relationships will greatly assist the illumination of the underlining cause(s) of heterotrophic ^{13}C enrichment.

Therefore, the objective of the present study was to gain insight into the longstanding puzzle of heterotrophic ^{13}C enrichment by jointly addressing the two deficiencies identified above. We conducted systematic and simultaneous analyses of carbon isotope ratios and N and P contents with excavated whole architectures of *Nitraria tangutorum* Bobrov, a C_3 shrub species endemic to northwestern deserts in China. These analyses were complemented with investigations of seasonal variations in leaf carbon isotope ratios on intact plants of the same species, thus enabling the analyses of carbon isotope compositions of different heterotrophic organs in a dynamic reference framework. *N. tangutorum* is interesting because it has an exceptional capability of controlling landscape evolution by fixing sands and building sand dunes known as nebkha or coppice dunes around its extensive shoot and root systems (Baas and Nield, 2007; Lang et al., 2013; Li et al., 2013). This characteristic makes it relatively easy to excavate the whole plant including roots for isotope and nutrient analyses, although to our knowledge, this species has never been investigated for heterotrophic ^{13}C enrichment.

We will report, for the first time, that variations in ^{13}C enrichment in different heterotrophic organs strongly depend on their N contents, indicating a role of a within-organ N-mediated process in heterotrophic ^{13}C enrichment. We will also show that the observed N–heterotrophic ^{13}C enrichment relationship is most parsimoniously explained through the respiratory CO_2 refixation by PEP carboxylase. Future studies on heterotrophic ^{13}C enrichment should investigate isotopic effects of N content and CO_2 refixation in different plant organs. Direct measurements of PEP carboxylase activity will be essential.

2 Materials and methods

2.1 Biological and environmental characteristics of *Nitraria tangutorum* Bobrov

Nitraria tangutorum Bobrov (Fig. 1) is a spiny shrub species in the *Nitraria* genus of the Zygophyllaceae family. Species in the *Nitraria* genus are generally xerophytes, widely distributed in the Middle East, Central Asia, and in the northwestern regions of China. *N. tangutorum*, however, is endemic to the northwestern regions of China, including northeastern Tibet, Gansu, Qinghai, Xinjiang, western Inner Mongolia, western Ningxia, and northern Shaanxi. It is a pioneer species and has high tolerance for drought, heat, and salts. *N. tangutorum* plays an important ecological role in combating desertification due to its exceptional capabilities in forming phytogenic nebkha dunes which prevent or slow down the movement of sands. According to Li and Jiang (2011) and Li et al. (2013), the process of forming a nebkha typically starts when occasional ample moisture allows a seed to germinate inside clay cracks in dried-up flat beds of previous rivers or lakes. As the resulting ortet grows, it intercepts aeolian sands and the plant enters into a clonal reproductive stage. When branches are buried by sands, layering occurs and adventitious roots are formed. Under appropriate sand-burial depth and sufficient moisture, ramets are developed from axillary buds in the layering and a clonal colony is formed. If aeolian sand supply is not interrupted, repetitive layering and ramet development will enlarge the colony and further increase its capacity to intercept aeolian sands; thereby, a phytogenic nebkha dune is formed (Fig. 1c).

The height of a *N. tangutorum* nebkha ranges from 1 to 3 m and some can reach 5 m. The base of a nebkha often has the shape of an ellipse with the major axis parallel to the local prevailing wind direction. The formation of nebkhas alters local micro-environments and provides habitats for other desert species. Li and Jiang (2011) described in detail the biological and environmental characteristics of species in the *Nitraria* genus with a focus on *N. tangutorum*.

Figure 1. Flowers (top, 10 June 2009, Minqin), fruits (middle, 18 July 2009, Minqin) and nebkha (bottom, 3 August 2010, Dengkou) of *Nitraria tangutorum* Bobrov. Pictures courtesy of Jianmin Chu, Research Institute of Forestry, Chinese Academy of Forestry.

2.2 Study sites

The field work was carried out at two desert locations. The first study site was within an experimental area (40°24′ N, 106°43′ E) managed by the Experimental Center of Desert Forestry of the Chinese Academy of Forestry. This site is located in Dengkou County, Inner Mongolia Autonomous Region, China. Dengkou County is at the junction between the Hetao Plain and Ulan Buh Desert of the Mongolian Plateau in the middle reaches of the Yellow River. The mean annual temperature is 8.84° and the mean annual precipitation is

147 mm, with 77.5 % of annual rainfall occurring from June to September (1983–2012 averages). The mean annual potential evaporation is 2381 mm (Li et al., 2013). The soils in the study region in general are sandy soil and gray-brown desert soil (Cambic Arenosols and Luvic Gypsisols in FAO (Food and Agriculture Organization of the United Nations) taxonomy). The *N. tangutorum* nebkhas at the study site are formed on clay soils deposited by the Yellow River. Although the plant community is dominated by *N. tangutorum*, xerophytic species such as semi-shrub *Artemisia ordosica*, perennial grass *Psammochloa villosa*, and annual species *Agriophyllum squarrosum* and *Corispermum mongolicum* can also be found.

The second study site was the Gansu Minqin Desert Ecosystem Research Station (38°34′ N, 102°58′ E), Minqin County, Gansu Province, China. Minqin County is located in the lower reaches of Shiyang River, surrounded by the Badain Jaran Desert in the west and north and the Tengger Desert in the east. The mean annual temperature is 8.87° and the mean annual precipitation is 117 mm, with 73.1 % of annual rainfall occurring from June to September (1983–2012 averages). The mean annual potential evaporation is 2643 mm (Du et al., 2010). Thus the second study site is somewhat drier than the first site, but with similar annual mean temperatures. The soil at the Minqin site is similar to that at the Dengkou site with sandy soil in the nebkhas and gray-brown desert soil between nebkhas. The native vegetation in the study area is usually dominated by shrubs and semi-shrubs with species such as *N. tangutorum* and *Calligonum mongolicum*. Experimental plots used in this study contained semi-fixed nebkha dunes developed by the growth of *N. tangutorum*. Typically in dry years, *N. tangutorum* is the only species growing in the nebkhas although in wet years, annual species such as *Agriophyllum squarrosum* and *Corispermum mongolicum* can also be found. Because the Minqin site is drier than the Dengkou site, the nebkhas at the Minqin site are generally smaller and less populated with plants than at the Dengkou site. The rooting depth is deeper at the Minqin site than at the Dengkou site (Table 1).

2.3 Excavation of *Nitraria tangutorum* nebkhas

In August 2012, we excavated three nebkhas at each study site. The geometrical and biometrical characteristics of the six nebkhas were summarized in Table 1. At the Dengkou site, the three nebkhas were excavated in a sampling area of 40 m × 40 m. At the Minqin site, nebkhas were generally much smaller. To ensure availability for analyses of sufficient biomass materials at this site, particularly the fine roots (see below), three sampling areas each with a dimension of 30 m × 30 m were established and three nebkhas from each sampling area were tentatively excavated. Two nebkhas from one sampling area and one from another were determined to have sufficient amount of fine roots for analyses and were therefore excavated fully.

We excavated the nebkhas by carefully teasing away the sands from the mounds to expose the root architecture of *N. tangutorum* with particular attention paid to the preservation of fine roots. The roots of a *N. tangutorum* can be found inside the sand mounds as well as inside the clay layer that generally forms a plain on which the sand mounds rest. We therefore also excavated any roots inside the clay layer to a depth until no more roots could be found.

We separated the whole plant biomass into leaves, stems, in-sand roots and below-plain roots. The in-sand roots, which were roots found inside the nebkha sands but above the plain formed by the underlying clay layer, were further separated into in-sand fine roots (diameter $\leq$ 2 mm) and in-sand coarse roots (diameter > 2 mm). The same root diameter threshold was used to separate the below-plain roots, which were found inside the clay layer under the nebkha sands. Furthermore, the below-plain fine and coarse roots were grouped in a 20 cm depth increment from the plain surface. We did not separate the in-sand fine and coarse roots into layers, because a nebkha has a cone shape on top, making a layer hard to define. Also we did not use a simple "below-ground" group, because "ground" is not well defined in a nebkha-populated landscape, and because there are large physical and chemical differences between sands and clay, which may affect the isotope compositions of roots growing in them. Litter was rarely found on the nebkhas, presumably because strong winds at the study sites can easily blow away any litter produced. However, woody debris from dead ramets was present inside the sand mounds and was collected during excavation. Thus for each nebkha, we differentiated the following categories of *N. tangutorum* biomass: the autotrophic organ of leaves, the heterotrophic organs of stems, in-sand fine roots (ISFR), in-sand coarse roots (ISCR), below-plain fine roots (BPFR) in 20 cm depth increments, and below-plain coarse roots (BPCR) in 20 cm increments, and the heterotrophic woody debris (WD). Nutrient contents and carbon isotope compositions were measured separately for each category.

2.4 Measurements of nutrient contents and carbon isotope compositions with excavated biomass

All categories of *N. tangutorum* biomass (leaves, stems, ISFR, ISCR, BPFR in 20 cm increments, BPCR in 20 cm increments, and WD) from each excavated nebkha were dried to constant weight (60°, 48 h). The dry weight of biomass was determined with 0.01 g accuracy on an analytical scale. The biomass carbon stocks were expressed relative to the base area of the nebkha which was assumed to be an ellipse. The fraction of each component was also calculated.

Dried materials were randomly selected from each biomass category and ground to 80 mesh. The resultant powder was separated into six duplicates. Three duplicates were analyzed for carbon (C), nitrogen (N) and phosphorous (P) contents and the remaining three for isotope compositions. The C, N, and P contents were measured in the

Table 1. Main geometrical and biometrical characteristics of the nebkhas excavated in this study.

Nebkha		Dengkou-1	Dengkou-2	Dengkou-3	Minqin-1	Minqin-2	Minqin-3
Major axis (m)		13.6	9.9	3.65	4	4.6	6.4
Minor axis (m)		8.38	5.9	3.24	3.5	2.9	4.6
Height (m)		2.02	1.38	0.57	0.35	0.44	0.8
Plant cover (%)		80	70	80	11	15	7
Below-plain rooting depth (cm)		< 60	< 40	< 40	< 80	< 80	< 80
Leaf biomass	($g\,C\,m^{-2}$ and %)	62.9 (10)	93.7 (12)	85.1 (11)	12.7 (6)	23.0 (11)	11.0 (9)
Stem biomass	($g\,C\,m^{-2}$ and %)	159.7 (25)	169.3 (22)	213.3 (28)	35.2 (16)	70.0 (34)	22.2 (19)
In-sand root biomass	($g\,C\,m^{-2}$ and %)	289.9 (45)	370.6 (47)	214.7 (28)	92.0 (41)	34.9 (17)	51.9 (44)
Blow-plain root biomass	($g\,C\,m^{-2}$ and %)	137.7 (21)	148.7 (19)	260.8 (34)	84.5 (38)	80.5 (39)	32.5 (28)
Total biomass	($g\,C\,m^{-2}$ and %)	650.2 (100)	782.3 (100)	773.9 (100)	224.4 (100)	208.3 (100)	117.6 (100)

EnvironmentalChemistry Analysis Laboratory in the Institute of Geographic Sciences and Natural Resources Research, the Chinese Academy of Sciences, Beijing, China. Total sample carbon and N were measured with the vario MACRO cube (Elementar Company, Germany). The analytical precision was better than 0.5 % relative standard deviation (RSD). Total P was measured with the ICP-OES OPTIMA 5300DV (PE, USA). The analytical precision was better than 2 % RSD.

The carbon isotope compositions were analyzed at the Stable Isotope Ratio Mass Spectrometer Laboratory of the Chinese Academy of Forestry (SIRMSL, CAF), Beijing, China. The instrument used was a Delta V Advantage Mass Spectrometer (Thermo Fisher Scientific, Inc., USA) coupled with an elemental analyzer (FlashEA 1112; HT Instruments, Inc., USA) in the continuous flow mode. Isotope compositions were expressed using the delta notation (δ) in parts per thousand (‰): $\delta^{13}C$ (‰) $= [(R_{sample}) / (R_{standard}) - 1] \times 1000$, where R is the ratio of ^{13}C to ^{12}C. The measurement applied the IAEA-600 standard (caffeine) relative to V-PDB (Vienna PeeDee formation belemnite limestone). The analytical precision was better than 0.1‰, based on replicate measurements of the reference standard.

2.5 Measurements of seasonal variations in leaf $\delta^{13}C$ and C_i / C_a ratio

Photosynthetic carbon isotope discrimination depends on environmental conditions (Farquhar et al., 1982; Farquhar and Cernusak, 2012; Gu and Sun, 2014); consequently, leaf carbon isotope ratio $\delta^{13}C$ may change seasonally, potentially making the autotrophic–heterotrophic differences in carbon isotope compositions time dependent. Thus in addition to the isotopic and nutrient analyses for samples from the excavated plant materials, we also measured seasonal variations in leaf carbon isotope compositions and ratios of leaf intercellular airspace (C_i) to ambient (C_a) CO_2 concentrations on nearby un-excavated nebkhas at both the Dengkou and Minqin study sites. Four samples of leaves were taken in each month from May to September of 2012 at both sites and analyzed for carbon isotope ratios at the SIRMSL of CAF. The seasonal variations in C_i / C_a ratios were measured with a Li-6400 portable photosynthetic system (Li-Cor Environmental Sciences, Lincoln, NE, USA) each month from June to September of 2012 at the Dengkou site with 24–28 samples per month and from July to September of 2011 at the Minqin site with 16 samples per month. The chamber environment (temperature, light, and relative humidity) was kept close to ambient conditions at the time of measurement. Seasonal variations in leaf nutrient contents were not measured. The measurements of seasonal variations in leaf $\delta^{13}C$ provide a dynamic reference framework for examining the $\delta^{13}C$ values of heterotrophic organs, while the independent measurements of seasonal variations in C_i / C_a ratios allow us to determine whether the seasonal patterns in leaf $\delta^{13}C$ are consistent with our current understanding of the photosynthetic carbon isotope discrimination (Farquhar et al., 1982).

2.6 Quantification of heterotrophic ^{13}C enrichment and statistical analyses

We quantified the difference in carbon isotope composition between the leaves (autotrophic) and a heterotrophic organ with the following expression:

$$\Delta^{13}C_{organ} = \left(\frac{R_{leaf}}{R_{organ}} - 1\right) \times 1000 \quad (1)$$
$$= \left(\frac{\delta^{13}C_{leaf}/1000 + 1}{\delta^{13}C_{organ}/1000 + 1} - 1\right) \times 1000$$
$$= \frac{\delta^{13}C_{leaf} - \delta^{13}C_{organ}}{1 + \delta^{13}C_{organ}/1000}.$$

Thus a value of $\Delta^{13}C_{organ} < 0$ indicates an enrichment of ^{13}C in a heterotrophic organ relative to the leaves while $\Delta^{13}C_{organ} > 0$ indicates heterotrophic depletion. The values of $\delta^{13}C_{Leaf}$ used to calculate $\Delta^{13}C_{organ}$ came from leaves harvested from *N. tangutorum* of the excavated nebkhas, not from those for seasonal patterns. The use of Δ in Eq. (1) makes the relationship between autotrophic and heterotrophic organs analogous to that between reactants and

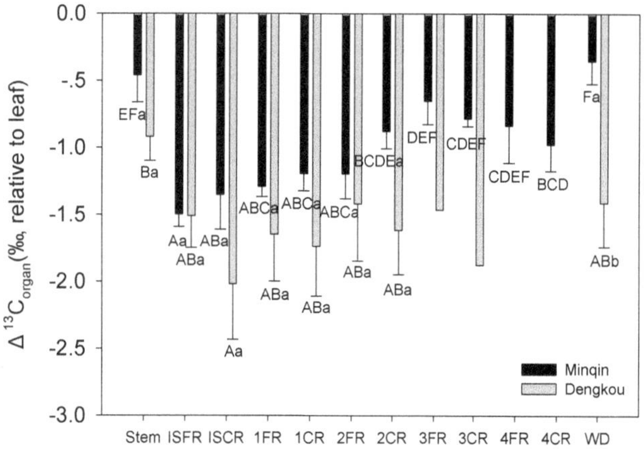

Figure 2. The difference in carbon isotope compositions between leaves and heterotrophic organs of *Nitraria tangutorum* Bobrov, which is measured by $\Delta^{13}C_{organ}$ in Eq. (1) and averaged across the nebkhas excavated at the same study site (Dengkou or Minqin). Negative values indicate ^{13}C enrichment in heterotrophic organs compared to leaves. Upper-case letters denote analysis-of-variance (ANOVA) results within a study site (i.e., comparing $\Delta^{13}C_{organ}$ among different organs at the same site) and lower case letters between the two sites (i.e., comparing $\Delta^{13}C_{organ}$ of the same organ between the two sites). IS stands for in-sand, FR for fine root, and for CR coarse root. 1, 2, 3, and 4 in front of FR or CR stand for 0–20, 20–40, 40–60, and 60–80 cm, respectively, below the plains on which nebkhas rest. Woody debris (WD) from dead ramets is also included in the figure. No ANOVA results for 3FR and 3CR at the Dengkou site as there was only one nebkha having roots between 40 and 60 cm. No roots were found below 60 cm at the Dengkou site.

products (Farquhar et al., 1989), which is appropriate for the purpose of this study. A great advantage of introducing $\Delta^{13}C_{organ}$ is that heterotrophic ^{13}C enrichment can be compared not only among the organs of the same plant, but also across different plants at the same site or at different sites which may differ in autotrophic isotopic signatures. Thus the use of $\Delta^{13}C_{organ}$ facilitates the identification of general patterns.

Two-way ANOVA analyses (organ by site) were performed with SPSS (Ver.17.0). C, N, and P contents, $\delta^{13}C$, $\Delta^{13}C_{organ}$, C / N ratios, N / P ratios, and C / P ratios were analyzed for differences between organs and between study sites. Tukey post hoc tests were used to determine pairwise differences for significant effects ($P < 0.05$). Regression analyses were used to determine the relationship between the heterotrophic ^{13}C enrichment and nutrient contents.

3 Results

3.1 Variations in $\Delta^{13}C_{organ}$ among plant organs and between study sites

At both the Dengkou and Minqin study sites, the values of $\Delta^{13}C_{organ}$ for all heterotrophic organs examined were significantly smaller than zero, indicating that without any exception, the heterotrophic organs were enriched in ^{13}C compared to the leaves (Fig. 2). However, there were considerable variations in $\Delta^{13}C_{organ}$ among the heterotrophic organs at both study sites and between the heterotrophic organs across the study sites. Stems were less enriched (closer to zero) than roots at both sites. At the Dengkou site, the most enriched organ was the coarse roots inside the nebkha sands. At the Minqin site, the most enriched part was the fine roots inside the nebkha sands although the difference between the coarse and fine roots inside the sands was not significant. At the Dengkou site, the coarse roots were consistently more enriched than the corresponding fine roots both inside the nebkha sands and below the plains. In contrast, at the Minqin site, the coarse roots were less enriched than the corresponding fine roots except for the roots deep into the plains (40–80 cm) where the coarse roots were more enriched. However at both sites, the statistical power of the coarse–fine root isotope differences were low as they were not significant at the significance level of 0.05. At the Dengkou site, the woody debris was more enriched than the stems but less enriched than the roots while at the Minqin site, it was less enriched than either the stems or the roots. In all biomass categories investigated, the Dengkou site was more enriched than the Minqin site, particularly in below-plain roots and in woody debris.

3.2 Variations in nutrient concentrations among plant organs and between sites

There are considerable variations in nutrient contents among plant organs and between sites (Fig. 3). At both the Dengkou and Minqin sites, leaves appeared to have the lowest C (Fig. 3a) but highest N (Fig. 3b) and P (Fig. 3c) contents. At both sites, stems tended to have lower N contents than roots either inside the sand dunes or below the plains under the sand dunes; in contrast, P contents in stems were within the variations of P contents in roots. At the Dengkou site, roots inside the sand dunes had lower N contents than roots below the plain; at the Minqin site, the coarse roots inside the sand dunes had lower N than either coarse or fine roots below the plain while the fine roots inside the sand dunes had N within the variations of those of coarse and fine roots below the plain. At the Dengkou site, the fine roots appeared to have higher P than coarse roots but the differences diminished from inside sands to below plain. There were no clear patterns on root P at the Minqin site. Woody debris had N contents similar to stems at both sites and tended to have

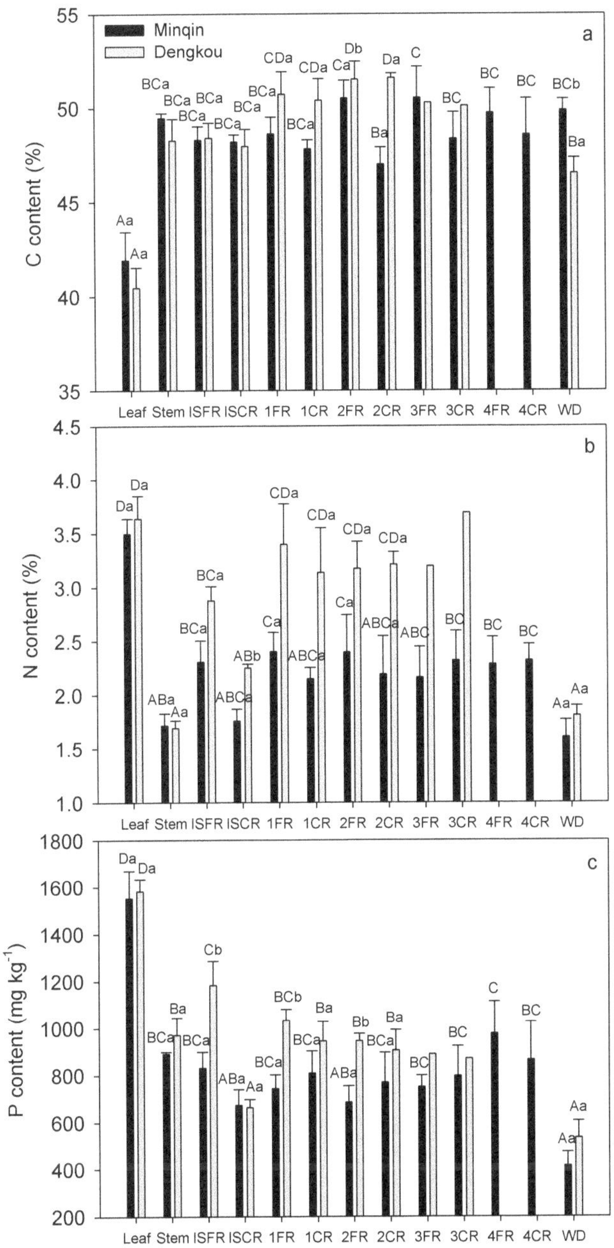

Figure 3. Carbon (C) **(a)**, nitrogen (N) **(b)** and phosphorous (P) content **(c)** of different organs of *Nitraria tangutorum* Bobrov, at the Dengkou and Minqin study sites. Symbols and letters denoting ANOVA results are explained in Fig. 2.

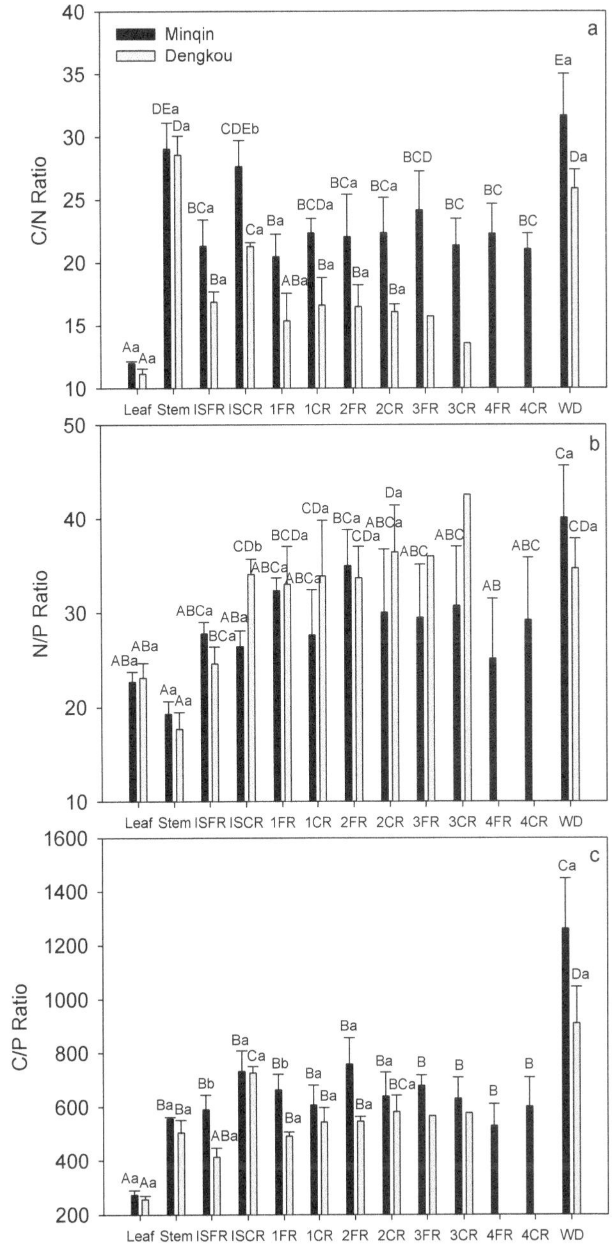

Figure 4. Carbon (C) to nitrogen (N) **(a)**, N to phosphorous (P) **(b)** and C to P mass ratios **(c)** of different organs of *Nitraria tangutorum* Bobrov, at the Dengkou and Minqin study sites. Symbols and letters denoting ANOVA results are explained in Fig. 2.

significantly less P contents than leaves, stems or roots. Between the two study sites, the leaves had lower C but higher N and P contents at the Dengkou site than at the Minqin site, but the difference is not significant at the significance level of 0.05. In contrast, heterotrophic organs at the Dengkou site tended to have significantly higher N and P contents than at the Minqin site. This contrast suggests that *N. tangutorum* may be able to maintain nutrient contents in leaves for photosynthesis at the expense of stems and roots.

Consistent with the variations in C, N and P contents, there were also substantial variations in the ratios of C / N (Fig. 4a), N / P (Fig. 4b) and C / P (Fig. 4c) among plant organs and between sites. For the live biomass (leaves, stems, and roots), the ratios of C / N ranged from about 11 to 30, N / P from 20 to 40 and C / P from 300 to 700. As expected, leaves had the lowest C / N and C / P ratios at both sites. Leaves also had the lowest N / P ratios except for stems. Overall, the Dengkou site had lower ratios of C / N and C / P but higher ratios of N / P than the Minqin site, particularly for roots below the plain.

3.3 Relationships between ^{13}C enrichment and nutrient contents

The observed large variations in ^{13}C enrichment and nutrient contents among heterotrophic organs and between study sites give us an opportunity to examine whether ^{13}C enrichment in heterotrophic organs relative to leaves could be affected by their nutrient contents. We found that across the two study sites and across the heterotrophic organs, $\Delta^{13}C_{organ}$ was significantly correlated with the N content (Fig. 5b), the C / N ratio (Fig. 5d), and the N / P ratio (Fig. 5e) in the heterotrophic organs. The correlation was negative for N content and N / P ratio but positive for C / N ratio, indicating that higher heterotrophic N contents resulted in larger heterotrophic ^{13}C enrichment relative to leaves. The C / N ratio explained a higher percentage (52 %) of variance in $\Delta^{13}C_{organ}$ than did the N content or the N / P ratio (44 and 42 %, respectively). No significant effect of heterotrophic organ C content (Fig. 5a), P content (Fig. 5c), or C / P ratio (Fig. 5f) on $\Delta^{13}C_{organ}$ were found.

We did not have enough independent samples to look at how leaf N contents might affect the heterotrophic ^{13}C enrichment. However, we examined the relationship between $\Delta^{13}C_{organ}$ and organ nutrient contents normalized by the corresponding leaf nutrient contents (i.e., the ratio of heterotrophic to corresponding leaf nutrient values). The normalized heterotrophic N contents explained somewhat less variance with reduced statistical power compared to the unnormalized values (compare Supplement Fig. S1 to Fig. 5), suggesting that it is the absolute N contents of the heterotrophic organs, not their relative departure from the corresponding leaf N contents, that affect the heterotrophic ^{13}C enrichment.

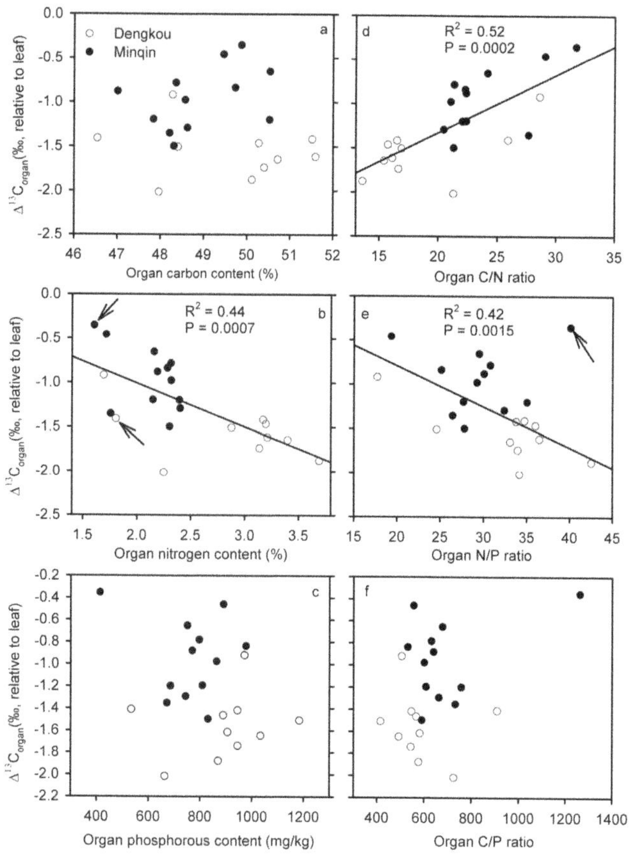

Figure 5. Nutrient dependence of the difference in carbon isotope compositions between leaves and heterotrophic organs of *Nitraria tangutorum* Bobrov, which is measured by $\Delta^{13}C_{organ}$ in Eq. (1) and averaged across the nebkhas excavated at the same study site. Negative values indicate ^{13}C enrichment in heterotrophic organs compared to leaves. Changes of $\Delta^{13}C_{organ}$ as a function of organ contents of carbon (C) **(a)**, nitrogen (N) **(b)** and phosphorous (P) **(c)** and of organ mass ratios of C to N **(d)**, N to P **(e)**, and C to P **(f)**. The two arrows in **(b)** indicate values for woody debris from dead ramets at each study site while in **(d)** indicates an outlier caused by measurements in phosphorous content (see the outlier in **c** and **f**).

3.4 Seasonal variations in leaf $\delta^{13}C$ and C_i/C_a ratios

At both Dengkou and Minqin sites, leaf $\delta^{13}C$ of *N. tangutorum* decreased from May to September (Fig. 6a), indicating progressive depletion in the heavier carbon isotope in leaves as the season progressed. Meanwhile, the C_i / C_a ratio increased from the early to late growing season (Fig. 6b). Thus the relationship between the seasonal patterns in leaf $\delta^{13}C$ and C_i / C_a ratios is consistent with the prediction by the leaf photosynthetic carbon isotope discrimination models (Farquhar et al., 1982; Farquhar and Cernusak, 2012; Gu and Sun, 2014). However, the differences in leaf $\delta^{13}C$ between the two sites cannot be entirely explained by the differences in the C_i / C_a ratios. In all months examined, the C_i / C_a ratios at the Dengkou site were consistently higher than at the Minqin site. If the C_i / C_a ratios were the only factor controlling the leaf $\delta^{13}C$, then the Dengkou site should have consistently lower leaf $\delta^{13}C$ (higher C_i / C_a ratios increase discrimination against ^{13}C during photosynthesis). To the contrary, the Dengkou site had higher leaf $\delta^{13}C$ than the Minqin site in May, June and July; only in August and September, the difference in leaf $\delta^{13}C$ was consistent with the effect of the difference in C_i / C_a ratios between the two sites (although the difference in leaf $\delta^{13}C$ between the two sites were still not significant).

Interestingly, the leaf $\delta^{13}C$ in May and June was close to the biomass-weighted average of root $\delta^{13}C$ at both study sites, suggesting that the initial building materials of new leaves might have largely come from stored carbon in roots.

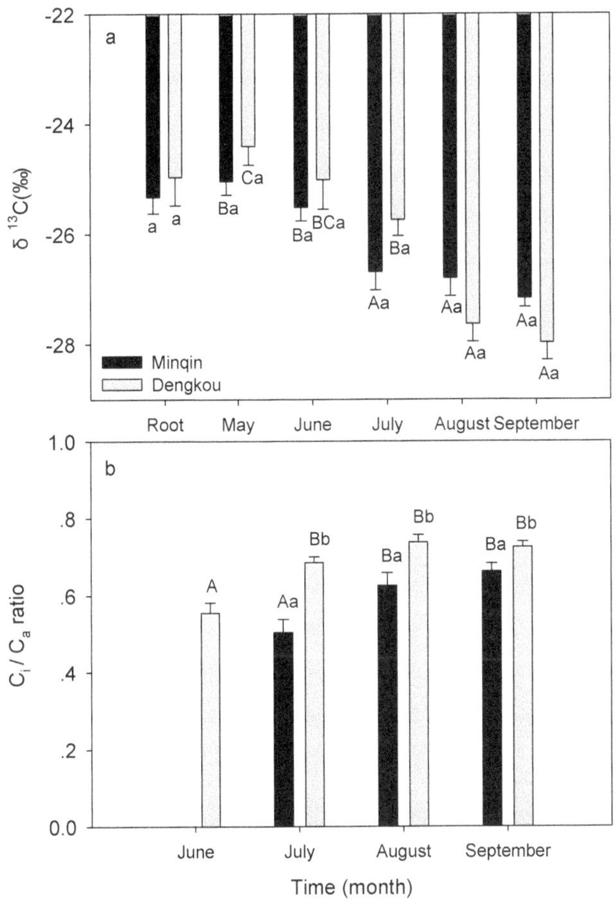

Figure 6. Seasonal changes in the ratios of leaf carbon isotopes (**a**) and intercellular (C_i) to ambient (C_a) CO_2 concentrations of *Nitraria tangutorum* Bobrov at the Dengkou and Minqin study sites (**b**). For comparison, the biomass-averaged isotope ratios of roots from the excavated nebkhas are also shown in (**a**).

4 Discussion

A major finding from this study is that the N content of a heterotrophic organ, expressed either as a fraction of total dry biomass or as a ratio of C to N or N to P, is strongly correlated with this organ's enrichment in ^{13}C relative to leaves with higher N concentrations corresponding to larger enrichment. Because this relationship is caused by variations among heterotrophic organs, and because normalizing the heterotrophic N content by the corresponding leaf N content did not improve or even worsen this relationship, the process responsible for it must reside inside the heterotrophic organs themselves. Further, this process must be mediated by N.

What N-mediated process could be responsible for the positive N–^{13}C enrichment relationship among heterotrophic organs? A parsimonious candidate is the respiratory CO_2 refixation by PEP carboxylase. CO_2 from the respiration of heterotrophic organs may dissolve into water and be hydrated into HCO_3^- which is then fixed by PEP carboxylase into oxaloacetate. Both the dissolution of CO_2 into water and the fixation of HCO_3^- by PEP carboxylase discriminate slightly against ^{13}C. However, the hydration process fractionates strongly in favor of ^{13}C and causes it to concentrate in HCO_3^-. Consequently, the CO_2 refixation by PEP carboxylase has a net fractionation of 5.7‰ in favor of ^{13}C relative to the gaseous CO_2 (Farquhar, 1983; Melzer and O'Leary, 1987; Farquhar et al., 1989). Thus the respiratory CO_2 refixation by PEP carboxylase should lead to a depletion of ^{13}C in CO_2 escaped to outside compared to the original substrates for respiration while heterotrophic organs should be ^{13}C-enriched due to the addition of organic materials from PEP carboxylase activities. Previous studies have reported high PEP carboxylase activities in heterotrophic organs of a variety of C_3 plant species (Melzer and O'Leary 1987; Berveiller and Damesin 2008; Gessler et al., 2009, 2014). If increased N content increases the respiratory CO_2 refixation in heterotrophic organs, then it should also increase ^{13}C enrichment in these organs. Berveiller et al. (2010) showed that CO_2 refixation rates of *Fagus sylvatica* stems increased as stem N content increased, which provides a direct support for the hypothesis that CO_2 refixation by PEP carboxylase is a process responsible for our observed positive relationship between N and ^{13}C enrichment in heterotrophic organs.

Observed respiration rates of leaves, stems and roots tend to increase with increased N contents (Reich et al., 2008). This does not necessarily contradict the PEP carboxylase hypothesis suggested above. The actual respiration rates of these organs may increase so much with increased N contents that the increase cannot be offset by the increased refixation rates by PEP carboxylase. Consequently, the observed rates of CO_2 evolved from heterotrophic organs may still increase even though the refixation rates have increased with increased N contents.

The PEP carboxylase hypothesis does imply that the CO_2 escaped to outside from the heterotrophic organs are depleted in ^{13}C compared to the substrates utilized for respiration. As summarized in the review of Ghashghaie and Badeck (2014), most isotopic studies on root respiration have found that CO_2 evolved from roots are depleted in ^{13}C compared with bulk root material, in contrast to leaf dark respiration, which is generally enriched. For stem respiration, however, more contradictory results have been reported. Wingate et al. (2010) showed that CO_2 evolved from stems of *Pinus pinaster* was more depleted in ^{13}C when compared to the currently measured net CO_2 flux by photosynthetic branches or with the phloem water-soluble organic matter and wood cellulose. Gessler et al. (2009) also found that the respiration of stems as well as roots of *Ricinus communis* was depleted in ^{13}C relative to the assumed respiratory substrates. This latter study was particularly relevant to this present study because the authors determined that the depletion was caused by a strong refixation of respiratory CO_2 catalyzed by PEP carboxylase. In contrast to these studies, Damesin and Lelarge (2003) reported that stem respiration of *Fagus sylvatica* was enriched in ^{13}C compared with the total organic matter, while Kodama et al. (2008) showed that CO_2 evolved from the stem of *Pinus sylvestris* had higher or similar $\delta^{13}C$ values compared to that of phloem exudate organic matter, depending on respiration rates. More studies are needed to determine whether carbon isotope fractionations of stem respiration depend on species, ages, or environments. Also, the dissolution and hydration of respiratory CO_2 may decouple in location from the fixation of HCO_3^- by PEP carboxylase if there is a strong transpiration stream in xylem, with isotopic consequences. For example, respiratory CO_2 can be dissolved and hydrated in roots and stems but the HCO_3^- molecules formed can be carried up in xylem transpiration streams (Aubrey and Teskey, 2009; Angert and Sherer, 2011; Bloemen et al., 2013; Trumbore et al., 2013) and fixed by PEP carboxylase in branches, which will serve to redistribute isotope signatures among different parts of the plant body.

Additional studies are also needed to determine whether there are other causes for the observed heterotrophic N–^{13}C enrichment relationship. For example, if different organ N contents are associated with chemical compounds with different isotope signatures or different "fragmentation fractionation" (enzymatic reaction of substrate molecules with heterogeneous ^{13}C distribution; Tcherkez et al., 2004; Hobbie and Werner 2004), one may expect organ N contents to be correlated with organ isotope signatures, potentially leading to the observed relationship. Another possibility to consider is that atmospheric $\delta^{13}C$ has been decreasing since the Industrial Revolution due to the emission of ^{13}C-depleted fossil CO_2. If a heterotrophic organ contains a higher fraction of carbon with an old age, then its bulk $\delta^{13}C$ would be higher. Stems and roots should contain more old carbon than leaves do. We do not have data to quantify this possibility. However, a qualitative reasoning led us to doubt that a general decreasing trend in atmospheric $\delta^{13}C$ can explain the observed heterotrophic N–^{13}C enrichment relationship. Although we do not know the ages of the six nebkhas excavated, atmospheric N deposition has probably been increasing during the lifetime of these nebkhas. Therefore, younger tissues should contain lower $\delta^{13}C$ and higher N, which would imply a negative N–^{13}C enrichment relationship, opposite to what we observed. Therefore the positive heterotrophic N–^{13}C enrichment relationship most likely has a phytogenic (rather than an atmospheric) origin.

It is important to clarify that our suggestion that the process responsible for the positive heterotrophic N–^{13}C enrichment relationship resides in heterotrophic organs does not imply that the cause(s) for heterotrophic enrichment of ^{13}C relative to leaves resides entirely in heterotrophic organs. In fact, to explain the full magnitude of the observed heterotrophic enrichment (2‰), about 35 % (100 × 2/5.7) of the carbon of heterotrophic organs has to have cycled through PEP carboxylase once, which appears to be surprisingly large for C_3 plants (Hobbie et al., 2003). Also, our finding that the $\delta^{13}C$ of leaves in the early growing season was close to the mean isotope ratio of roots, but decreased as the season progressed, indicates that processes inside leaves must also contribute to the overall isotope differences between leaves and heterotrophic organs, if the leaf samples for reference are from middle to late growing seasons. The reference leaf samples in our calculation of $\Delta^{13}C_{organ}$ were from middle growing seasons (August). Therefore, the progressive seasonal depletion in foliar ^{13}C increased the magnitude of the obtained $\Delta^{13}C_{organ}$. Furthermore, processes such as preferential loading into phloem of the heavier isotope and loss of depleted outer bark materials should also affect the overall autotrophic–heterotrophic isotope differences (Cernusak et al., 2009; Ghashghaie and Badeck, 2014). While these processes may boost the overall magnitude of heterotrophic ^{13}C enrichment, they cannot explain its relationship with N content among heterotrophic organs.

It is likely that leaf N also plays an important role in determining ^{13}C enrichment in heterotrophic organs relative to leaves. We do not have enough leaf-level data to examine this issue in depth but findings from previous studies allow us to speculate about what this role might be. As discussed early, leaf N content is positively correlated with leaf $\delta^{13}C$ because higher leaf N increases leaf photosynthetic capacity, which results in decreased C_i / C_a ratios and thus reduced discrimination against ^{13}C during photosynthesis (Sparks and Ehleringer, 1997; Livingston et al., 1999; Duursma and Marshall, 2006; Cernusak et al., 2007, 2013). However, a positive relationship between leaf N and leaf $\delta^{13}C$ does not necessarily mean that higher leaf N will reduce the degree of heterotrophic enrichment in ^{13}C compared to leaves as heterotrophic organs use photosynthetic products from leaves. An interesting pathway for leaf N to influence heterotrophic ^{13}C enrichment may lie in the relationship between leaf N and dark respiration. It is known that leaf dark respiration scales with leaf N (Reich et al., 2008). It is also known that leaf dark respiration is enriched in ^{13}C, contrary to respirations of stems and roots (Ghashghaie and Badeck, 2014). Thus higher leaf N may actually increase the depletion of ^{13}C in leaves relative to heterotrophic organs. Consequently one may expect that N in autotrophic and heterotrophic organs of plants contributes to the isotope difference between these two types of organs in the same direction but through fundamentally different mechanisms.

Our analyses benefitted from the large variations in nutrient contents and heterotrophic ^{13}C enrichment both across plant organs and between sites, allowing any relationship (if exists) between these two sets of variables to be seen clearly. The large variations across plant organs are a validation of our systemic, whole-plant sampling strategy. The large between-site differences in organ nutrient contents likely reflect a site difference in soil fertility. The soil of the vegetated area at Dengkou contained 0.024 ± 0.006 % N (Jia, 2010) while at Minqin the value was 0.01 ± 0.001 % (Song et al., 2012), explaining the generally higher plant organ N contents at Dengkou than at Minqin. Soil P contents have not been measured at either site. However, we suspect that soil at Dengkou was also richer in P than at Minqin, as plant organs generally contained higher P contents at the former than latter site. The cross-organ variations in nutrient contents were larger at Dengkou than at Minqin, possibly because poorer soil nutrient availability limited organ nutrient content variations at the latter site. Correspondingly, the range of heterotrophic ^{13}C enrichment was also wider at Dengkou than at Minqin. Both the cross-organ and between-site variations contributed the observed relationship between the N content and heterotrophic ^{13}C enrichment. However, even within the same site, a pattern between N content and heterotrophic ^{13}C enrichment can be clearly seen, particularly at the Dengkou site. Further, the patterns of the two sites appear to be consistent with each other and form a single relationship. This consistency suggests that the same mechanism operates at the two sites to generate a unified dependence of ^{13}C enrichment on N content across heterotrophic plant organs.

The lack of a clear relationship between P content and heterotrophic ^{13}C enrichment (Fig. 5c and Fig. S1c) is interesting. In plants, proteins, which are rich in N, must be maintained with an allocation of a certain fraction of total body P to ribosomal ribonucleic acid (rRNA) (Niklas et al., 2005; Elser et al., 2010). Thus the N and P contents are generally positively correlated, and the measurements from Minqin and Dengkou are not exceptions (Fig. S2). So why is there is a clear dependence of heterotrophic ^{13}C enrichment on N but not on P? It could be that the relationship of heterotrophic ^{13}C enrichment with P is considerably weaker than that with N, and our data were not sensitive enough to detect it.

The relationship of heterotrophic ^{13}C enrichment with the N / P ratio (Fig. 5e and S1e) is broadly similar to that with N (Fig. 5b and S1b), suggesting that the relationship of heterotrophic ^{13}C enrichment with the N / P ratio is largely due to the effect of N rather than to the ratio itself. However, some level of direct dependence of the enrichment on the N / P ratio cannot be ruled out. Niklas et al. (2005) and Elser et al. (2010) integrated biological stoichiometry and metabolic scaling theories, which led them to suggest that growth rates and plant sizes should be related to N / P ratios. These authors' analyses focused on individual plants while our study is on plant organs. However, if the N / P ratio affects fractionating metabolic processes of plant organs, it is conceivable that the N / P ratio can also affect the ^{13}C enrichment (or depletion) of this organ relative to leaves.

5 Conclusions

We conclude that heterotrophic ^{13}C enrichment is affected jointly by fractionation processes occurring within heterotrophic organs and within leaves. Processes taking place between heterotrophic organs and leaves (e.g., preferential phloem loading of ^{13}C-enriched sugars) may also contribute to this phenomenon. A nitrogen-mediated process, hypothesized to be the CO_2 refixation by PEP carboxylase, may be responsible for variations in ^{13}C enrichment within heterotrophic organs, while processes within leaves or between leaves and heterotrophic organs may determine the overall magnitude of heterotrophic ^{13}C enrichment. We suggest that future efforts should focus on the roles of nitrogen and refixation of respiratory CO_2 by PEP carboxylase in carbon isotope fractionation processes both within leaves and within heterotrophic organs as well as in between them. The findings of this study may have implications beyond isotope ecology. There has been a general lack of studies of refixation of respiratory CO_2 by PEP carboxylase in C_3 plant species. To our knowledge, no current terrestrial carbon cycle models consider this post-photosynthetic process. If PEP carboxylase can significantly affect carbon isotope compositions in heterotrophic organs of C_3 plant species, it may very well have strong influence on post-photosynthetic plant carbon budget and therefore terrestrial ecosystem carbon balance.

Acknowledgements. Field work, data acquisition and analyses were conducted at the Institute of Desertification Studies, Chinese Academy of Forestry with support in part by grants from the Forestry Public Welfare Scientific Research Funding (201104077), the Program of Introducing Advanced Forestry Technologies from Other Countries (2008-4-47), the Science and Technology Foundation (CAFYBB2007008) and the Lecture and Study Program for Outstanding Scholars from Home and Abroad of the Chinese Academy of Forestry, and the National Key Technology R&D Program of the Ministry of Science and Technology of China (2012BAD16B01). Data analyses and manuscript writing were partly carried out at Oak Ridge National Laboratory (ORNL) with support from US Department of Energy, Office of Science, Biological and Environmental Research Program, Climate and Environmental Sciences Division. It also received support from the ORNL's Laboratory Directed Research and Development Program. ORNL is managed by UT-Battelle, LLC, for the US Department of Energy under contract DE-AC05-00OR22725.

Edited by: M. Bahn

References

Angert, A. and Sherer, Y.: Determining the relationship between tree–respiration and CO_2 efflux by δO_2/Ar measurements, Rapid Commun. Mass Spectrom., 25, 1752–1756, 2011.

Aubrey, D. P. and Teskey, R. O.: Root-derived CO2 efflux via xylem stream rivals soil CO_2 efflux, New Phytol., 184, 35-40, 2009.

Baas, A. C. W. and Nield, J. M.: Modelling vegetated dune landscapes, Geophys. Res. Lett., 34, L06405, doi:10.1029/2006GL029152, 2007.

Badeck, F. W., Tcherkez, G., Noguès, S., Piel, C., and Ghashghaie, J.: Postphotosynthetic fractionation of stable carbon isotopes between plant organs – a widespread phenomenon, Rapid Commun. Mass Spectrom., 19, 1381–1391, 2005.

Berveiller, D. and Damesin, C.: Carbon assimilation by tree stems: potential involvement of phosphoenolpyruvate carboxylase, Trees, 22, 149–157, 2008.

Berveiller, D., Fresneau, C., and Damesin, C.: Effect of soil N supply on carbon assimilation by tree stems, Ann. Forest Sci., 67, 609, doi:10.1051/forest/2010022, 2010.

Bloemen, J., McGuire, M. A., Aubrey, D. P., Teskey, R. O., and Steppe, K.: Transport of root-respired CO_2 via the transpiration stream affects aboveground carbon assimilation and CO_2 efflux in trees, New Phytol., 197, 555–565, 2013.

Bowling, D. R., Pataki, D. E., and Randerson, J. T.: Carbon isotopes in terrestrial ecosystem pools and CO_2 fluxes, New Phytol., 178, 24–40, 2008.

Cernusak, L. A., Winter, K., Aranda, J., Turner, B. L., and Marshall, J. D.: Transpiration efficiency of a tropical pioneer tree (Ficus insipida) in relation to soil fertility, J. Experim. Bot., 58, 3549–3566, 2007.

Cernusak, L. A., Tcherkez, G., Keitel, C., Cornwell, W. K., Santiago, L. S., Knohl, A., Barbour, M. M., Williams, D. G., Reich, P. B., Ellsworth, D. S., Dawson, T. E., Griffiths, H. G., Farquhar, G. D., and Wright, I. J.: Viewpoint: why are non-photosynthetic tissues generally ^{13}C enriched compared with leaves in C_3 plants?, Review and synthesis of current hypotheses, Funct. Plant Biol., 36, 199–213, 2009.

Cernusak, L. A., Ubierna, N., Winter, K., Holtum, J. A. M., Marshall, J. D., and Farquhar, G. D.: Environmental and physiological determinants of carbon isotope discrimination in terrestrial plants, New Phytol., 200, 950–965, 2013.

Craig, H.: The geochemistry of the stable carbon isotopes, Geochim. Cosmochim. Ac., 3, 53–92, 1953.

Damesin, C. and Lelarge, C.: Carbon isotope composition of current-year shoots from *Fagus sylvatica* in relation to growth, respiration and use of reserves, Plant, Cell Environ., 26, 207–219, 2003.

Dawson, T. E., Mambelli, S., Plamboeck, A. H., Templer, P. H., and Tu, K. P.: Stable isotopes in plant ecology, Annu. Rev. Ecol. Sys., 33, 507–559, 2002.

Domingues, T. F., Meir, P., Feldpausch, T. R., Saiz, G., Veenendaal, E. M., Schrodt, F., Bird, M., Djagbletey, G., Hien, F., Compaore, H., Diallo, A., Grace, J., and Lloyd, J.: Co-limitation of photosynthetic capacity by N and phosphorus in West Africa woodlands, Plant Cell Environ., 33, 959–980, 2010.

Du, J. H., Yan, P., and Dong, Y. X.: Phenological response of Nitraria tangutorum to climate change in Minqin County, Gansu Province, northwest China, Internat. J. Biometeorol., 54, 583–593, 2010.

Duursma, R. A. and Marshall, J. D.: Vertical canopy gradients in d13C correspond with leaf N content in a mixed-species conifer forest, Trees-Structure and Function, 20, 496–506, 2006.

Ehleringer, J. R., Comstock, J. P., and Cooper, T. A.: Leaf-twig carbon isotope ratio differences in photosynthetic-twig desert shrubs, Oecologia, 71, 318–320, 1987.

Elser, J. J., Fagan, W. F., Kerkhoff, A. J., Swenson, N. G., and Enquist, B. J.: Biological stoichiometry of plant production: netabolism, scaling and ecological response to global change, New Phytol., 186, 593–608, 2010.

Farquhar, G. D.: On the nature of carbon isotope discrimination in C_4 species, Aust. J. Plant Physiol., 10, 205–2634, 1983.

Farquhar, G. D., O'Leary, M. H., and Berry, J. A.: On the relationship between carbon isotope discrimination and the intercellular carbon dioxide concentration in leaves, Aust. J. Plant Physiol., 9, 121–137, 1982.

Farquhar, G. D., Ehleringer, J. R., and Hubick, K. T.: Carbon isotope discrimination and photosynthesis, Annu. Rev. Plant Physiol. Plant Mol. Biol., 40, 503–537, 1989.

Farquhar, G. D. and Cernusak, L. A.: Ternary effects on the gas exchange of isotopologues of carbon dioxide, Plant, Cell Environ., 35, 1221–1231, 2012.

Field, C. and Mooney, H. A.: The photosynthesis – N relationship in wild plants, in: On the economy of plant form and function, edited by: Givnish, T., New York, NY, USA, Cambridge University Press, 25–55, 1986.

Gessler, A., Tcherkez, G., Karyanto, O., Keitel, C., Ferrio, J. P., Ghashghaie, J., Kreuzwieser, J., and Farquhar, G. D.: On the metabolic origin of the carbon isotope composition of CO_2 evolved from darkened light-acclimated leaves in *Ricinus communis*, New Phytol., 181, 374–386, 2009.

Gessler, A., Ferrio, J. P., Hommel, R., Treydte, K., Werner, R. A., and Monson, R. K.: Stable isotopes in tree rings: towards a mechanistic understanding of isotope fractionation and mixing processes from the leaves to the wood, Tree Physiol., 1–23, doi:10.1093/treephys/tpu040, 2014.

Ghashghaie, J. and Badeck, F. W.: Opposite carbon isotope discrimination during dark respiration in leaves versus roots – a review, New Phytologist, 201, 751–769, 2014.

Gu, L. H. and Sun, Y.: Artifactual responses of mesophyll conductance to CO_2 and irradiance estimated with the variable *J* and online isotope discrimination methods, Plant. Cell Environ., 37, 1231–1249, 2014.

Hobbie, E. A. and Werner, R. A.: Intramolecular, compound-specific, and bulk carbon isotope patterns in C_3 and C_4 plants: a review and synthesis, New Phytol., 161, 371–385, 2004.

Hobbie, E. A., Watrud, L. S., Maggard, S., Shiroyama, T., Paul, T., and Rygiewicz, P. T.: Carbohydrate use and assimilation by litter and soil fungi assessed by carbon isotopes and BIOLOG® assays, Soil Biol. Biogeochem., 35, 303–311, 2003.

Jia, Z. Y.: Responses of soil respiration in a desert *Nitraria tangutorum* ecosystem to simulated rainfaill, Dissertation, Chinese Academy of Forestry, 2010 (in Chinese with English abstract).

Kodama, N., Barnard, R. L., Salmon, Y., Weston, C., Ferrio, J. P., Holst, J., Werner, R. A., Saurer, M., Rennenberg, H., Buchmann, N., and Gessler, A.: Temporal dynamics of the carbon isotope composition in a *Pinus sylvestris* stand: from newly assimilated organic carbon to respired carbon dioxide, Oecologia, 156, 737–750, 2008.

Lang, L. L., Wang, X. M., Hasi, E., and Hua, T.: Nebkha (coppice dune) formation and significance to environmental change reconstructions in arid and semiarid areas, J. Geograph. Sci., 23, 344–358, 2013.

Leavitt, S. W. and Long, A.: Evidence for ^{13}C / ^{12}C fractionation between tree leaves and wood, Nature, 298, 742–744, 1982.

Leavitt, S. W. and Long, A.: Stable-carbon isotope variability in tree foliage and wood, Ecology, 67, 1002–1010, 1986.

Li, Q. H. and Jiang, Z. P.: Research on Plant Species of Genus *Nitraria* L. (238 pp., China Forestry Press, Beijing, China), 2011.

Li, Q. H., Xu, J., Li, H. Q., Wang, S. X., Yan, X., Xin, Z. M., Jiang, Z. P., Wang, L. L., and Jia, Z. Q.: Effects of aspect on clonal reproduction and biomass allocation of layering modules of *Nitraria tangutorum* in Nebkha dunes, PLOS One, 8, e79927, doi:10.1371/journal.pone.0079927, 2013.

Livingston, N. J., Guy, R. D., Sun, Z. J., and Ethier, G. J.: The effects of N stress on the stable carbon isotope composition, productivity and water use efficiency of white spruce (*Picea glauca* (Moench) Voss) seedlings, Plant Cell Environ., 22, 281–289, 1999.

Melzer, E. and O'Leary, M. H.: Anapleurotic CO_2 fixation by phosphoenolpyruvate carboxylase in C_3 plants, Plant Physiol., 84, 58–60, 1987.

Niklas, K. J., Owens, T., Reich, P. B., and Cobb, E. D.: Nitrogen/phosphorus leaf stoichiometry and the scaling of plant growth, Ecol. Lett., 8, 636–642, 2005.

Reich, P. B., Tjoelker, M. G., Pregitzer, K. S., Wright, I. J., Oleksyn, J., and Machado, J.,L.: Scaling of respiration to N in leaves, stems and roots of higher land plants, Ecol. Lett., 11, 793–801, 2008.

Song, W. M., Chen, S. P., Wu, B., Zhu, Y. J., Zhou, Y. D., Li, Y. H., Cao, Y. L., Lu, Q., and Lin, G. H.: Vegetation cover and rain timing co-regulate the responses of soil CO_2 efflux to rain increases in an arid desert ecosystem, Soil Biol. Biochem., 49, 114–123, 2012.

Sparks, J. P. and Ehleringer, J. R.: Leaf carbon isotope discrimination and N content for riparian trees along elevational transects, Oecologia, 109, 362–367, 1997.

Tcherkez, G., Farquhar, G., Badeck, F., and Ghashghaie, J.: Theoretical considerations about carbon isotope distribution in glucose of C_3 plants, Funct. Plant Biol., 31, 857–877, 2004.

Tcherkez, G., Mahé, A., and Hodges, M.: ^{12}C / ^{13}C fractionations in plant primary metabolism, Trends Plant Sci., 16, 499–506, 2011.

Trumbore, S. E., Angert, A., Kunert, N., Muhr, J., and Chambers, J. Q.: What's the flux? Unraveling how CO_2 fluxes from trees reflect underlying physiological processes, New Phytol., 197, 353–355, 2013.

Wingate, L., Ogee, J., Burlett, R., Bosc, A., Devaux, M., Grace, J., Loustau, D., and Gessler, A.: Photosynthetic carbon isotope discrimination and its relationship to the carbon isotope signals of stem, soil and ecosystem respiration, New Phytol., 188, 576–589, 2010.

2

Deploying four optical UAV-based sensors over grassland: challenges and limitations

S. K. von Bueren[1,*], A. Burkart[2,*], A. Hueni[3], U. Rascher[2], M. P. Tuohy[1], and I. J. Yule[1]

[1]Institute of Agriculture & Environment, Massey University, Palmerston North, New Zealand
[2]Institute of Bio- and Geosciences, IBG-2: Plant Sciences, Forschungszentrum Jülich GmbH, Jülich, Germany
[3]Remote Sensing Laboratories, University of Zurich, Zurich, Switzerland
[*]These authors contributed equally to this work.

Correspondence to: A. Burkart (an.burkart@fz-juelich.de)

Abstract. Unmanned aerial vehicles (UAVs) equipped with lightweight spectral sensors facilitate non-destructive, near-real-time vegetation analysis. In order to guarantee robust scientific analysis, data acquisition protocols and processing methodologies need to be developed and new sensors must be compared with state-of-the-art instruments. Four different types of optical UAV-based sensors (RGB camera, converted near-infrared camera, six-band multispectral camera and high spectral resolution spectrometer) were deployed and compared in order to evaluate their applicability for vegetation monitoring with a focus on precision agricultural applications. Data were collected in New Zealand over ryegrass pastures of various conditions and compared to ground spectral measurements. The UAV STS spectrometer and the multispectral camera MCA6 (Multiple Camera Array) were found to deliver spectral data that can match the spectral measurements of an ASD at ground level when compared over all waypoints (UAV STS: $R^2 = 0.98$; MCA6: $R^2 = 0.92$). Variability was highest in the near-infrared bands for both sensors while the band multispectral camera also overestimated the green peak reflectance. Reflectance factors derived from the RGB ($R^2 = 0.63$) and converted near-infrared ($R^2 = 0.65$) cameras resulted in lower accordance with reference measurements. The UAV spectrometer system is capable of providing narrow-band information for crop and pasture management. The six-band multispectral camera has the potential to be deployed to target specific broad wavebands if shortcomings in radiometric limitations can be addressed. Large-scale imaging of pasture variability can be achieved by either using a true colour or a modified near-infrared camera. Data quality from UAV-based sensors can only be assured, if field protocols are followed and environmental conditions allow for stable platform behaviour and illumination.

1 Introduction

In the last decade, the use of unmanned aerial vehicles (UAVs) as remote sensing platforms has become increasingly popular for a wide range of scientific disciplines and applications. With the development of robust, autonomous and lightweight sensors, UAVs are rapidly evolving into standalone remote sensing systems that deliver information of high spatial and temporal resolution in a non-invasive manner. UAV systems are particularly promising for precision agriculture where spatial information needs to be available at high temporal frequency and spatial resolution in order to identify in-field variability (Stafford, 2000; Seelan et al., 2003; Lelong et al., 2008; Nebiker et al., 2008; Link et al., 2013). Zhang and Kovacs (2012) provide a comprehensive review of unmanned aerial systems applied in precision agriculture.

Precision agriculture aims at identifying crop and soil properties in near-real-time (Lebourgeois et al., 2012; Primicerio et al., 2012a) and at delivering results to farmers and decision makers with minimum delay to enable management decisions based on current crop and soil status. The use of input resources such as fertilizers, herbicides or water (Van Alphen and Stoorvogel, 2000; Carrara et al., 2004; Chávez et al., 2010) are matched to the current demand by the crops, leading to an economical use of resources. The use of UAV-based sensors to detect water stress and quantify biomass and nitrogen content in crops and grasses has been demonstrated (Berni et al., 2008, 2009; Kawamura et al., 2011). Yield forecasting in wheat (Jensen et al., 2007) and rice (Swain et al., 2010), rangeland management (Rango et al., 2009), leaf area index (LAI) and green normalized difference vegetation index (NDVI) estimation in winter wheat (Hunt et al., 2010) and site-specific vineyard management (Turner, 2011; Primicerio et al., 2012b) have been accomplished using unmanned aerial platforms.

Proximal remote sensing methods can be used to detect pasture and crop biophysical parameters such as biomass, dry matter, fibre content, organic matter digestibility and macronutrient availability (nitrogen, phosphorus and potassium). Pasture monitoring approaches capable of measuring biophysical variables over the whole farm at a high spatial resolution allow for site-specific management decisions and optimum nutrient management (Sanches et al., 2012). While vegetation indices have been frequently applied for biomass and dry matter estimation (Mutanga, 2004; Duan et al., 2011; Vescovo et al., 2012), waveband-specific algorithms have been developed to estimate macronutrients (Mutanga and Skidmore, 2007; Pullanagari et al., 2012a, b).

In a pasture management context in New Zealand, where air- and spaceborne remote sensing methods are often limited by frequent cloud cover, UAV-based remote sensing can potentially overcome some of those limitations. Recent developments in commercially available lightweight and small digital cameras and multispectral sensors support precision nutrient management. However, these sensors need to be characterized and validated against state-of-the-art reference instruments. The extraction of quantitative information relies on thorough calibration procedures, good instrument characterization and a high standard of field operation.

Various studies have specifically evaluated multispectral sensors and consumer-grade digital cameras and assessed their potential for vegetation monitoring. The use of a conventional, ground-based broadband digital RGB camera has shown limited success in estimating green biomass on short-grass prairie, suggesting that narrow-band sensors are more promising for application over such complex ecosystems (Vanamburg et al., 2006). An image processing workflow for three consumer digital cameras has been developed by Lebourgeois et al. (2012) and they have suggested that the cameras have a high potential for terrestrial remote sensing of vegetation due to their versatility and multispectral capabilities. Vegetation indices derived from visible and near-infrared imagery acquired by two compact digital cameras were found to generate strong relationships with crop biophysical parameters and to be practical for monitoring of temporal changes in crop growth (Sakamoto et al., 2012). Kelcey and Lucieer (2012) developed a processing chain to improve the imagery acquired with the same six-band multispectral sensor that was used in the current study. They showed that image quality can be improved through application of sensor correction techniques to facilitate subsequent image analysis. A novel, UAV-based lightweight high-resolution spectrometer, which was tested in the field for the first time in the current study, was introduced by Burkart et al. (2013). Nijland et al. (2014) evaluated the use of near-infrared (NIR) and RGB cameras for the use of vegetation monitoring and plant phenology trend detection and found that the NIR-converted cameras were outperformed by standard RGB cameras. Poor band separation and the limited dynamic range of the NIR camera system limited the use of the sensors for vegetation monitoring in a controlled laboratory and in a field experiment.

Studies usually deploy a single UAV sensing system over an area of interest. But because different agricultural applications and environmental frameworks demand specific capabilities of an UAV remote sensing system, the current study uses four different sensors over the same experimental area to evaluate each sensor's suitability for applied grassland monitoring. From preliminary experiments, it was evident that the UAV system, including platform and sensor, need to be specifically matched to the vegetation parameter to be investigated. The present study used two compact digital cameras (RGB and NIR), a six-band multispectral camera (visible/near-infrared – VNIR) and a high-resolution spectrometer (VNIR) mounted on two different UAV platforms to acquire spectral information over dairy pastures in order to characterise each instrument in terms of radiometric quality and accuracy of spectral information obtainable, as compared to a ground reference instrument. Handling and limitations of the UAVs, flight planning, field procedures and the capabilities of the different sensors are discussed as a prospective guideline for upcoming UAV sensor-based research. Results are evaluated with a focus on inter-sensor comparability, aspects of field data collection using UAVs and the sensor's capabilities for monitoring green vegetation.

1.1 Experimental site

The experimental flight campaign was conducted in February 2013 on a Massey University dairy farm near Palmerston North, New Zealand, (No. 1 Dairy, located at lat. −40.376, long. 175.606). No. 1 Dairy is a fully operational dairy farm with an effective area of 119.7 ha. UAV flights were performed over four different paddocks with distinct characteristics from bare soil to dry and irrigated ryegrass pasture. At the time of data acquisition between 11:00 and 15:00 LT no clouds were visible.

1.2 UAV systems

As shown in Table 1, two different UAV systems were used: a QuadKopter (MikroKopter), owned and operated by Massey University, and a Falcon-8 (AscTec (Ascending Technologies), Krailing, Germany), from the Research Centre Jülich,

Table 1. UAV platforms.

Name	QuadKopter	Falcon-8
Manufacturer	MikroKopter	Ascending Technologies
Weight [g]	1900	1800
Max. Payload [g]	1000	500
Power source	LiPo, 4200 mAh, 14.8 V	Lipo, 6400 mAh, 11.1 V
Endurance [min]	12	15
GPS navigation	Ublox LEA 6s GPS chip	Ublox LEA 6T
Features	Open Source Gyro-stabilized camera mount	Stabilized camera mount, live video link, motor redundancy
Sensors	MCA6	UAV STS, RGB, Canon IR

Germany. The Falcon-8 uses the AscTec Autopilot Control V1.68 software. It has two identical exchangeable gimbals manufactured by AscTec, one for the Sony camera the other one for the spectrometer and Canon camera. Both gimbals are dampened and actively stabilized in pitch and roll. The MikroKopter UAV was fitted with an AV130 Standard Gimbal produced by Photo Higher. The gimballed camera mounts levelled out any platform movement to ensure the sensors were pointing in nadir direction to the ground at all times during the flight. The main difference between the Falcon-8 and the MikroKopter platforms is the payload restriction, which precludes the Falcon-8 from lifting sensors heavier than 0.5 kg, thus making it necessary to use the MikroKopter UAV to lift the Mini-MCA6 sensor. Both UAVs with their payloads were intensively tested on multiple flights before the study.

Figure 1. UAV-based sensors: **(a)** Sony Nex5n RGB camera **(b)** Canon PowerShot IR camera **(c)** MCA6 multispectral camera **(d)** Spectrometer (UAV STS).

1.3 UAV sensors

Four UAV sensors (Fig. 1) were tested and compared in terms of their ability to produce reflectance data over pastures. All of the sensors were lighter than 1 kg including batteries and were either modified or specifically designed for use on remotely controlled platforms. The sensors share a spectral range in the VNIR which is considered the most relevant region of the electromagnetic spectrum for agricultural research applications (Lebourgeois et al., 2008). In terms of spatial and spectral resolution (Fig. 2), the sensors differ significantly. Table 2 lists their relevant properties.

Mini-MCA6 (MCA6): the Mini-MCA6 (Multispectral Camera Array) is a six-band multispectral camera (Tetracam, Chatsworth, CA, USA) that can acquire imagery in six discrete wavebands. A camera-specific image alignment file is provided by the manufacturer. Exchangeable filters in the range of 400 to 1100 nm can be fitted to six identical monochromatic cameras. Table 3 lists the filter setup used during the study. The camera firmware allows pre-setting all imaging related parameters such as exposure time, shutter release interval and image format and size. Six two gigabyte CompactFlash memory cards store up to 800 images (10 bit RAW format, full resolution). With an opening angle of 38.3° × 31.0°, the camera has a relatively narrow field of view as opposed to the Canon and Sony cameras. The camera was set to a 2 ms exposure time and was run on a 2 s shutter release interval with images saved in the 10 bit RAW format. Positioning of the camera was achieved by hovering the UAV over the vegetation target for at least 30 s per waypoint.

STS spectrometer (UAV STS): the spectrometer was adapted for UAV-based remote sensing at the Research Centre Jülich. Its design is based on the STS VIS spectrometer (Ocean Optics, Dunedin, FL, USA) with the addition of a micro-controller to enable remote triggering and saving of spectral data. The spectrometer operated on an independent power source and its low weight and fine spectral resolution made it ideal for use on an UAV. The full specifications, calibration efforts and validation of the STS spectrometer are presented in Burkart et al. (2013). An identical spectrometer, on the ground, acquired spectra of incoming radiance every time the airborne sensor was triggered. Spectra were saved on a micro SD card.

Sony RGB camera: a SONY Nex5n (Sony Corporation, Minato, Japan) modified by AscTec was attached to the Falcon-8 using a specially designed camera mount. A live video feed from the camera to the UAV operator and remote triggering were available. Spectral sensitivity was given by

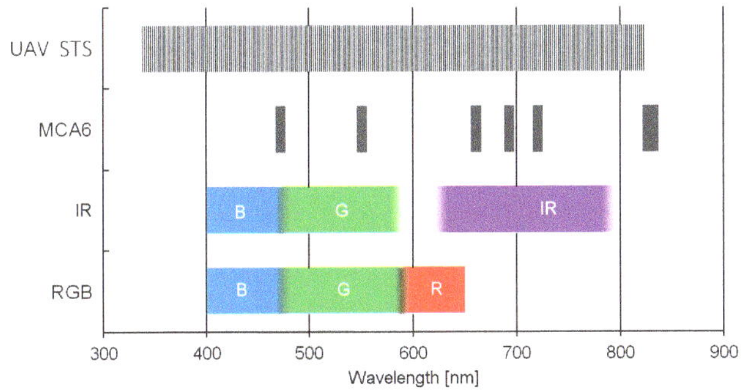

Figure 2. Spectral sensitivity of the four sensors. Spectral bands are indicated by different colours.

the common Bayer matrix (Bayer, 1976; Hirakawa et al., 2007) and hot mirror used in consumer digital cameras.

Canon PowerShot camera: the Canon PowerShot SD780 IS is a consumer digital camera that has been professionally (LDP LLC, Carlstadt, US) converted to acquire near-infrared imagery. The near-infrared filter has been replaced with a red-light-blocking filter. Again, the spectral response of the camera is based on the Bayer pattern colour filter array. Customized CHDK (Canon Hack Development Kit) firmware allows running the camera in a continuous capture mode at specific time intervals (2 s, user defined). Camera acquisition was set to automatic as time constraints and UAV batteries did not allow for accurate manual configuration of white balance, aperture, ISO and shutter speed. Images were saved as JPEGs. A live video link from the UAV's on-board camera enabled precise positioning of the RGB and infrared cameras over the ryegrass pastures. The main difference to the MCA6 is the inability to adjust filter settings and the camera's bandwidths. According to manufacturer information each band has an approximate width of 100 nm.

1.4 Ground-based sensors

ASD HandHeld 2 ground-based reference sensor: ground-based spectral measurements were acquired with an ASD HandHeld 2 portable spectroradiometer (Analytical Spectral Devices, Inc., Boulder, Colorado, US). The device covers a spectral range from 325 nm to 1075 nm which makes it suitable for comparison with all UAV sensors flown in this study. At 700 nm the device has a spectral resolution of 3 nm and the field of view equates to 25°. A Spectralon® panel (Spectralon®, Labsphere, Inc., North Sutton, NH, USA) was used to acquire white reference measurements before each target measurement. Each target was measured 10 times from 1 m distance while moving over the area of interest.

1.5 Flight planning and data acquisition procedure

Taking into account the operational requirements of each sensor and flying platform, a detailed flight plan was developed. Eight sampling locations defined by waypoints were selected from overview images and supported by an in situ visual assessment of the paddock. A focus was put on covering a wide range of pasture qualities from dry to fully irrigated ryegrass pastures. Waypoints were selected in paddock areas with homogeneous pasture cover. This ensured that each waypoint can be considered representative for the area of the paddock it is located in, and it aided dealing with the different sensor footprint sizes (Table 4).

Each sampling location was marked with a tarpaulin square, which was clearly visible in all spectral bands of the aerial images. In order to avoid interference effects of the markers with the UAV STS measurements, they were removed before acquisition of spectra. Next to the first waypoint, a calibration site with coloured tarpaulin squares was set-up and measured with the ASD HandHeld 2.

The sensors were flown over the targets in the following order: (1) RGB camera for an overview shot, (2) IR camera for an overview shot, (3) MCA6 over calibration sites (black, grey, white and red tarpaulins black foam material, bare soil) and waypoints and (4) UAV spectrometer over waypoints. Overview images cover all sampling locations in an area with a single shot from 100 to 150 m flying height. MCA6 images were taken from 25 m above the ground. UAV STS data were collected from a height of 10–15 m, and 15 spectra were taken over each waypoint. During the experiment, the Falcon-8 was flown in semi-autonomous GPS mode. Coordinates of the sampling locations were recorded with a low-accuracy GPS (Legend, HTC, Taoyuan, Taiwan). The Falcon-8 used those coordinates to autonomously reach the marker locations. Over each sampling location, the flight mode was then switched to manual and the UAV was positioned over the target as accurately as possible using a live video link. The UAV STS and the live camera were on the same stabilized gimbal and aligned in a way, that the centre of the FPV camera approximates the UAV STS's field of view. The QuadKopter was flown in manual mode during the entire experiment. In test flights preceding this experiment, it was found that the GPS on board of the MikroKopter was not accurate enough to position the sensor over a waypoint.

Flights were conducted consecutively to minimize variability due to changing illumination and vegetation status. Figure 3 depicts raw data from the imaging sensors before any processing has been applied. Before the flight of the UAV spectrometer, ASD ground reference measurements were taken at each waypoint.

1.6 Data processing

Data from each sensor underwent calibration and correction procedures.

MCA6: a proprietary software package (PixelWrench2 by Tetracam) that was delivered with the Tetracam was used to transfer images from the CompactFlash memory cards to the computer. Each RAW band was processed to a TIFF (Tagged Image File Format) image in order to identify all images that

Table 2. Sensor properties.

Name	Sony Nex5n RGB	Canon Powershot IR	MCA6	STS
Company	Sony – modified	Canon – modified	Tetracam	Ocean Optics – modified
Type	RGB camera integrated	VIS + Infrared camera in the Falcon-8 UAV	Multispectral Imager with 6 bands of 10 nm width	Spectroradiometer with additional electronics for remote control
Field of View	73.7° × 53.1°	57.2° × 40°	38.3° × 31.0°	12°
Spectral bands	3	3	6	256
Spectral range	Blue, Green, Red	Blue, Green, IR	450–1000 nm	338–824 nm
Image size	4912 × 3264	4000 × 3000	1280 × 1024	n/a
Image format	JPEG	JPEG	RAW	n/a
Dynamic Range	8 bit	8 bit	10 bit	14 bit
Weight [g]	500	100	790	216
Handling	Wireless trigger, live view	Interval mode	Interval mode	Wireless trigger, live view

Table 3. MCA6 filter specifications.

	Slave 1	Master	Slave 2	Slave 3	Slave 4	Slave 5
Centre wavelength FWHM (nm)	473	551	661	693	722	831
Bandwidth FWHM (nm)	9.26	9.72	9.73	9.27	9.73	17.81
Peak transmission (%)	64.37	72.54	61.4	66.89	63.63	65.72

show the target area. As a result, between 6 and 15 images per target were found to be suitable for further image processing (total of 109 images) and two images showing the tarpaulin areas and bare soil were selected for reflectance factor calibration. From there, RAW image processing was done in Matlab (The MathWorks Inc., 2011). Both the calibration images and the vegetation target images were noise corrected and vignetting effects were removed for each of the six cameras (Yu, 2004; Olsen et al., 2010; Kelcey and Lucieer, 2012). A sensor correction factor was applied to each filter based on filter sensitivity factory information (Kelcey and Lucieer, 2012).

UAV STS: as described in Burkart et al. (2013) a temperature-based dark current correction (Kuusk, 2011) and an inter-calibration of the air- and ground-based spectrometer were applied before derivation of reflectance factors.

Sony RGB Camera: the red, green and blue bands were calibrated to a reflectance factor with the empirical line method (Smith and Milton, 1999; Baugh and Groeneveld, 2008) relating the ASD reflectance over the coloured reference tarpaulins (Fig. 3) to real reflectance (Aber et al., 2006).

Canon infrared camera: the camera was corrected using the same method as for the RGB camera, but with the centre wavelengths adapted to the infrared sensitive pixels.

The images that show the tarpaulin and the bare soil were selected as calibration images and processed separately. The white and the red tarpaulins were excluded from analysis due to pixel saturation and high specular reflection. For each of the calibration surfaces (black, grey, black foam and bare soil) a subset image area was defined from which the pixel values for the empirical line method were derived.

For each calibration target, ten ASD reference spectra were convolved to the spectral response of the Mini-MCA6 (see Spectral Convolution). The empirical line method was applied to establish band-specific calibration coefficients. Using those coefficients, the empirical line method was applied to each vegetation target image on a pixel-by-pixel basis, thus converting digital numbers of the image pixels to a surface reflectance factor.

In order to extract the footprint area over which ground ASD and UAV spectrometer data had been acquired, the relevant image area was identified and extracted from each image by identifying the markers in the image. Footprints were matched between sensors by defining a 0.3 by 0.3 m area below the waypoint marker as the region of interest. An average reflectance factor was calculated for each footprint resulting in between 6 and 15 values per sample location for the MCA6 images. Standard deviations, mean and median were calculated for each waypoint.

ASD HandHeld 2 ground reference sensor: ASD HandHeld 2 spectral binary files were downloaded and converted to reflectance using the HH2Sync software package (Version 1.30, ASD Inc.). Spectral data were then imported into the spectral database SPECCHIO (Hueni et al., 2009).

Spectral Convolutions: in order to synthesize STS spectrometer data from ground-based ASD data, a discrete spectral convolution was applied (Kenta and Masao, 2012). Each STS band was convolved by applying Eq. (1), using a Gaussian function to represent the spectral response function of each STS band. These spectral response functions (SRFs) were parameterized by the calibrated centre wavelengths of the STS instrument and by a nominal FWHM (full width at half maximum) of 3 nm for all spectral bands. The discrete

Table 4. Optical sensor footprint properties.

	UAV STS	MCA6	Canon IR	Sony RGB	ASD
Footprint shape	Circular	Rectangular	Rectangular	Rectangular	Circular
Footprint size [Sensor height (m)]	Ø 2.1 m [10]	17.3 × 13.9 m [25]	109.0 × 72.8 m [100]	149.9 × 99.9 m [100]	Ø 0.44 m [1]
Number of pixels	n/a	1280 × 1024	4000 × 3000	4912 × 3264	n/a
Ground resolution (m)	n/a	0.0135	0.0273	0.0305	n/a

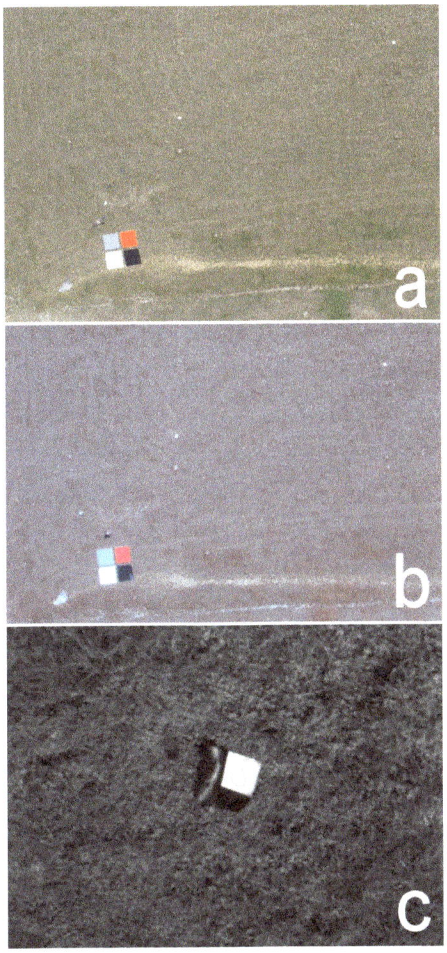

Figure 3. Raw data from the imaging sensors **(a)** RGB camera at 100 m altitude, **(b)** IR camera at 100 m altitude, **(c)** MCA6 at 25 m altitude (red band). The images show the region of interest cropped from a larger image. White points represent the tarpaulin waypoint markers.

convolution range (nm) of each band was based on $\pm 3\sigma$ of the Gaussian function and applied at the wavelength positions where an ASD band occurred, i.e. at every nanometre. It must be noted that the results of this convolution cannot truly emulate the actual system response of the STS as the ASD sampled input spectra are already a discrete representation of the continuous electromagnetic spectrum and are hence already inherently smoothed by the measurement process of the ASD.

In a similar manner, MCA6 bands were simulated, but having replaced the Gaussian assumption of the SRFs with the spectral transmission values (Table 3) digitized from analogue figures supplied by the filter manufacturer (Andover Corporation, Salem, US).

$$R_k = \frac{\sum_{j=n}^{m} c_j R_j}{\sum_{j=n}^{m} c_j}, \quad (1)$$

where R_k = reflectance factor of Ocean Optics band k, R_j = reflectance factor of ASD band j, c_j = weighting coefficient based on the Ocean Optics STS, spectral responsivity at wavelength of ASD band j, $n : m$ = convolution range of Ocean Optics band k.

2 Results

MCA6 and UAV STS: calibrated reflectance factors of the UAV spectrometer and the MCA6 were compared to calculated ASD reflectance values using linear regression analysis. The UAV STS and the ASD HandHeld 2 were compared over the whole STS spectrum, while the MCA6 was compared to the ASD in its six discrete bands.

Figure 4 shows the spectral information derived from both the STS spectrometer and MCA6 in direct comparison with the convolved ASD-derived reflectance spectra for two distinctively different waypoints in terms of ground biomass cover and greenness of vegetation. Waypoint 2 is a recently grazed pasture with a high percentage of dead matter and senescent leaves. Soil background reflectance was high and the paddock was very dry, with no irrigation scheme operating. Pasture at waypoint 8 had not been grazed recently and therefore vegetation cover was dense with a mix of ryegrass pastures and clover. The paddock undergoes daily irrigation and no soil background signal was detectable. The data indicated that the MCA6 estimates higher reflectance factors than the UAV spectrometer and the ASD for the blue, green and the lowest red band. In the far-red and NIR bands, values were consistently lower than those derived from the ASD but still higher than reflectance measured by the UAV STS. While the ASD detected a steep increase in reflectance in the

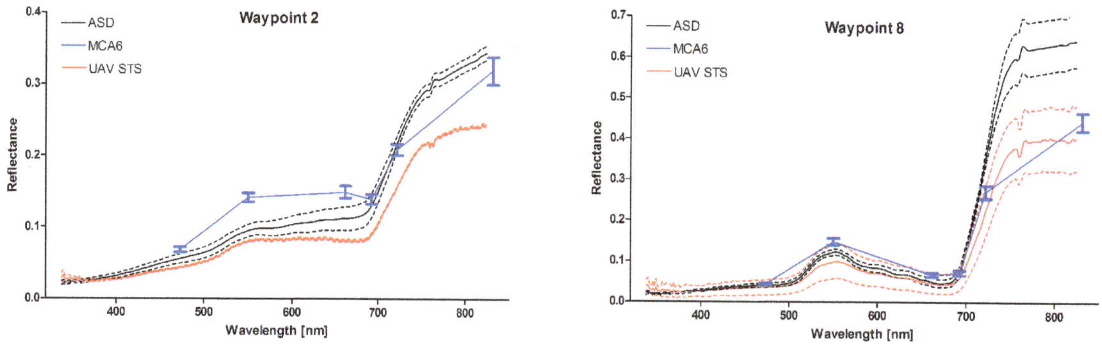

Figure 4. Reflectance of the spectral sensors ASD (black), MCA6 (blue) and UAV STS (red) as measured over the exemplary waypoints 2 and 8. SD in dotted lines for the ASD and UAV STS and with error bars for the 6 bands of the MCA6.

Table 5. Correlation matrix of the optical sensors (R^2). Values were calculated for corresponding bands of each sensor pair over all waypoints. Number of images (n) is given in brackets.

	RGB	IR	MCA6	UAV STS
RGB	1			
IR	0.913 (16)	1		
MCA6	0.377 (16)	0.945 (16)	1	
UAV STS	0.681 (24)	0.891 (24)	0.826 (48)	1
ASD	0.674 (24)	0.647 (24)	0.924 (48)	0.978 (3856)

red edge, both UAV sensors detected a lower signal in the same region of the spectrum.

The mean MCA6-derived spectra showed an increase in reflectance in the green peak region of the vegetation spectrum that is approximately 0.05 % higher than in the same region of the UAV spectrometer. The slope between the green and the red bands is positive for both sensors demonstrating the dried, stressed state of the vegetation at waypoint 2. While MCA6 bands show low correlations with the UAV STS and the ASD for the 551 nm and the 661 nm bands, its values are in line with the other sensors in the red-edge region of the spectra.

The MCA6 correlates significantly with ASD-derived reflectance (R^2 0.92, Fig. 5, Table 5) when compared over all eight waypoints and over all six-bands ($n = 48$). Shortcomings of spectral accuracy of the MCA6 are revealed when comparing band reflectance values over different sample locations and per waypoint (Fig. 6). The green band (551 nm) achieves lowest correlations with ASD convolved reflectance values ($R^2 = 0.68$), with MCA6 reflectance factors overestimated for all waypoints. The remaining five bands show correlations with R^2 between 0.70 (722 nm) and 0.97 (661 nm). Overall, the MCA6 overestimates bands below the red edge, while it shows low deviations from the STS- and the ASD-derived reflectance values for the red-edge bands. Due to the low number of waypoints, the blue-, green- and red-band correlations need to be interpreted with caution. With an

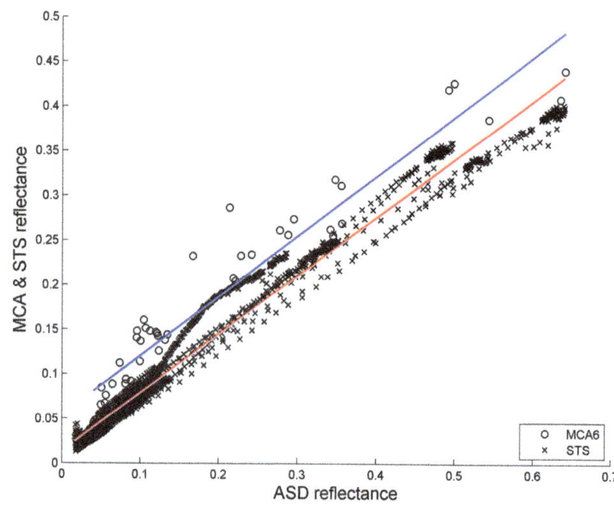

Figure 5. Reflectance comparison of UAV-based sensors to convolved ASD-derived reflectance showing data over all eight sample locations and spectra (MCA6 $n = 48$, STS $n = 120$). MCA6 vs. ASD (blue): $R^2 = 0.92$, slope of linear regression: 0.6691, offset: 0.0533. STS vs. ASD (red): $R^2 = 0.98$, slope of linear regression: 0.6522, offset: 0.0142.

R^2 of 0.98, the UAV spectrometer strongly correlates to the reflectance derived from the ASD when compared over all waypoints (Table 4). Even though the trend of the spectra is similar to the ASD ground truth, differences are visible in the magnitude of the reflectance mainly in the near-infrared.

RGB and NIR camera: as can be seen in Table 4, the correlation between the RGB and IR cameras results in an R^2 of 0.91, whereas the correlations to the high-resolution spectrometers are as low as 0.65 between the NIR camera and the ASD. The RGB camera and MCA6 are poorly correlated with a R^2 of 0.38.

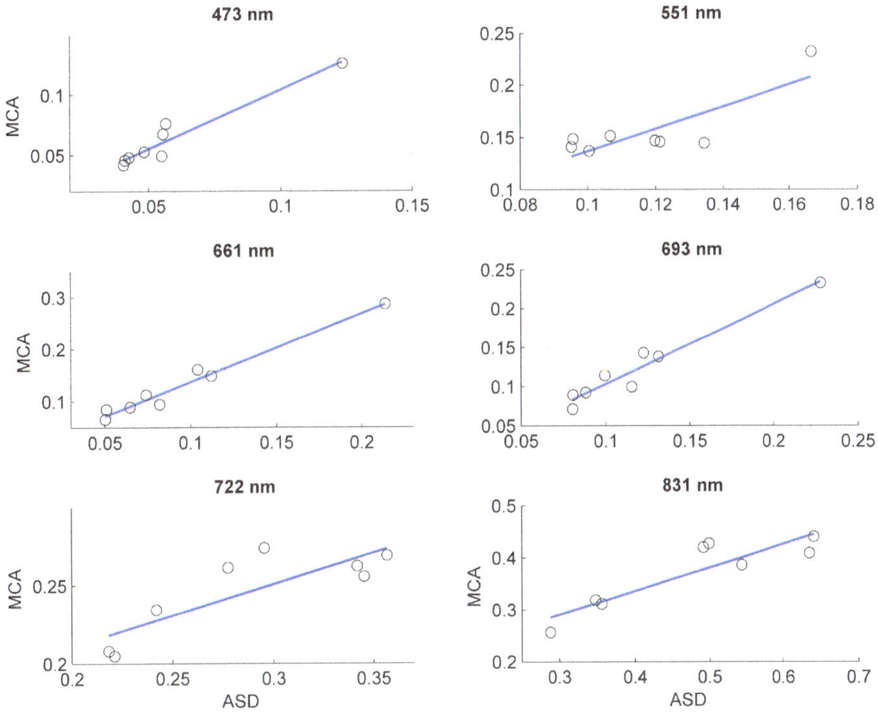

Figure 6. Comparison of reflectance values between MCA6 and convolved ASD reflectance for each MCA6 band. 473 nm: $R^2 = 0.93$, regression slope (RS): 0.9783; 551 nm: $R^2 = 0.68$, RS: 1.0654; 661 nm: $R^2 = 0.97$, RS: 1.311; 693 nm: $R^2 = 0.95$, RS: 1.0225; 722 nm: $R^2 = 0.7$, RS: 0.4009; 831 nm: $R^2 = 0.8$, RS: 0.4516.

3 Discussion

MCA6: when compared to the UAV spectrometer and the ground reference data, the MCA6 filters performed well in the red-edge region of the electromagnetic spectrum. This observation is supported by the CMOS sensor relative sensitivity which is over 90 % in the red-edge and the near-infrared bands according to factory information (Tetracam Inc.). The largest deviations were observed in the green band, where the MCA6 consistently overestimates vegetation reflectance factors. In sample locations with low biomass cover and/or stressed pastures, this results in a negative slope between the red bands. The sensor's performance is further impaired when high soil background reflectance is present, as is the case for the first three waypoints and the bare soil calibration target. While the green peak in the UAV STS and ASD measurements is barely visible over waypoint 2 but pronounced for waypoint 8, the MCA does not pick up on that feature. Green-band reflectance is overestimated for the drier pasture, while deviations from the other sensors' measurements over irrigated, greener pasture are lower. Those differences must be put down to radiometric inconsistencies in the MCA6 and potential calibration issues and it suggests that with the current filter setup, the MCA6 cannot be regarded as suitable for remote sensing of biochemical constituents and fine-scale monitoring of vegetation variability. Another complexity can be seen in the near-infrared regions of the derived spectra. For the UAV STS, MCA6 and the ASD, the variability of measured reflectance factors increases. This discrepancy is likely to arise from a combination of areas of different spatial support in terms of the sensor's field-of view (FOV) and calibration biases (sensor and reflectance calibration). Further investigation into sensor performance over targets with complex spectral behaviour must be conducted in order to evaluate the spectral performance of those bands. The number of waypoints visited was not high enough to fully assess the performance of the four lower MCA6 bands as can be seen in Fig. 6. Due to the statistical distribution of the data points, a definite statement on the performance of those bands is not possible. The empirical line method used for reflectance calibration introduces further errors because only one calibration image was acquired over the entire measurement procedure. Reflectance factor reliability can be improved by more frequent acquisition of calibration images.

UAV STS: the UAV STS-delivered spectra with strong correlations to the ASD measurements. The calculation of narrow-band indices or spectral fitting algorithms is thus possible. However, depending on the status of the vegetation target the ASD-derived reflectance factors can be up to 1.5 times (Fig. 4) higher than the UAV STS measurements. This result, particularly striking in the NIR, is below expectations, as Burkart et al. (2013) compared the Ocean Optics spectrometer (UAV STS) to an ASD Field Spec 4 and reported good agreements between the two instruments. The main source of discrepancies between the ASD and STS measurements can be attributed to inconsistencies in footprint matching due to using a live feed from a camera that can only approximate the spectrometer's field of view. By choosing homogeneous surfaces and averaging over multiple measurements, parts of the problems arising from footprint were addressed in this study. However the matching of the footprint of two different spectrometers can go beyond comparing circles and rectangles due their optical path as recently shown by MacArthur et al. (2012). A more thorough inter-comparison of the ASD and the particular Ocean Optics device employed on the UAV will be required in the future.

RGB and NIR cameras: an empirical line calibration was used for the reflectance factor estimation of both consumer RGB and infrared-modified cameras. Although correlations between the digital cameras and the high-resolution spectrometers exist, they must be treated with caution. This is due to the unknown radiometric response of the cameras, band overlaps and the inherent differences between simple digital cameras and numerical sensors. Both cameras provide imagery with high on-ground resolution, thus enabling identification of in-field variations. In terms of the NIR camera, the wide bandwidth and limited information on the spectral response call for cautious use and further evaluation if the camera is to be used for quantitative vegetation monitoring. At this stage, this study can only suggest that the sensor might be used for support of visual paddock assessment and broadband vegetation indices. Nevertheless, the results demonstrate the opportunities these low-budget sensors offer for simple assessment of vegetation status over large areas using UAVs. If illumination conditions enable an empirical line calibration, reasonable three-band reflectance results can be calculated. Further improvements of radiometric image quality can be expected from fixed settings of shutter speed, ISO, white balance and aperture, as well as for the use of the RAW format. A calibration of lens distortion and vignetting parameters could further increase the quality, especially in the edges of the image (Yu, 2004). However, operational efficiency increases with automatic camera settings which only varied minimally due to the stable illumination conditions at the time of the study.

The empirical line method that was used for reflectance calibration was based on some simplifications. Variations in illumination and atmospheric conditions require frequent calibration image acquisition in order to produce accurate radiometric calibration results. Due to the conservative management of battery power and thus relatively short flight times, only one MCA6 flight was conducted to acquire an image of the calibration tarpaulins and the bare soil. The same restriction applies to the quality of the radiometric calibration of the RGB and IR camera. The use of colour tarpaulin surfaces as calibration targets has implications on the quality of the achieved reflectance calibration in this study. Although they provide low-cost and easy-to-handle calibration surfaces, they are not as spectrally flat as would be needed for a sensor calibration with minimum errors. Moran et al. (2001, 2003) have investigated the use of chemically treated canvas tarpaulins and painted targets in terms of their suitability as stable reference targets for image calibration to reflectance and introduce measures to ensure optimum calibration results. They concluded that specially painted tarps could provide more suitable calibration targets for agricultural applications.

Discrepancies in measured reflectance factors between the UAV STS, the MCA6 and the ASD arise from a combination of factors. Foremost, inherent differences in their spectral and radiometric properties lead to variations in measured reflectance factors. Deviations in footprint matching between the STS spectrometer and the ground measurements, although kept to a minimum, lead to areas of different spatial support and cannot be fully eliminated. Another dimension to this complexity is added by the UAVs and the camera gimbals. Although platform movements were minimal due to the stable environmental conditions and the compensation of any small platform instabilities by the camera gimbals, a small variability in measured radiant flux must be attributed to uncertainties in sensor viewing directions. For a complete cross-calibration between the UAV-based and ground sensors, these potential error sources need to be quantified. Within the context of evaluating sensors for their usability and potential for in-field monitoring of vegetation, those challenges were not addressed in the current study.

In-field data acquisition and flight procedures, one of the key challenges in accommodating four airborne sensors over the same area of interest is accurate footprint matching and minimizing any errors that are introduced by this complexity. Camera gimbals, on board GPS software, piloting skills and waypoint selection maximized footprint matching between sensors. The Falcon-8 UAV was capable of a very stable hover flight over the area of interest while the MikroKopter UAV required manual piloting to ensure that it hovered over the area of interest. The tarpaulin markers were invaluable as a visual aid both during piloting of the UAVs and during subsequent image processing for identifying the footprint areas in each image. Because of the need to select waypoints that were representative for a large area of the paddock, the stable hovering behaviour of the Falcon-8 ensured that the UAV spectrometer's footprint was comparable to the other sensors' field of view. Although the described measures and precautions enabled confident matching of footprints, they can only be applied when working in homogeneous areas of pasture and vegetation cover. Confounding factors, such as soil background influence, large variations in vegetation cover inside the footprint area and strong winds that destabilise the platform, will compromise accurate footprint matching.

When acquiring data with UAVs, responses to changes in environmental conditions, such as increasing wind speeds and cloud presence, need to be immediate. Although specifications from UAV manufacturers attest that the flying vehicles are able to cope with winds of up to 30 km h^{-1}, in reality the wind speed at which a flight must be interrupted is considerably lower. Platform stability, altitude control and footprint matching accuracy between sensors are compromised under high winds. The fact that two different UAV platforms had been used potentially introduces more variability that cannot be quantified. However, the aforementioned payload restrictions make the use of two different platforms inevitable. Due to the fast progress in UAV platform development, this intricacy is likely to be irrelevant in the future as platforms become more versatile and adaptable to accommodate various sensor requirements.

Technical specifications of UAVs: both UAVs were powered with lithium polymer (LiPo) batteries. A fully charged battery enabled flying times of approximately 10 min for the payload carried. With only four batteries available for each UAV, this lead to a data acquisition time frame of about 40 min per flying platform. However, because turbulence, unplanned take offs and landings and inaccurate GPS positions frequently required revisiting a waypoint, the total number of sample locations that could be investigated between 11:00 and 15:00 LT when illumination conditions were most favourable, was low. This makes thorough flight planning, marking of waypoints and efficient collection of ground reference data essential. Due to the non-availability of power outlets and the time it takes to fully recharge a LiPo battery, battery life limits the time frame in which airborne data can be collected. At the time of the study, higher powered LiPo batteries were still too heavy, thus neutralizing a gain in flight time due to the high weight of the more powerful battery. Those restrictions can slow down data acquisition considerably and the number of ground sampling locations is limited. In the future, improvements in platform stability and electronics as well as higher powered batteries will enable larger ground coverage by UAVs. Using in-field portable charging options such as powered from car batteries can significantly enhance the endurance of rotary wing UAVs.

The evaluated UAV sensors differ in their suitability for deployment in vegetation monitoring and more specifically pasture management applications. While high spectral accuracy is essential for quantifying parameters such as nutritional status in crops and pastures, the high spatial resolution imaging ability of digital cameras can be used to assess paddocks and fields with regard to spatial variations that may not be visible to a ground observer.

Usability of sensors: the UAV STS spectrometer with its high spectral resolution can be used to derive narrowband vegetation indices such as the PRI (photochemical reflectance index) (Suarez et al., 2009) or TSAVI (transformed soil adjusted vegetation index) (Baret et al., 1989). Furthermore, its narrow bands facilitate identification of wavebands that are relevant for agricultural crop characterization (Thenkabail et al., 2002). Once those centre wavelengths have been identified, a more broadband sensor such as the MCA6 could target crop and pasture characteristics with specific filter setups provided the MCA6 performance can be enhanced in terms of radiometric reliability. The consumer digital cameras seem to be useful for derivation of broadband vegetation indices such as the green NDVI (Gitelson et al., 1996) or the GRVI (Motohka et al., 2010). Identification of wet and dry areas in paddocks and growth variations are further applications that such cameras can cover. Imaging sensors that identify areas in a paddock that need special attention are extremely useful, and although they do not provide the high spectral resolution of the UAV STS spectrometer, they do give a visual indication of vegetation status.

Challenges and limitations: deploying UAVs is a promising new approach to collect vegetation data. As opposed to ground-based proximal sensing methods, UAVs offer nondestructive and efficient data collection and less accessible areas can be imaged relatively easy. Moreover, UAVs can potentially provide remote sensing data when aircraft sensors and satellite imagery are unavailable. However, three main factors can cause radiometric inconsistencies in the measurements: sensors, flying platforms and the environment.

The sensors mounted on the UAVs introduce the largest level of uncertainty in the data. Radiometric aberrations across the camera lenses can be addressed by a flat field-correction of the images.

Further factors are camera settings. In this study, shutter speed, exposure time and ISO were set on automatic because of the clear sky and stable illumination conditions. However, to facilitate extraction of radiance values and quantitative information on the vegetation, these settings need to be fixed for all the flights in order to make the images comparable.

The RAW image format is recommended when attempting to work with absolute levels of radiance as it applies the least alterations to pixel digital numbers.

Furthermore, footprint matching between sensors with different sizes and shapes is challenging. While it is straightforward for imaging cameras with rectangular shaped footprints, matching measurements between the UAV STS, ASD and the imaging sensors can only be approximated. While footprint shape is fixed, the size can be influenced by the flying altitude above ground.

However, it is also important to be aware of any bidirectional effects that are introduced as a result of the camera lens' view angle and illumination direction (Nicodemus, 1965).

Although UAV platforms are equipped with gyrostabilization mechanisms, GPS chips and camera gimbals, an uncertainty remains of whether the camera is in fact pointing nadir and at the target. Slight winds or a motor imbalance can destabilise the UAV system enough to cause the sensor field of view to be misaligned. For imaging sensors this is less of an issue as it is for numerical sensors such as the UAV STS. The live view will only ever be an approximation of the sensor's actual FOV. Careful setting up of the two systems on the camera gimbal and periodical measurement of known targets to align the spectrometer's FOV to the live view camera can help to minimise deviations between FOVs.

The environment also needs to be considered for the collection of robust radiometric data. Even if all other factors are perfect, winds or wobbling of the platform caused by, e.g., a motor imbalance or a bad GPS position hold can cause the sensor to direct its FOV to the wrong spot. In terms of the imaging cameras this is again simple to check after image download whereas the UAV STS data might possibly not show any deviations in the data.

With a good knowledge of the sensors characteristics and the necessary ground references an UAV operator will be able to acquire satisfying data sets, if the environmental conditions are opportune. Based on a tested UAV with known uncertainties in GPS and gimbal accuracy the data set can be quality flagged and approved for further analysis.

4 Conclusions

UAVs are rapidly evolving into easy-to-use sensor platforms that can be deployed to acquire fine-scale vegetation data over large areas with minimal effort. In this study, four optical sensors, including the first available UAV-based micro-spectrometer were flown over ryegrass pastures and cross-compared. Overall, the quality of the reflectance measurements of the UAV sensors is dependent on thorough data acquisition processes and accurate calibration procedures. The novel high-resolution STS spectrometer operates reliably in the field and delivers spectra that show high correlations to ground reference measurements. For vegetation analysis, the UAV STS holds potential for feature identification in crops and pastures as well as the derivation of narrow-band vegetation indices. Further investigations and cross-calibrations are needed, mainly with regard to the near-infrared measurements in order to establish a full characterization of the system. It was also demonstrated that the six-band MCA6 camera can be used as a low spectral resolution multispectral sensor with the potential to deliver high-resolution multispectral imagery. In terms of its poor radiometric performance in the green and near-infrared filter regions, it is evident that the sensor needs further testing and correction efforts to eliminate the error sources of these inconsistencies. Over sample locations with low vegetation cover and strong soil background interference, the MCA6 image data needs to be processed with caution. Individual filters must be assessed further, with a focus on the green and NIR regions of the electromagnetic spectrum. Any negative effects that depreciate data quality, such as potentially unsuitable calibration targets (coloured tarpaulins) need to be identified and further examined in order to guarantee high quality data. If those issues can be addressed and the sensor is equipped with relevant filters, the MCA6 can become a useful tool for crop and pasture monitoring. The modified Canon infrared and the RGB Sony camera have proven to be easy-to-use sensors that deliver instant high-resolution imagery covering a large spatial area. No spectral calibration has been performed on those sensors, but factory spectral information allowed converting digital numbers to a ground reflectance factor. Near-real-time assessment of variations in vegetation cover and identification of areas of wetness/dryness as well as calculation of broad-band vegetation indices can be achieved using these cameras. A number of issues have been identified during the field experiments and data processing. Exact footprint matching between the sensors was not achieved due to differences in the FOVs of the sensors, instabilities in UAV platforms during hovering and potential inaccuracies in viewing directions of the sensors due to gimbal movements. Although those differences in spatial scale reduce the quality of sensor inter-comparison, it must be stated that under field conditions a complete match of footprints between sensors is not achievable. For the empirical line calibration method that was applied to the MCA6 and the digital cameras, we propose the use of spectrally flat painted panels for radiometric calibration rather than tarpaulin surfaces. To reduce complexity of the experiment and keep the focus on the practicality of deploying multiple sensors on UAVs, the influence of directional effects has been neglected.

The field protocols developed allow for straightforward field procedures and timely coordination of multiple UAV-based sensors as well as ground reference instruments. The more autonomously the UAV can fly, the more focus can be put on data acquisition. Piloting UAVs in a field where obstacles such as power lines and trees are present requires the full concentration of the pilot and at least one support person to observe the flying area. Due to technical restrictions, the total area that can be covered by rotary wing UAVs is still relatively small, resulting in a point sampling strategy. Higher powered, lightweight batteries on UAVs can allow for more frequent calibration image acquisition and the coverage of natural calibration targets, thus improving the radiometric calibration. Differences in UAV specifications and capabilities lead to the UAVs having a specific range of applications that they can undertake reliably.

As shown in this study even after calibration efforts, biases and uncertainties remain and must be carefully evaluated in terms of their effects on data accuracy and reliability. Restrictions and limitations imposed by flight equipment must be carefully balanced with scientific data acquisition protocols. The different UAV platforms and sensors each have their strengths and limitations that have to be managed by matching platform and sensor specifications and limitations to data acquisition requirements. UAV-based sensors can be quickly deployed in suitable environmental conditions and thus enable the timely collection of remote sensing data. The specific applications that can be covered by the presented UAV sensors range from broad visual identification of paddock areas that require increased attention to the identification of waveband-specific biochemical crop and pasture properties on a fine spatial scale. With the development of sensor-specific data processing chains, it is possible to generate data sets for agricultural decision making within a few hours of data acquisition and thus enable the adjustment of management strategies based on highly current information.

Acknowledgements. The research was supported by a Massey University doctoral scholarship granted to S. von Bueren and a travel grant from COST ES0903 EUROSPEC to A. Burkart. The authors acknowledge the funding of the CROP.SENSe.net project in the context of Ziel 2-Programmes NRW 2007–2013 "Regionale Wettbewerbsfähigkeit und Beschäftigung (EFRE)" by the Ministry for Innovation, Science and Research (MIWF) of the state of North Rhine–Westphalia (NRW) and European Union Funds for regional development (EFRE) (FKZ 005-1012-0001) while collaborating on the preparation of the manuscript.

All of us were shocked and saddened by the tragic death of Stefanie von Bueren on 25 August. We remember her as an enthusiastic adventurer and aspiring researcher.

Edited by: M. Rossini

References

Aber, J. S., Aber, S. W., Pavri, F., Volkova, E., and Penner II, R. L.: Small-format aerial photography for assessing change in wetland vegetation, Cheyenne Bottoms, Kansas, Transactions of the Kansas Academy of Science, 109, 47–57, doi:10.1660/0022-8443(2006)109[47:sapfac]2.0.co;2, 2006.

Baret, F., Guyot, G., and Major, D. J.: TSAVI: A Vegetation Index Which Minimizes Soil Brightness Effects On LAI And APAR Estimation, Geoscience and Remote Sensing Symposium, 1989, IGARSS'89, 12th Canadian Symposium on Remote Sensing, International, 1355–1358, 1989.

Baugh, W. M. and Groeneveld, D. P.: Empirical proof of the empirical line, Int. J. Remote Sens., 29, 665–672, doi:10.1080/01431160701352162, 2008.

Bayer, B. E.: Color imaging array, 1976.

Berni, J., Zarco-Tejada, P., Surez, L., González-Dugo, V., and Fereres, E.: Remote sensing of vegetation from uav platforms using lightweight multispectral and thermal imaging sensors, The International Archives of the Photogrammetry, Remote Sensing and Spatial Information Sciences, XXXVII, 2008.

Berni, J. A. J., Zarco-Tejada, P. J., Suarez, L., and Fereres, E.: Thermal and Narrowband Multispectral Remote Sensing for Vegetation Monitoring From an Unmanned Aerial Vehicle, Ieee T. Geosci. Remote, 47, 722–738, doi:10.1109/Tgrs.2008.2010457, 2009.

Burkart, A., Cogliati, S., Schickling, A., and Rascher, U.: A novel UAV-based ultra-light weight spectrometer for field spectroscopy, Sensors Journal, IEEE, 14, 62–67, doi:10.1109/jsen.2013.2279720, 2013.

Carrara, M., Comparetti, A., Febo, P., and Orlando, S.: Spatially variable rate herbicide application on durum wheat in Sicily, Biosys. Eng., 87, 387–392, 2004.

Chávez, J. L., Pierce, F. J., Elliott, T. V., Evans, R. G., Kim, Y., and Iversen, W. M.: A remote irrigation monitoring and control system (RIMCS) for continuous move systems. Part B: Field testing and results, Precis. Agric., 11, 11–26, 2010.

Duan, M., Gao, Q., Wan, Y., Li, Y., Guo, Y., Ganzhu, Z., Liu, Y., and Qin, X.: Biomass estimation of alpine grasslands under different grazing intensities using spectral vegetation indices, Can. J. Remote Sens., 37, 413–421, 2011.

Gitelson, A. A., Kaufman, Y. J., and Merzlyak, M. N.: Use of a green channel in remote sensing of global vegetation from EOS-MODIS, Remote Sens. Environ., 58, 289–298, 1996.

Hirakawa, K., Wolfe, P. J., and Ieee: Spatio-spectral color filter array design for enhanced image fidelity, in: 2007 Ieee International Conference on Image Processing, Vols 1–7, IEEE International Conference on Image Processing ICIP, 645–648, 2007.

Hueni, A., Nieke, J., Schopfer, J., Kneubühler, M., and Itten, K. I.: The spectral database SPECCHIO for improved long-term usability and data sharing, Comput. Geosci., 35, 557–565, doi:10.1016/j.cageo.2008.03.015, 2009.

Hunt, E. R., Hively, W. D., Fujikawa, S. J., Linden, D. S., Daughtry, C. S. T., and McCarty, G. W.: Acquisition of NIR-Green-Blue Digital Photographs from Unmanned Aircraft for Crop Monitoring, Remote Sens., 2, 290–305, doi:10.3390/Rs2010290, 2010.

Jensen, T., Apan, A., Young, F., and Zeller, L.: Detecting the attributes of a wheat crop using digital imagery acquired from a low-altitude platform, Comput. Electron. Agr., 59, 66–77, doi:10.1016/j.compag.2007.05.004, 2007.

Kawamura, K., Sakuno, Y., Tanaka, Y., Lee, H.-J., Lim, J., Kurokawa, Y., and Watanabe, N.: Mapping herbage biomass and nitrogen status in an Italian ryegrass (Lolium multiflorum L.) field using a digital video camera with balloon system, J. Appl. Remote Sens., 5, 053562, doi:10.1117/1.3659893, 2011.

Kelcey, J. and Lucieer, A.: Sensor Correction of a 6-Band Multispectral Imaging Sensor for UAV Remote Sensing, Remote Sens., 4, 1462–1493, 2012.

Kenta, T. and Masao, M.: Radiometric calibration method of the general purpose digital camera and its application for the vegetation monitoring, Land Surf. Remote Sens., 8524, doi:10.1117/12.977211, 2012.

Kuusk, J.: Dark Signal Temperature Dependence Correction Method for Miniature Spectrometer Modules, J. Sens., 2011, 608157, 2011.

Lebourgeois, V., Bégué, A., Labbé, S., Mallavan, B., Prévot, L., and Roux, B.: Can commercial digital cameras be used as multispectral sensors? A crop monitoring test, Sensors, 8, 7300–7322, 2008.

Lebourgeois, V., Begue, A., Labbe, S., Houles, M., and Martine, J. F.: A light-weight multi-spectral aerial imaging system for nitrogen crop monitoring, Precis. Agric., 13, 525–541, doi:10.1007/s11119-012-9262-9, 2012.

Lelong, C. C. D., Burger, P., Jubelin, G., Roux, B., Labbe, S., and Baret, F.: Assessment of unmanned aerial vehicles imagery for quantitative monitoring of wheat crop in small plots, Sensors, 8, 3557–3585, doi:10.3390/S8053557, 2008.

Link, J., Senner, D., and Claupein, W.: Developing and evaluating an aerial sensor platform (ASP) to collect multispectral data for deriving management decisions in precision farming, Comput. Electron. Agr., 94, 20–28, doi:10.1016/j.compag.2013.03.003, 2013.

MacArthur, A., MacLellan, C. J., and Malthus, T.: The fields of view and directional response functions of two field spectroradiometers, IEEE Geosci. Remote Sens., 50, 3892–3907, 2012.
Moran, M. S., Bryant, R. B., Clarke, T. R., and Qi, J. G.: Deployment and calibration of reference reflectance tarps for use with airborne imaging sensors, Photogram. Eng. Rem. S., 67, 273–286, 2001.
Moran, S., Fitzgerald, G., Rango, A., Walthall, C., Barnes, E., Bausch, W., Clarke, T., Daughtry, C., Everitt, J., Escobar, D., Hatfield, J., Havstad, K., Jackson, T., Kitchen, N., Kustas, W., McGuire, M., Pinter, P., Sudduth, K., Schepers, J., Schmugge, T., Starks, P., and Upchurch, D.: Sensor development and radiometric correction for agricultural applications, Photogram. Eng. Rem. S., 69, 705–718, 2003.
Motohka, T., Nasahara, K. N., Oguma, H., and Tsuchida, S.: Applicability of green-red vegetation index for remote sensing of vegetation phenology, Remote Sensing, 2, 2369–2387, 2010.
Mutanga, O.: Hyperspectral remote sensing of tropical grass quality and quantity, Hyperspectral remote sensing of tropical grass quality and quantity, x + 195 pp.-x + 195 pp., 2004.
Mutanga, O. and Skidmore, A. K.: Red edge shift and biochemical content in grass canopies, ISPRS J. Photogram., 62, 34–42, doi:10.1016/j.isprsjprs.2007.02.001, 2007.
Nebiker, S., Annen, A., Scherrer, M., and Oesch, D.: A Lightweight Multispectral Sensor for Micro UAV – Opportunities for Very High Resolution Airborne Remote Sensing, XXI ISPRS Congress. Beijing, China, 2008.
Nijland, W., de Jong, R., de Jong, S. M., Wulder, M. A., Bater, C. W., and Coops, N. C.: Monitoring plant condition and phenology using infrared sensitive consumer grade digital cameras, Agr. Forest Meteorol., 184, 98–106, doi:10.1016/j.agrformet.2013.09.007, 2014.
Olsen, D., Dou, C., Zhang, X., Hu, L., Kim, H., and Hildum, E.: Radiometric Calibration for AgCam, Remote Sens., 2, 464–477, doi:10.3390/rs2020464, 2010.
Primicerio, J., Di Gennaro, S. F., Fiorillo, E., Genesio, L., Lugato, E., Matese, A., and Vaccari, F. P.: A flexible unmanned aerial vehicle for precision agriculture, Precis. Agric., 13, 517–523, doi:10.1007/s11119-012-9257-6, 2012a.
Primicerio, J., Di Gennaro, S. F., Fiorillo, E., Genesio, L., Lugato, E., Matese, A., and Vaccari, F. P.: A flexible unmanned aerial vehicle for precision agriculture, Precis. Agric., 13, 1–7, 2012b.
Pullanagari, R. R., Yule, I. J., Hedley, M. J., Tuohy, M. P., Dynes, R. A., and King, W. M.: Multi-spectral radiometry to estimate pasture quality components, Precis. Agric., 13, 442–456, doi:10.1007/s11119-012-9260-y, 2012a.
Pullanagari, R. R., Yule, I. J., Tuohy, M. P., Hedley, M. J., Dynes, R. A., and King, W. M.: In-field hyperspectral proximal sensing for estimating quality parameters of mixed pasture, Precis. Agric., 13, 351–369, doi:10.1007/s11119-011-9251-4, 2012b.
Rango, A., Laliberte, A., Herrick, J. E., Winters, C., Havstad, K., Steele, C., and Browning, D.: Unmanned aerial vehicle-based remote sensing for rangeland assessment, monitoring, and management, J. Appl. Remote Sens., 3, 033542, doi:10.1117/1.3216822, 2009.
Sakamoto, T., Gitelson, A. A., Nguy-Robertson, A. L., Arkebauer, T. J., Wardlow, B. D., Suyker, A. E., Verma, S. B., and Shibayama, M.: An alternative method using digital cameras for continuous monitoring of crop status, Agr. Forest Meteorol., 154, 113–126, doi:10.1016/j.agrformet.2011.10.014, 2012.
Sanches, I. D., Tuohy, M. P., Hedley, M. J., and Mackay, A. D.: Seasonal prediction of in situ pasture macronutrients in New Zealand pastoral systems using hyperspectral data, Int. J. Remote Sens., 34, 276–302, doi:10.1080/01431161.2012.713528, 2012.
Seelan, S. K., Laguette, S., Casady, G. M., and Seielstad, G. A.: Remote sensing applications for precision agriculture: A learning community approach, Remote Sens. Environ., 88, 157–169, 2003.
Smith, G. M. and Milton, E. J.: The use of the empirical line method to calibrate remotely sensed data to reflectance, Int. J. Remote Sens., 20, 2653–2662, doi:10.1080/014311699211994, 1999.
Stafford, J. V.: Implementing precision agriculture in the 21st century, J. Agr. Eng. Res., 76, 267–275, 2000.
Suarez, L., Zarco-Tejada, P. J., Berni, J. A. J., Gonzalez-Dugo, V., and Fereres, E.: Modelling PRI for water stress detection using radiative transfer models, Remote Sens. Environ., 113, 730–744, doi:10.1016/j.rse.2008.12.001, 2009.
Swain, K. C., Thomson, S. J., and Jayasuriya, H. P. W.: Adaption of an unmanned helicopter for low altitude remote sensing to estimate yield and total biomass of a rice crop, Trans. ASABE, 53, 21–27, 2010.
Thenkabail, P. S., Smith, R. B., and De Pauw, E.: Evaluation of narrowband and broadband vegetation indices for determining optimal hyperspectral wavebands for agricultural crop characterization, Photogram. Eng. Rem. S., 68, 607–621, 2002.
Turner, D. J.: Development of an Unmanned Aerial Vehicle (UAV) for hyper-resolution vineyard mapping based on visible, multispectral and thermal imagery, School of Geography & Environmental Studies Conference 2011, 2011.
Van Alphen, B. and Stoorvogel, J.: A methodology for precision nitrogen fertilization in high-input farming systems, Precis. Agric., 2, 319–332, 2000.
Vanamburg, L. K., Trlica, M. J., Hoffer, R. M., and Weltz, M. A.: Ground based digital imagery for grassland biomass estimation, Int. J. Remote Sens., 27, 939–950, doi:10.1080/01431160500114789, 2006.
Vescovo, L., Wohlfahrt, G., Balzarolo, M., Pilloni, S., Sottocornola, M., Rodeghiero, M., and Gianelle, D.: New spectral vegetation indices based on the near-infrared shoulder wavelengths for remote detection of grassland phytomass, Int. J. Remote Sens., 33, 2178–2195, doi:10.1080/01431161.2011.607195, 2012.
Yu, W.: Practical anti-vignetting methods for digital cameras, IEEE Transactions on Consumer Electronics, 50, 975–983, 2004.
Zhang, C. and Kovacs, J. M.: The application of small unmanned aerial systems for precision agriculture: a review, Precis. Agric., 13, 693–712, doi:10.1007/s11119-012-9274-5, 2012.

3

Atmospheric inversion of surface carbon flux with consideration of the spatial distribution of US crop production and consumption

J. M. Chen[1,2], J. W. Fung[2], G. Mo[2], F. Deng[3], and T. O. West[4]

[1]International Institute of Earth System Science, Nanjing University, Nanjing, Jiangsu, China
[2]Department of Geography and Program in Planning, University of Toronto, Toronto, Ontario, M5S 3G3, Canada
[3]Department of Physics, University of Toronto, Toronto, Ontario, M5S 3G3, Canada
[4]Joint Global Change Research Institute, Pacific Northwest National Laboratory, College Park, Maryland, USA

Correspondence to: J. M. Chen, (chenj@geog.utoronto.ca)

Abstract. In order to improve quantification of the spatial distribution of carbon sinks and sources in the conterminous US, we conduct a nested global atmospheric inversion with detailed spatial information on crop production and consumption. County-level cropland net primary productivity, harvested biomass, soil carbon change, and human and livestock consumption data over the conterminous US are used for this purpose. Time-dependent Bayesian synthesis inversions are conducted based on CO_2 observations at 210 stations to infer CO_2 fluxes globally at monthly time steps with a nested focus on 30 regions in North America. Prior land surface carbon fluxes are first generated using a biospheric model, and the inversions are constrained using prior fluxes with and without adjustments for crop production and consumption over the 2002–2007 period. After these adjustments, the inverted regional carbon sink in the US Midwest increases from 0.25 ± 0.03 to 0.42 ± 0.13 Pg C yr^{-1}, whereas the large sink in the US southeast forest region is weakened from 0.41 ± 0.12 to 0.29 ± 0.12 Pg C yr^{-1}. These adjustments also reduce the inverted sink in the west region from 0.066 ± 0.04 to 0.040 ± 0.02 Pg C yr^{-1} because of high crop consumption and respiration by humans and livestock. The general pattern of sink increases in crop production areas and sink decreases (or source increases) in crop consumption areas highlights the importance of considering the lateral carbon transfer in crop products in atmospheric inverse modeling, which provides a reliable atmospheric perspective of the overall carbon balance at the continental scale but is unreliable for separating fluxes from different ecosystems.

1 Introduction

Human activities have greatly modified the global carbon cycle through fossil fuel consumption, cement production, and land use (Canadell et al., 2007; Le Quéré et al., 2013). The airborne fraction of these carbon sources has been highly variable from year to year, mostly due to large variations in the terrestrial carbon sink (Le Quéré et al., 2009, 2013). Due to the complexity and heterogeneity of land cover, it has been a challenge to estimate the spatial distribution and magnitude of terrestrial carbon sources and sinks. It has been more reliable thus far to derive the terrestrial sink as a residual of the global carbon budget than to estimate it using land-based data (Le Quéré et al., 2013). Our ability to project the carbon cycle and estimate its influence on climate will remain limited if we cannot resolve the current carbon source and sink distribution patterns and provide plausible mechanistic explanations for the patterns. In this regard, regional studies that focus on the spatial distribution of carbon dynamics would be a useful direction for improving our understanding of the global carbon cycle.

With regard to regional carbon cycles, North America may be one of the most closely observed and studied regions in the world. However, the magnitude and the spatial distribution of carbon sinks and sources over the continent are still highly uncertain (Huntzinger et al., 2012; Hayes et al., 2012). In terms of the sink magnitude, atmospheric inversion studies (referred to as "top-down") performed for different time periods from 1992 to 2007 suggest that North America has been a sink ranging from 0.54 to 1.06 Pg C yr^{-1} (Gurney et al., 2002, 2003, 2004; Rödenbeck et al., 2003; Baker et al., 2006; Peters et al., 2007; Deng et al., 2013; Peylin et al., 2013), while 19 biospheric models (referred to as "bottom-up") produced an average sink of 0.6 $Pg\,C\,yr^{-1}$, with a range from −0.7 to 2.2 $Pg\,C\,yr^{-1}$, for North America over the period from 2000 to 2005 (Huntzinger et al., 2012). Using a biospheric model, Turner et al. (2013) estimated that the net ecosystem productivity (NEP) over North America in 2004 was 1.73 ± 0.37 $Pg\,C\,yr^{-1}$, and this NEP value is reduced by 0.616 $Pg\,C\,yr^{-1}$ for the carbon loss due to harvested product emission, river/stream evasion, and fire emission in order to estimate the total land sink. These top-down and bottom-up estimates at the continental scale broadly agree, giving us confidence that North America is a large and important contributor to the global terrestrial carbon sink.

With respect to the spatial distribution of the North American carbon sink, results become more uncertain at higher spatial resolutions. Disaggregation of the sink between North American boreal and temperate regions is plagued with uncertainties. For the boreal region, an inversion study (Fan, 1998) produced a sink of 0.2 ± 0.4 $Pg\,C\,yr^{-1}$ in 1988–1992, while a TransCom 3 experiment (Gurney et al., 2003) showed a source of 0.26 ± 0.39 $Pg\,C\,yr^{-1}$ in 1992–1996. When the TransCom 3 experiment was repeated with updated data, it became a sink of 0.003 ± 0.28 $Pg\,C\,yr^{-1}$ in 1992–1996 (Yuen et al., 2005). Using forest inventory data, Canadian forests were found to be a carbon source of 0.069 $Pg\,C\,yr^{-1}$ in 1985–1989 (Kurz and Apps, 1999), while (Pan et al., 2011a) estimated that forests in Canada and Alaska were a carbon sink of 0.26 ± 0.06 $Pg\,C\,yr^{-1}$ in 1990–2008. A biospheric model calculated a weak sink of 0.05 $Pg\,C\,yr^{-1}$ for Canadian forests during the 1980s to 1990s (Chen et al., 2003). For the conterminous US, with approximately 32 % of the total North America area, estimates of the sink from a set of bottom-up and top-down methods fall in a range from 0.30 to 0.58 $Pg\,C\,yr^{-1}$ in 1980–1989 (Pacala et al., 2001), while TransCom 3 experiments inferred the sink in 1992–1996 to be 0.89 ± 0.22 and 0.82 ± 0.40 $Pg\,C\,yr^{-1}$ for inversions at monthly and annual time steps, respectively (Gurney et al., 2003, 2004; Baker et al., 2006). These differences suggest that uncertainty due to the temporal resolution of inverse modeling is considerable. Peters et al. (2007) developed a carbon assimilation system implemented at weekly time steps, referred to as CarbonTracker (CT), and showed that the sink in 2000–2005 in temperate North America was 0.50 ± 0.60 $Pg\,C\,yr^{-1}$. With a simple atmospheric budgeting approach applied to CO_2 measurements in the inflows and outflows through the troposphere over the conterminous US, Crevoisier et al. (2010) deduced a sink of 0.5 ± 0.4 $Pg\,C\,yr^{-1}$ in 2004–2006. Seven other inversion studies, on average, indicate that temperate North America was a sink of 0.685 ± 0.574 $Pg\,C\,yr^{-1}$ in 2000–2006 (see summary in Hayes et al., 2012). Although these estimates have large uncertainties, they generally indicate that the major sink in North America is located in the temperate region while the boreal region is either a small sink or source.

The locations of carbon sinks and sources within temperate North America are highly uncertain. The 19 biospheric models employed by Huntzinger et al. (2012) generated very different sink and source spatial patterns over the conterminous US, although the average sink mostly appears in forested areas in southeast, northeast and northwest regions. Another bottom-up estimate using long-term modeling and recent remote sensing inputs suggested that forests in the southeast region are large sinks because of their predominant mid-age structure (Zhang et al., 2012). Based on forest inventory data, Williams et al. (2012) deduced that forests along the east and west coasts of the US were large sinks in 2005–2006, with most areas in the southeast region (i.e., Arkansas, Louisiana, Mississippi, and Alabama) having sinks in the range of 110–140 $g\,C\,m^{-2}\,yr^{-1}$. However, CarbonTracker (Peters et al., 2007) repeatedly produces large sinks in cropland and adjacent grassland areas, while the southeast region varied between being a small source and a small sink. Inversion studies that divide North America into 30 regions (11 in the conterminous USA; Deng et al., 2007; Deng and Chen, 2011) indicate broad patterns of the sink distribution in both the southeast forest and Midwestern crop regions. In these two studies, the seasonal variations of the carbon flux from various terrestrial ecosystems modeled by a biospheric model and neutralized at the annual time step were used as the prior flux to constrain the inversion. Under the neutralized flux constraint, the inverted sink may be more or less regarded as the atmospheric signal, although the inversion results are inevitably influenced by errors in transport modeling and other prior fluxes such as fossil fuel, biomass burning, and ocean–atmospheric exchange. Inversion studies by Peters et al. (2007), Deng and Chen (2011), Lauvaux et al. (2012), and Schuh et al. (2013) indicate that the midwest croplands persistently behave as a large regional sink, but this sink generally does not accumulate locally in soil and vegetation due to lateral transfer of agricultural products and therefore could not be estimated from the local carbon stock changes. From the atmospheric perspective, crop production during the growing season results in large uptake of CO_2 from the atmosphere, while crop consumption after harvest takes place in areas outside of croplands. Gourdji et al. (2012) found a large discrepancy between inverse and biospheric model results over crop production and consumption areas in both growing and dormant seasons and attributed this discrepancy to the lack of lateral transfer of agricultural carbon included in biospheric models, although poor biospheric model performance for crops (Huntzinger et al., 2012) could also be a reason for the discrepancy. Similarly, lateral transfer of carbon also occurs with forest products (Hayes et al., 2012). Hayes et al. (2012) successfully reconciled a large portion of this discrepancy using crop and forest product information for North America. Based on this study, it can be inferred that these lateral transports of carbon

Table 1. Summary of background prior fluxes and their uncertainties.

Background flux	Data source	Temporal variability	Uncertainty
Fossil fuel	CDIAC (Marland et al., 2009) + EDGAR 4 database (Olivier and Aardenne, 2005)	Interannual	n/a
Fire	GFEDv2 (Randerson et al., 2007)	Interannual	n/a
Biosphere	BEPS model (Chen et al., 2012; Ju et al., 2006)	Interannual, seasonal, diurnal	2.0 Pg C yr^{-1} (Gurney et al., 2003) distributed globally over land surfaces regions based on spatial pattern of the GPP
Ocean	OPA-PISCES-T model (Buitenhuis et al., 2006)	Seasonal	0.67 Pg C yr^{-1} distributed over ocean regions (Deng and Chen, 2011)

need to be considered in both bottom-up and top-down modeling in order for them to converge on similar spatial patterns of the carbon sink and source distribution.

The necessity of including lateral transfer of carbon in the prior flux for constraining inverse modeling can be questioned because it could be argued that atmospheric CO_2 measurements have already integrated the outcome of all carbon cycle processes including the lateral movement of carbon both at the surface and in the atmosphere. Theoretically, this argument is well grounded if we have sufficient atmospheric CO_2 measurements, and it could possibly hold true for North America, which is one of the most densely observed regions in the world with respect to atmospheric CO_2. However, the extent to which the carbon source and sink distribution over North America is determined by atmospheric CO_2 measurements has not been systematically assessed. The Bayesian synthesis inversion framework, in which a prior flux is used to constrain the inversion (Enting, 2002), provides an ideal tool for this assessment. The extent to which the inverted flux distribution is influenced by the inclusion of lateral carbon transfer in the prior flux could be an indicator of the strength of existing atmospheric CO_2 measurements on the carbon cycle relative to the prior flux.

In this study, we attempt first to include crop production and consumption information in the prior flux for constraining our existing inverse modeling system (Deng and Chen, 2011) and then to assess the necessity of this inclusion for determining the carbon source and sink distribution over North America. The consumption of crop products by livestock and humans is about twice the consumption of forest products (West et al., 2011; Hayes et al., 2012). Unlike forest products with a large range of residence times, crop products can be assumed to be consumed within a year of harvest (West et al., 2011). The spatial distributions of crop production and consumption at the county level (West et al., 2011) provide a sufficient resolution for use in atmospheric inverse modeling. The specific objectives of our study are: (1) to investigate the changes in the inverted carbon source and sink distribution after considering the spatial patterns of crop production and consumption, (2) to explore whether these changes improve our understanding of the carbon source and sink distribution within the conterminous US, (3) to evaluate the impact of the crop data on the inverted carbon balance for the conterminous US and other regions of the globe, and (4) to assess the relative importance of atmospheric CO_2 data and the prior flux in determining the spatial pattern of the carbon sink in the conterminous US.

2 Atmospheric CO_2 inversion methodology

The Bayesian synthesis inversion method (Enting and Trudinger, 1995) is used in this study. This method includes forward modeling of atmospheric CO_2 concentration using a transport model with prior fluxes and inverse modeling of the surface CO_2 flux based on the difference between modeled and observed CO_2 concentrations.

2.1 Forward modeling

2.1.1 Prior fluxes and their uncertainties

The a priori fluxes needed in the Bayesian synthesis inversion include sources from fossil fuel emissions, fire emissions, net carbon exchange between atmosphere and land, and net carbon exchange between atmosphere and ocean. These fluxes for the time period from 2000 and 2007 used in this study are the same as those used in Deng and Chen (2011) (Table 1). The fossil fuel emission field used in this study is constructed based on the fossil fuel CO_2 emission inventory from 1871 to 2006 from the Carbon Dioxide Information Analysis Center (CDIAC) (Marland et al., 2009) and the EDGAR 4 databases on a $1° \times 1°$ grid (Olivier and Aardenne, 2005). The grid point fire emission field used in this research is from the Global Emissions Fire Database version 2 (GFEDv2) (Randerson et al., 2007; van der Werf et al., 2006).

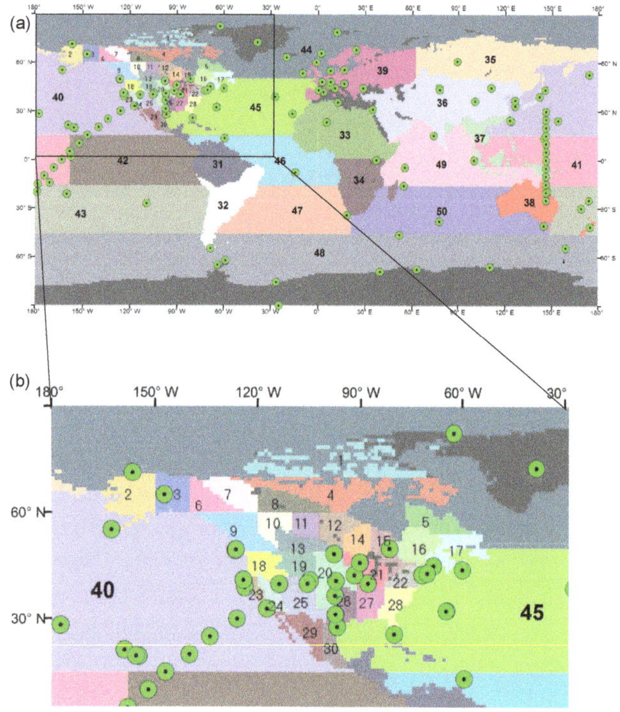

Figure 1. **(a)** A nested inversion system with 30 regions for North America and 20 regions for the remainder of the globe. The 210 measurement sites are indicated as circles. **(b)** An enlarged portion for North America.

The Boreal Ecosystem Productivity Simulator (BEPS) is employed to produce seasonally varying net ecosystem exchange (NEE) fluxes in hourly time steps globally (Chen et al., 1999, 2012; Ju et al., 2006). Developed based on FOREST Biogeochemical Cycles (FOREST-BGC) (Running and Coughlan, 1988), BEPS was originally intended for modeling the Canadian forest carbon cycle (Chen et al., 2007; Ju et al., 2006; Liu et al., 1999, 2002), but it has been extended to temperate and tropical ecosystems (Higuchi et al., 2005; Matsushita and Tamura, 2002; Sun et al., 2004; Chen et al., 2012). It uses remotely sensed leaf area index (LAI) (Deng et al., 2006), land cover type from Global Land Cover (GLC2000), meteorology from NCEP-reanalysis (Kalnay et al., 1996), and soil textural data (Webb et al., 1991). A unique feature of BEPS is the separation of sunlit and shaded components in the canopy using not only LAI but also a foliage clumping index (Chen et al., 2005; He et al., 2012) when calculating photosynthesis (Liu et al., 2002; Chen et al., 2012). Given that soil carbon pools are often not well modeled by terrestrial biosphere models, large uncertainties exist in the modeled annual carbon fluxes. However these models are still useful in estimating seasonal and diurnal patterns in response to changes in environmental conditions. Therefore, in most atmospheric inversion studies the prior annual mean NEE from land surfaces at each grid is set to zero (Gurney et al., 2004; Rödenbeck et al., 2003). The use of seasonally and diurnally varying biospheric fluxes is essential for the forward modeling to include the covariance of atmospheric transport and the surface flux (Denning et al., 1995; Gurney et al., 2004; Randerson et al., 1997; Deng and Chen, 2011).

In this study, two sets of control terrestrial biosphere fluxes from BEPS are prepared: (1) annually balanced but seasonally and diurnally varying NEE fluxes, and (2) annually, seasonally, and diurnally varying NEE fluxes. The annually balanced NEE fluxes are prepared by forcing the annual mean NEE to be zero, resulting in no interannual variability, while the seasonal and diurnal variability is retained. In fulfilling the objectives of this research, cropland carbon adjustments are made to these prior NEE fluxes (see Sect. 3).

The main processes responsible for ocean CO_2 uptake is the partial pressure difference between the sea surface and the overlying air which in part depends on the seasonal growth of phytoplankton in the oceans. The daily air–sea CO_2 fluxes across the sea surface in this research are simulated by the OPA-PISCES-T model, which is a global circulation model (OPA) (Madec et al., 1998) coupled with an ocean biochemistry model (PISCES-T) (Aumont, 2003; Buitenhuis et al., 2006). This coupled model is forced by daily wind stress, heat, and water fluxes from NCEP reanalysis (Kalnay et al., 1996). The ocean–atmosphere flux – along with the land–atmosphere flux, fossil emissions, and fire emissions – from 2000 to 2007 were fitted to a $1° \times 1°$ spatial scale at hourly time steps and aggregated to $3° \times 2°$ for North America and $6° \times 4°$ for the rest of the globe as inputs for forward modeling.

2.1.2 Atmospheric transport

The atmospheric transport model chosen for this research is the transport-only version of the global chemistry transport model (Krol et al., 2005) version 5 (TM5), which is an offline model driven by meteorological data from the European Centre for Medium-Range Weather Forecast (ECMWF) model. In this study, we define a global grid of $6° \times 4°$ with nested grids focusing on North America at $3° \times 2°$ based on Peters et al. (2005). The model consists of 25 vertical layers: 5 layers in the boundary layer, 10 in the free troposphere, and 10 in the stratosphere. In this study, the prior hourly fluxes are input into TM5 to generate forward simulations of hourly concentrations at 210 CO_2 observation sites across the globe for emissions from fossil fuel consumption (c_{ff}), fire (c_{fire}), terrestrial biosphere (c_{bio}), and oceans (c_{oce}).

2.1.3 CO_2 observations

The atmospheric CO_2 concentrations used in this study are the monthly CO_2 observation data from 2000 to 2007 compiled in the GLOBALVIEW-CO_2 2008 database. These concentrations are not actual data but rather baseline conditions which are derived using data integration techniques described by Masarie and Tans (1995). The data set is compiled with different data types, including surface flask, tower, aircraft, and ship measurements. In this study, selected months that used measurement-based data from 210 stations are taken to compile the CO_2 concentration matrix, and they consist of 12 181 station measurements during the 8-year period from 2000 to 2007 (Fig. 1).

In order to find the concentration corresponding to the biases in the surface carbon flux to be adjusted through inverse modeling, simulated concentrations corresponding to the prior fluxes need to be subtracted from CO_2 measurements. For continental tower sites, the GLOBALVIEW-CO_2 data set contains weekly averages of measurements in only afternoon hours to capture the well-mixed condition within the planetary boundary layer, and therefore the monthly simulated concentrations at these sites are also taken as average values in the same afternoon hours. For non-tower sites, GLOBALVIEW-CO_2 provides a summary of the sample collection times for discrete observations. The simulated concentrations are also sampled at the same times to obtain the monthly mean values, so as to be consistent with the GLOBALVIEW-CO_2 data set. These monthly averaged simulated concentrations are then subtracted from the corresponding 12 181 monthly CO_2 measurements ($\boldsymbol{c}_{\text{obs}}$) from GLOBALVIEW-$CO_2$ to produce the residual concentration ($\mathbf{C}$), expressed as follows:

$$\boldsymbol{c} = \boldsymbol{c}_{\text{obs}} - \boldsymbol{c}_{\text{ff}} - \boldsymbol{c}_{\text{fire}} - \boldsymbol{c}_{\text{bio}} - \boldsymbol{c}_{\text{oce}}. \tag{1}$$

This residual concentration is used as input to the inversion system to optimize the surface carbon flux.

2.2 Inverse modeling

2.2.1 Inversion regions and concentration locations

The atmospheric CO_2 concentrations are used for inversion of 50 global regions, including 30 regions in North America (Fig. 1), following Deng et al. (2007) and Deng and Chen (2011). The 30 regions in North America are delineated based on a 1 km resolution land cover map from AVHRR (Advanced Very High Resolution Radiometer) data (DeFries and Townshend, 1994) and provincial and state boundaries. This nested inversion system allows for a reduction of errors due to spatial aggregation over the focused region of North America and in the meantime does incur excessive computation. Although the spatial resolution over North America is still relatively low, the number of regions is adequate to capture the atmospheric signal and to show the broad source and sink patterns.

2.2.2 Time-dependent Bayesian synthesis approach

In the time-dependent Bayesian synthesis approach (Enting, 2002) used in our inverse modeling, a linear combination of source and sink terms is formulated to match with CO_2 concentration observations:

$$\boldsymbol{c} = Gf + A\boldsymbol{c}_0 + \varepsilon, \tag{2}$$

where $\boldsymbol{c}_{m\times 1}$ is a vector of m atmospheric CO_2 observations at given space and times; $\boldsymbol{\varepsilon}_{m\times 1}$ is a random error vector with a zero mean and a covariance matrix $cov(\boldsymbol{\varepsilon}) = \mathbf{R}_{m\times m}$; $\mathbf{G}_{m\times(n-1)}$ is a given matrix representing a transport (observation) operator, where $n-1$ is the number of fluxes to be determined; $\mathbf{A}_{m\times 1}$ is a unity vector (filled with 1) that relates to the assumed initially well-mixed atmospheric CO_2 concentrations ($\boldsymbol{c}_0$); and $f_{(n-1)\times 1}$ is an unknown vector of carbon fluxes of all studied regions. In this research, $m = 12\,181$ (the number of measurements as mentioned in Sect. 2.1.3) and $(n-1) = 4800$ (50 regions × 8 years × 12 months).

After combining matrixes $\mathbf{G}$ and $\mathbf{A}$ into $\mathbf{M}_{m\times n} = (\mathbf{G}, \mathbf{A})$ and vectors $\boldsymbol{f}$ and $\boldsymbol{c}_0$ as $s_{n\times 1} = (\boldsymbol{f}^T, \boldsymbol{c}_0)^T$, Eq. (2) is rewritten as

$$\boldsymbol{c} = \mathbf{M}s + \boldsymbol{\varepsilon}. \tag{3}$$

while Eq. (3) can be solved for s by the conventional least-squared technique, the problem is poorly constrained. The Bayes approach (Tarantola, 2005) is generally used for ill-constrained problems through introducing a priori information in the inversion process. The best a priori information for this purpose is a prior estimate of the surface flux. The a posteriori flux is obtained by minimizing the following cost function J:

$$J = \frac{1}{2}(\mathbf{M}s - \boldsymbol{c})^T R^{-1}(\mathbf{M}s - \boldsymbol{c}) + \frac{1}{2}(s - s_{\text{p}})^T \mathbf{Q}^{-1}(s - s_{\text{p}}), \tag{4}$$

where $s_{\text{p}_n\times 1}$ is the a priori estimate of s (set to zero after subtracting its contribution to concentration from the atmospheric CO_2 observation); the covariance matrix $\mathbf{Q}_{n\times n}$ represents the uncertainty in the a priori estimate; and $R_{m\times m}$ is the model–data mismatch error covariance. Through minimizing this cost function in Eq. (4), the posterior best estimate of s (Enting, 2002) is defined as

$$\hat{s} = (\mathbf{M}^T R^{-1}\mathbf{M} + \mathbf{Q}^{-1})^{-1}(\mathbf{M}^T R^{-1}\boldsymbol{c} + \mathbf{Q}^{-1}s_{\text{p}}), \tag{5}$$

with the posterior uncertainty expressed as follows:

$$\hat{Q} = (\mathbf{Q}^{-1} + \mathbf{M}^T R^{-1}\mathbf{M})^{-1}. \tag{6}$$

2.2.3 Transport (observation) operator, model–data mismatch, and prior uncertainties

The matrices for transport (observation) operator, $\mathbf{M}$; the model–data mismatch, R; and the prior uncertainties, $\mathbf{Q}$, are taken from Deng and Chen (2011). The transport (observation) operator, $\mathbf{M}$, is formed from 4800 forward simulations (8 years and 50 regions). An initial flux of 1 Pg C for each month and region was prescribed in the TM5 transport model to find the monthly contribution of each region to CO_2

concentration at each observation site. The model–data mismatch, R, reflects the difference between the modeled and the observed CO_2 concentrations, which include both errors from transport modeling and measurement (such as instrument errors). The observation sites were divided into five categories, each with its own constant portion (σ_{const}) and variable portion (GVsd) that is computed monthly from the standard deviation data given in GLOBALVIEW-CO_2 2008 variation (var) files. The constant portion is defined under the following categories: Antarctic sites (0.15), oceanic sites (0.30), land and tower sites (1.25), mountain sites (0.90), and aircraft samples (0.75). The variable portion is the statistical summary of average atmospheric variability for each measurement record. Therefore, the covariance matrix, $\mathbf{R}$, is given as a diagonal matrix that contains the error for each month i:

$$\mathbf{R}_{ii} = \sigma^2_{\text{const}} + \text{GVsd}^2. \tag{7}$$

Additionally, a weighting factor, $\mathbf{W}$, is inserted to the cost function in Eq. (4) to account for the vertical correlation between measurements at different levels of the same tower and aircraft sites (i.e., smaller weights are given to each of the measurements at the same site) (Deng and Chen, 2011).

$$J = \frac{1}{2}((\mathbf{M}s - \mathbf{c})W)^T R^{-1}((\mathbf{M}s - \mathbf{c})W) + \frac{1}{2}(s - s_{\text{p}})^T \mathbf{Q}^{-1}(s - s_{\text{p}}) \tag{8}$$

The weight, W, is a diagonal matrix with the diagonal terms given by

$$w_{ii} = 1/(1 + 0.6(n-1)), \tag{9}$$

where n is the number of observations at different levels at the same site.

While the a priori fluxes are set to zero after subtracting their contributions toward the measured CO_2 concentrations, the a priori uncertainties, $\mathbf{Q}$, are important in forcing the spatial distribution of the inverted fluxes (Table 1). The uncertainties for fossil fuel of ±6 % (Marland et al., 2009) and for fire fluxes of ±20 % (van der Werf et al., 2010) are not included in $\mathbf{Q}$, and hence, it is important to note that the inverted fluxes are subject to these additional uncertainties.

The χ^2 test (Gurney et al., 2003) is employed to test the consistency of the fit to CO_2 data and the prior flux estimate simultaneously:

$$\chi^2 = \frac{J_{\min}}{N_{\text{obs}}} = \frac{\sum\limits_{m=1}^{m} \frac{(\mathbf{M}s-\mathbf{c})^2}{R^2} + \sum\limits_{n=1}^{n} \frac{(s-s_{\text{p}})^2}{\mathbf{Q}^2}}{N_{\text{obs}}}, \tag{10}$$

where N_{obs} is the number of degrees of freedom and $J_{\min}$ is the cost function from Eq. (4). The consistency of the fit is highest when the value of χ^2 equals unity.

3 US cropland carbon integration methodology

During the growing season, cropland in the US Midwest represents a strong regional carbon sink. However, a large portion of crop biomass is removed during harvest and transported to other regions for processing and consumption. When the crop products are consumed by humans and livestock, CO_2 is released back into the atmosphere at geographic locations that differ from the origin of production. In this study, these lateral redistributions of carbon are considered in the prior fluxes used in atmospheric CO_2 inversion studies in order to investigate their influence on the inverted carbon source and sink distribution.

3.1 US cropland carbon budget based on inventory data

The regional patterns of CO_2 uptake and release in US croplands based on agricultural statistics (West et al., 2011) are used to adjust the prior biosphere flux used in our inversion. The data include county-level net primary productivity (NPP), harvest, and changes in soil carbon from 1990 to 2008, as well as human and livestock crop consumption from 2000 to 2008 (CDIAC, 2014). The national crop carbon budget for the USA from 2000 to 2008 is balanced within 0.3 to 6.1 % yr^{-1} based on the study from West et al. (2011). Although many other important components are included in the overall US cropland carbon budget, such as exports and crop carbon used for fuel, the vertical net carbon exchange (NCE) into the atmosphere is given by the sum of net carbon uptake from NPP, net change in soil carbon, and the release of carbon from biomass decomposition, human consumption, and livestock consumption.

3.1.1 Crop NPP, harvest, biomass decomposition, and changes in soil carbon pool

The cropland NPP used in West et al. (2011) is calculated based on annual statistics of crop production (P) in units of tons of biomass and harvested crop area (HA) reported by the US Department of Agriculture (USDA) National Agricultural Statistics Service. County-level statistics are gap-filled using district-level data and then converted to county-level NPP in units of carbon using Eq. (11), which is documented in earlier studies (Hicke, 2004; Hicke and Lobell, 2004; Prince et al., 2001; West et al., 2010):

$$\text{NPP}_{\text{crops}} = \sum_i \frac{P_i \times (1 - \text{MC}_i) \times C}{\text{HI}_i \times f_{\text{AG},i} \times \text{HA}_i}, \tag{11}$$

where P is the reported crop production; MC is the harvest moisture content; C is the conversion factor from biomass to carbon (0.45 g of C per g of dry mass); HI is the harvest index, i.e., the ratio of yield mass to aboveground biomass; f_{AG} is the fraction of production allocated aboveground; and the subscript i indicates 17 different crops (corn, soybean,

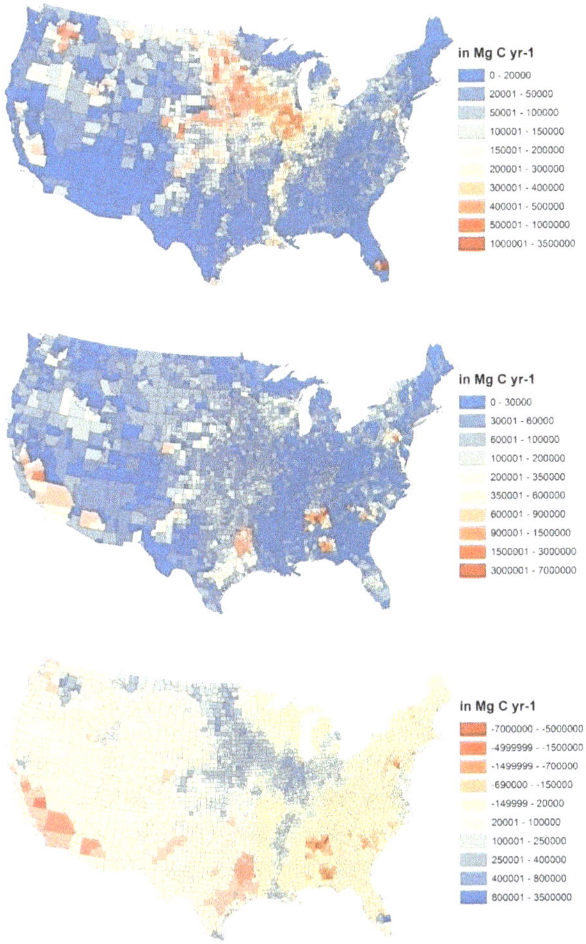

Figure 2. Spatial distributions of annual crop production (top); crop consumption (middle); and crop NEP (bottom) for 2003. Data from CDIAC (2014).

oats, barley, wheat, sunflower, hay, sorghum, cotton, rice, peanuts, potatoes, sugar beets, sugarcane, tobacco, rye, and beans) representing the majority of the crops grown in the conterminous US (West et al., 2010, 2011). The harvested amount is calculated based on crop yields. Carbon released from biomass decomposition is calculated from NPP by subtracting the amount that is harvested. The remainder of crops, i.e., the amount not harvested, is left as residue and is either sequestered into the soil or decomposed within the same year.

Changes in soil carbon are calculated based on empirical relationships between land management practices and soil carbon change (West et al., 2008). In order to capture the long-term impacts on soil carbon pools from a 20-year history of changes in crop rotation and tillage intensity, the soil carbon change was calculated from 1980 to 2008.

3.1.2 Human and livestock consumption

Human crop carbon consumption data provided by West et al. (2011) were calculated based on the per capita food consumption and the US population census data. The livestock consumption data were calculated in a similar way but also considered different feed consumption rates of different animal species. The consumed amount was assumed to be released back into the atmosphere within the same year as CO_2 through respiration, excretion, and flatus (West et al., 2009). Excretion typically entered the waste treatment facilities within the county, and the emissions were taken into account in the consumption term above.

3.2 Prior flux adjustments for US croplands

One way to integrate the carbon exchange of crops into the inversion model is to adjust the prior NEP modeled by BEPS. As part of the crop carbon exchange, the harvested amount is taken away from the field and respired back to the atmosphere when consumed by humans, $R_{\text{human\_consumption}}$, and livestock, $R_{\text{livestock\_consumption}}$. Therefore, crop NEP is given by

$$\text{NEP}_{\text{crop}} = \text{NPP}_{\text{crop}} - R_{\text{h, residue}} - R_{\text{human\_consumption}} - R_{\text{livestock\_consumption}} + \Delta c_{\text{soil}}. \quad (12)$$

Since the residue amount, including both the remaining aboveground and belowground biomass, is the difference between the total crop biomass and the amount harvested, Eq. (12) can be rewritten as:

$$\text{NEP}_{\text{crop}} = \text{NPP}_{\text{harvested}} + \Delta c_{\text{soil}} - R_{\text{human\_consumption}} - R_{\text{livestock\_consumption}} = \text{production-consumption}, \quad (13)$$

where

$$\text{production} = \text{NPP}_{\text{harvested}} + \Delta c_{\text{soil}}, \quad (14)$$

$$\text{consumption} = R_{\text{human\_consumption}} + R_{\text{livestock\_consumption}}. \quad (15)$$

The spatial patterns of crop production, consumption, and NEP are shown in Fig. 2. The spatial distributions of crop production and consumption are calculated monthly and are used to adjust the monthly NEP distributions modeled by BEPS over the conterminous US. The production and consumption terms are adjusted separately due to their different seasonal patterns.

3.2.1 Production adjustment

The simulated terrestrial NPP by BEPS is adjusted to integrate cropland production over the contiguous US using the following equation:

$$\begin{aligned}\text{NPP}_{\text{adjusted}} &= \text{NPP}_{\text{biosphere}} - r_{\text{crop\_area}} \cdot (\text{NPP}_{\text{biosphere}}) \\ &+ \text{production} = (1 - r_{\text{crop\_area}}) \cdot \text{NPP}_{\text{biosphere}} \\ &+ \text{production} = (1 - r_{\text{crop\_area}}) \cdot \text{NPP}_{\text{biosphere}} \\ &+ \text{NPP}_{\text{harvested}} + \Delta c_{\text{soil}}, \end{aligned} \quad (16)$$

where $\text{NPP}_{\text{biosphere}}$ is the $1° \times 1°$ NPP output from BEPS and $r_{\text{crop\_area}}$ is the ratio of harvested crop area within the $1° \times 1°$

grid over the total area of the grid. The county-level cropland production is first extrapolated into a mean value for each 1° × 1° grid and then adjusted using Eq. (16). The ratio r_{crop_area} is calculated based on the harvested area data. The basic idea of Eq. (16) is to replace BEPS NPP with more accurate crop production data for the portion of land area that is used for crop production, while BEPS NPP is unchanged for the remaining area. In this way, both the productive and non-productive areas in a grid are included in the prior, and the simulation of non-crop area by BEPS may be influenced by adjacent crop areas. BEPS uses GLC2000 as the land cover data in NPP simulations, but these data are not as accurate as the agricultural crop area statistics within each grid. Chan and Lin (2011) cautioned researchers against the direct comparison of carbon accounting based on agricultural census data and fluxes simulated by biospheric models due to large differences of different land cover classifications used by the models. For this reason, we choose to use the crop area ratio method for the production adjustment.

Since the prior surface CO_2 fluxes enter into the atmospheric transport model on hourly time steps, the annual crop carbon production data are interpolated into the seasonal and diurnal patterns simulated by BEPS. Firstly, the annual $NPP_{biosphere}$ is converted to annual $NPP_{adjusted}$, and the ratio between the two is taken and multiplied by the hourly $NPP_{biosphere}$ fluxes, resulting in hourly adjusted fluxes from 2000 to 2008 for each 1° × 1° grid.

3.2.2 Consumption adjustment

The consumption terms are integrated into BEPS-simulated R_h over the contiguous US using the following equation:

$$\begin{aligned} R_{h,\,adjusted} &= R_{h,\,biosphere} - r_{crop\_area} \cdot (R_{h,\,biosphere}) \\ &+ \text{consumption} = (1 - r_{crop\_area}) \cdot R_{h,\,biosphere} \\ &+ \text{consumption} = (1 - r_{crop\_area}) \cdot R_{h,\,biosphere} \\ &+ R_{human\_consumption} + R_{livestock\_consumption}, \end{aligned} \tag{17}$$

where $R_{h,\,biosphere}$ is the 1° × 1° R_h output from BEPS and r_{crop_area} is the ratio of harvested crop area within the 1° × 1° grid over the total area of the grid. The county-level cropland consumption data are first resampled into each 1° × 1° grid, and the mean value of each grid is used to adjust R_h using Eq. (17). However, unlike the production adjustment, the temporal patterns of CO_2 release from human and livestock consumptions do not follow the seasonal and diurnal patterns simulated for the biosphere. We therefore assume constant hourly release of CO_2 from crop consumption throughout the year. In this way, the annual consumption amount is divided equally into the hourly values and added to the hourly simulated $R_{h,\,biosphere}$ from BEPS for the time period from 2000 to 2008 at each 1° × 1° grid.

3.3 Schemes of inversion experiments

Two sets of inversion experiments (Table 2) are designed to test for the impact of integrating the cropland carbon data into the prior terrestrial flux. In the first set (Experiments 1a, 1b, 1c), the monthly terrestrial biosphere NEP modeled by BEPS at hourly time steps is first neutralized on an annual basis, meaning that the annual mean terrestrial flux is 0 for each grid cell but its seasonal variation is retained. The neutralized flux is then adjusted for crop production and consumption in Experiments 1b and 1c. After these adjustments, the annual prior flux in Experiments 1b and 1c (Table 3) differs from zero. The prior terrestrial surface flux in many atmospheric inversion studies only contains seasonal instead of interannual variations (Gurney, 2004; Rödenbeck et al., 2003) in order to minimize the influence of the errors in the prior flux on the inverted annual flux so that the results are mostly based on the "atmospheric view". In the second set (Experiments 2a, 2b, 2c), the same monthly BEPS NEP is used without annual neutralization (Table 3) so that both seasonal and interannual variations in NEP are retained and the best estimate of the biospheric carbon flux is used to constrain the inversion. Experiments 1a–c are therefore designed to explore the strength of the carbon cycle signal in the atmospheric CO_2 measurement, while Experiments 2a–c are considered to be the best final estimates by integrating ecosystem modeling results and crop statistics. In each set of experiments, there is a control run (Experiments 1a, 2a), which is used to assess influences of the production adjustment (Experiments 1b, 2b) and influences of the combined production and consumption adjustments (Experiments 1c, 2c) on the inverted flux.

The a priori uncertainty matrix **Q** described in Sect. 2.2.3 is adjusted to take into account uncertainties from the cropland production and consumption data. Uncertainties in the agricultural production data consist of parameter uncertainties associated with the harvest index (HI) and reported crop production (P) (Chan and Lin, 2011). Bolinder et al. (2007) and Prince et al. (2001) show the standard deviation of HI, the dominant source of uncertainty towards NPP, to be 10 %. Therefore, experiments that adjust for production only are prescribed a 10 % uncertainty in annual NPP distributed over the production regions and weighted with the original matrix **Q**. Experiments that adjust for both production and consumption use the above method to include production uncertainty as well as an additional 1 % of annual total human consumption (West et al., 2009) and 20 % of the annual total livestock consumption distributed over consumption areas (Ciais et al., 2007). Additionally, the χ^2 test (Eq. 12) is employed to show the consistency of the fit to data and prior flux estimates simultaneously. The χ^2 test indicates that experiments 1c and 2c, having the χ^2 values closest to unity, have the greatest consistency.

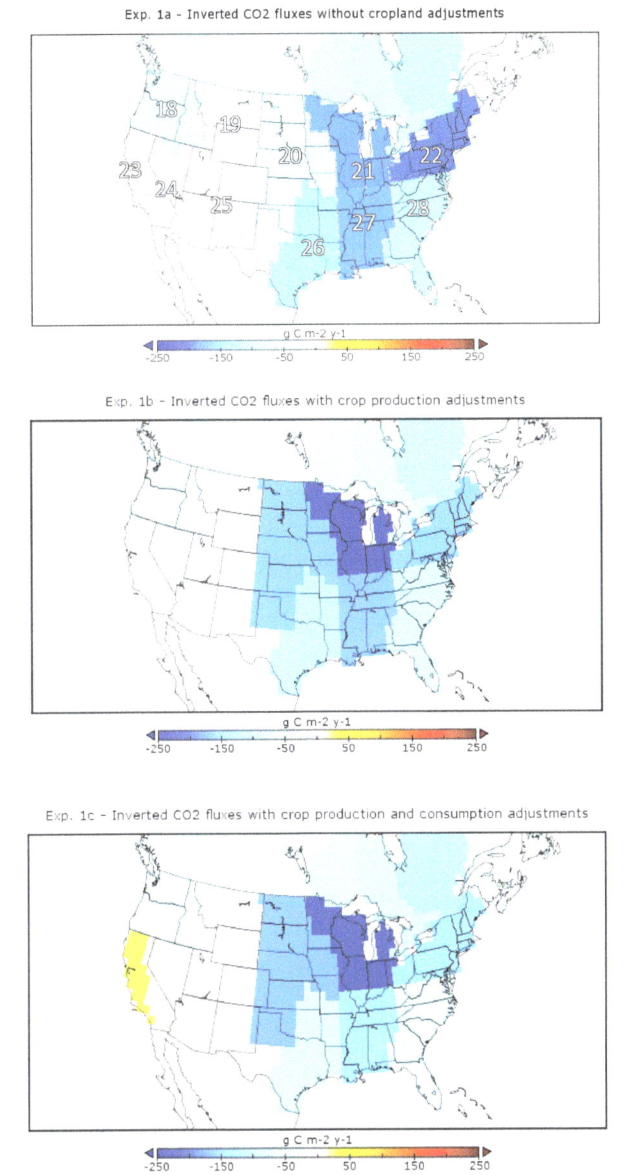

Figure 3. Map of mean annual inverted CO_2 flux for experiment 1 during 2002 to 2007 over contiguous US regions. Negative values represent carbon uptake.

4 Results and discussion

The inversion experiments described in Sect. 3.3 are conducted to test the impacts of cropland inventory on the inverted CO_2 fluxes. The impacts are evaluated based on the multi-year mean annual values and seasonal variations.

4.1 Multi-year regional carbon budget

4.1.1 Average annual flux

The average annual inverted CO_2 fluxes over the conterminous US are reported for experiment 1 (Fig. 3) and experiment 2 (Fig. 4). Table 4 summarizes the mean inverted CO_2 fluxes (μ) and errors (ε), as well as the percentage changes ($\Delta\%$) from the control case calculated by

$$\Delta\% = \frac{\mu - \mu_{\text{control}}}{\mu_{\text{control}}}, \tag{18}$$

where μ_{control} is the annual mean inverted flux for the corresponding control experiment.

To evaluate the impact of integrating crop production and consumption data into the prior fluxes, comparisons of the inverted fluxes can be made between the experiments with and without crop adjustments. However, in order for these comparisons to be meaningful, a signal-to-noise ratio (SNR) is calculated for each region, using the following equation:

$$\text{SNR} = \frac{|\mu - \mu_{\text{control}}|}{\varepsilon}, \tag{19}$$

where ε is the posterior uncertainty in the inverted flux for a region. Note that the "signal" is defined here as the mean difference in the inverted fluxes between the control and a crop-adjusted experiment, in order to represent the signal introduced by the adjustment. If SNR is less than unity, the resulting change due to an agricultural adjustment is less than the uncertainty, and hence the signal is within the noise level of the results. In Table 4, regions with SNR greater than unity are denoted with an asterisk.

Comparing Fig. 3a and b, it can be seen that the production adjustment redistributed the carbon sink from forested regions in the southeast (regions 26 to 28) and northwest (regions 18, 19) to cropland area in the Midwest (regions 20 and 21). SNR for experiment 1b for region 20 is greater than 1 (Table 4), indicating that the sink increase in this region is beyond inversion uncertainty. The increase in the sink size in the US Midwest can be attributed to the large CO_2 uptake during the growing season, but much of the carbon is released through crop consumption in other geographic locations. Seasonal results for these regions are shown in Sect. 4.1.3.

The impact of crop consumption adjustment on the inversion result can be assessed by comparing Fig. 3b and c. In region 28 with large crop consumption, the inverted carbon sink (Fig. 3c) is reduced by 68 % from the case with crop production adjustment only (Fig. 3b) and by 76 % from the control case (Fig. 3a), which is a sink of 69 Tg C yr^{-1} (Table 4). In regions 23 and 24 with large crop consumption and low vegetation growth, the crop consumption adjustment turns weak carbon sinks (Fig. 3a and b) into sources (Fig. 3c). SNR for experiment 1c for region 24 (California) is greater than 1, confirming that this shift from sink to source is beyond the inversion uncertainty. In general, crop consumption adjustment in regions with considerable consumptions (regions 18, 19, 23, 24, 25, 26, 27, 28) induces noticeable reduction in the carbon sink. It would also be interesting to note that the crop production adjustment (experiment 1b) increases the USA sink by 44 % but decreases the

Canadian sink by 15 %, and the crop production and consumption adjustments (experiment 1c) reduce the sink by 1.4 and 12.1 % for USA and Canada, respectively, relative to the control case (experiment 1a, Table 4). These results suggest that the crop information used in the inversion affects not only US carbon sinks but also neighboring regions (such as Canada). The inverted results from both crop production and consumption adjustments (experiment 1c, Fig. 3c) show that the US Midwest sink in 2002–2007 is $425 \pm 129\,\mathrm{Tg\,C\,yr^{-1}}$. This sink value obtained in experiment 1c is smaller than that ($522\,\mathrm{Tg\,C\,yr^{-1}}$) in experiment 1b but considerably larger than that ($189\,\mathrm{Tg\,C\,yr^{-1}}$) in the control case, confirming the importance of considering the crop production and consumption data for atmospheric inversion.

Experiments 1a–c analyzed above are designed to accentuate the information content of atmospheric CO_2 measurements for the surface carbon flux by neutralizing the annual biospheric flux before making the crop consumption adjustments. However, it could be argued that the annually neutralized fluxes may not be the best prior information for constraining the inversion. Experiments 2a–c are therefore conducted with the best prior estimates possible based on a biospheric model and crop data. The prior biospheric carbon flux used in experiment 2 differs from that in experiment 1 in the following ways: (1) the annual net carbon flux modeled by BEPS is used without neutralization, (2) the interannual variations in the prior flux are considered, and (3) the interannual variations in crop production and consumption are also considered. Although errors in the annual mean prior fluxes would have imprints on the inverted results in experiment 2, the unneutralized fluxes may nudge the inverted results closer to reality as they integrate prior knowledge on the carbon source and sink distribution and the interannual variability.

Figure 4 shows inverted results under experiment 2 with comparisons to the prior estimates. The production adjustment (experiment 2b) leads to significant increases in the CO_2 sink in the Midwest crop area (region 20). Regions 19, 21, and 25 also gain noticeable sink increases, but sinks in other regions in the US decrease in compensation for these gains. Relative to experiment 2b, the additional consumption adjustment made in experiment 2c significantly weakened the sinks in the west region and noticeably in the southeast. These results are consistent with the findings from experiment 1. Crop adjustments in experiment 2 greatly enhanced the sinks in the cropland in the US Midwest (regions 20, 21), mostly at the expense of forested areas in the southeast (regions 26, 27, 28). This is also similar to experiment 1, which only takes into account the seasonal variations and not the interannual variations. For the purpose of comprehensive evaluation of how crop adjustments spatially redistribute the carbon flux, Table 5 provides a matrix of correlation coefficients of the change in the inverted flux from experiment 2a to experiment 2c among the 11 regions in the US, where positive correlations indicate that crop adjustments make the inverted flux increase or decrease in the same direction. Region 21, for example, is negatively correlated with most regions except regions 20, 23, and 24, suggesting that its increase in carbon sink after the crop adjustments is mostly balanced by decreases in other regions except regions 20, 23, and 24. Another interesting example is region 28, which is strongly and negatively correlated with the Midwest region 21 and strongly and positively correlated to the west regions 18, 19, and 25. A large part of these correlations is due to adjustments to the priors (Table 3), as the change in the inverted flux for a region is highly correlated with the adjustment to the region (Fig. 6). However, atmospheric circulation patterns, such as the westerly and the monsoon air flows (air mass moving from the southeast to the Midwest), could also have impacts on these correlations.

The spatial distributions of the prior and posterior fluxes for the three pairs shown in Fig. 4 are visually well correlated. As an example, the correlation between the prior and posterior fluxes of experiment 2a is shown in Fig. 5a. It is highly significant ($r^2 = 0.92$), meaning that the inverted flux distribution among the regions is highly correlated with the prior flux used to constrain the inversion. It is interesting to note that the posterior fluxes of the various regions are about 1.6 times their prior fluxes, suggesting that atmospheric inversion forces the fluxes in all regions in the conterminous US to increase in about the same proportion, which is perhaps determined according to the balance of the US with other large regions. This means that the overall magnitude of the US sink is more or less determined globally while the spatial sink pattern within the US is correlated with the prior. Figure 5b further affirms this point by demonstrating that the changes in the posterior flux in experiment 2c are also highly correlated with the adjustments made to the prior flux. It is interesting to note that the changes are also about 1.6 times the adjustments, indicating the same strength of the influence of the adjustments to the prior on the inversion. Strong correlations between the prior and posterior exist in all experiments ($r^2 = 0.85$–0.95). However, these strong correlations should not be taken as an indication of the sole importance of the prior in determining the posterior as it can be argued that the correlation is good because the prior flux distribution matches with the atmospheric signal. Figure 5c is shown to demonstrate that the posterior flux of experiment 2c is significantly correlated ($r^2 = 0.63$) with the posterior flux of experiment 1a which is obtained with a neutralized prior and arguably represents the atmospheric signal. This correlation means that the atmospheric signal is responsible for about 63 % of the spatial distribution of the inverted flux. Figure 5d shows that the posterior flux from experiment 2a constrained with a prior without the crop adjustments is highly correlated with the posterior flux from experiment 1a. This high correlation indicates that the unneutralized prior flux does not provide much additional information for inverting the spatial distribution of the carbon flux. It also indicates that the atmospheric signal is not strong enough to sense all the sinks in crop production areas. These correlations demonstrate that

Table 2. Schemes of experiments designed to test for the impact of integrating information from crop production and consumption data into the prior fluxes.

Experiment 1.	Annually balanced terrestrial biosphere	χ^2
(a)	Neutralized terrestrial prior fluxes from BEPS model (annual mean = 0)	1.09
(b)	Neutralized terrestrial prior fluxes (annual mean = 0) adjusted for crop production	1.12
(c)	Neutralized terrestrial prior fluxes (annual mean = 0) adjusted for crop production and crop consumption (added to the prior fluxes as a source)	0.97
Experiment 2.	**Interannual variability in terrestrial biosphere**	χ^2
(a)	Terrestrial prior fluxes from BEPS model (annual mean $\neq$ 0)	1.11
(b)	Terrestrial prior fluxes (annual mean $\neq$ 0) adjusted for crop production	1.16
(c)	Terrestrial prior fluxes (annual mean $\neq$ 0) adjusted for crop production and crop consumption	0.97

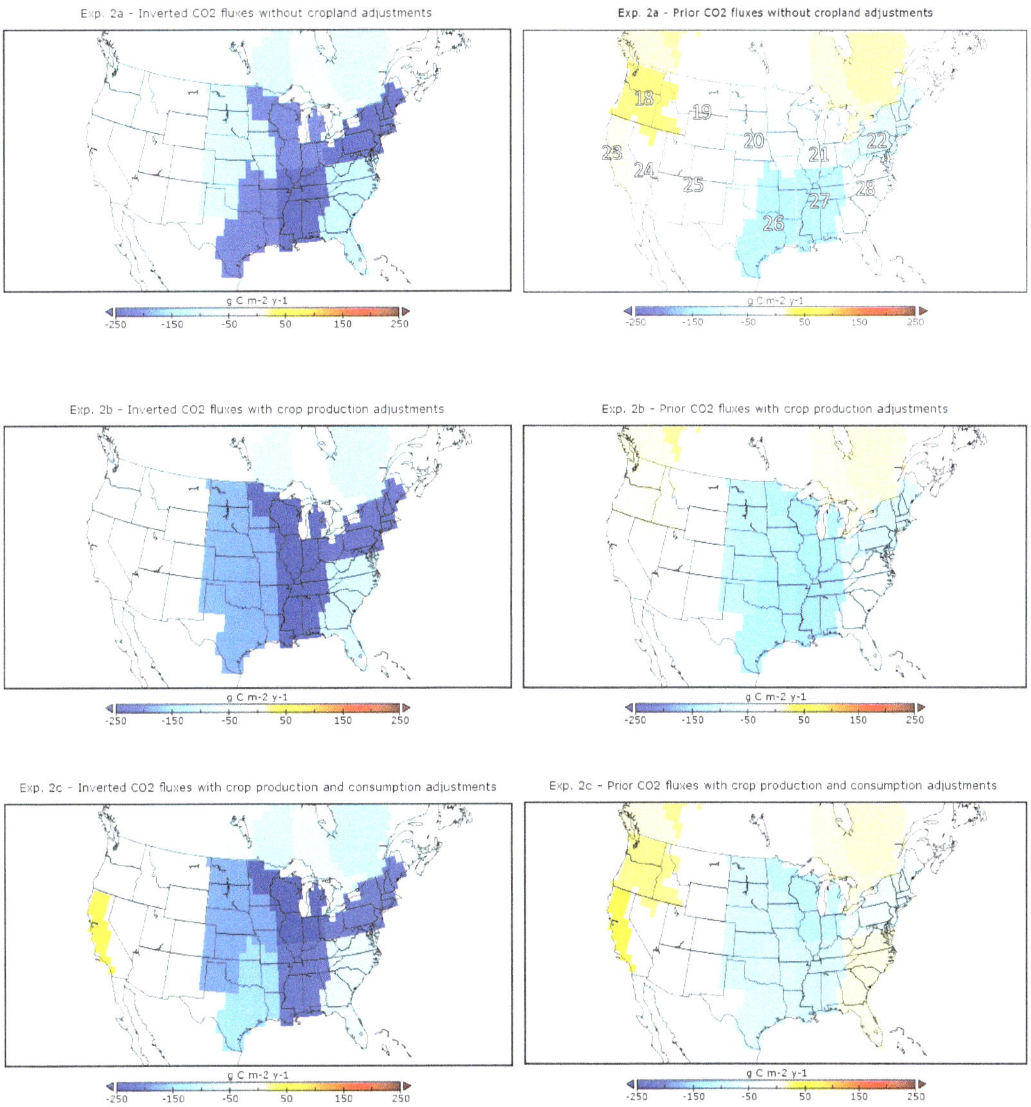

Figure 4. Map of mean annual inverted (left) and prior (right) CO_2 flux for experiment 2 during 2002 to 2007 over the contiguous US. Negative values represent carbon uptake.

Table 3. Prior fluxes used in the two sets of inversion experiments

Region	Experiment 1			Experiment 2		
	a	b	c	a	b	c
18	0 ± 6.5	−8.91 ± 8.3	−1.94 ± 9.5	15.88 ± 6.5	6.97 ± 8.3	13.94 ± 9.5
19	0 ± 2.9	−6.35 ± 3.7	−1.96 ± 4.4	−5.69 ± 2.9	−12.04 ± 3.7	−7.65 ± 4.4
20	0 ± 8.6	−75.83 ± 27.4	−37.47 ± 32.9	−66.51 ± 8.6	−142.34 ± 27.4	−103.98 ± 32.9
21	0 ± 8.5	−51.75 ± 18.2	−34.97 ± 20.3	−34.71 ± 8.5	−86.46 ± 18.2	−69.69 ± 20.3
22	0 ± 9.6	−0.42 ± 11.6	17.56 ± 13.6	−49.03 ± 9.6	−49.45 ± 11.6	−31.47 ± 13.6
23	0 ± 3.0	−1.93 ± 3.9	5.38 ± 6.2	2.75 ± 3.0	0.82 ± 3.9	8.12 ± 6.2
24	0 ± 0.7	−1.14 ± 1.8	3.80 ± 5.5	−2.95 ± 0.7	−4.09 ± 1.8	0.85 ± 5.5
25	0 ± 3.0	−3.97 ± 3.8	3.39 ± 5.1	−16.78 ± 3.0	−20.75 ± 3.8	−13.40 ± 5.1
26	0 ± 12.5	3.02 ± 15.1	48.67 ± 23.0	−105.11 ± 12.5	−102.09 ± 15.1	−56.45 ± 23.0
27	0 ± 14.9	−4.86 ± 20.0	36.57 ± 26.6	−88.63 ± 14.9	−93.49 ± 20.0	−52.06 ± 26.6
28	0 ± 15.6	−4.69 ± 18.7	29.87 ± 24.9	−21.50 ± 15.6	−26.19 ± 18.7	8.37 ± 24.9
West	0 ± 4.2	−22.29 ± 5.4	8.66 ± 7.2	−6.79 ± 4.2	−29.09 ± 5.4	1.86 ± 7.2
Midwest	0 ± 8.6	−127.56 ± 23.3	−72.43 ± 27.3	−101.22 ± 8.6	−228.79 ± 23.3	−173.66 ± 27.3
Northeast	0 ± 9.6	−0.35 ± 11.6	17.56 ± 13.6	−49.03 ± 9.6	−49.45 ± 11.6	−31.47 ± 13.6
Southeast	0 ± 14.3	−6.53 ± 18.1	115.11 ± 24.9	−215.24 ± 14.3	−221.77 ± 18.1	−100.14 ± 24.9
US	0 ± 9.9	−156.83 ± 16.1	68.87 ± 20.0	−372.28 ± 9.9	−529.11 ± 16.1	−303.41 ± 20.0
Canada	0 ± 5.9	0.00 ± 5.9	0.00 ± 5.9	50.75 ± 5.9	50.75 ± 5.9	50.75 ± 5.9
NA	0 ± 9.1	−156.83 ± 14.8	68.87 ± 18.5	−363.99 ± 9.1	−520.81 ± 14.8	−295.12 ± 18.5

Table 4. Mean inverted CO_2 flux (μ), error (ε) in $Tg\,C\,yr^{-1}$,and the percentage change ($\Delta\%$) for global regions from 2002 to 2007.

Region	Experiment 1					Experiment 2				
	(a)	(b)		(c)		(a)	(b)		(c)	
	$\mu \pm \varepsilon$	$\mu \pm \varepsilon$	Δ%	$\mu \pm \varepsilon$	Δ%	$\mu \pm \varepsilon$	$\mu \pm \varepsilon$	Δ%	$\mu \pm \varepsilon$	Δ%
18	−43.27 ± 33.1	−38.45 ± 25.7	−11.13	−30.41 ± 28.4	−29.71	−26.48 ± 33.1	−18.95 ± 25.7	−28.43	−14.74 ± 28.4	−44.32
19	−10.80 ± 20.3	−17.94 ± 15.8	66.08	−9.49 ± 17.7	−12.07	−15.22 ± 20.3	−17.18 ± 15.8	12.88	−13.41 ± 17.7	−11.87
20	−71.59 ± 56.9	−271.27 ± 91.4	278.94*	−223.82 ± 102.4	212.67*	−113.85 ± 56.9	−227.33 ± 91.4	99.67*	−231.71 ± 102.4	103.52*
21	−117.07 ± 59.8	−251.38 ± 72.7	114.73*	−201.33 ± 78.3	71.98	−132.20 ± 59.8	−192.76 ± 72.7	45.81	−188.63 ± 78.3	42.68
22	−104.91 ± 57.7	−97.89 ± 46.1	−6.69	−64.04 ± 51.9	−38.95	−145.62 ± 57.7	−124.70 ± 46.1	−14.36	−113.33 ± 51.9	−22.17
23	−0.79 ± 8.9	−10.69 ± 7	1257.11*	23.21 ± 10.2	−3044.92*	1.66 ± 8.9	0.34 ± 7	−79.75	7.23 ± 10.2	334.19
24	−0.05 ± 0	−2.26 ± 0	4169.81*	6.71 ± 3.5	−12754.7*	−3.00 ± 0	−4.93 ± 0	64.71*	−0.12 ± 3.5	−96.03
25	−7.95 ± 14.2	−7.69 ± 10.9	−3.29	−4.29 ± 14.2	−45.94	−23.48 ± 14.2	−26.09 ± 10.9	11.12	−19.70 ± 14.2	−16.1
26	−92.88 ± 59.8	−79.49 ± 47.5	−14.42	−36.65 ± 63.8	−60.54	−160.95 ± 59.8	−138.33 ± 47.5	−14.06	−113.01 ± 63.8	−29.79
27	−111.07 ± 78.7	−76.22 ± 64.6	−31.38	−63.91 ± 82.9	−42.46	−175.18 ± 78.7	−167.96 ± 64.6	−4.12	−141.68 ± 82.9	−19.12
28	−69.17 ± 67.5	−52.21 ± 51	−24.52	−16.86 ± 64.5	−75.68	−76.25 ± 67.5	−63.05 ± 51	−17.3	−38.29 ± 64.5	−49.78
West	−62.85 ± 39.1	−77.03 ± 30.4	22.55	−14.29 ± 32.7	−77.26*	−66.50 ± 40.2	−66.81 ± 20.6	0.46	−40.75 ± 23.4	−38.73*
Midwest	−188.65 ± 80.7	−522.65 ± 116.8	177.04*	−425.15 ± 128.8	125.36*	−246.05 ± 82.3	−420.09 ± 115.5	70.73*	−420.33 ± 128.2	70.83*
Northeast	−104.91 ± 57.7	−97.89 ± 46.1	−6.69	−64.04 ± 51.9	−38.95	−145.62 ± 57.7	−124.70 ± 46.1	−14.36	−131.31 ± 51.9	−9.82
Southeast	−273.13 ± 118.2	−207.92 ± 94.2	−23.87	−117.42 ± 119.3	−57.01	−412.38 ± 115.5	−369.34 ± 93.1	−10.44	−292.98 ± 119.4	−28.95
US	−629.54 ± 157.7	−905.48 ± 153.9	43.83	−620.91 ± 172.7	−1.37	−870.55 ± 156.3	−980.94 ± 152.2	12.68	−885.37 ± 182.4	1.7
Canada	−295.59 ± 142.8	−340.58 ± 139.5	−15.22	−259.87 ± 139.7	−12.08	−255.50 ± 137.2	−219.74 ± 136.7	−14	−217.45 ± 136.3	−14.89
NA	−932.61 ± 205.9	−1222.9 ± 195.9	31.12	−883.58 ± 211.4	−5.26	−1139.67 ± 189.9	−1206.04 ± 181.8	5.82	−1110.17 ± 200.3	−2.59

* Represents regions with SNR > 1, and positive percentage change in Δ% represents increase in uptake. West includes regions 18, 19, 23, 24, and 25; Midwest includes regions 20 and 21; northeast includes region 22; and southeast includes regions 26, 27, and 28.

the atmospheric CO_2 measurements are able to differentiate about two-third of the differences in fluxes among the regions and the inverted flux is highly sensitive to the prior flux.

Even though the nested regions in North America are fairly large, the atmospheric signal from the CO_2 observation network is blurred substantially among the regions due to efficient atmospheric mixing. Schuh et al. (2010) conducted a high-resolution inversion over North America and found that the inverted flux spatial distribution in 40 km grids differs greatly from that of CarbonTracker because the prior fluxes used in their system and CT are quite different. The high-resolution posterior maps of the carbon source and sink distributions from these two systems resemble closely the spatial details in the prior maps, demonstrating that the atmosphere measurements could only adjust the broad spatial patterns but not the spatial details in the inversion process. In their study, a decorrelation distance of 600 km was judged to be suitable to produce the posterior flux distribution, meaning that the ability of the atmospheric measurements to differentiate flux spatial patterns deteriorates greatly at spatial

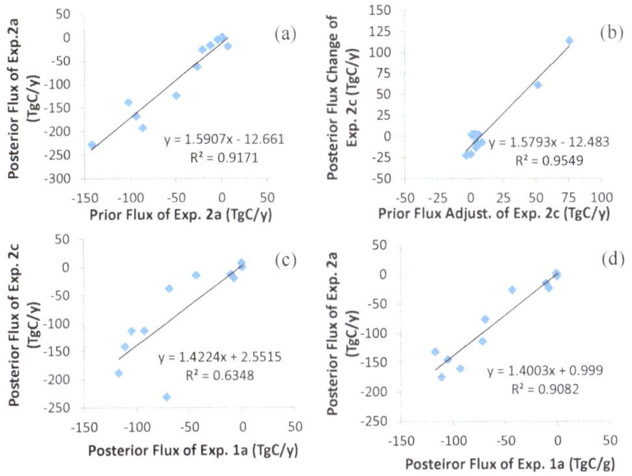

Figure 5. Dependence of the posterior flux from inversions on the prior flux for regions 18–28. **(a)** Correlation between posterior fluxes from experiment 2a and their prior fluxes, **(b)** correlation between changes in posterior fluxes from experiment 2c and adjustments to their prior fluxes, **(c)** correlation of the posterior fluxes between experiment 1a and experiment 2c, and **(d)** correlation of the posterior fluxes between experiment 1a and experiment 2a.

scales smaller than 600 km. Our nested region size is about the same as this decorrelation distance, and therefore our inverted regional fluxes do not lose much spatial information contained in the atmospheric measurements.

The result of experiment 2a for region 20 is comparable to the sink of 0.11 $\mathrm{Pg\,C\,yr^{-1}}$ for cropland in about the same region from 2001 to 2005 reported by Peters et al. (2007), in which the biospheric flux was not adjusted for crop production and consumption. The comparison suggests that both sets of inversions captured about the same strength of the atmospheric sink signal in this region. After crop adjustments in experiment 2c, the sink magnitude in region 20 is approximately doubled with a total of 0.23 $\mathrm{Pg\,C\,yr^{-1}}$ (Table 4) or 298 $\mathrm{g\,C\,m^{-2}\,yr^{-1}}$. The corresponding sink magnitude in region 21 is 0.19 $\mathrm{Pg\,C\,yr^{-1}}$ or 181 $\mathrm{g\,C\,m^{-2}\,yr^{-1}}$. In the Mid-Continent Intensive (MCI) study (Schuh et al., 2013), three inversion models produced an average sink of about 0.15 $\mathrm{Pg\,C\,yr^{-1}}$ or 150 $\mathrm{g\,C\,m^{-2}\,yr^{-1}}$. This result is comparable to the value of 0.183 $\mathrm{Pg\,C\,yr^{-1}}$ or 183 $\mathrm{g\,C\,m^{-2}\,yr^{-1}}$ produced in another inversion study (Lauvaux et al., 2012) for the same MCI area over a growing season from June to December (the dormant season in other months presumably has small fluxes). The MCI study area of 1000 km × 1000 km covers 52 % of regions 20 and 21, with areas of 1275 and 631 $\mathrm{km^2}$, respectively. The area-weighted flux density of these two regions is 219 $\mathrm{g\,C\,m^{-2}\,yr^{-1}}$, which is about 15–40 % larger than previous inversion results from Schuh et al. (2013) and Lauvaux et al. (2012). The difference of our inversion results from these MCI studies is within the posterior uncertainty of our inversion, but it also indicates the need for further evaluation of our inversion results. In these regions, the major crops are corn and soybean. Using eddy covariance systems at three local sites in region 20, Verma et al. (2005) measured NEP of irrigated corn, irrigated corn–soybean rotation, and rainfed corn–soybean rotation in 2001–2004 to be 441, 351, and 296 $\mathrm{g\,C\,m^{-2}\,yr^{-1}}$, respectively. The inverted sink per unit land area is much smaller than these site-level sink values because productive crops only occupy about 50 % of the area in these regions.

Table 4 also provides aggregated results for four large US census regions – west, Midwest, northeast, and southeast – as well as the US and Canada. Crop consumption adjustments make significant (SNR > 1) changes to the inverted carbon flux in the west region in both experiments 1 and 2. This is mostly due to the large crop consumption in California. In experiment 2b with the non-neutralized and interannually variable prior biospheric flux, the crop production adjustment alone does not greatly affect the inverted sink in the west region, but it does significantly increase the sink in the Midwest region with high crop production. Similar to experiment 1, the total sinks in North America and Canada from experiment 2 show large changes in sink sizes after the production adjustment. This suggests that in atmospheric inversion estimates the changes in the a priori flux affect the inverted results not only locally but also globally (Gurney et al., 2004). The overall US and North American carbon sinks are similar to the control case when the crop consumption adjustment is also made (experiments 2b), where the percentage changes are only 1.7 and −2.6 % (Table 3), respectively. This result reinforces our original assumption that the long-term cropland carbon budget is approximately balanced between the crop production and consumption (West et al., 2011).

4.1.2 Multi-year global carbon budget

Although the cropland carbon adjustments were only made for the US regions, the CO_2 flux is inverted globally using the nested inversion system, which not only avoids the need to set up boundary conditions for North America but also produces results for other regions for comprehensive analysis. Table 6 provides the average annual inverted CO_2 fluxes globally from 2002 to 2007 for experiments 1 and 2. It also summarizes the mean inverted CO_2 fluxes (μ) and errors (ε), as well as percentage changes ($\Delta\%$) from the control case for the large regions outside of North America. In the results for both experiments 1 and 2, agricultural adjustments affect the two North America regions (NA-S and NA-N) similarly, indicating that these two regions are closely linked as many CO_2 stations within and around North America are affected by fluxes from both regions. With the crop adjustments, changes in the inverted flux are observed in many regions outside the US, although no adjustments to the prior flux were made outside of the US. A plausible explanation for these results is that any changes in the total sinks in the conterminous US must be compensated for by other regions around the globe since the global carbon

Table 5. Correlation coefficients of the change in the inverted flux from experiments 2a to that of experiment 2c among 11 regions in the US, calculated based on inverted annual fluxes in 2000–2007.

Region	18	19	20	21	22	23	24	25	26	27	28
18	1	0.88	−0.57	−0.56	0.33	0.13	−0.16	0.55	0.09	0.16	0.75
19	0.88	1	−0.54	−0.69	0.43	0.29	0.02	0.72	0.17	0.16	0.70
20	−0.57	−0.54	1	0.06	−0.36	−0.34	0.12	−0.01	−0.46	−0.21	−0.19
21	−0.56	−0.69	0.06	1	−0.48	0.10	0.10	−0.78	−0.44	−0.25	−0.71
22	0.33	0.43	−0.36	−0.48	1	−0.18	−0.45	0.12	0.61	0.09	−0.01
23	0.13	0.29	−0.34	0.10	−0.18	1	0.87	0.41	−0.14	0.67	0.23
24	−0.16	0.02	0.12	0.10	−0.45	0.87	1	0.41	−0.27	0.63	0.16
25	0.55	0.72	−0.01	−0.78	0.12	0.41	0.41	1	−0.03	0.46	0.86
26	0.09	0.17	−0.46	−0.44	0.61	−0.14	−0.27	−0.03	1	0.29	−0.03
27	0.16	0.16	−0.21	−0.25	0.09	0.67	0.63	0.46	0.29	1	0.37
28	0.75	0.70	−0.19	−0.71	−0.01	0.23	0.16	0.86	−0.03	0.37	1

budget is constrained by mean changes in the atmospheric CO_2 concentration. In an atmospheric inversion study, Gurney et al. (2004) found that contributions from land fluxes appeared in the adjacent ocean regions, and they described this phenomenon as flux "leakage". In our case, the compensating effect seen in other regions outside of the conterminous US may also be considered as a leakage. This leakage puts into question the reliability of the atmospheric CO_2 measurements in optimizing the local flux if the prior information is biased, given the fact that North America is one of the continents with strong data constraints. We would therefore expect that similar adjustments in other regions of the globe with fewer observation stations can cause larger flux leakages toward regions with weaker atmospheric constraints.

The inverted total global land sink from experiment 2 is larger than that from experiment 1 (Table 5) because the prior land sink in experiment 2 is larger than that in experiment 1. To compensate for this influence of the prior flux over land, the inverted ocean sink is smaller in experiment 2 than in experiment 1. The total land and ocean sinks do not show large changes with the crop production or crop consumption adjustments in the US in both experiments (Table 5). Relative to experiment 1, experiment 2 shows a slightly larger decrease in the terrestrial sink and a slightly larger increase in the ocean sink. The overall crop adjustments allocate a slightly smaller sink to North America, while most regions outside of North America adjust accordingly to balance the global carbon budget due to weak data constraints in other regions.

4.1.3 Seasonal variation

The timing of the CO_2 uptake and release from croplands exerts great influences on the atmospheric CO_2 (Corbin et al., 2010), as crops in different regions differ in their growth patterns. The seasonal variation pattern in the surface flux may coincide with that in the strength of atmospheric boundary layer mixing, causing the rectifier effect on atmospheric transport (Denning et al., 1995; Gurney, 2004). In this research, the a priori seasonal variation for cropland is based on BEPS, which considers the seasonal variations in vegetation structure using remotely sensed LAI and meteorology. Figure 6 shows the monthly a priori and a posteriori CO_2 fluxes for the period of 2002–2007 from experiment 2 for regions of high crop production and high crop consumption in the US. Results from experiment 1 are not shown, but they are similar in the seasonal variation pattern.

In regions 20 and 21 of high crop production, the inverted monthly fluxes with agricultural adjustments show higher uptake during the peak growing season of June, July, and August and a stronger release of carbon during the non-growing season in October and November than the prior fluxes. March and April show a net uptake in the CO_2 inverted fluxes in agriculturally adjusted experiments in region 20 instead of near carbon neutrality shown in the control experiment. In regions 23 and 24 of high crop consumption with low vegetation, the impacts of crop consumption adjustment are mostly shown as persistent decreases in sinks throughout the year, while those with the production adjustment have persistent increases. In the US southeast, a region with mixed forests and moderate crop production and consumption (regions 26, 27, 28), the crop production adjustment (experiment 2b) and the crop production and consumption adjustments (experiment 2c) cause discernible to considerable impacts on the magnitude of the CO_2 uptake during the growing season and release during non-growing seasons.

The changes in the monthly inverted fluxes after adjusting for crop production and consumption are mainly observed as changes in the magnitude instead of the timing of the growing seasons. This means that the seasonal pattern of the inverted CO_2 flux generally follows that of the a priori flux. In making our US cropland carbon adjustments, annual cropland census data are used instead of actual cropland seasonal variations, while the seasonal growth pattern is modeled based on remotely sensed LAI and meteorology. Therefore, these results are constrained by remote sensing

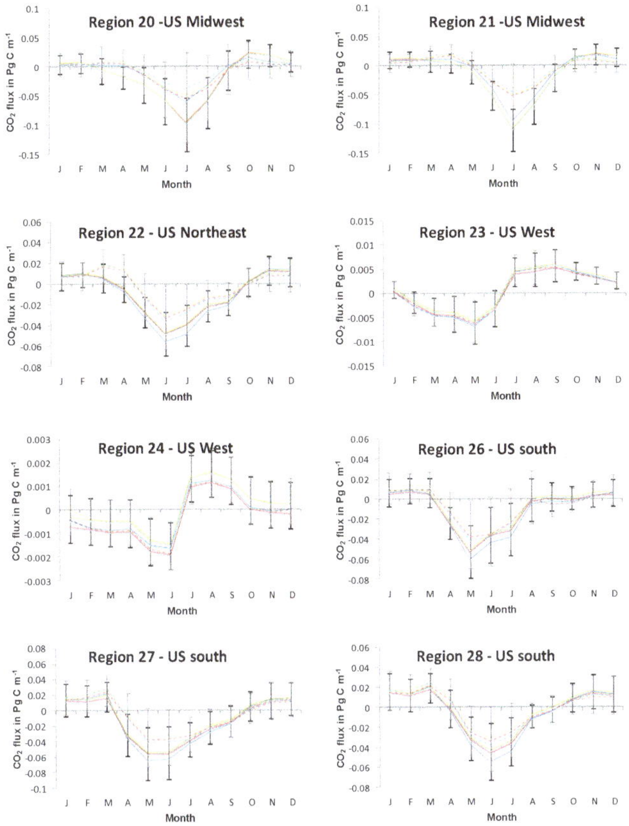

Figure 6. Monthly inverted (solid lines) and prior (dashed lines) CO_2 flux (negative values are sinks) averaged over 2002 to 2007 for Experiments 2 – annually varying terrestrial prior fluxes with **(a)** no adjustments – blue; **(b)** production adjustments – red; **(c)** crop production and consumption adjustments – green.

information, although they are also subject to certain uncertainties due to errors in retrieved LAI and the model.

4.2 Discussion

After integrating US crop production and consumption data into the prior flux, the US Midwest becomes a large regional CO_2 sink from 2002 to 2007 in our inverse modeling. This is mainly due to the large uptake of CO_2 during the growing season (Peters et al., 2007). Because of the release of the CO_2 in regions of sizable crop consumption by humans and livestock, a weakening of the US southeast forest sink is found in our inversion. With the neutralization of the annual prior flux before adjusting for crop consumption (experiment 1), the US Midwest is perceived by the "atmospheric view" as the dominant sink in North America. While croplands in the US Midwest uptake large amounts of carbon during the growing season, some of this carbon is decomposed on-site, while the remainder is transported off-site and respired back to the atmosphere elsewhere. Therefore, lateral transport of carbon needs to be considered in order to reconcile atmospheric inversion results with ground-based measurements (Hayes et al., 2012; Gourdji et al., 2012). On the west coast, including California, where there are high levels of human and livestock consumption, an overall CO_2 source is obtained from the inversion. With the incorporation of US cropland carbon data, the main US sink is redistributed from the US southeast and west forested regions to the US Midwest cropland in the inversion process. In order to assess the possible range of errors in our inversion system, this redistribution is shown in the results from two sets of experiments. Experiment 1, which gives larger weights to the influence of atmospheric CO_2 observation than to the prior terrestrial flux, shows the sink in the US Midwest to be larger than that in the US southeast. Similar results are also obtained in experiment 2, which gives larger weights to the prior terrestrial flux as an opposite case. In both experiments, the inclusion of agricultural inventory data in the priors significantly altered the inversion results for several regions in the west and Midwest regions. Although the atmospheric CO_2 measurements contain much information on the spatial distribution of the carbon sink over the continent, our analysis reveals that they are insufficient for differentiating large flux differences among productive and non-productive regions (Fig. 5). The inverted flux distribution is mostly determined by the prior distribution. This point may be no better demonstrated than the MCI study of Schuh et al. (2013), in which three inversion studies show different inversion results but all closely resemble the prior distributions, although there are 16 CO_2 observation sites within the MCI study region.

In experiment 1, based on an annually neutralized prior ecosystem flux, the overall magnitudes of the US and North American sinks are weakened by ~ 1.4 and ~ 5.3 % after the cropland production and consumption adjustments are made, respectively. Furthermore, regions outside of the US are also affected by these adjustments, and the Canadian carbon sink shows noticeable weakening (12.1 %) (Table 4). Many regions outside of North America, which are not as well constrained by CO_2 observations, also show small changes when these adjustments are made (Table 6). While these changes may have been affected by the additional errors introduced to the uncertainty matrix of the prior flux, the leakage of the CO_2 flux from the adjusted regions into other regions with poor constraints may be the main reason for these changes (Gurney et al., 2004). While not conclusive, this phenomenon of leakage points to the importance of the prior flux for constraining atmospheric inversion for the carbon flux of small regions with sparse atmospheric CO_2 measurements. Since the unneutralized flux is closer to reality than the neutralized flux in terms of the annual sink magnitude, the following further discussion is therefore based on results from experiment 2.

During the period from 2002 to 2007, the annual mean crop production and consumption over the conterminous US based on West et al. (2011) were ~ 0.27 and $\sim 0.24\,\mathrm{Pg\,C\,yr^{-1}}$, respectively. The consumption is smaller than production by about $0.03\,\mathrm{Pg\,C\,yr^{-1}}$, which is close to

Table 6. Mean inverted CO_2 flux (μ), error (ε) in $Pg\,C\,yr^{-1}$, and the percentage change ($\Delta\%$) for global regions from 2002 to 2007.

Region	Experiment 1					Experiment 2				
	(a)	(b)		(c)		(a)	(b)		(c)	
	$\mu \pm \varepsilon$	$\mu \pm \varepsilon$	$\Delta\%$	$\mu \pm \varepsilon$	$\Delta\%$	$\mu \pm \varepsilon$	$\mu \pm \varepsilon$	$\Delta\%$	$\mu \pm \varepsilon$	$\Delta\%$
NA-N	−0.28 ± 0.14	−0.24 ± 0.14	−13.01	−0.26 ± 0.14	−7.57	−0.22 ± 0.14	−0.18 ± 0.14	−14.89	−0.20 ± 0.14	−7.83
NA-S	−0.65 ± 0.16	−0.74 ± 0.16	14.07	−0.63 ± 0.16	−2.95	−0.92 ± 0.16	−1.02 ± 0.16	10.65	−0.91 ± 0.19	−1.37
31	0.36 ± 0.41	0.35 ± 0.41	−0.61	0.34 ± 0.41	−5.59	0.34 ± 0.41	0.39 ± 0.41	14.34	0.37 ± 0.41	9.12
32	−0.07 ± 0.27	−0.07 ± 0.27	−3.62	−0.06 ± 0.27	−10.04	−0.11 ± 0.27	−0.09 ± 0.27	−16.89	−0.08 ± 0.27	−21.02
33	−0.20 ± 0.28	−0.19 ± 0.28	−5.45	−0.22 ± 0.28	8.30	−0.28 ± 0.28	−0.19 ± 0.28	−33.92	−0.22 ± 0.28	−24.07
34	−0.86 ± 0.27	−0.86 ± 0.27	−0.15	−0.86 ± 0.27	−0.63	−0.93 ± 0.27	−0.91 ± 0.27	−2.33	−0.91 ± 0.27	−2.78
35	−0.90 ± 0.26	−0.87 ± 0.26	−2.97	−0.89 ± 0.26	−0.40	−1.00 ± 0.26	−0.96 ± 0.26	−3.73	−0.99 ± 0.26	−1.42
36	−0.63 ± 0.24	−0.61 ± 0.24	−3.34	−0.64 ± 0.24	0.57	−0.52 ± 0.24	−0.54 ± 0.24	4.97	−0.57 ± 0.24	9.75
37	−0.54 ± 0.24	−0.54 ± 0.24	0.57	−0.54 ± 0.24	0.78	−0.63 ± 0.24	−0.59 ± 0.24	−7.49	−0.59 ± 0.24	−7.32
38	0.06 ± 0.08	0.06 ± 0.08	1.05	0.06 ± 0.08	1.91	0.04 ± 0.08	0.04 ± 0.08	−2.77	0.04 ± 0.08	−1.58
39	−0.21 ± 0.21	−0.18 ± 0.21	−14.44	−0.19 ± 0.21	−7.45	−0.06 ± 0.21	−0.01 ± 0.21	−88.22	−0.02 ± 0.21	−64.84
40	−0.58 ± 0.15	−0.58 ± 0.15	1.22	−0.59 ± 0.15	3.00	−0.62 ± 0.15	−0.63 ± 0.15	1.22	−0.64 ± 0.15	2.87
41	−0.06 ± 0.12	−0.07 ± 0.12	10.86	−0.06 ± 0.12	2.11	0.00 ± 0.12	−0.01 ± 0.12	138.30	0.00 ± 0.12	−45.39
42	0.74 ± 0.14	0.74 ± 0.14	−0.72	0.74 ± 0.14	−0.14	0.80 ± 0.14	0.79 ± 0.14	−0.90	0.79 ± 0.14	−0.36
43	−0.58 ± 0.18	−0.58 ± 0.18	0.25	−0.58 ± 0.18	−0.35	−0.51 ± 0.18	−0.51 ± 0.18	0.77	−0.51 ± 0.18	0.08
44	−0.27 ± 0.07	−0.28 ± 0.07	2.85	−0.27 ± 0.07	1.07	−0.27 ± 0.07	−0.28 ± 0.07	3.52	−0.28 ± 0.07	1.76
45	−0.34 ± 0.12	−0.34 ± 0.12	−0.84	−0.36 ± 0.12	4.30	−0.31 ± 0.12	−0.30 ± 0.12	−3.08	−0.32 ± 0.12	2.52
46	0.10 ± 0.12	0.10 ± 0.12	−0.60	0.10 ± 0.12	−0.41	0.15 ± 0.12	0.15 ± 0.12	1.02	0.15 ± 0.12	1.14
47	−0.32 ± 0.14	−0.32 ± 0.14	−1.28	−0.32 ± 0.14	−1.52	−0.33 ± 0.14	−0.33 ± 0.14	−0.94	−0.33 ± 0.14	−1.18
48	0.15 ± 0.06	0.15 ± 0.06	−1.04	0.15 ± 0.06	−0.77	0.14 ± 0.06	0.14 ± 0.06	−0.73	0.14 ± 0.06	−0.45
49	0.06 ± 0.16	0.05 ± 0.16	−6.59	0.06 ± 0.16	−2.21	0.14 ± 0.16	0.14 ± 0.16	2.54	0.15 ± 0.16	4.36
50	−0.56 ± 0.13	−0.56 ± 0.13	−0.64	−0.56 ± 0.13	−0.87	−0.53 ± 0.13	−0.53 ± 0.13	−0.42	−0.53 ± 0.13	−0.66
Land	−3.93 ± 0.67	−3.89 ± 0.67	−0.83	−3.90 ± 0.65	−0.72	−4.30 ± 0.66	−4.07 ± 0.65	−5.39	−4.07 ± 0.66	−5.29
Oceans	−1.67 ± 0.37	−1.69 ± 0.37	1.43	−1.70 ± 0.37	1.70	−1.37 ± 0.38	−1.38 ± 0.37	0.98	−1.39 ± 0.37	1.31
Total	−5.59 ± 0.75	−5.59 ± 0.75	−0.16	−5.59 ± 0.74	0.01	−5.66 ± 0.75	−5.45 ± 0.74	−3.85	−5.46 ± 0.75	−3.70

* represents regions with SNR > 1, and positive percentage change in $\Delta\%$ represents increase in uptake/release.

the value of $0.047\,Pg\,C\,yr^{-1}$ for the net US agricultural carbon export in 2008 estimated by West et al. (2011). For the US Midwest, the inverted results with agricultural adjustments show a large sink of $0.42 \pm 0.13\,Pg\,C\,yr^{-1}$ (experiment 2c). The value is reduced to $0.24\,Pg\,C\,yr^{-1}$ if these adjustments are not made (experiment 2a). These adjustments make a difference of approximately 67 % of the total annual crop production in the US. This suggests that the NEP simulations from the Boreal Ecosystem Productivity Simulator underestimate the processes that relate to cropland CO_2 sinks because it assumes the consumption (respiration) occurs at the same location as production.

The US west and southeast, being regions of large population and crop consumption, show a combined $0.145\,Pg\,C\,yr^{-1}$ weakening of the sink after the crop adjustments (experiment 2c). This sink reduction is about 60 % of the total annual crop consumption averaged over 2002 to 2007 ($\sim 0.24\,Pg\,C\,yr^{-1}$, based on West et al., 2011), indicating the importance of using the crop consumption data for these two regions in the inversion.

Using an atmospheric inversion technique over North America, Peters et al. (2007) showed the US Midwest to be the major sink and the US southeast forest region to be either carbon-neutral or a source between 2002 and 2007. In our inversion constrained by the annually neutralized a priori flux adjusted for crop production and consumption (experiment 1c, Table 4), the US Midwest is also shown to be a large CO_2 sink, but the US southeast forest region remains a sizable carbon sink of $0.117 \pm 0.12\,Pg\,C\,yr^{-1}$ in the same period. Our result differs from that of Peters et al. (2007) for the US southeast, mostly because of different prior fluxes used. Our inversion with the unneutralized prior flux modeled by BEPS (experiment 2a) produced a large sink of $0.413 \pm 0.12\,Pg\,C\,yr^{-1}$ for the southeast (Table 4), and, with crop production and consumption adjustments to the unneutralized prior flux (experiment 2c), the southeast still remains a strong sink of $0.293 \pm 0.12\,Pg\,C\,yr^{-1}$. From the neutralized experiments, we can infer that the atmospheric signal is forcing the inversion in the sink direction for the US southeast. Although the sink is reduced by 42 % when crop production and consumption in this region are considered, this atmospheric signal seems to be robust in indicating this region to be a large sink. The results from experiment 2 show that this sink signal is considerably strengthened if the prior flux is a sink, as we would expect from bottom-up modeling (Zhang et al., 2012) for forests considering the predominant mid-age forest structure (Pan et al., 2011b). The effect of the crop consumption adjustment taken as the difference between experiments 2b and 2c give us confidence that the crop consumption in the US southeast is not large enough to offset the forest sink in this region.

The main components of the US cropland carbon budget presented by West et al. (2011) are the crop production and consumption. However, there is also a non-negligible portion of the crop products being exported outside of the US. Ciais et al. (2007) showed the importance of this lateral transport of carbon internationally. In our study, the lateral international transport of crop carbon is not considered, but its effects on the US carbon cycle might have been included to the first order by the difference between crop production of $\sim 0.27\,\mathrm{Pg\,C\,yr^{-1}}$ and crop consumption of $\sim 0.24\,\mathrm{Pg\,C\,yr^{-1}}$ within the USA. This small imbalance has equivalent effects on the distribution of carbon sinks between Canada and the US, and among continents.

While the use of census-based cropland carbon data is cautioned due to their potentially large uncertainties (Chan and Lin, 2011), our research shows that agricultural statistical data give valuable information on the spatial patterns and the magnitude of the surface CO_2 flux. Cropland carbon and atmospheric CO_2 data, when used in combination, are shown to be able to optimize the prior terrestrial CO_2 flux simulated by a biospheric model and can suggest areas of improvement in the model. This research shows that for managed land cover types such as croplands, where the carbon uptake and release are not simply determined by biospheric processes, biospheric models should not be the only tool used for generating prior terrestrial surface CO_2 fluxes (Ciais et al., 2007). Inventory data, such as agricultural statistics, are shown in this research to provide the missing spatial information useful for regional carbon budget estimation using atmospheric inversion techniques. Since the atmospheric mixing is efficient and CO_2 concentration measurements are sparse, we rely on an accurate prior to obtain a reliable estimate of the regional carbon sink and source distribution.

5 Conclusions

Through adjusting the prior terrestrial carbon flux from a biospheric model using crop production and consumption data at the county level over the conterminous US, we conducted two sets of atmospheric inversion experiments for the purpose of investigating the influence of the crop data on the spatial distribution of inverted carbon flux over North America from 2002 to 2007. One set of experiments utilizes the prior flux that is annually neutralized for each region before the adjustments, and the other set utilizes the prior flux without the neutralization. These two sets of experiments allow us to explore the strength of the atmospheric signal on the carbon source and sink distribution and the possible ranges of the influences of crop data on the inverted carbon flux. The main findings of this research are summarized as follows.

1. The spatial distribution of the inverted carbon flux among 30 regions in North America is sensitive to changes in the prior flux after the crop production and consumption adjustments. Similar to many other biospheric models, the model used in this study provides a prior carbon flux without considering the lateral transport of carbon and therefore does not capture the large consumption sources and underestimates the large production sinks. With the annually neutralized prior, atmospheric CO_2 measurements detected strong sink signals in crop production regions but not the weak source signals in crop consumption regions. While we agree with Gurney et al. (2003), who suggested that inverted fluxes are insensitive to small changes in the a priori fluxes for regions well constrained by atmospheric data, our results show that lateral carbon transport adjustments to the a priori flux are large enough to significantly alter the spatial distribution of the inverted flux. In some crop production and consumption regions, changes in the inverted flux are larger than the uncertainty estimate. Although West et al. (2011) showed the annual cropland carbon budget to be approximately balanced over the conterminous US, the spatial distributions of crop production and consumption are highly uneven, and it would be useful to integrate this information into the prior flux in atmospheric inversion systems. We therefore agree with the suggestion of Ciais et al. (2007) that simply forcing the prior fluxes in inversions to follow spatial patterns of NPP is not compatible with regional patterns of CO_2 sources and sinks due to crop product displacement.

2. Integration of crop data in the prior fluxes has helped identify some robust atmospheric signals in some source and sink regions. Similar to Peters et al. (2007), we also find the US Midwest to be a large CO_2 sink due to crop production, whether or not the production and consumption adjustments are made or the prior flux is neutralized. Dissimilar to Peters et al. (2007), we find the forest regions in the US southeast to be a considerable sink. The neutralization of the prior flux and the adjustments for crop consumption make the sink smaller, but these regions remain sizable sinks throughout the inversion period from 2002 to 2007. This robust signal confirms the finding of Zhang et al. (2012) that the US southeast is a large sink because of the large fraction of mid-age, actively growing forests. The usage of the crop data has helped reconcile top-down and bottom-up results, in agreement with Hayes et al. (2012).

3. The crop adjustments made to the prior carbon flux over the conterminous US also altered the inverted flux outside the US when the biospheric flux was neutralized on an annual basis. This seems to be a leakage of the flux to regions outside of the adjusted regions, a similar case to that previously found by Gurney et al. (2004). Without sufficient atmospheric CO_2 data to constrain each region, the alteration of the flux in one region

is mostly compensated for by other regions, especially those in the vicinity and that are least constrained. This result confirms that our current ground-based CO_2 observation network is not dense enough in any region in the world to conduct inversions without the additional prior flux constraint. Therefore, more attention should be given to obtaining accurate prior fluxes for inversion purposes or greatly expanding the current CO_2 observation network.

Acknowledgements. This research is supported by a research grant (2010CB950704) under the Global Change Program of the Chinese Ministry of Science and Technology, an individual grant (GR-646) from the previous Canadian Foundation of Climate and Atmospheric Sciences (CFCAS), and a Canada Research Chair program.

Edited by: A. Michalak

References

Aumont, O.: An ecosystem model of the global ocean including Fe, Si, P colimitations, Global Biogeochem. Cy., 17, doi:10.1029/2001GB001745, 2003.

Baker, D., Law, R. M., Gurney, K. R., Rayner, P. J., Peylin, P., Denning, A. S., Bousquet, P., Bruhwiler, L., Chen, Y. H., Ciais, P., Fung, I. Y., Heimann, M., John, J., Maki, T., Maksyutov, S., Masarie, K., Prather, M., Pak, B., Taguchi, S., and Zhu, Z.: TransCom 3 inversion intercomparison: Impact of transport model errors on the interannual variability of regional CO_2 fluxes, 1988–2003, Global Biogeochem. Cy., 20, 1988–2003, doi:10.1029/2004GB002439, 2006.

Bolinder, M. A., Janzen, H. H., Gregorich, E. G., Angers, D. A., and VandenBygaart, A. J.: An approach for estimating net primary productivity and annual carbon inputs to soil for common agricultural crops in Canada, Agr. Ecosys. Environ., 18, 29–42, doi:10.1016/j.agee.2006.05.013, 2007.

Buitenhuis, E. T., Le Quéré, C., Aumont, O., Beaugrand, G., Bunker, A., Hirst, A., Ikeda, T., O'Brien, T., Piontkovski, S., and Straile D.: Biogeochemical fluxes through mesozooplankton, Global Biogeochem. Cy., 20, 1–18, doi:10.1029/2005GB002511, 2006.

Canadell, J. G., Le Quéré, C., Raupach, M. R., Field, C. B., Buitenhuis, E. T., Ciais, P., Conway, T. J., Gillett N. P., Houghton R. A., and Marland G.: Contributions to accelerating atmospheric CO_2 growth from economic activity, carbon intensity, and efficiency of natural sinks, P. Natl. Acad. Sci. USA, 104, 18866–18870, doi:10.1073/pnas.0702737104, 2007.

CDIAC: Carbon Sources and Sinks Associated with US Cropland Production, US Department of Energy, Carbon Dioxide Information Analysis Center, Oak Ridge, Tennessee, USA, available at: http://cdiac.ornl.gov/carbonmanagement (last access: 12 January 2015), 2014.

Chan, E. C. and Lin, J. C.: What is the value of agricultural census data in carbon cycle studies?, J. Geophys. Res., 116, 1–14, doi:10.1029/2010JG001617, 2011.

Chen, B., Chen, J. M., Mo, G., Yuen, C.-W., Margolis, H., Higuchi, K., and Chan, D.: Modeling and Scaling Coupled Energy, Water, and Carbon Fluxes Based on Remote Sensing: An Application to Canada's Landmass, J. Hydrometeorol., 8, 123–143, doi:10.1175/JHM566.1, 2007.

Chen, J. M., Liu, J., and Cihlar, J.: Daily canopy photosynthesis model through temporal and spatial scaling for remote sensing applications, Ecol. Model., 124, 99–119, doi:10.1016/S0304-3800(99)00156-8, 1999.

Chen, J. M., Ju, W., Cihlar, J., Price, D., Liu, J., Chen, W., Pan, J., Black, T. A., and Barr, A.: Spatial distribution of carbon sources and sinks in Canada's forests, Tellus, 55, 622–641, 2003.

Chen, J. M., Menges, C. H., and Leblanc, S. G.: Global derivation of the vegetation clumping index from multi-angular satellite data, Remote Sens. Environ., 97, 447–457, 2005.

Chen, J. M., Mo, G., Pisek, J., Deng, F., Ishozawa, M., and Chan, D.: Effects of foliage clumping on global terrestrial gross primary productivity, Global Biogeochem. Cy., 26, GB1019, doi:10.1029/2010GB003996, 2012.

Ciais, P., Bousquet, P., Freibauer, A., and Naegler, T.: Horizontal displacement of carbon associated with agriculture and its impacts on atmospheric CO_2, Global Biogeochem. Cy., 21, 1–12, doi:10.1029/2006GB002741, 2007.

Corbin, K. D., Denning, A. S., Lokupitiya, E. Y., Schuh, A. E., Miles, N. L., Davis, K. J., Richardson, S., and Baker, I. T.: Assessing the impact of crops on regional CO_2 fluxes and atmospheric concentrations, Tellus, 62, 521–532, doi:10.1111/j.1600-0889.2010.00485.x, 2010.

Crevoisier, C., Sweeney, C., Gloor, M., Sarmientob, J. L., and Tans, P. P.: Regional US carbon sinks from three-dimensional atmospheric CO_2 sampling, P. Natl. Acad. Sci. USA, 107, 18348–18353, doi:10.1073/pnas.0900062107, 2010.

DeFries, R. and Townshend, J. R.: NDVI-derived land cover classifications at a global scale, Int. J. Remote Sens., 15, (May 2012), 37–41, 1994.

Deng, F. and Chen, J. M.: Recent global CO_2 flux inferred from atmospheric CO_2 observations and its regional analyses, Biogeosciences, 8, 3263–3281, doi:10.5194/bg-8-3263-2011, 2011.

Deng, F., Chen, J. M., and Plummer, S.: Algorithm for global leaf area index retrieval using satellite imagery, IEEE T. Geosci. Remote, 44, 2219–2229, 2006.

Deng, F., Chen, J. M., Ishizawa, M., Yuen, C.-W., Mo, G., Higuchi, K., Chan, D., Chen, B., and Maksyutov, S.: Global monthly CO_2 flux inversion with a focus over North America, Tellus, 59, 179–190, doi:10.1111/j.1600-0889.2006.00235.x, 2007.

Deng, F., Chen, J. M., Pan, Y., Peters, W., Birdsey, R., McCullough, K., and Xiao, J.: The use of forest stand age information in an atmospheric CO_2 inversion applied to North America, Biogeosciences, 10, 5335–5348, doi:10.5194/bg-10-5335-2013, 2013.

Denning, A. S., Fung, I. Y., and Randall, D.: Latitudinal gradient of atmospheric CO_2 due to seasonal exchange with land biota, Nature, 76, 240–243, doi:10.1038/376240a0, 1995.

Enting, I. G.: Inverse Problems, Carbon, 392 pp., Cambridge University Press, Atmospheric Constituent Transport, doi:10.1017/CBO9780511535741, 2002.

Enting, I. and Trudinger, C.: A synthesis inversion of the concentration and δ; ^{13}C of atmospheric CO_2, Tellus, B47, 35–52, doi:10.1034/j.1600-0889.47.issue1.5.x, 1995.

Fan, S.: A Large Terrestrial Carbon Sink in North America Implied by Atmospheric and Oceanic Carbon Dioxide Data and Models, Science, 282, 442–446, doi:10.1126/science.282.5388.442, 1998.

Gourdji, S. M., Mueller, K. L., Yadav, V., Huntzinger, D. N., Andrews, A. E., Trudeau, M., Petron, G., Nehrkorn, T., Eluszkiewicz, J., Henderson, J., Wen, D., Lin, J., Fischer, M., Sweeney, C., and Michalak, A. M.: North American CO_2 exchange: inter-comparison of modeled estimates with results from a fine-scale atmospheric inversion, Biogeosciences, 9, 457–475, doi:10.5194/bg-9-457-2012, 2012.

Gurney, K. R., Law, R. M., Denning, A. S., Rayner, P. J., Baker, D., Bousquet, P., Bruhwiler, L., Chen, Y. H., Ciais, P., Fan, S., Fung, I. Y., Gloor, M., Heimann, M., Higuchi, K., John, J., Maki, T., Maksyutov, S., Masarie, K., Peylin, P., Prather, M., Pak, B. C., Randerson, J., Sarmiento, J., Taguchi, S., Takahashi, T., and Yuen, C. W.: Towards robust regional estimates of CO_2 sources and sinks using atmospheric transport models, Nature, 415, 626–30, doi:10.1038/415626a, 2002.

Gurney, K. R., Law, R. M., Denning, A. S., Rayner, P. J., Baker, D., Bousquet, P., Bruhwiler, L., Chen, Y. H., Ciais, P., Fan, S., Fung, I. Y., Gloor, M., Heimann, M., Higuchi, K., John, J., Kowalczyk, E., Maki, T., Maksyutov, S., Peylin, P., Prather, M., Pak, B. C., Sarmiento, J., Taguchi, S., Takahashi, T., and Yuen, C. W.: TransCom 3 CO_2 inversion intercomparison: 1. Annual mean control results and sensitivity to transport and prior flux information, Tellus, 55, 555–579, 2003.

Gurney, K. R., Law, R. M., Denning, A. S., Rayner, P. J., Pak, B. C., Baker, D., Bousquet, P., Bruhwiler, L., Chen, Y.-H., Ciais, P., Fung, I. Y., Heiman, M., John, J., Maki, T., Maksyutov, S., Peylin, P., Prather, M., and Taguchi, S.: Transcom 3 inversion intercomparison: Model mean results for the estimation of seasonal carbon sources and sinks, Global Biogeochem. Cy., 18, GB1010, doi:10.1029/2003GB002111, 2004.

Hayes, D. J., Turner, D. P., Stinson, G., McGuire, a. D., Wei, Y., West, T. O., Heath, L. S., de Jong, B., McConkey, B. G., Birdsey, R. A., Kurz, W. A., Jacobson, A. R., Huntzinger, D. N., Pan, Y., Post, W. M., and Cook, R. B: Reconciling estimates of the contemporary North American carbon balance among terrestrial biosphere models, atmospheric inversions, and a new approach for estimating net ecosystem exchange from inventory-based data, Glob. Change Biol., 18, 1282–1299, doi:10.1111/j.1365-2486.2011.02627.x, 2012.

He, L., Chen, J. M., Pisek, J., Schaaf, C. B., and Strahler, A. H.: Global clumping index map derived from the MODIS BRDF product, Remote Sens. Environ., 119, 118–130, 2012.

Hicke, J. A.: Spatiotemporal patterns of cropland area and net primary production in the central United States estimated from USDA agricultural information, Geophys. Res. Lett., 31, 1–5, doi:10.1029/2004GL020927, 2004.

Hicke, J. A. and Lobell, D.: Cropland area and net primary production computed from 30 years of USDA agricultural harvest data, Earth Interact., 8, 1–20, 2004.

Higuchi, K., Shashkov, A., Chan, D., Saigusa, N., Murayama, S., Yamamoto, S., Kondo, H., Chen, J. M., Liu, J., and Chen, B.: Simulations of seasonal and inter-annual variability of gross primary productivity at Takayama with BEPS ecosystem model, Agr. Forest Meteorol., 134, 143–150, doi:10.1016/j.agrformet.2005.08.018, 2005.

Huntzinger, D. N., Post, W. M., Wei, Y., Michalak, a. M., West, T. O., Jacobson, a. R., Baker, I. T., Jacobson, A. R., Baker, I. T., Chen, J. M., Davis, K. J., Hayes, D. J., Hoffman, F. M., Jain, A. K., Liu, S., McGuire, A. D., Neilson, R. P., Potter, C., Poulter, B., Price, D., Raczka, B. M., Tian, H. Q., Thornton, P., Tomelleri, E., Viovy, N., Xiao, J., Yuan, W., Zeng, N., Zhao, M., and Cook, R.: North American Carbon Program (NACP) regional interim synthesis: Terrestrial biospheric model intercomparison, Ecol. Model., 232, 144–157, doi:10.1016/j.ecolmodel.2012.02.004, 2012.

Ju, W., Chen, J. M., Black, T. A., Barr, A. G., Liu, J., and Chen, B.: Modelling multi-year coupled carbon and water fluxes in a boreal aspen forest, Agr. Forest Meteorol., 140, 136–151, doi:10.1016/j.agrformet.2006.08.008, 2006.

Kalnay, E., Kanamitsu, M., Kistler, R., Collins, W., Deaven, D., Gandin, L., Iredell, M., Saha, S., White, G., Woollen, J., Zhu, Y., Leetmaa, A., Reynolds, B., Chelliah, M., Ebisuzaki, W., Higgins, W., Janowiak, J., Mo, K. C., Ropelewski, C., Wang, J., Jenne, R., and Joseph, D.: The NCEP/NCAR 40-year reanalysis project, B. Am. Meteorol. Soc., 77, 437–471, 1996.

Kurz, W. A. and Apps, M. J.: A 70-year retrospective analysis of carbon fluxes in the Canadian forest sector, Ecol. Appl., 9, 526–547, 1999.

Krol, M., Houweling, S., Bregman, B., van den Broek, M., Segers, A., van Velthoven, P., Peters, W., Dentener, F., and Bergamaschi, P.: The two-way nested global chemistry-transport zoom model TM5: algorithm and applications, Atmos. Chem. Phys., 5, 417–432, doi:10.5194/acp-5-417-2005, 2005.

Lauvaux, T., Schuh, A. E., Uliasz, M., Richardson, S., Miles, N., Andrews, A. E., Sweeney, C., Diaz, L. I., Martins, D., Shepson, P. B., and Davis, K. J.: Constraining the CO_2 budget of the corn belt: exploring uncertainties from the assumptions in a mesoscale inverse system, Atmos. Chem. Phys., 12, 337–354, doi:10.5194/acp-12-337-2012, 2012.

Le Quéré, C., Raupach, M. R., Canadell, J. G., Marland, G., Bopp, L., Ciais, P., Conway, T. J., Doney, S. C., Feely, R. A., Foster, P. N., Friedlingstein, P., Gurney, K., Houghton, R. A., House, J. I., Huntingford, C., Levy, P. E., Lomas, M. R., Majkut, J., Metzl, N., Ometto, J. P., Peters, G. P., Prentice, I. C., Randerson, J. T., Running, S. W., Sarmiento, J. L., Schuster, U., Sitch, S., Takahashi, T., Viovy, N., van der Werf, G. R. and Woodward, F. I.: Trends in the sources and sinks of carbon dioxide, Nat. Geosci., 2, 831–836, doi:10.1038/ngeo689, 2009.

Le Quéré, C., Andres, R. J., Boden, T., Conway, T., Houghton, R. A., House, J. I., Marland, G., Peters, G. P., van der Werf, G. R., Ahlström, A., Andrew, R. M., Bopp, L., Canadell, J. G., Ciais, P., Doney, S. C., Enright, C., Friedlingstein, P., Huntingford, C., Jain, A. K., Jourdain, C., Kato, E., Keeling, R. F., Klein Goldewijk, K., Levis, S., Levy, P., Lomas, M., Poulter, B., Raupach, M. R., Schwinger, J., Sitch, S., Stocker, B. D., Viovy, N., Zaehle, S., and Zeng, N.: The global carbon budget 1959–2011, Earth Syst. Sci. Data, 5, 165–185, doi:10.5194/essd-5-165-2013, 2013.

Liu, J., Chen, J. M., Cihlar, J., and Chen, W.: Net primary productivity distribution in the BOREAS region from a process model using satellite and surface data, J. Geophys. Res., 104, 27735–27754, 1999.

Liu, J., Chen, J. M., Cihlar, J., and Chen, W.: Net primary productivity mapped for Canada at 1 km resolution, Glob. Ecol. Biogeogr., 11, 115–129, 2002.

Madec, G., Delecluse, P., Imbard, M., and Levy, C.: OPA 8.1 Ocean General Circulation Model reference manual, Notes du pôle de modélisation Institut Pierre Simon Laplace IPSL France, 11, 91 pp., 1998.

Marland, G., Hamal, K., and Jonas, M.: How Uncertain Are Estimates of CO_2 Emissions?, J. Ind. Ecol., 13, 4–7, doi:10.1111/j.1530-9290.2009.00108.x, 2009.

Masarie, K. and Tans, P. P.: Extension and integration of atmospheric carbon dioxide data into a globally consistent measurement record, J. Geophys. Res., 100, 11593–11, 1995.

Matsushita, B. and Tamura, M.: Integrating remotely sensed data with an ecosystem model to estimate net primary productivity in East Asia, Remote Sens. Environ., 81, 58–66, 2002.

Olivier, J. and Aardenne, J. V.: Recent trends in global greenhouse gas emissions: regional trends 1970–2000 and spatial distribution of key sources in 2000, Environ. Sci., 2, 81–99, 2005.

Pacala, S., Hurtt, G., Baker, D., and Peylin, P.: Consistent land- and atmosphere-based US carbon sink estimates, Science, 292, 2316–2320, 2001.

Pan, Y., Birdsey, R. A., Fang, J., Houghton, R. a., Kauppi, P. E., Kurz, W. A., Phillips, O. L., Shvidenko, A., Lewis, S. L., Canadell, J. G., Ciais, P., J ackson, R. B., Pacala, S., McGuire, A. D., Piao, S., Rautiainen, A., and Sitch, S.: A large and persistent carbon sink in the World's forests, Science, 333, 988–993, 2011a.

Pan, Y., Chen, J. M., Birdsey, R., McCullough, K., He, L., and Deng, F.: Age structure and disturbance legacy of North American forests, Biogeosciences, 8, 715–732, doi:10.5194/bg-8-715-2011, 2011b.

Peters, W., Miller, J. B., Whitaker, J., Denning, A. S., Hirsch, A., Krol, M. C., Zupanski, D., Bruhwiler, L., and Tans, P. P.: An ensemble data assimilation system to estimate CO_2 surface fluxes from atmospheric trace gas observations, J. Geophys. Res., 110, D24304, doi:10.1029/2005JD006157, 2005.

Peters, W., Jacobson, A. R., Sweeney, C., Andrews, A. E., Conway, T. J., Masarie, K., Miller, J. B., Bruhwiler, L. M. P., Petron, G., Hirsch, A. I., Worthy, D. E. J., van der Werf, G. R., Randerson, J. T., Wennberg, P. O., Krol, M. C., and Tans, P. P.: An atmospheric perspective on North American carbon dioxide exchange: CarbonTracker, P. Natl. Acad. Sci. USA, 104, 18925–18930, doi:10.1073/pnas.0708986104, 2007.

Peylin, P., Law, R. M., Gurney, K. R., Chevallier, F., Jacobson, A. R., Maki, T., Niwa, Y., Patra, P. K., Peters, W., Rayner, P. J., Rödenbeck, C., van der Laan-Luijkx, I. T., and Zhang, X.: Global atmospheric carbon budget: results from an ensemble of atmospheric CO_2 inversions, Biogeosciences, 10, 6699–6720, doi:10.5194/bg-10-6699-2013, 2013.

Prince, S. D., Haskett, J., Steininger, M., Strand, H., and Wright, R. Net primary production of US Midwest croplands from agricultural harvest yield data, Ecol. Appl., 11, 1194–1205, 2001.

Randerson, J. T., Thompson, V., Conway, T. J., Fung, I. Y., and Field, B. The contribution of terrestrial sources and sinks to trends in the seasonal cycle of atmospheric carbon dioxide, Global Biogeochem. Cy., 11, 535–560, 1997.

Randerson, J. T., Van Der Werf, G. R., Giglio, L., Collatz, G. J., and Kasibhatla, P. S.: Global Fire Emissions Database, Version 2 (GFEDv2.1), Data set, Oak Ridge National Laboratory Distributed Active Archive Center Oak Ridge Tennessee USA, 2007.

Running, S. W. and Coughlan, J. C.: A general model of forest ecosystem processes for regional applications I, Hydrologic balance, canopy gas exchange and primary production processes, Ecol. Model., 42, 125–154, 1988.

Rödenbeck, C., Houweling, S., Gloor, M., and Heimann, M.: CO_2 flux history 1982–2001 inferred from atmospheric data using a global inversion of atmospheric transport, Atmos. Chem. Phys., 3, 1919–1964, doi:10.5194/acp-3-1919-2003, 2003.

Schuh, A. E., Denning, A. S., Corbin, K. D., Baker, I. T., Uliasz, M., Parazoo, N., Andrews, A. E., and Worthy, D. E. J.: A regional high-resolution carbon flux inversion of North America for 2004, Biogeosciences, 7, 1625–1644, doi:10.5194/bg-7-1625-2010, 2010.

Schuh, A. E., Lauvaux, T., West, T. O., Denning, S., Davis, K., Miles, N., Richardson, S., Uliasz, M., Lokupitiya, E., Cooley, S., Andrews, A., and Ogle, S. Evaluating atmospheric CO_2 inversions at multiple scales over a highly inventoried agricultural landscape, Glob. Change Biol., 19, 1424–1439, doi:10.1111/gcb.12141, 2013.

Sun, R., Chen, J., Zhu, Q., Zhou, Y., and Liu, J.: Spatial distribution of net primary productivity and evapotranspiration in Changbaishan Natural Reserve, China, using Landsat ETM+ data, Can. J. Remote Sens., 30, 731–742, 2004.

Tarantola, A.: Inverse Problem Theory, published by SIAM: Society for Industrial and Applied Mathematics, Physica B, 130, 77–78, doi:10.1137/1.9780898717921, 2005.

Turner, D. P, Jacobson, A. R., Ritts, W. D., Wang W. L., and Nemani, R.: A Large Proportion of North American Net Ecosystem Production is Offset by Emissions from Harvested Products, River/Stream Evasion, and Biomass Burning, Glob. Change Biol., Glob. Change Biolo., 19, C3528, doi:10.1111/gcb.12313, 2013.

Webb, R. S., Rosenzweig, C. E., and Levine, E. R.: A global data set of soil particle size properties, NASA Tech. Memo., TM-4286, 40 pp., 1991.

West, T. O., Brandt, C. C., Wilson, B. S., Hellwinckel, C. M., Tyler, D. D., Marland, G., De La Torre Ugarte, D. G., Larson, J. A., and Nelson, R. G.: Estimating Regional Changes in Soil Carbon with High Spatial Resolution, Soil Sci. Soc. Am. J., 72, 285. doi:10.2136/sssaj2007.0113, 2008.

West, T. O., Marland, G., Singh, N., Bhaduri, B. L., and Roddy, A. B.: The human carbon budget: an estimate of the spatial distribution of metabolic carbon consumption and release in the United States, Biogeochemistry, 94, 29–41, doi:10.1007/s10533-009-9306-z, 2009.

West, T. O., Brandt, C. C., Baskaran, L. M., Hellwinckel, C. M., Mueller, R., Bernacchi, C. J., Bandaru, V., Yang, B., Wilson, B. S., Marland, G., Nelson, R. G., De la Torre Ugarte, D. G., and Post, W. M.: Cropland carbon fluxes in the United States: increasing geospatial resolution of inventory-based carbon accounting, Ecol. Appl., 20, 1074–86, 2010.

West, T. O., Bandaru, V., Brandt, C. C., Schuh, A. E., and Ogle, S. M.: Regional uptake and release of crop carbon in the United States, Biogeosciences, 8, 2037–2046, doi:10.5194/bg-8-2037-2011, 2011.

Williams, C. A., Collatz, G. J., Masek, J., and Goward, S. N.: Carbon consequences of forest disturbance and recovery across the conterminous United States, Global Biogeochem. Cy., 26, GB1005, doi:10.1029/2010GB003947, 2012.

Yuen, C. W., Higuchi, K., Baker, D., Bousquet, P., Bruhwiler, L., Chen, Y. H., Ciais, P., Denning, A. S., Fan, S., Fung, I., Gloor, M., Gurney, K. R., Heimann, M., John, J., Law, R. M., Maki, T., Maksyutov, S., Pak, B., Peylin, P., Prather, M., Rayner, P., Sarmiento, J., Taguchi, S. and Takahashi, T.: : Impact of Fraserdale CO_2 observations on annual flux inversion of the North American boreal region, Tellus, B75, 203–209, 2005.

van der Werf, G. R., Randerson, J. T., Giglio, L., Collatz, G. J., Kasibhatla, P. S., and Arellano Jr., A. F.: Interannual variability of global biomass burning emissions from 1997 to 2004, Atmos. Chem. Phys. Discuss., 6, 3175–3226, doi:10.5194/acpd-6-3175-2006, 2006.

van der Werf, G. R., Randerson, J. T., Giglio, L., Collatz, G. J., Mu, M., Kasibhatla, P. S., Morton, D. C., DeFries, R. S., Jin, Y., and van Leeuwen, T. T.: Global fire emissions and the contribution of deforestation, savanna, forest, agricultural, and peat fires (1997–2009), Atmos. Chem. Phys., 10, 11707–11735, doi:10.5194/acp-10-11707-2010, 2010.

Verma S. B., Dobermann A., Cassman K. G., Walters D. T., Knops J. M., Arkebauer T. J., Suyker A. E., Burba G. G., Amos B., Yang H., Ginting D., Hubbard K. G., Gitelson A. A., and Walter-Shea E. A.: Annual carbon dioxide exchange in irrigated and rainfed maize-based agroecosystems, Agr. Forest Meteorol., 131, 77–96, 2005.

Zhang, F. M., Chen, J. M., Birdsey, R. A., Pan, Y., Shen, S., Ju, W. J., and He, L.: Attributing carbon changes in conterminous US forests to disturbance and non-disturbance factors from 1901 to 2010, J. Geophys. Res.-Biogeo., 117, G02021, doi:10.1029/2011JG001930, 2012.

4

CO_2 fluxes and ecosystem dynamics at five European treeless peatlands – merging data and process oriented modeling

C. Metzger[1,2], P.-E. Jansson[1], A. Lohila[3], M. Aurela[3], T. Eickenscheidt[4], L. Belelli-Marchesini[5], K. J. Dinsmore[6], J. Drewer[6], J. van Huissteden[5], and M. Drösler[4]

[1]Royal Institute of Technology, Department of Land and Water Resources Engineering, Stockholm, Sweden
[2]Technische Universität München, Chair of Restoration Ecology, Freising, Germany
[3]Finnish Meteorological Institute, Atmospheric Composition Research, Helsinki, Finland
[4]University of Applied Sciences Weihenstephan-Triesdorf, Chair of Vegetation Ecology, Freising, Germany
[5]VU University, Department of Earth Sciences, Amsterdam, the Netherlands
[6]Centre for Ecology and Hydrology, Penicuik, Midlothian, UK

Correspondence to: C. Metzger (cmetzger@kth.se)

Abstract. The carbon dioxide (CO_2) exchange of five different peatland systems across Europe with a wide gradient in land use intensity, water table depth, soil fertility and climate was simulated with the process oriented CoupModel. The aim of the study was to find out whether CO_2 fluxes, measured at different sites, can be explained by common processes and parameters or to what extend a site specific configuration is needed. The model was calibrated to fit measured CO_2 fluxes, soil temperature, snow depth and leaf area index (LAI) and resulting differences in model parameters were analyzed. Finding site independent model parameters would mean that differences in the measured fluxes could be explained solely by model input data: water table, meteorological data, management and soil inventory data.

Seasonal variability in the major fluxes was well captured, when a site independent configuration was utilized for most of the parameters. Parameters that differed between sites included the rate of soil organic decomposition, photosynthetic efficiency, and regulation of the mobile carbon (C) pool from senescence to shooting in the next year.

The largest difference between sites was the rate coefficient for heterotrophic respiration. Setting it to a common value would lead to underestimation of mean total respiration by a factor of 2.8 up to an overestimation by a factor of 4. Despite testing a wide range of different responses to soil water and temperature, rate coefficients for heterotrophic respiration were consistently the lowest on formerly drained sites and the highest on the managed sites. Substrate decomposability, pH and vegetation characteristics are possible explanations for the differences in decomposition rates.

Specific parameter values for the timing of plant shooting and senescence, the photosynthesis response to temperature, litter fall and plant respiration rates, leaf morphology and allocation fractions of new assimilates, were not needed, even though the gradient in site latitude ranged from 48° N (southern Germany) to 68° N (northern Finland) differed largely in their vegetation. This was also true for common parameters defining the moisture and temperature response for decomposition, leading to the conclusion that a site specific interpretation of these processes is not necessary. In contrast, the rate of soil organic decomposition, photosynthetic efficiency, and the regulation of the mobile carbon pool need to be estimated from available information on specific soil conditions, vegetation and management of the ecosystems, to be able to describe CO_2 fluxes under different conditions.

1 Introduction

In recent years, many data sets have been collected from a number of sites and across multiple years, containing detailed and high-resolution measurements of carbon (C) fluxes, plant and soil characteristics, meteorological and water table data (Baldocchi et al., 2001; Baldocchi, 2007).

Several of the measured sites are peatlands, which have accumulated a vast amount of C since the last deglaciation. Under drained conditions, peatlands have a high carbon dioxide (CO_2) emission potential (e.g., van den Bos, 2003; Lohila, 2004; Drösler et al., 2008; Maljanen et al., 2010). Understanding the processes driving CO_2 emissions is essential in the development of management practices to reduce greenhouse gas emissions.

Direct comparison of measured data can be used to explore the effect of single variables if the site conditions are similar or differ only in few variables, e.g., in manipulation experiments (Chivers et al., 2009; Ward et al., 2013) or different vegetation types at the same site (e.g., Chojnicki et al., 2010). However, the sites in this study have very different characteristics with respect to climate, hydrology, current and former land management, vegetation and soils. Direct site comparisons of measured flux data (e.g., Alm et al., 1999; Humphreys et al., 2006; Lund et al., 2009; Drewer et al., 2010) are often uninformative when trying to distinguish between responses of several individual factors. Typically, multiple factors are linked and interact with each other complicating the analysis. Therefore, important drivers at one site might not play a significant role on another site (e.g., Lafleur et al., 2005). Process oriented modeling provides a method to identify to what extent observations at different sites can be described by the same processes, while accounting for such interactions.

Process oriented modeling requires (1) that the model can describe the observations and (2) that the parameters used in the model to describe the observations can be estimated from available data. Typically, studies focus on demonstrating how well the model can describe a certain set of data (e.g., van Huissteden et al., 2009; Calanca et al., 2007; Frolking et al., 2001; St-Hilaire et al., 2010). In contrast, the focus of this study was exploring differences between the sites while model performance was subordinate. Process oriented models often require a large number of input parameters which are usually difficult to estimate based on available data from less intensively investigated sites (Juston et al., 2010). Parameters may interact with each other and the available information does not allow for a single or unambiguous mathematical solution (Beven and Freer; 2001, Beven, 2006; van Oijen et al., 2013). However, for all sites in this study, accurate gas flux measurements in combination with detailed measurements of soil and plant conditions were available. Such extensive measurements have been demonstrated to be useful in identifying the governing properties for specific sites. For example, the modeling of CO_2 from forest sites has shown that dynamics of CO_2 fluxes are restricted to a certain range of parameter values (Wu and Jansson, 2013; Wu et al., 2013a).

A systematic evaluation of one model against data from multiple sites with a common set of parameters will allow for a better understanding of processes not only at the individual sites but also on the site specific differences which control the resulting fluxes (e.g., Calanca et al., 2007; van Huissteden et al., 2006; van Huissteden et al., 2009). This is a necessary precondition for accurate predictions of CO_2 fluxes under different climate scenarios or at different locations. On peatlands, some attempts have been made to consider site differences using simplified process models on national (e.g., ECOSSE; Bell et al., 2012) and global scales (e.g., InTec; Ju and Chen, 2005; McGill; St-Hilaire et al., 2010) and up to a millennial timescale (Schuldt et al., 2013). However, we are not aware of any studies comparing differences in parameter distributions of CO_2-related processes between treeless peatland sites using an uncertainty-based approach and a detailed process oriented model running on site scale.

Many carbon ecosystem models are available for site-scale application such as Biome-BC (Feng et al., 2011), DNDC (Li et al., 1992a, b; Dietiker et al., 2010), PaSim (Calanca et al., 2007), PIXGRO (Adiku et al., 2006), CANDY (Franko et al., 1997), or DAYCENT (CENTURY) (Del Grosso et al., 2005). Some models were explicitly created or adapted to peatlands such as PDM (Frolking et al., 2001), PCARS (Frolking et al., 2002), CASA (Potter et al., 2001), NASA-CASA (Del Grosso et al., 2005), ecosys (Grant et al., 2012), wetland-DNDC (Zhang et al., 2002), peatland DOS-TEM (Fan et al., 2013), PEATLAND-VU (van Huissteden et al., 2006) or GUESS-ROMUL (Yurova et al., 2007).

In this work the CoupModel was used, which is a detailed process oriented model coupling heat and mass transfer for soil–plant–atmosphere systems (Jansson and Karlberg, 2010). The CoupModel was chosen for the following reasons: the model was designed for a wide range of soil types and different ecosystems and applications (see Jansson, 2012 for review) which might be useful as some of the sites in this study are already quite degraded and might not respond like a typical, intact peatland anymore. The model has been shown to be capable of simulating all three main greenhouse gases from peatlands: CO_2 (Klemedtsson et al., 2008), nitrous oxide (N_2O) (Norman et al., 2008) and methane (CH_4) (Ravina, 2007). Furthermore, the CoupModel includes detailed submodules for the most relevant processes in the carbon cycle: it predicts plant growth, plant transpiration and autotrophic respiration, soil nitrogen (N) and C processes, energy and heat fluxes, soil temperature, soil frost and snow depth. It supports an hourly time step for input and output data and can run in even finer time resolution, which is necessary for analyzing e.g., chamber flux data. The user can select between different submodels, different equations and different complexities and easily access all parameters via a user interface. Calibration procedures with randomized parameter values and methods for visualization and detailed analysis of the model output are supported. An extensive model description can be found in Jansson and Karlberg (2010). The model and its documentation as well as several tutorials for its application can be downloaded from the CoupModel home page (CoupModel, 2014).

Table 1. Site characteristics.

Code	Lom	Amo	Hor	FsA and FsB
Country	Finland	UK	Netherlands	Germany
Site name	Lompolojänkkä	Auchencorth Moss	Horstermeer	Freisinger Moos
Area (m^2)	120 000	250 000	120 000	400
Latitude, longitude	67°59′83″ N, 24°12′55″ E	55°47′34″ N, 3°14′35″ W	52°14′25″ N, 5°4′17″ E	48°22′50″ N, 11°41′12″ E
Peatland type	fen	bog	fen	fen
Dominant vegetation	mosses, sedges, shrubs	grasses, sedges, soft rush, mosses	grass, reeds, small shrubs	sedges, herbs, grasses (A), tall sedges (B)
Land use and management	natural mire	restored; grazed	restored; nature reserve	drained, 1 cut a^{-1}
Mean temperature/range[a] (°C)	−1.4/−15–13	10/4–15	9.8/3–17	7.5/−2–17
Mean water table (cm)	+1.2	−12.5	∼ −10	−25 (A) −20 (B)
Annual precipitation (mm)	484	1155	797	788
N deposition (kg ha^{-1} a^{-1})	8.13	1.59		7.1
Peat depth (m)	2–3	0.5–5	2	3
pH	5.5–6.0	4.4	4.8–6.0	5.5–6.7

[a] Annual range of mean monthly temperatures.

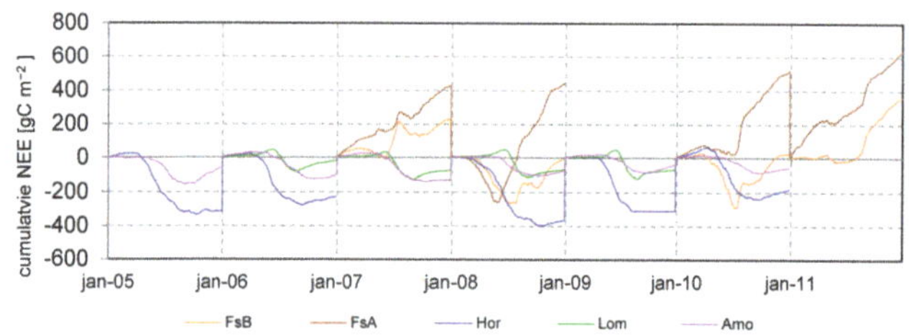

Figure 1. Measured NEE from gap-filled time series. Positive values indicate CO_2 emission, negative CO_2 uptake by the ecosystem.

The main aim of this study was to find out to what extend the large differences in measured CO_2 fluxes between five data rich European flux measurement sites can be solely explained by the differences in meteorology, water table and management. Therefore, the process oriented CoupModel was applied using an uncertainty-based Monte Carlo approach. Specific objectives were

I. to identify differences and similarities between various sites in CO_2-related processes, corresponding parameters and responses to forcing data;

II. to identify and discuss the impact of available data for estimating key parameters in CO_2 flux models in general;

III. to identify problems related to the model representation of the different ecosystem processes for open peatlands.

2 Methods

2.1 Description of sites and investigations

The CoupModel was applied to five treeless peatland sites with a wide gradient in land use intensity, water level, soil nutrient status and mean annual temperature (Table 1). Together with the climatic gradient from northern Finland to southern Germany and a different growing season, this leads to great differences in amplitude and dynamics of gross primary productivity (GPP), ecosystem respiration (R_{eco}) and different amounts of biomass. This is reflected in the annual accumulated net ecosystem exchange (NEE) based on measurements, ranging from −395 to 636 g C m^{-2} (Fig. 1).

Dynamic forcing data for model input (water table and meteorology) were available from measurements at all sites (Table S1). Data used for model parameter constraint included measurements of LAI, soil temperature and NEE (Table S2). Measured NEE was partitioned into R_{eco} and GPP by the use of empirical models. At sites where the eddy covariance method was used, R_{eco} was derived from nighttime NEE, otherwise it was taken from opaque chamber measurements.. The empirical R_{eco} models are based on temperature (Lloyd and Taylor, 1994), while light-level-based functions were used for GPP according to Falge et al. (2001). Corrections and gap filling at flux tower sites was done according to the methods described in Reichstein et al. (2005). A detailed description is given in the references listed in Table S2. Though R_{eco} and GPP are not explicitly measured, this will be called measured data in the following for simple distinction from the simulated fluxes by the CoupModel.

The northernmost site, Lompolojänkkä fen (Lom), located in Finland is a nutrient-rich natural mire with sedges, shrubs and mosses. Mean air temperature from 2006 to 2010 was −1.4 °C and the mean groundwater table during the snow-free season was close to the peat surface. Data for model calibration were available from 2006 to 2010 and consisted of eddy covariance (EC) and automatic chamber data of CO_2 fluxes, snow depth and leaf area index (LAI) measurements. A detailed description of the site and measurement methods can be found in Aurela et al. (2009), Drewer et al. (2010) and Lohila et al. (2010).

The Scottish site, Auchencorth Moss (Amo) is an ombrotrophic bog, with vegetation consisting of grasses, sedges and soft rushes, covering a primarily Sphagnum base layer. The site is managed for low-intensity sheep grazing with less than one livestock unit per hectare, but this was not accounted for in the model. Amo encompasses a small area of peat extraction in the southwest of the catchment, which is unlikely to fall within the flux footprint of the EC system. The site was drained over a century ago, however, the drains are no longer considered to be in operation. The mean water table was −12.5 cm between 2006 and 2010. Mean temperature during this period was 10 °C, CO_2 data from EC during the same period was used for model calibration. A detailed description of the site and measurements can be found in Helfter et al. (2014), Drewer et al. (2010) and Dinsmore et al. (2010).

Horstermeer fen (Hor) is located in the Netherlands in a drained natural lake. It used to be agricultural land but was abandoned more than 15 years ago. The water table was raised during restoration leading to a mean value of −10 cm during the simulation period from 2004 to 2010. It became a seminatural grassland, a nature reserve without any mowing management. The vegetation is very heterogeneous with reeds, grasses and small shrubs (Hendriks, 2009). The mean temperature during the simulation period was 10 °C. CO_2 fluxes were measured half-hourly by EC and biweekly with opaque chambers between 2004 and 2010. A detailed description of the site and measurement methods can be found in Hendriks et al. (2007).

Freisinger Moos (FsA and FsB) is a drained nutrient-rich fen in the south of Germany. The two sites FsA and FsB lie next to each other in a drained sedge meadow which was cut once per year. The mean annual hay yield was 4.19 or 4.07 t dry weight $ha^{-1}\,a^{-1}$ for FsA and 5.67 or 6.17 t dry weight $ha^{-1}\,a^{-1}$ for FsB for the years 2010 and 2011, respectively. FsB is located in a small depression with a mean water level of −20 cm compared to −25 cm for FsA during the years 2007–2011. Mean temperature during this period was 7.5 °C. FsB contains mainly tall sedges with little reed while FsA is vegetated by a mixture of sedges, grasses and herbs. Manual transparent and opaque chamber data of CO_2 fluxes ($n = 3$ for each plot), measured several times a day every 3–4 weeks and half-hourly meteorological data were available for the time period of 2007–2011. A detailed description of the chamber configuration, measurement technique and empirical model approach were given in Drösler (2005), Beetz et al. (2013) and Leiber-Sauheitl et al. (2014). The measured R_{eco} and empirical modeled GPP during the measurement period of each measurement day were used for parameter constraint, empirically modeled values between measurement days were only used for visualization and comparison.

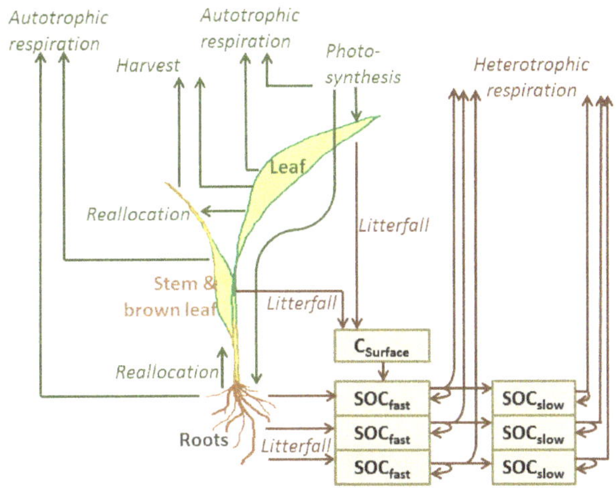

Figure 2. Scheme of carbon fluxes and pools in the current CoupModel setup.

2.2 Model description

CoupModel v4 from 12 April 2013 was used for simulations. The current version can be downloaded from the CoupModel home page (CoupModel, 2014). A detailed description can be found in Jansson and Karlberg (2010). The model represents the ecosystem by a description of C and N fluxes in the soil and in the plant. It includes all main abiotic fluxes, such as soil heat and water fluxes, that represent the major drivers for regulation of the biological components of the ecosystem. The most important equations with the corresponding parameters and switches differing from the default setup in the used version can be found in Tables S3, S4, S5 and S6. The major model assumptions relating to the model application to peatlands are described below. Figure 2 shows a scheme of the main carbon fluxes and pools in the current CoupModel setup.

Table 2. Abbreviations and symbols.

Abbreviation	Description
C	carbon
CO_2	carbon dioxide
EC	eddy covariance
GHG	greenhouse gas
GPP	gross primary production
k_{gresp}	growth respiration coefficient
k_h	rate coefficient for the decay of the slow C pool
k_l	rate coefficient for the decay of the fast C pool
k_{tot}	total rate of decomposition calculated from k_h, k_l and SOC of the corresponding pools in the upper 30 cm
$k_{mrespleaf}$	maintenance respiration coefficient for leaves
k_{rn}	extinction coefficient in the Beer law used to calculate the partitioning of net radiation between plant canopy and soil surface
LAI	leaf area index
ME	mean error
m_{retain}	coefficient for determining allocation to mobile internal storage pool
N	nitrogen
NSE	Nash–Sutcliffe efficiency
p_{ck}	speed at which the maximum surface cover of plants is reached
$p_{\theta p}$	power coefficient in the response function of microbial activity in dependency of soil moisture
$p_{\theta Satact}$	activity under saturated conditions in the soil moisture response function for microbial activity, mineralization–immobilization, nitrification and denitrification
$p_{\theta Upp}$	water content interval in the soil moisture response function for microbial activity, mineralization–immobilization, nitrification and denitrification
R^2	coefficient of determination
R_{eco}	ecosystem respiration
SOC	soil organic carbon
T_{amean}	assumed value of mean air temperature for the lower boundary condition for heat conduction
T_{aamp}	assumed value of the amplitude of the sine curve , representing the lower boundary condition for heat conduction
$T_{MatureSum}$	temperature sum beginning from grain filling stage for plant reaching maturity stage
T_{DormTh}	critical air temperature that must be undershot for temperature sum calculation
$T_{EmergeSum}$	air temperature sum that is the threshold for start of plant development
$T_{EmergeTh}$	critical air temperature that must be exceeded for temperature sum calculation
t_{Q10}	response to a 10 °C soil temperature change on the microbial activity, mineralization–immobilization, nitrification, denitrification and plant respiration
t_{Q10bas}	base temperature for the microbial activity, mineralization–immobilization, nitrification and denitrification at which the response is 1
ε_L	radiation-use efficiency

2.2.1 Meteorological driving variables and integration time step of the model

Hourly values of global radiation, relative humidity, precipitation, wind speed, and air temperature, measured at each site were used as input. Data was gap filled by simple linear interpolation for gaps <6 h. Larger gaps were filled by values from other adjacent climate stations. At Hor the station used for gap filling provided only daily values. Hourly values were retrieved assuming uniform distribution over 24 h for precipitation, wind speed and relative humidity and sinusoidal distribution for temperature and global radiation.

Model performance was only evaluated for the years when meteorological data was available. The simulations were started 2 years prior to the evaluation period, so the system (in particular the plant) could adapt to the site conditions and become more independent of initial values. Data from the available years was copied to previous years if not available from an adjacent climate station.

The model's internal time step was half-hourly for abiotic processes and hourly for nitrogen- and carbon-related processes.

2.2.2 Dynamic coupled heat and water model for above soil surface conditions

An interception model for both, radiation and precipitation, a snow model and a surface pool model was used to provide boundary conditions at the soil surface. Interception and plant evaporation was dependent on the simulated leaf area index of the plant as well as the degree of coverage, while transpiration depended additionally on the simulated water uptake of the plant. Cloud fraction was calculated from global radiation input and latitude. Incoming radiation was partitioned between one part which was absorbed by the plant canopy and another part which reached the soil. Surface temperature was simulated based on an energy balance approach, where the radiation reaching the soil equals the sum of sensible and latent heat flux to the air and heat flux to the soil. Soil evaporation was derived from an iterative solution of the soil surface energy balance of the soil surface, using an empirical parameter for estimating the vapor pressure and temperature at the soil surface. Vapor pressure deficit was calculated from the relative humidity input. Snowfall was simulated from precipitation and air temperature, snowmelt from global radiation, air temperature and simulated soil heat flux. Surface runoff was controlled by a surface pool of water that covers various fractions of the soil surface. Under oversaturated periods the flow of water in the upper soil compartment could be directed upwards, towards the surface pool. Surface runoff was calculated as a function of the amount of water in the surface pool.

2.2.3 Dynamic heat and water model for the soil

The soil profiles were divided into 12 layers with an increasing layer depth from 5 cm for the upper layer to 100 cm in the lowest layer. Heat flow between adjacent soil layers were calculated based on thermal conductivity functions accounting for the content of ice and water. The heat flow equation is based on a coupled equation accounting for the freezing and thawing in the soil (Jansson and Halldin, 1979). Convection was not accounted for. The lower boundary was calculated as temperature based on a sine variation at the soil surface and a damping depth for the whole soil profile as well as a parameter for the annual mean temperature T_{amean} and annual amplitude of temperature T_{aamp} at the site (a list of symbols and abbreviations can be found in Table 2).

Soil water depended on infiltration to the soil, soil evaporation, water uptake by plant roots and groundwater flow. Soil moisture, represented as liquid water content, was calculated based on the water storage and temperature. Water flows between adjacent soil layers were calculated according to the Richards equation (1931), considering hydraulic conductivity, water potential gradient and vapor diffusion. Soil water characteristics were described by the Brooks and Corey (1964) equation between two threshold water tensions, while a log-linear expression was applied at higher water tensions and a linear expression at water contents close to saturation. Unsaturated conductivity was simulated according to Mualem (1976), additionally accounting for the conductivity in macro pores. The groundwater level was defined by assuming a continuous zone of saturation from the water table level down to the lower boundary of the considered soil profile. To force saturation at the measured groundwater level, water was added to or removed from the corresponding layer.

2.2.4 Vegetation

Vegetation was simulated according to the explicit big leaves concept (e.g., Dai et al., 2004) but only one plant canopy layer, representing the complete plant community, was defined. Albedo, LAI, vegetation height and vegetation cover were simulated. Permanent, perennial vegetation was configured with maximal plant height of 0.6 m, a lowest root depth of −0.6 m and a maximal plant cover of 100 %. Grain development was assumed to play a minor role and was therefore disabled. Plant respiration was assumed to be dependent on growth and maintenance (e.g., Hansen and Jensen, 1977).

For leaf assimilation, the light-use efficiency approach (Monteith, 1972; Monteith and Moss, 1977, see e.g., Hilker et al., 2008 for review) was used, at which total plant growth is proportional to the global radiation absorbed by canopy but limited by unfavorable temperature and limited soil water. For simplicity, plant assimilation was simulated independent of dynamics in N availability. This might be justified as none of the sites were fertilized in the recent years and the vegetation community was assumed to be adapted to the nutrient conditions at each site. Differences in N availability between sites are included in the radiation efficiency (ε_{L}). Plants were assumed to be well adapted to wet conditions (Keddy, 1992, Steed et al., 2002), including aerenchyma to tolerate water saturated soil conditions (Jackson and Armstrong, 1999). Plant stress due to high water saturation was therefore disabled.

Plant development started every spring when the accumulated sum of air temperatures above a threshold value ($T_{\mathrm{EmergeTth}}$) reached the value of $T_{\mathrm{EmergeSum}}$. Both parameters were calibrated (Table S4). The accumulation of temperatures started when the day length exceeded 10 h. Snow cover hindered shooting by reducing the radiation passing through to the plant, while low soil temperatures reduced plant water uptake.

Besides a small amount of litter fall occurring during the whole plant growth period (Robson, 1973; Duru and Ducrocq, 2000; Fulkerson and Donaghy, 2001), senescence was assumed to start after the plant reached maturity and therefore depended on growth stage (e.g., Thomas and Stoddart, 1980) and temperature sums (e.g., Davidson and Campbell, 1983). As this was not yet directly supported by the model, the stem pool was used for brown, senescent, standing biomass. Therefore new assimilates were constantly allocated to roots and leafs only, while existing leaf biomass was reallocated after maturity to the stem pool. A third stage of litter fall was configured depending on a minimum threshold temperature sum: five consecutive days in the autumn with day lengths shorter than 10 h and with temperatures below T_{DormTth} terminated the growing season and plants went into dormancy.

During litter fall part of the C is stored in a mobile pool, which can be then reused for regrowth in spring in the next year (e.g., White, 1973; Wingler, 2005).

Harvest took place at FsA and FsB. Based on observations in the field, 85 % of the aboveground plant material was removed at harvest. Harvest dates were known and implemented in the model. After harvest the growth stage was allowed to be reset to a lower value (e.g., Thomas and Stoddart, 1980). Reallocation of C from roots to leaves could take place as reported for, e.g., *Festuca pratensis* (Johansson, 1992; 1993).

2.2.5 Soil carbon and nitrogen

The organic substrate was represented by two C and N pools for each of the 12 soil layers: one with a slow and one with a high turnover rate coefficient. Decomposition products from the fast pools are partitioned into CO_2 which is released to the atmosphere and C which is partly moved to the slow pools and partly returned to the fast pools. Decomposition products from the slow pools are partly released as CO_2 and partly returned to the slow pools. The initial values for the amount of C and N per layer was given by measurements and partitioned into the two pools for each layer according to the measured C : N ratio as described in Sect. 2.2.5 and Table 3. Beside the turnover rate coefficients and amount of substrate in each pool per layer, decomposition rates depended on the response to soil moisture and temperature in the corresponding layer.

As the rate coefficients for decomposition were expected to strongly affect each other, only the coefficient for the fast decomposition pools were calibrated. The coefficient for the slow pools (k_h) was kept constant at a low value of $2 \times 10^{-8}\,\text{day}^{-1}$ during the calibration runs, which might be justified as decomposition of resistant carbon is less responsible for the variation in soil respiration (e.g., Whalen et al., 2000).

Nitrogen- and methane-related processes were considered by a model including the most important pathways and fluxes. However no emphasize on the calibration of these processes was made in this study since the current objective was on CO_2 fluxes from the peatlands.

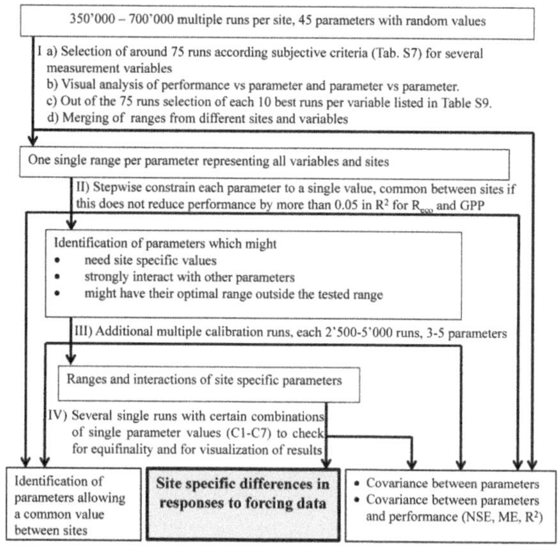

Figure 3. Stepwise parameter calibration. Boxes show the outcome of each step. Description for scenarios C1–C7 can be found in Table S8.

2.2.6 Independent approach to find values of site specific parameters

Dry and wet N deposition, latitude and thickness of the organic layer were used as constant site specific input.

Water retention parameters were assigned to each soil layer according to soil data from each site. However, at Amo and Lom, water retention and, at all sites, unsaturated conductivity was assigned from the CoupModel soil database as suggested by Lundmark (2008) for peat soils. Measured total soil organic carbon (SOC) per layer was partitioned to the two SOC pools per layer on the basis of the measured total C : N ratio per layer whereas the initial C : N ratios of the slow decomposing pools were assumed to be 10, while for the fast pool's 27.5 was chosen according to measured C : N of leaf tissues at FsA and FsB (Table 3).

2.3 Parameter calibration approach

The aim of the calibration was to find out to what extent the same parameter values could be used for all sites compared to a site specific representation. A stepwise approach was carried out starting with finding the best site specific parameter representations and then trying to merge them to common values valid for all sites. Finally, the common representation was revised to some few parameters showing great site specific effects on model performance. An overview of the different steps can be found in Fig. 3, details on the calibration procedures are presented as supplementary material.

For the basic calibration (step I, Fig. 3), 350 000–700 000 runs were performed for each site. 45 parameters, which were suspicious of eventually being site specific, were selected and calibrated with an assumed uniform random range

Table 3. Partitioning of measured SOC to the pools. The data in the table is aggregated into 3 soil layers, however 12 layers were used in the model.

	depth (m)	Lom	Amo	Hor	FsA	FsB
Measured total C ($kg\,m^{-3}$)	0–0.1	24	190	72	107	88
	0.1–0.3	30	187	79	104	90
	>0.3	51	175	156	70	61
Measured C : N ($kg\,m^{-3}$)	0–0.1	27	23	13	11	12
	0.1–0.3	20	22	13	14	13
	>0.3	20	21	22	17	17
Estimated fraction of fast pool/total C	0–0.1	95 %	72 %	18 %	3 %	9 %
	0.1–0.3	56 %	73 %	20 %	20 %	16 %
	>0.3	55 %	68 %	62 %	35 %	41 %
Dry bulk density ($g\,cm^{-3}$)	0–0.1	0.06[a]	0.39	0.35	0.59	0.33
	0.1–0.3	0.06	0.37	0.48	0.29	0.52
	>0.3	0.10	0.37[b]	0.50	0.18	0.17

[a] No data available, value from lower layer used. [b] No data available, value from upper layer used.

(Table S4). Parameter ranges were then constrained based on selected runs (steps I and II; Fig. 3), showing acceptable performance to multiple variables (Table S7), measured at the sites.

Several additional multiple calibration runs were performed, with few selected parameters each, to unravel parameter interactions (step III; Fig. 3). A number of simulations were also made by single value representations of parameters (step IV; Fig. 3) to visualize the impact of certain parameter values on interacting parameters and on performance. These runs are called single runs in the following, numbered C1–C7 and described in Table S8.

Selection of runs and evaluation of performance was based on three indices: coefficient of determination (R^2) assesses how well the dynamics in the measurement derived values are represented by the model. Mean error (ME), also called *y* intercept (Willmott, 1982) indicates a lag or lead between model predictions and measured data (Moriasi et al., 2007). Nash–Sutcliffe efficiency (NSE) (Nash and Sutcliffe, 1970) accounts for both deviation of dynamics and magnitude. It ranges from $-\infty$ to 1, whereas 1 means the best fit of modeled to measured data and values <0 indicates that the mean measured value is a better predictor than the simulated value, which indicates an unacceptable performance (Moriasi et al., 2007).

3 Results

3.1 Model performance – results of basic calibration and selected common configuration

Model performance showed distinct differences between the sites, depending on the investigated variable and on the number of considered runs (Table 4). Figure 4 shows the differences between measurements and model C1.

3.1.1 Fluxes

At all sites the dynamics in R_{eco} fluxes were simulated considerably better than GPP (Table 4). Performances for NEE were worse as simulation errors in GPP and R_{eco} are summed up.

With respect to R_{eco} and GPP, the selected single runs represent a parameter configuration close to the best ones possible in the tested range: their R^2 value did not differ by more than 0.05 from the best result achieved in the multiple calibration, while ME values were smaller than $0.1\,g\,C\,m^{-2}\,day^{-1}$. Clearly, the lower R^2 and higher ME values in single runs for biomass and LAI simulation indicate that none of the runs could give the best results for all variables at the same time. For example, the best values for GPP can only be achieved if poorer performance would have been accepted for other parameters such as winter R_{eco} or LAI (see criteria for accepted runs in Table S7).

The ME values in Table 4 show a clear overestimation of winter fluxes by 3.21 and $2.11\,g\,C\,m^{-2}\,day^{-1}$ for the single runs at FsA and FsB, respectively, and a weaker overestimation for the accepted runs. The overestimation was less pronounced at Amo ($0.13\,g\,C\,m^{-2}\,day^{-1}$) and Lom ($0.01\,g\,C\,m^{-2}\,day^{-1}$). At Hor, winter fluxes were underestimated with a ME of $-0.26\,g\,C\,m^{-2}\,day^{-1}$. This was reflected in the accumulated NEE (Fig. 4) leading to a much higher CO_2 loss compared to the CO_2 balance estimated by the empirical model approach at FsA and FsB. At Lom, higher accumulated NEE due to the overestimation of winter R_{eco} was visible in the first months of each year. It was nearly compensated due to the underestimated spring R_{eco}

Table 4. Highest achieved values for selected performance indices.

Variable	Index	Lom all/selected runs	Lom single run	Amo all/selected runs	Amo single run	Hor all/selected runs	Hor single run	FsA all/selected runs	FsA single run	FsB all/selected runs	FsB single run
NEE	R^2	0.61/0.60	0.59	0.59/0.58	0.55	0.53/0.51	0.48	0.20/0.16	0.15	0.25/0.21	0.19
	ME	0.00	0.05	0.00	0.04	0.00	0.02	0.00	1.43	0.00	−0.05
GPP	R^2	0.66/0.66	0.65	0.68/0.68	0.66	0.58/0.57	0.55	0.38/0.35	0.34	0.40/0.39	0.35
	ME	0.00	0.05	0.00	−0.09	0.00	0.04	0.00	0.06	0.00	−0.03
R_{eco}EC	R^2	0.79/0.74	0.69	0.71/0.71	0.66	0.78/0.77	0.75	NA	NA	NA	NA
	ME	0.00	0.00	0.00	−0.05	0.00	−0.06	NA	NA	NA	NA
R_{eco}	R^2	0.73/0.71	0.64	0.67/0.57	0.38	0.52/0.48	0.45	0.73/0.66	0.69	0.87/0.81	0.85
Chamber	ME	0.00	−0.06	0.00	0.04	0.00/−4.74	−5.38	0.00	−0.01	0.00	−0.08
R_{eco}	R^2	0.67/0.63	0.63	0.14/0.08	0.06	0.28	0.28	0.51/0.43	0.32	0.92/0.89	0.89
winter	ME	0.00	0.01	0/0.04	0.13	0.00	−0.26	0/1.60	3.21	0.00/0.73	2.11
Upper soil	R^2	0.88/0.87	0.87	0.86	0.84	0.92	0.91	0.88/0.86	0.84	0.88/0.86	0.84
temperature	ME	0.00	−0.01	−0.03	−0.08	−1.37/−1.51	−1.77	0.00/0.58	0.35	0/1.20	0.35
Lower soil	R^2	0.95	0.95	0.90	0.89	0.89	0.89	0.97/0.96	0.94	0.92/0.91	0.94
temperature	ME	0.00	−0.03	0.00	0.02	0.00	−0.08	0.00	−0.15	0.00	−0.15
Snow depth	R^2	0.75	0.75	NA	NA	NA	NA	NA	NA	NA	NA
	ME	−0.1	−0.06	NA	NA	NA	NA	NA	NA	NA	NA
LAI	R^2	0.65/0.51	0.53	NA	NA	0.36/0.31	0.33	0.75/0.69	0.61	0.82/0.76	0.61
	ME	0.00	0.11	NA	NA	0.00/−0.61	−1.49	0.00	0.12	0.00	0.05
Aboveground	R^2	NA	NA	NA	NA	0.02/0.00	0.00	0.31/0.26	0.24	0.47/0.43	0.32
living biomass	ME	NA	NA	NA	NA	0	−112	0/−20	−21	0/−36	−48
Root biomass	R^2	NA	NA	NA	NA	0.28/0.07	0.01	NA	NA	NA	NA
	ME	NA	NA	NA	NA	0.00	−282	NA	NA	NA	NA

NA: not available.

or overcompensated due to GPP overestimation as, e.g., in summer 2006, which was very dry.

3.1.2 Explanatory variables

Of all variables, the highest R^2 values were achieved for soil temperature at all sites. Temperatures in deeper soil layers (−50 or −60 cm) had better fits than in upper layers with R^2 values close to 0.9 or higher and a maximum mean deviation of 0.15 °C. The fit of modeled vs. measured snow depth, which was only available at Lom, had a R^2 value of 0.75 with a mean error of less than 10 cm.

Simulation of LAI represented the measurements quite well with R^2 values of between 0.53 and 0.76 and a mean error of maximum 0.12 $m^2\,m^{-2}$. An exception was Hor, where LAI was underestimated by a ME of −0.61 and −1.49 $m^2\,m^{-2}$ in the accepted 75 runs and in the selected single run C1, respectively. At Hor, root biomass was underestimated in the single run by a ME of −281 $g\,C\,m^{-2}$ and living leaf biomass by −122 $g\,C\,m^{-2}$.

In most of the runs of the basic calibration at Hor, either GPP was overestimated or leaf biomass and LAI were underestimated. Therefore, beside the common configuration C1, a different configuration was tested where plant respiration and litter fall parameters for Hor were set to much lower values than in the tested range to fit to GPP and LAI at the same time. However, this reduced performance for R_{eco} R^2 to 0.66 compared to 0.75 in C1 and led to an overestimation of winter R_{eco} with a ME of 0.75 $g\,C\,m^{-2}\,day^{-1}$.

3.2 Parameter constraint

Site specific calibration was needed for the speed at which the maximum surface cover is reached (p_{ck}), the mean value in the analytical air temperature function (T_{amean}), temperature sum for reaching plant maturity ($T_{MatureSum}$), coefficient for determining allocation to mobile internal storage pool (m_{retain}), decomposition rate of the fast SOC pools (k_l) and radiation-use efficiency (ε_L).

Activity under saturated conditions ($p_{\theta Satact}$), threshold temperature for plant dormancy (T_{DormTh}), response to a 10 °C soil temperature change on the microbial activity (t_{Q10}) and base temperature for the microbial activity (t_{Q10bas}) covary with performance indices but showed different patterns for different validation variables and for different sites.

Most of the parameters did not show any influence on performance indices within the tested range (Fig. S1), demonstrating that either the relatively low effect of the parameter was overcompensated by the effect of more sensitive parameters, or the range used for calibration is sufficiently constraining. Each of these parameters did not reduce model performance indicated by R^2 by more than 0.05 for GPP or R_{eco} after setting them to a common value.

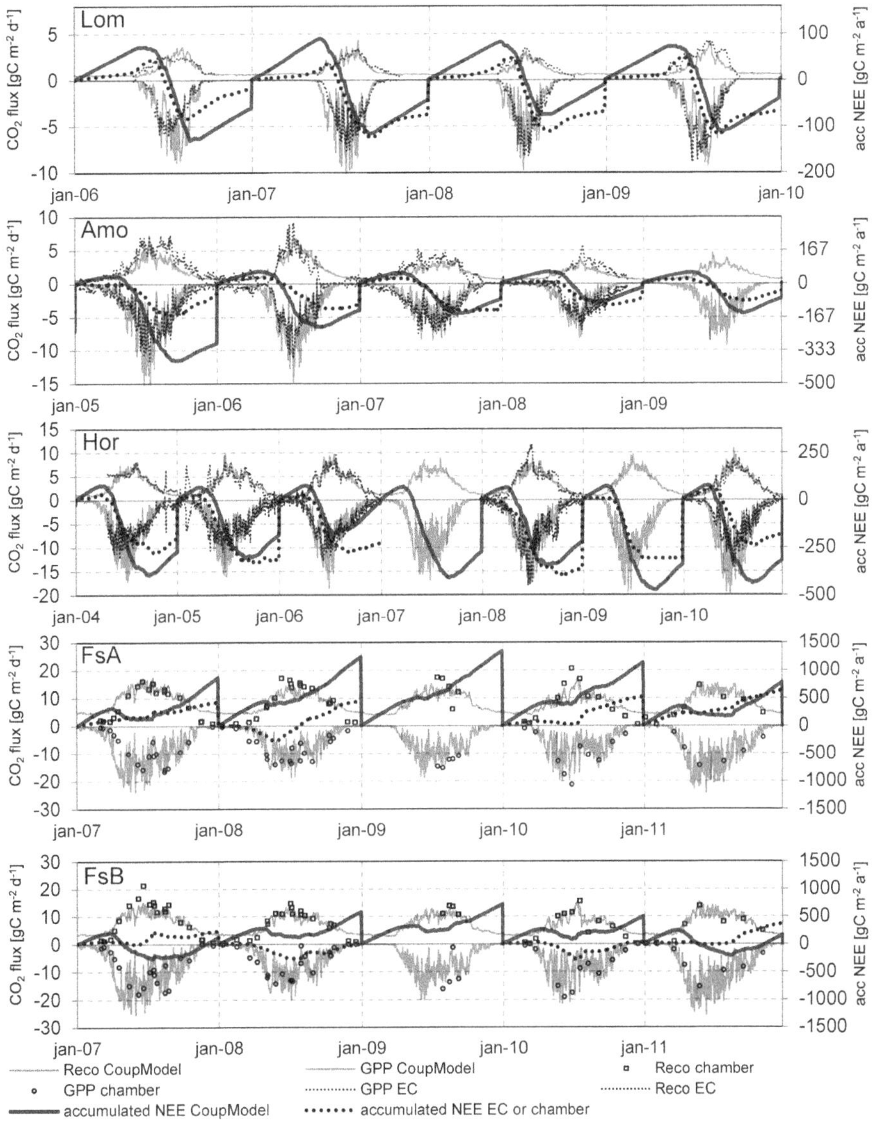

Figure 4. Simulated and measured R_{eco} (positive) and GPP (negative) fluxes and accumulated NEE for one selected set of parameter values (C1) common between all sites. Note the different scales.

3.3 Correlations between parameters

In the basic calibration, the following parameters were identified to interact with other parameters: p_{ck} covaried with the extinction coefficient in the Beer law (k_{rn}) which is used to calculate the partitioning of net radiation between canopy and soil surface. Strong, linear, negative correlation between coefficients for growth (k_{gresp}) and maintenance respiration ($k_{mrespleaf}$) was detected.

The effect of the different parameters in the water response function, $p_{\theta Satact}$, $p_{\theta Upp}$ and $p_{\theta p}$, compensated each other. They could not be constrained without a very high measurement resolution of fluxes and water table combined with high water table fluctuation at the same time. Therefore, $p_{\theta Upp}$ and $p_{\theta p}$ were set to default values and $p_{\theta Satact}$ was constrained by additional multiple runs together with k_l. Differences between sites in k_l are reduced with higher $p_{\theta Satact}$ values (Fig. 5), however, higher $p_{\theta Satact}$ values increase overestimation of winter R_{eco} at FsA and FsB (Figs. 6, 7d). A wider range of $p_{\theta Satact}$ values was acceptable for summer R_{eco} (Fig. 6).

Besides moisture response, decomposition rate (k_l) and temperature response (t_{Q10}, t_{Q10bas}) control soil respiration. The effect on R_{eco} was cofounded by plant respiration. Different patterns for different sites and variables for each of the parameters were even more pronounced when only k_l, t_{Q10} and $k_{mrespleaf}$ were in calibration (Fig. 6).

Single runs with different configurations (Fig. 7) revealed that higher plant respiration as well as steeper temperature response can lead to less overestimation of respiration in winter

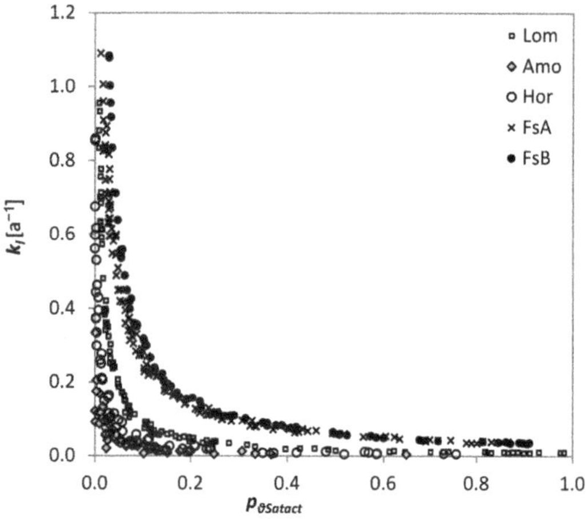

Figure 5. Dependencies between the parameters for decomposition rate and saturation activity for the different sites, based on additional multiple runs.

(Fig. 7d) but lead to reduced performance (Fig. 7c). In all single runs, despite the different configurations, FsA always showed the highest k_l, while Amo had the lowest (Fig. 7a). A higher saturation activity reduces the difference in k_l values, but leads to higher overestimation of winter fluxes.

4 Discussion

Despite not being specifically developed for peatlands, the CoupModel was able to simulate measured fluxes quite well. The model was run in a simple configuration with only two SOC pools per layer, no explicit representation of microbes, and only one plant layer. Even though the CoupModel was capable of adequately reproducing the measurements. Several points were identified where further peatland specific processes or more detailed representations might improve the model. Those are discussed in the following subsections.

From the 45 calibrated parameters, 8 parameters could be identified to actually need a site specific representation to achieve acceptable performance. Those parameters are discussed in Sects. 4.3–4.10. The remaining 37 parameters were not sensitive in the tested ranges, even though site specific values could have been expected; for example, it is known that grassland species differ in their assimilation and growth response to temperatures (Billings et al., 1978; Wohlfahrt et al., 1999). Plant respiration rates in graminoids differ between species (Poorter et al., 1991; Scheurwater et al., 1998; van der Werf et al., 1988) and depend, among other factors, on light (Rovira, 1969; Bahn et al., 2013), nutrient (Paterson and Sim, 2000) and moisture conditions (Crow and Wieder, 2005) as well as on cutting regime (Bokhari, 1977). Allocation fractions to different plant parts differ between species and depend on nutrient conditions (Aerts et al., 1991; Gong et al., 2014) as well as shading (Bahn et al., 2013). Values for specific leaf areas are species specific (e.g., Poorter et al., 1990; Reich et al., 1998) and differ in response to nutrient availability (Meziane and Shipley, 1999). Leaf lifetime (e.g., Ryser and Urbas, 2000) as well as leaf and root turnover rates (Schläpfer and Ryer, 1996) vary between graminoid species.

The five peatland sites largely differed in their vegetation composition, plant life-forms and species. Nevertheless, common values for all sites could be applied for parameters related to these processes, without reducing model performance on R_{eco} and GPP in R^2 values by more than 0.05. That shows that either the studied sites on a vegetation community level did not differ much in these processes, or that the impact of those parameters is subordinate compared to the impact of other parameters, meteorological input and other site conditions.

Therefore, models with a focus on multiple-year carbon fluxes do not need a site specific interpretation of these processes.

4.1 Model initialization

Many models use spin-up routines of many years until SOC pools reach a steady state (e.g., Dimitrov et al., 2010; Smith et al., 2010; Thornton and Rosenbloom, 2005). Here, measured C : N values were used to partition the SOC between pools, while ranges for parameter values were chosen in a way that the amount of carbon in the soil pools did not change very drastically. However, no further effort was made to force the pools to be in equilibrium. It was assumed that this might not be the case in the real world either: drainage ditches at FsA and FsB are still maintained, leading to high carbon losses and changes in substrate quality. Land use at Hor was quite recently changed from a fertilized and deeply drained cropland to a nature reserve with a restored water table. Additionally, Amo used to be more intensively managed and drained, but the drainage system was not maintained. Land use history was not known and SOC measurements were available from only one date per site. The measured carbon fluxes were therefore the only indication about carbon loss or addition to the complete system, while changes in relative pool sizes were not known. The partitioning of the SOC has implications on the parameter distribution for the rate coefficient for decomposition, which is discussed in Sect. 4.10.

4.2 Model performance

The best performance achieved highly differed between the different validation variables and between the different sites. This was not only caused by the model's ability to simulate the different output parameters but also due to measurement quality, measurement uncertainty, measurement methods (temporal and spatial resolution) and heterogeneity of the sites.

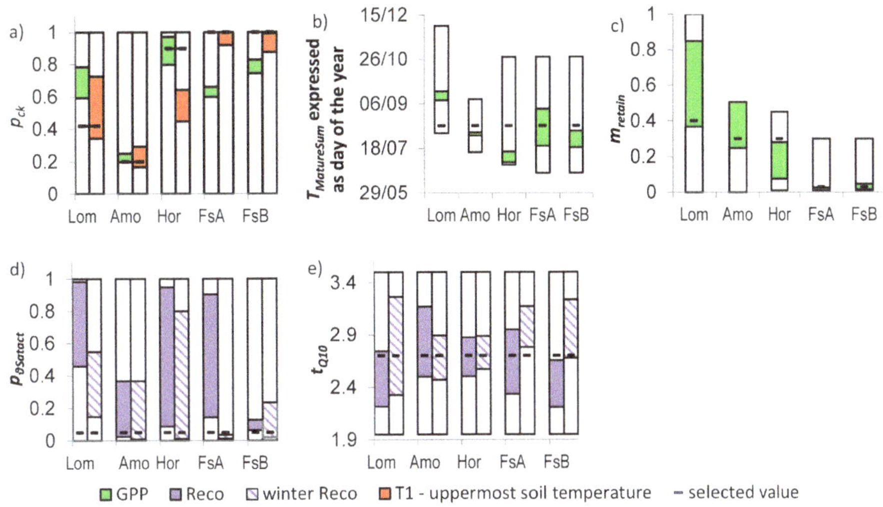

Figure 6. Obtained distributions of parameter values as constrained by additional multiple runs (calibration step III). Ranges for k_{l1} and ε_L are not shown due to their interactions with several parameters. Colored bars show the range of the 10 runs with the best performance for each validation variable. Prior ranges are indicated by the frame around the bar. Black dash is the value chosen for the common configuration C1.

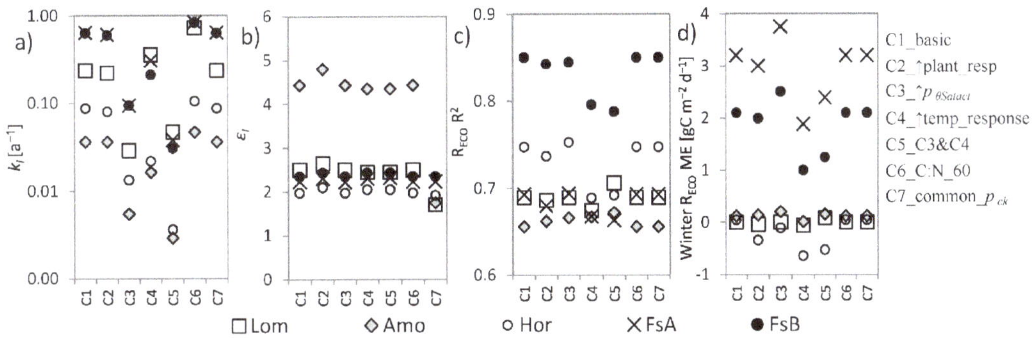

Figure 7. Values for the parameters of decomposition rate **(a)** light-use efficiency **(b)** and resulting model performance **(c, d)** when applying various single-value representations of parameters (C1–C7; see Table S8).

GPP was simulated markedly poorer as compared to R_{eco} at all sites and not only in the single runs, but also in the complete set of performed multiple runs. An explanation might be that in the model the whole plant community consisting of different individuals, species and even functional types, with different life cycles and adaptations to light availability and temperature was simplified to only one plant. Especially mosses, which differ largely from vascular plants in respect to their ecology and response to water, temperature and light conditions (Gaberščik and Martinčič, 1987; Harley et al., 1989; Murray et al., 1993; Turetsky, 2003), might be important at the moss-rich Lom and Amo sites. The vegetation at Hor consists of species with very different strategies and requirements for nutrients and water. At FsB, reeds, which are known for a late emergence, were well present in some of the years and hardly appeared in other years. FsA is relatively species-rich and several of these species are abundant only during parts of the vegetation period. Also, using a more complex photosynthesis model such as in e.g., Farquhar et al. (1980, 2001) and testing a wider range of parameters might lead to a better fit. Including plant stress due to high water levels and nutrient limitation might improve the performance on some sites. For example, Sagerfors et al. (2008) found photosynthesis to be limited also by too high water levels, so that the McGill wetland model assumed reduced photosynthesis outside a water level range of −10 to −20 cm (Wu et al., 2013b). Furthermore, GPP cannot be measured directly neither by the chamber nor the EC method. Instead it was derived from NEE and R_{eco} or nighttime NEE, including the uncertainty of two different measurements and empirical modeling.

Heterogeneity of vegetation was very distinct at Hor, which might explain the difficulties to simulate the right amounts of GPP and biomass at the same time. The biomass and LAI taken into account for this study might not be fully representative of the whole EC fetch for all wind directions.

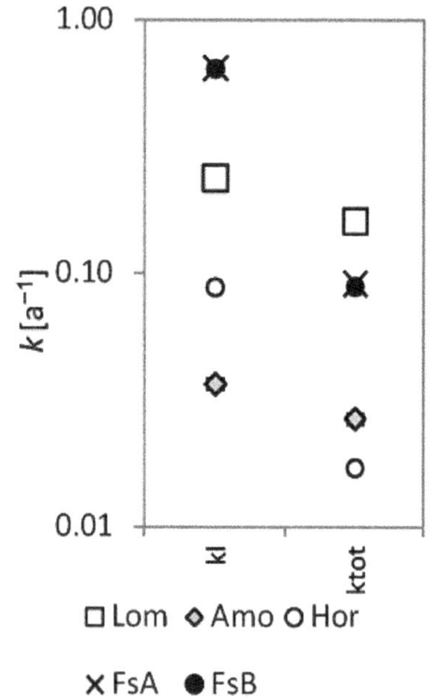

Figure 8. Decomposition rates of fast pools (k_l) and calculated rates of total organic matter decomposition if only one pool was used (k_{tot}) for each site and each layer.

Hor is also a site which deviates strongly with respect to other sites, with recent large changes in management. It is in successional transition from intensively used dairy farming meadow (approximately 20 years ago) towards reed fen with willow thickets. Soil and vegetation still show the imprint of high nutrient levels derived from manuring practices (e.g., patches with abundant *Urtica dioca*). This likely still affects GPP. These features could be a better explanation of the deviating GPP than the additionally tested configuration with strongly reduced litter fall and plant respiration rates.

Even though the winter fluxes are small compared to the summer fluxes, they have a marked role in the annual NEE balances (Fig. 4). Overestimation of winter R_{eco} in combination with slightly underestimated winter GPP leads to high overestimation of annual accumulated NEE, emphasizing the importance of winter flux dynamics in the annual balances. At all sites except Hor, winter R_{eco} was overestimated in the selected single run. For FsB and especially FsA this was also true for all multiple runs. As R_{eco} at Lom and Amo are typically relative low, the effect was less pronounced.

Several different reasons for the winter R_{eco} overestimation are possible: explanations due to model setup and parameterization are discussed in the Sects. 4.7, 4.8 and 4.9. Additionally, gases might be trapped within the snow and under the ice (Bubier et al., 2002; Maljanen et al., 2010) and therefore captured by the measurement instruments only in springtime, when they are released. A gastight ice cover was not realized in the current model setup. Frozen or ice covered soils are quite common at the boreal Lom, but also at FsA and FsB which have a more continental climate than the other sites.

The ability of the model to simulate soil moisture could not be evaluated, as this variable was measured only at Lom, where the soil was close to saturation throughout the year. Therefore, and as groundwater level was used as input, hydraulic properties could not be constrained. Furthermore, swelling, shrinking and hysteresis effects which are important factors in hydraulic characteristics of peat soils (e.g., Kellner and Halldin, 2002) were not accounted for. This could have an effect on model performance and parameter values, especially those related to the soil moisture response.

4.3 Soil temperature dynamics

Due to the isolating impact of the snow cover (e.g., Zhang, 2005), the value of mean annual soil temperature (T_{amean}) was expected to be slightly higher than the mean annual air temperature. Constrained values of soil temperature were 1.5–5 °C higher than the mean annual air temperature at all sites. If the model was run under different conditions without further fitting, factors causing differences between mean annual soil temperatures and corresponding air temperature need to be considered.

4.4 The role of soil temperature and GPP to constrain the plant cover

Accepted fits for soil temperature in the uppermost measured soil layer led to p_{ck} values close to the measured coverage of vascular plants for each site. Therefore, the measured coverage could directly be used in the configuration C1 (Fig. 6a). Setting p_{ck} to a common value of 100 % reduced the differences in ε_L between the site's C7 (Fig. 6e), but led to underestimation of soil temperature in the uppermost soil layer by at most −0.45 °C in ME at Amo. An explanation could be that mosses are contributing to the plant coverage in respect to GPP but not to temperature, especially at sites where they are the main peat-forming material.

4.5 Start of senescence

Site specific calibration was needed for the temperature sum initiating the start of senescence ($T_{MatureSum}$). However, if the resulting day of the year was plotted instead, the differences between sites became small (Fig. 6) and setting it to the mean value of all sites did not reduce model performance in GPP R^2 by more than 0.05. Induction of senescence with graminoids is known to depend on both temperature and day length (Nuttonson, 1958; Proebsting et al., 1976; Thomas and Stoddart, 1980; Davidson and Campbell, 1983). However, the differences between the sites in this study could be explained solely by the relative day length.

4.6 Seasonal and management control of mobile plant pool for regrowth

The proportion of C in the plant which does not become litter, but instead is stored for shooting in the next year (m_{retain}), differed largely between sites. At Lom, a value of at least 40 % led to and acceptable performance while a maximum of 3 % was found for FsA and FsB; a mean value of 20 % would reduce R^2 of GPP by at least 0.04 for these sites. At Amo and Hor neither a value of 3 % nor 40 % reduced R^2 of GPP by more than 0.01. An explanation for the low m_{retain} at FsA and FsB could be that the same pool is used for regrowth after being cut and therefore not available for shooting anymore, as the regrowth rate in both early spring and after being cut depend on the carbohydrate reserve (White, 1973; Davies, 1988; Klimeš and Klimešová, 2002). Steele et al. (1984) conclude that defoliation late in the year will affect spring regrowth.

At Lom, high m_{retain} might be an adaption to the short vegetation period (Kistritz et al., 1983). Evergreen parts of the vegetation like dwarf shrubs, lower leaf parts of graminoids and mosses were not accounted for, which also affects regrowth in spring. Saarinen (1998) found that 60–70 % of shoots and 20 % of green biomass in a *Carex rostrata* fen survived the winter and, therefore, hypothesized based on comparisons with other studies that the proportion increases with increasing latitude.

The storage pool is an important parameter needing site specific calibration but can be fitted if several measurements during spring and early summer of either GPP, biomass or LAI are available.

4.7 Radiation-use efficiency

As plants were not nutrient limited in the model setup, the lowest values for ε_L were expected under the most nutrient-poor conditions (Longstreth and Nobel, 1980; Reich et al., 1994; Haxeltine and Prentice, 1996; Gamon et al., 1997; Wohlfahrt et al., 1999). The opposite was true if site specific values were used for p_{ck}. However, a common value for p_{ck} reduced the differences in ε_L and led to low ε_L at the ombrotrophic Amo site, but to an even lower value at the minerotrophic Lom site. The nutrient status of the soil can therefore not explain the differences in ε_L. The assumption of plants being well adapted to nutrient and water stress might not be true for the restored Hor site, where parts of the vegetation still consist of species which are not typical for wetlands. This might explain the low productivity at that site, but could only be covered by a model if site specific plant responses to high water levels would be applied. Additionally, ε_L is known to be species specific (Sinclair and Horie, 1989; Reich et al., 1998; Wohlfahrt et al., 1999).

Radiation-use efficiency is an important parameter needing site specific calibration. If common values were used for ε_L, p_{ck} and m_{retain}, mean GPP would be underestimated by a factor of 2.4 (FsB) or overestimated by a factor of 3 (Lom). If site specific values were used for p_{ck} and m_{retain}, the discrepancy would be even higher. However ε_L can easily be fitted if either GPP, biomass or LAI is known.

4.8 The control of decomposition and plant respiration by soil temperature

The annual R_{eco}, which was dominated by summer R_{eco}, could be described by a single temperature response function at all sites. However, it was not possible to find an equally good fit to both summer and winter R_{eco} using the same t_{Q10} value. Higher t_{Q10} values would decrease overestimation of winter R_{eco} especially at the southern sites FsA and FsB, but also reduce model performance for the annual R_{eco}. Different temperature responses for different sites (e.g., Jacobs et al., 2007), seasons (e.g., Lipson et al., 2002) and temperature ranges (e.g., Lloyd and Taylor, 1994; Paul, 2001; Atkin et al., 2003) are reported in the literature. This is partly explained by multiplicative effects of several temperature sensitive processes (Davidson et al., 2006; Kirschbaum, 2006) but, still, a constant t_{Q10} might be a wrong assumption (Atkin et al., 2005).

More sophisticated temperature responses like the Ratkowsky function (Ratkowsky et al., 1982) might improve the performance for individual sites. This might also be true for separate temperature response functions for plants and soil, as summer R_{eco} includes autotrophic and heterotrophic respiration, while winter R_{eco} is strongly dominated by heterotrophic respiration.

4.9 The control of decomposition by soil moisture

The activity under saturated conditions in respect to unsaturated conditions is described by $p_{\theta Satact}$ and was strongly negatively correlated with decomposition rate k_l. Patterns for $p_{\theta Satact}$ differed between sites and variables. At all sites a minimum value of around 5 % led to the acceptable performance in annual R_{eco}, while also quite high values did not reduce the performance except at FsB. At Lom only winter R_{eco} was considered, as conditions were always saturated during summer. For acceptable winter R_{eco}, $p_{\theta Satact}$ needed to be very low. This was not true for Lom, where water in the upper soil layer partly froze in the model and led to high winter respiration.

As the soil at FsA and FsB was saturated during winter, a common lower value for $p_{\theta Satact}$ would decrease overestimation of winter fluxes. However, it would also reduce model performance at all sites and increase the site specific differences in k_l (Fig. 7).

Permanently saturated soils contain less O_2 than temporally saturated ones (e.g., Kettunen et al., 1999), which affects decomposition (e.g., Reddy and Patrick, 1975; DeLaune et al., 1981; Holden et al., 2004). Therefore a lower $p_{\theta Satact}$ would be justified for wetter sites. If k_l was constant between sites and instead $p_{\theta\mathrm{Satact}}$ was fitted, this would lead the value of $p_{\theta\mathrm{Satact}}$ to decrease in the order FsB > FsA > Lom > Hor > Amo (Fig. 5), which cannot be justified by the differences in water levels which increase in the order FsA < FsB << Amo < Hor << Lom. Therefore, a different $p_{\theta\mathrm{Satact}}$ cannot explain differences in soil respiration between sites. However, the amount of aerenchymatous plants, leading to soil aeration (e.g., Armstrong, 1980; Bendix et al., 1994; Grosse et al., 1996) was not taken into account. It reaches its highest coverage at FsB (90 %), followed by FsA (62 %), Hor (50 %), Lom (around 10 %) and Amo (around 6 %). Modeling water response depending on soil O_2 and redox potential, including O_2 conductance from plants, might help us to analyze the differences in decomposition rate and reduce winter overestimation. For example, in the Wetland-DNDC model, the water response function depends on redox potential: decomposition under saturated condition is reduced by a factor of only 0.6 if redox potential is high, but by a factor of 0.2 if redox potential is low (Zhang et al., 2002).

4.10 The control of decomposition by substrate

The largest differences of parameters between sites appeared for the maximum decay rate of the fast C pools, k_l. Setting it to a common value would lead to an underestimation of mean R_eco by a factor of 2.8 at FsB or an overestimation by a factor of 4 at Amo.

Despite different temperature and water response curves being tested, k_l values at FsA and FsB are substantially higher than at Amo (Figs. 5, 7). Higher $t_{\mathrm{Q}10}$ values lead to two groups of k_l values: similar high ones for Lom, FsA and FsB and substantially lower ones for Hor and Amo (Fig. 7).

The partitioning into SOC pools strongly effects the differences, as can be shown by calculating decomposition rates for the total SOC (k_tot) based on k_l, k_h and SOC in the pools of the upper 30 cm as used in the C1 scenario (Fig. 8). However, FsB and FsA still have much higher rates than Amo. The resulting values and ranges of k_tot (0.02–0.16 a^{-1}) are comparable with the reported values from laboratory incubation studies of peat cores (0.03–1.66 a^{-1}, Moore and Dalva, 1997; 0.01–0.35 a^{-1}, Glatzel et al., 2004; 0.008 a^{-1}, Kechavarzi et al., 2010; a SOC content of 30 % was assumed for conversion from dry mass).

Lower decomposability is often associated with higher C : N ratios (e.g., Zeitz and Velty, 2002; Limpens and Berendse, 2003; Bragazza et al., 2006; Zhang et al., 2008), which might be important especially for the moss-rich Amo and Lom. Assuming a C : N ratio of 60 for the fast pools (Fig. 7, C6) leads to a decomposition rate at Lom which is close to those at FsA and FsB, while those of Hor and Amo remain substantially lower.

Low pH might be one reason for the low k_l at Amo (e.g., DeLaune et al., 1981; Bergman et al., 1999). Despite being nutrient-rich and having a high pH and high biomass production, leading to large amounts of labile carbon added to the soil, k_l values at Hor were very low. This might be connected to land use history and the origin of the peat from partly clayey lake sediment. Most of the labile C in the parent peat in the upper, formerly drained soil layers might have been decomposed before and therefore became stabilized.

In the current setup the slow pools were almost inert. A higher decay rate for the slow pools would result in a lower k_l for sites with high C stock in the slow pools (cf. Table 3). This would decrease the differences between FsA and FsB compared to Lom and Amo, but increase the differences between FsA compared to FsB and Hor.

Substrate quality is known to affect decomposition rates (e.g., Raich and Schleisinger, 1992; Belyea, 1996; Fang and Moncrieff, 2005; Yeloff and Mauquoy, 2006). Therefore, many other SOC models use several different SOC pools (e.g., Franko et al., 1997; Smith et al., 1997; Cui et al., 2005; Del Grosso et al., 2005; van Huissteden et al., 2006) to account for differences in substrate quality. This leads to the problem of partitioning total SOC into the pools (e.g., Helfrich et al., 2007; Zimmermann et al., 2007). In some models, the various SOC pools differ also in their response functions (e.g., Smith et. al, 2010).

The highest decomposition rates occurred at sites with highest biomass production. A correlation of productivity with soil respiration was found in several comparison studies (e.g., Janssens et al., 2001; Reichstein et al., 2003). Fresh material provided by the plants might lead to higher microbial activity and priming effect (e.g., Kuzyakov, 2002; Fontaine et al., 2007). A higher plant to soil respiration ratio reduced the differences in k_l between the sites and lowered winter R_eco, especially at the highly productive FsA and FsB sites, but also reduced the model performance at all sites except Amo.

Vegetation at Amo and Lom consist largely of mosses which are more resistant to decomposition than vascular plants (Rudolph and Samland, 1985; Verhoeven and Toth, 1995; Limpens and Berendse, 2003; Moore et al., 2007) and might further explain the low k_l value at Amo. Despite the lower biomass production, higher moss cover and higher C : N ratio compared to Hor, FsA or FsB, Lom has a relatively high decomposition rate. This can be explained by the very low dry bulk density, resulting in a low amount of C in the upper soil layers (Table 3), which are most exposed to decomposition (e.g., Fang and Moncrieff, 2005). Also, a low dry bulk density accompanies the low degree of degradation and therefore high amounts of labile carbon (e.g., Grosse-Brauckmann, 1990).

Despite the large differences in accumulated NEE (Fig. 1) between FsA and FsB, these sites have nearly identical decomposition rates. This confirms the expectations that the differences in NEE between FsA and FsB can be fully explained by the differences in the water table and the biomass and carbon stocks.

5 Conclusions

Differences between sites with respect to CO_2 fluxes could be explained if, in addition to air temperature, water table and soil C and N stocks, site specific plant productivity and decomposition rates were also taken into account. Differences in nutrient availability and soil wetness could not explain the differences in plant productivity between the sites. Substrate quality and litter input, as well as pH values, are likely explanations for the differences in decomposition rates. A site specific interpretation was not needed for processes related to plant phenology, their response to temperature, allocation of new assimilates and plant respiration and litter fall rates.

The model parameters which strongly affected model performance were successfully constrained by the available long-term measurement data on NEE, partitioned into GPP and R_{eco}, LAI and biomass, including rooting depth and root biomass at one site, water table, soil temperature and soil C and N stocks, as well as meteorological data and snow data at one site. It would have been useful if additional information about root biomass, litter fall and soil water content to validate the model performance in the corresponding processes was available for all sites. A second measurement of C and N stocks, several years after the first, as well as information about the degree of decomposition on all sites would have been very helpful to constrain decomposition rates and partitioning between SOM pools.

Some improvements in the model and its configuration were identified to obtain a better performance for simulations of GHG fluxes from treeless peatlands. Examples include separate temperature responses for plant and soil heterotrophic respiration. The static response to water saturated conditions needs to be replaced by a function that considers the change of O_2 in the soil.

Acknowledgements. This study was financed by GHG-Europe – Greenhouse gas management in European land use systems – supported by the European Commission in the 7th Framework Programme (grant agreement no. 244122) and by the Formas project LAGGE – Landscape Greenhouse Gas Exchange – Integration of Terrestrial and Freshwater sources and sinks, coordinated by Leif Klemedtsson.

We gratefully acknowledge experimentalists and students for their hard work collecting, processing and providing the field data. The measurements were funded within the projects GHG-Europe (grant agreement no. 244122), NitroEurope IP (017841) and Carbo Europe (contract no. GOCE-CT2003-505572) by the European Commission, the CEH UK Carbon Catchment network and the joint research project Organic soils – Acquisition and development of methods, activity data and emission factors for the climate reporting under LULUCF/AFOLU, founded by the Thünen Institute. Eddy and soil data at Amo and Lom were provided by the NitroEurope database (http://nitroeuropedata.ceh.ac.uk, V: 0.3.3.0; Copyright 2014 CEH) and Eddy data at Hor by the European Fluxnet Database Cluster (http://gaia.agraria.unitus.it).

Furthermore we thank Wolfram Adelmann for the flux data from 2007–2009 at FsA and FsB, the Royal Netherlands Meteorological Institute for providing meteorological data which was used for gap filling at the Dutch site, Päivi Mäkiranta from METLA for additional manual chamber and LAI data at Lom, Michel Bechtold from the Thünen Institute for water table and water retention values at FsA and FsB, Michael Sommer from ZALF for soil C, N and dry bulk density at FsA and FsB, and Stephan Raspe from the Bavarian State Institute of Forestry for providing the N deposition data for FsA and FsB.

We would also like to thank Michel Bechtold from the Thünen Institute, Magnus Svensson from KTH and the editor Bärbel Tiemeyer for helpful comments on the manuscript.

The simulation files with the setup according the common configuration together with the respective version of the CoupModel can be provided upon request.

Edited by: B. Tiemeyer

References

Adiku, S. G. K., Reichstein, M., Lohila, A., Dinh, N. Q., Aurela, M., Laurila, T., Lueers, J., and Tenhunen, J. D.: PIXGRO: A model for simulating the ecosystem CO_2 exchange and growth of spring barley, Ecol. Model., 190, 260–276, doi:10.1016/j.ecolmodel.2005.04.024, 2006.

Aerts, R., Boot, R. G. A., and van der Aart, P. J. M.: The relation between above- and belowground biomass allocation patterns and competitive ability, Oecologia, 87, 551–559, 1991.

Alm, J., Saarnio, S., Nykänen, H., Silvola, J., and Martikainen, P.: Winter CO_2, CH_4 and N_2O fluxes on some natural and drained boreal peatlands, Biogeochemistry, 44, 163–186, 1999.

Armstrong, W.: Aeration in Higher Plants, Adv. Bot. Res., 7, 225–332, doi:10.1016/S0065-2296(08)60089-0, 1980.

Atkin, O. K., Tjoelker, M. G., and Atkin, O.: Thermal acclimation and the dynamic response of plant respiration to temperature, Trends Plant Sci., 8, 343–351, doi:10.1016/S1360-1385(03)00136-5, 2003.

Atkin, O. K., Bruhn, D., Hurry, V. M., and Tjoelker, M. G.: Evans Review No. 2: The hot and the cold: unravelling the variable response of plant respiration to temperature, 32, 87–105, doi:10.1071/FP03176, 2005.

Aurela, M., Lohila, A., Tuovinen, J.-P., Hatakka, J. R. T., and Laurila, T.: Carbon dioxide exchange on a northern boreal fen, Boreal Environ. Res., 14, 699–710, 2009.

Bahn, M., Lattanzi, F. A., Hasibeder, R., Wild, B., Koranda, M., Danese, V., Brüggemann, N., Schmitt, M., Siegwolf, R., and Richter, A.: Responses of belowground carbon allocation dynamics to extended shading in mountain grassland, New Phytologist, 198, 116–126, doi:10.1111/nph.12138, 2013.

Baldocchi, D.: Forward, Glob. Change Biol., 13, 547, doi:10.1111/j.1365-2486.2007.01345.x, 2007.

Baldocchi, D., Falge, E., Gu, L., Olson, R., Hollinger, D., Running, S., Anthoni, P., Bernhofer, C., Davis, K. J., Evans, R., Fuentes,

J., Goldstein, A., Katul, G., Law, B., Lee, X., Malhi, Y., Meyers, T., Munger, W., Oechel, W., Paw, K. T., Pilegaard, K., Schmid, H. P., Valentini, R., Verma, S., Vesala, T., Wilson, K., and Wofsy, S.: FLUXNET: A New Tool to Study the Temporal and Spatial Variability of Ecosystem–Scale Carbon Dioxide, Water Vapor, and Energy Flux Densities, Bull. Amer. Meteor. Soc, 82, 2415–2434, 2001.

Beetz, S., Liebersbach, H., Glatzel, S., Jurasinski, G., Buczko, U., and Höper, H.: Effects of land use intensity on the full greenhouse gas balance in an Atlantic peat bog, Biogeosciences, 10, 1067–1082, doi:10.5194/bg-10-1067-2013, 2013.

Bell, M. J., Jones, E., Smith, J., Smith, P., Yeluripati, J. B., Augustin, J., Juszczak, R., Olejnik, J., and Sommer, M.: Simulation of soil nitrogen, nitrous oxide emissions and mitigation scenarios at 3 European cropland sites using the ECOSSE model, Nutr. Cycl. Agroecosyst., 92, 161–181, doi:10.1007/s10705-011-9479-4, 2012.

Belyea, L. R.: Separating the Effects of Litter Quality and Microenvironment on Decomposition Rates in a Patterned Peatland, Oikos, 77, 529–539, 1996.

Bendix, M., Tornbjerg, T., and Brix, H.: Internal gas transport in Typha latifolia L. and Typha angustifolia L. 1. Humidity-induced pressurization and convective through flow, Aquat. Bot., 49, 75–89, doi:10.1016/0304-3770(94)90030-2, 1994.

Bergman, I., Lundberg, P., and Nilsson, M.: Microbial carbon mineralisation in an acid surface peat: effects of environmental factors in laboratory incubations, Soil Biol. Biochem., 31, 1867–1877, doi:10.1016/S0038-0717(99)00117-0, 1999.

Beven, K.: A manifesto for the equifinality thesis, J. Hydrol., 320, 18–36, doi:10.1016/j.jhydrol.2005.07.007, 2006.

Beven, K. and Freer, J.: Equifinality, data assimilation, and uncertainty estimation in mechanistic modelling of complex environmental systems using the GLUE methodology, J. Hydrol., 249, 11–29, doi:10.1016/S0022-1694(01)00421-8, 2001.

Billings, W. D., Peterson, K. M., and Shaver, G. R.: Growth, turnover, and respiration rates of roots and tillers in tundra graminoids, in: Vegetation and production ecology of an Alaskan arctic tundra, edited by: Tieszen, L. L., Springer-Verlag, 29, 415–434, doi:10.1007/978-1-4612-6307-4, 1978.

Bokhari, U. G.: Regrowth of western wheatgrass utilizing 14C-labeled assimilates stored in belowground parts, Plant Soil, 48, 115–127, 1977.

Bragazza, L., Freeman, C., Jones, T., Rydin, H., Limpens, J., Fenner, N., Ellis, T., Gerdol, R., Hájek, M., Tomáš, H., Lacumin, P., Kutnar, L., Tahvanainen, T., and Toberman, H.: Atmospheric nitrogen deposition promotes carbon loss from peat bogs, PNAS, 103, 19386–19389, 2006.

Brooks, R. and Corey, A.: Hydraulic Properties of Porous Media, Hydrology Papers 3, Colorado State University, Fort Collins, 1964.

Bubier, J., Crill, P. M., and Mosedale, A.: Net ecosystem CO_2 exchange measured by autochambers during the snow-covered season at a temperate peatland, Hydrol. Process., 16, 3667–3682, doi:10.1002/hyp.1233, 2002.

Calanca, P., Vuichard, N., Campbell, C. L., Viovy, N., Cozic, A., Fuhrer, J., and Soussana, J. F.: Simulating the fluxes of CO_2 and N_2O in European grasslands with the Pasture Simulation Model (PaSim), Agr. Ecosyst. Environ., 121, 164–174, doi:10.1016/j.agee.2006.12.010, 2007.

Chivers, M. R., Turetsky, M. R., Waddington, J. M., Harden, J. W., and McGuire, A. D.: Effects of Experimental Water Table and Temperature Manipulations on Ecosystem CO_2 Fluxes in an Alaskan Rich Fen, Ecosystems, 12, 1329–1342, doi:10.1007/s10021-009-9292-y, 2009.

Chojnicki, B. H., Michalak, M., Acosta, M., Juszczak, R., Augustin, J., Drösler, M., and Olejnik, J.: Measurements of Carbon Dioxide Fluxes by Chamber Method at the Rzecin Wetland Ecosystem, Poland, Polish J. of Environ. Stud., 19, 283–291, 2010.

CoupModel: Current version of COUP model for download, available at: http://www.coupmodel.com (last access: 11 November 2014), 2014.

Crow, S. E. and Wieder, R. K.: Sources of CO_2 emission from a northern peatland: root respiration, exudation and decomposition, Ecology, 86, 1825–1834, 2005.

Cui, J., Li, C., and Trettin, C. C.: Analyzing the ecosystem carbon and hydrologic characteristics of forested wetland using a biogeochemical process model, Glob. Change Biol., 11, 278–289, doi:10.1111/j.1365-2486.2005.00900.x, 2005.

Dai, Y., Dickinson, R. E., and Wang, Y.-P.: A Two-Big-Leaf Model for Canopy Temperature, Photosynthesis, and Stomatal Conductance, J. Climate, 17, 2281–2299, doi:10.1175/1520-0442(2004)017<2281:ATMFCT>2.0.CO;2, 2004.

Davidson, E. A., Janssens, I. A., and Luo, Y.: On the variability of respiration in terrestrial ecosystems: moving beyond Q10, Glob. Change Biol., 12, 154–164, doi:10.1111/j.1365-2486.2005.01065.x, 2006.

Davidson, H. R. and Campbell, C. A.: The effect of temperature, moisture and nitrogen on the rate of development of spring wheat as measured by degree days, Can. J. Plant Sci., 63, 833–846, 1983.

Davies, A.: The Regrowth of Grass Swards, in: The Grass Crop, edited by: Jones, M. B. and Lazenby, A., The Grass Crop, Chapman and Hall, London, 85–127, UK, 1988.

Del Grosso, S., Parton, W. J., Mosier, A., Holland, E., Pendall, E., Schimel, D., and Ojima, D.: Modeling soil CO_2 emissions from ecosystems, Biogeochemistry, 73, 71–91, doi:10.1007/s10533-004-0898-z, 2005.

DeLaune, R. D., Reddy, C. N., and Patrick Jr, W. H.: Organic matter decomposition in soil as influenced by pH and redox conditions, Soil Biol. Biochem., 13, 533–534, 1981.

Dietiker, D., Buchmann, N., and Eugster, W.: Testing the ability of the DNDC model to predict CO_2 and water vapour fluxes of a Swiss cropland site, Agr. Ecosyst. Environ., 139, 396–401, doi:10.1016/j.agee.2010.09.002, 2010.

Dimitrov, D. D., Grant, R. F., Lafleur, P. M., and Humphreys, E. R.: Modeling the effects of hydrology on ecosystem respiration at Mer Bleue bog, J. Geophys. Res.-Biogeosci. (2005–2012), 115, G04, doi:10.1029/2010JG001312, 2010.

Dinsmore, K. J., Billett, M. F., Skiba, U. M., Rees, R. M., Drewer, J., and Helfter, C.: Role of the aquatic pathway in the carbon and greenhouse gas budgets of a peatland catchment, Glob. Change Biol., 16, 2750–2762, doi:10.1111/j.1365-2486.2009.02119.x, 2010.

Drewer, J., Lohila, A., Aurela, M., Laurila, T., Minkkinen, K., Penttilä, T., Dinsmore, K. J., McKenzie, R. M., Helfter, C., Flechard, C., Sutton, M. A., and Skiba, U. M.: Comparison of greenhouse gas fluxes and nitrogen budgets from an ombotrophic bog in

Scotland and a minerotrophic sedge fen in Finland, Eur. J. Soil Sci., 61, 640–650, doi:10.1111/j.1365-2389.2010.01267.x, 2010.

Drösler, M.: Trace gas exchange and climatic relevance of bog ecosystems, Southern Germany, Ph.D. thesis, Chair of Vegetation Ecology, Department of Ecology, Technical University Munich, 179 pp., 2005.

Drösler, M., Freibauer, A., Christensen, T. R., and Friborg, T.: Observations and Status of Peatland Greenhouse Gas Emissions in Europe, in: The continental-scale greenhouse gas balance of Europe, edited by: Dolman, A. J., Freibauer, A., Valentini, R., Ecological studies, 203, Springer, New York, 243–261, 2008.

Duru, M. and Ducrocq, H.: Growth and Senescence of the Successive Grass Leaves on a Tiller, Ontogenic development and effect of temperature, Ann. Bot., 85, 635–643, doi:10.1006/anbo.2000.1116, 2000.

Falge, E., Baldocchi, D., Olson, R., Anthoni, P., Aubinet, M., Bernhofer, C., Burba, G., Ceulemans, R., Clement, R., Dolman, H., Granier, A., Gross, P., Grünwald, T., Hollinger, D. Y., Jensen, N.-O., Katul, G., Keronen, P., Kowalski, A., Lai, C. T., Law, B. E., Meyers, T., Moncrieff, J., Moors, E., Munger, W. J., Pilegaard, K., Rannik, Ü., Rebmann, C., Suyker, A., Tenhunen, J., Tu, K., Verma, S., Vesala, T., Wilson, K., and Wofsy, S.: Gap filling strategies for defensible annual sums of net ecosystem exchange, Agr. Forest Meteorol., 107, 43–69, doi:10.1016/s0168-1923(00)00225-2, 2001.

Fan, Z., David McGuire, A., Turetsky, M. R., Harden, J. W., Michael Waddington, J., and Kane, E. S.: The response of soil organic carbon of a rich fen peatland in interior Alaska to projected climate change, Glob. Change Biol., 19, 604–620, doi:10.1111/gcb.12041, 2013.

Fang, C. and Moncrieff, J. B.: The variation of soil microbial respiration with depth in relation to soil carbon composition, Plant Soil, 268, 243–253, doi:10.1007/s11104-004-0278-4, 2005.

Farquhar, G. D., von Caemmerer, S. von, and Berry, J. A.: A biochemical model of photosynthetic CO_2 assimilation in leaves of C3 species, Planta, 149, 78–90, 1980.

Farquhar, G. D., Caemmerer, S. von, and Berry, J. A.: Models of photosynthesis, Plant Physiol., 125, 42–45, 2001.

Feng, L., Rui, S., Tinglong, Z., Bo, H., and Tang, Y.: Simulation of carbon dioxide fluxes in agroecosystems based on BIOME-BGC model, in: Geoscience and Remote Sensing Symposium (IGARSS), Vancouver, BC, Canada, 24–29 July 2011, 3327–3329, 2011

Fontaine, S., Barot, S., Barré, P., Bdioui, N., Mary, B., and Rumpel, C.: Stability of organic carbon in deep soil layers controlled by fresh carbon supply, Nature, 450, 277–280, doi:10.1038/nature06275, 2007.

Franko, U., Crocker, G., Grace, P., Klír, J., Körschens, M., Poulton, P., and Richter, D.: Simulating trends in soil organic carbon in long-term experiments using the CANDY model, Geoderma, 81, 109–120, doi:10.1016/S0016-7061(97)00084-0, 1997.

Frolking, S. E., Roulet, N. T., Moore, T. R., Richard, P. J. H., Lavoie, M., and Muller, S. D.: Modeling Northern Peatland Decomposition and Peat Accumulation, Ecosystems, 4, 479–498, doi:10.1007/s10021-001-0105-1, 2001.

Frolking, S. E., Roulet, N. T., Moore, T. R., Lafleur, P. M., Bubier, J. L., and Crill, P. M.: Modeling seasonal to annual carbon balance of Mer Bleue Bog, Ontario, Canada, Glob. Biogeochem. Cy., 16, 4-1–4-21, doi:10.1029/2001GB001457, 2002.

Fulkerson, W. J. and Donaghy, D. J.: Plant-soluble carbohydrate reserves and senescence - key criteria for developing an effective grazing management system for ryegrass-based pastures: a review, Aust. J. Exp. Agr., 41, 261–275, 2001.

Gaberščik, A. and Martinčič, A.: Seasonal Dynamics of Net Photosynthesis and Productivity of Sphagnum papillosum, Lindbergia, 13, 105–110, 1987.

Gamon, J. A., Serrano, L., and Surfus, J. S.: The photochemical reflectance index: an optical indicator of photosynthetic radiation use efficiency across species, functional types, and nutrient levels, Oecologia, 112, 492–501, 1997.

Glatzel, S., Basiliko, N., and Moore, T.: Carbon dioxide and methane production potentials of peats from natural, harvested and restored sites, Eastern Quebec, Canada, Wetlands, 24, 261–267, 2004.

Gong, X., Berone, G., Agnusdei, M., Rodríguez Palma, R., Schäufele, R., and Lattanzi, F.: The allocation of assimilated carbon to shoot growth: in situ assessment in natural grasslands reveals nitrogen effects and interspecific differences, Oecologia, 174, 1085–1095, doi:10.1007/s00442-013-2838-x, 2014.

Grant, R. F., Desai, A. R., and Sulman, B. N.: Modelling contrasting responses of wetland productivity to changes in water table depth, Biogeosciences, 9, 4215–4231, doi:10.5194/bg-9-4215-2012, 2012.

Grosse, W., Jovy, K., and Tiebel, H.: Influence of plants on redox potential and methane production in water-saturated soil, Hydrobiol., 93–99, 1996.

Grosse-Brauckmann, G.: Ablagerungen der Moore, in: Moor-und Torfkunde, 3rd ed., edited by: Gottlich, K., Schweizerbart'sche Verlagsbuchhandlung, Stuttgart, 17–236, 1990.

Hansen, G. K. and Jensen, C. R.: Growth and Maintenance Respiration in Whole Plants, Tops, and Roots of Lolium multiflorum, Physiol. Plant., 39, 155–164, 1977.

Harley, P. C., Tenhunen, J. D., Murray, K. J., and Beyers, J.: Irradiance and temperature effects on photosynthesis of tussock tundra Sphagnum mosses from the foothills of the Philip Smith Mountains, Alaska, Oecologia, 79, 251–259, doi:10.1007/BF00388485, 1989.

Haxeltine, A. and Prentice, C. I.: A General Model for the Light-Use Efficiency of Primary Production, British Ecological Society, 10, 551–561, 1996.

Helfrich, M., Flessa, H., Mikutta, R., Dreves, A., and Ludwig, B.: Comparison of chemical fractionation methods for isolating stable soil organic carbon pools, Eur. J. Soil Sci., 58, 1316–1329, doi:10.1111/j.1365-2389.2007.00926.x, 2007.

Helfter, C., Campbell, C., Dinsmore, K. J., Drewer, J., Coyle, M., Anderson, M., Skiba, U., Nemitz, E., Billett, M. F., and Sutton, M. A.: Drivers of long-term variability in CO_2 net ecosystem exchange in a temperate peatland, Biogeosciences Discuss., 11, 14981–15018, doi:10.5194/bgd-11-14981-2014, 2014.

Hendriks, D. M. D., van Huissteden, J., Dolman, A. J., and van der Molen, M. K.: The full greenhouse gas balance of an abandoned peat meadow, Biogeosciences, 4, 411–424, doi:10.5194/bg-4-411-2007, 2007.

Hendriks, D.: Vegetation as indicator for methane emissions, carbon dioxide fluxes and greenhouse gas balances from peat land, in: Integrated observations of greenhouse gas budgets at the ecosystem level: Changing environment and management practices in peat meadows, edited by: Hendriks, D., Ph.D. thesis, VU-University, Amsterdam, 133–166, 2009.

Hilker, T., Coops, N. C., Wulder, M. A., Black, A. T., and Guy, R. D.: The use of remote sensing in light use efficiency based models of gross primary production: A review of current status and future requirements, Sci. Total Environ., 404, 411–423, doi:10.1016/j.scitotenv.2007.11.007, 2008.

Holden, J., Chapman, P. J., and Labadz, J. C.: Artificial drainage of peatlands: hydrological and hydrochemical process and wetland restoration, Prog. Phys. Geogr., 28, 95–123, doi:10.1191/0309133304pp403ra, 2004.

Humphreys, E. R., Lafleur, P. M., Flanagan, L. B., Hedstrom, N., Syed, K. H., Glenn, A. J., and Granger, R.: Summer carbon dioxide and water vapor fluxes across a range of northern peatlands, J. Geophys. Res., 111, G04011, doi:10.1029/2005JG000111, 2006.

Jackson, M. B. and Armstrong, W.: Formation of Aerenchyma and the Processes of Plant Ventilation in Relation to Soil Flooding and Submergence, Plant Biol., 1, 274–287, doi:10.1111/j.1438-8677.1999.tb00253.x, 1999.

Jacobs, C. M. J., Jacobs, A. F. G., Bosveld, F. C., Hendriks, D. M. D., Hensen, A., Kroon, P. S., Moors, E. J., Nol, L., Schrier-Uijl, A., and Veenendaal, E. M.: Variability of annual CO2 exchange from Dutch grasslands, Biogeosciences, 4, 803–816, doi:10.5194/bg-4-803-2007, 2007.

Janssens, I. A., Lankreijer, H., Matteucci, G., Kowalski, A. S., Buchmann, N., Epron, D., Pilegaard, K., Kutsch, W., Longdoz, B., Grünwald, T., Montagnani, L., Dore, S., Rebmann, C., Moors, E. J., Grelle, A., Rannik, Ü., Morgenstern, K., Oltchev, S., Clement, R., Guðmundsson, J., Minerbi, S., Berbigier, P., Ibrom, A., Moncrieff, J., Aubinet, M., Bernhofer, C., Jensen, N. O., Vesala, T., Granier, A., Schulze, E. D., Lindroth, A., Dolman, A. J., Jarvis, P. G., Ceulemans, R., and Valentini, R.: Productivity overshadows temperature in determining soil and ecosystem respiration across European forests, Glob. Change Biol., 7, 269–278, 2001.

Jansson, P. and Halldin, S.: Model for the annual water and energy flow in a layered soil, Comparison of forest and energy exchange models, Society for Ecol. Model., Copenhagen, 145–163, 1979.

Jansson, P.-E.: COUPModel: Model use, calibration and validation, Transactions of the ASABE, 55, 1335–1344, 2012.

Jansson, P.-E. and Karlberg, L.: Coupled heat and mass transfer model for soil–plant–atmosphere systems, Royal Institute of Technology, Stockholm, 484 pp., available at: http://www2.lwr.kth.se/VaraDatorprogram/CoupModel/coupmanual.pdf (last access: 15 June 2012), 2010.

Johansson, G.: Release of organic c from growing roots of meadow fescue (Festuca pratensis L.), Soil Biol. Biochem., 24, 427–433, doi:10.1016/0038-0717(92)90205-C, 1992.

Johansson, G.: Carbon distribution in grass (Festuca pratensis L.) during regrowth after cutting—utilization of stored and newly assimilated carbon, Plant Soil, 151, 11–20, 1993.

Ju, W. and Chen, J. M.: Distribution of soil carbon stocks in Canada's forests and wetlands simulated based on drainage class, topography and remotely sensed vegetation parameters, Hydrol. Process., 19, 77–94, doi:10.1002/hyp.5775, 2005.

Juston, J., Andrén, O., Kätterer, T., and Jansson, P.-E.: Uncertainty analyses for calibrating a soil carbon balance model to agricultural field trial data in Sweden and Kenya, Ecol. Model., 221, 1880–1888, doi:10.1016/j.ecolmodel.2010.04.019, 2010.

Kechavarzi, C., Dawson, Q., Bartlett, M., and Leeds-Harrison, P. B.: The role of soil moisture, temperature and nutrient amendment on CO_2 efflux from agricultural peat soil microcosms, Geoderma, 154, 203–210, 2010.

Keddy, P. A.: Assembly and response rules: two goals for predictive community ecology, J. Veg. Sci., 3, 157–164, doi:10.2307/3235676, 1992.

Kellner, E. and Halldin, S.: Water budget and surface layer water storage in a Sphagnum bog in central Sweden, Hydrol. Process., 16, 87–103, 2002.

Kettunen, A., Kaitala, V., Lehtinen, A., Lohila, A., Alm, J., Silvola, J., and Martikainen, P. J.: Methane production and oxidation potentials in relation to water table fluctuations in two boreal mires, Soil Biol. Biochem., 31, 1741–1749, doi:10.1016/S0038-0717(99)00093-0, 1999.

Kirschbaum, M.: The temperature dependence of organic-matter decomposition—still a topic of debate, Soil Biol. Biochem., 38, 2510–2518, doi:10.1016/j.soilbio.2006.01.030, 2006.

Kistritz, R. U., Hall, K. J., and Yesaki, I.: Productivity, detritus flux, and nutrient cycling in a Carex lyngbyei tidal marsh, Estuaries, 6, 227–236, 1983.

Klemedtsson, L., Jansson, P.-E., Gustafsson, D., Karlberg, L., Weslien, P., Arnold, K., Ernfors, M., Langvall, O., and Anders, L.: Bayesian calibration method used to elucidate carbon turnover in forest on drained organic soil, Biogeochemistry, 89, 61–79, doi:10.1007/s10533-007-9169-0, 2008.

Klimeš, L. and Klimešová, J.: The effects of mowing and fertilization on carbohydrate reserves and regrowth of grasses: do they promote plant coexistence in species-rich meadows?, Evolutionary Ecology, 363–382, doi:10.1007/978-94-017-1345-0_8, 2002.

Kuzyakov, Y.: Separating microbial respiration of exudates from root respiration in non-sterile soils: a comparison of four methods, Soil Biol. Biochem., 34, 1621–1631, doi:10.1016/S0038-0717(02)00146-3, 2002.

Lafleur, P., Moore, T., Roulet, N., and Frolking, S. E.: Ecosystem Respiration in a Cool Temperate Bog Depends on Peat Temperature But Not Water Table, Ecosystems, 8, 619–629, doi:10.1007/s10021-003-0131-2, 2005.

Leiber-Sauheitl, K., Fuß, R., Voigt, C., and Freibauer, A.: High CO_2 fluxes from grassland on histic Gleysol along soil carbon and drainage gradients, Biogeosciences, 11, 749–761, doi:10.5194/bg-11-749-2014, 2014.

Li, C., Frolking, S. E., and Frolking, T. A.: A model of nitrous oxide evolution from soil driven by rainfall events: 1. Model structure and sensitivity, J. Geophys. Res., 97, 9759–9776, 1992a.

Li, C., Frolking, S. E., and Frolking, T. A.: A model of nitrous oxide evolution from soil driven by rainfall events: 2. Model applications, J. Geophys. Res., 97, 9777–9783, 1992b.

Limpens, J. and Berendse, F.: How litter quality affects mass loss and N loss from decomposing Sphagnum, Oikos, 103, 537–547, 2003.

Lipson, D., Schadt, C., and Schmidt, S.: Changes in Soil Microbial Community Structure and Function in an Alpine Dry Meadow Following Spring Snow Melt, Microb. Ecol., 43, 307–314, doi:10.1007/s00248-001-1057-x, 2002.

Lloyd, J. and Taylor, J. A.: On the temperature dependence of soil respiration, Funct. Ecol., 8, 315–323, 1994.

Lohila, A.: Annual CO_2 exchange of a peat field growing spring barley or perennial forage grass, J. Geophys. Res., 109, D18166, doi:10.1029/2004JD004715, 2004.

Lohila, A., Aurela, M., Hatakka, J., Pihlatie, M., Minkkinen, K., Penttilä, T., and Laurila, T.: Responses of N_2O fluxes to temperature, water table and N deposition in a northern boreal fen, Eur. J. Soil Sci., 61, 651–661, doi:10.1111/j.1365-2389.2010.01265.x, 2010.

Longstreth, D. J. and Nobel, P. S.: Nutrient Influences on Leaf Photosynthesis: Effects of nitrogen, phosphorus and potassium for gossypium hirsutum L., Plant Physiol., 65, 541–543, 1980.

Lund, M., Lafleur, P. M., Roulet, N. T., Anders, L., Christensen, T. R., Aurela, M., Chojnicki, B. H., Lawrence, F. B., Humphreys, E. R., Laurila, T., Oechel, W. C., Olejnik, J., Rinne, J., Schubert, P. E. R., and Nilsson, M. B.: Variability in exchange of CO_2 across 12 northern peatland and tundra sites, Glob. Change Biol., 16, 2436–2448, doi:10.1111/j.1365-2486.2009.02104.x, 2009.

Lundmark, A.: Monitoring transport and fate of de-icing salt in the roadside environment: modelling and field measurements, Ph.D. thesis, KTH, Land and Water Resources Engineering, Stockholm, Sweden, 44 pp., 2008.

Maljanen, M., Sigurdsson, B. D., Gudmundsson, J., Óskarsson, H., Huttunen, J. T., and Martikainen, P. J.: Greenhouse gas balances of managed peatlands in the Nordic countries – present knowledge and gaps, Biogeosciences, 7, 2711–2738, doi:10.5194/bg-7-2711-2010, 2010.

Meziane, D. and Shipley, B.: Interacting determinants of specific leaf area in 22 herbaceous species: effects of irradiance and nutrient availability, Plant Cell Environ., 22, 447–459, 1999.

Monteith, J. L.: Solar radiation and productivity in tropical ecosystems, J. Appl. Ecol., 9, 747–766, 1972.

Monteith, J. L. and Moss, C. J.: Climate and the Efficiency of Crop Production in Britain [and Discussion], Philos. T. Roy. Soc. B, 281, 277–294, doi:10.1098/rstb.1977.0140, 1977.

Moore, T. R. and Dalva, M.: Methane and carbon dioxide exchange potentials of peat soils in aerobic and anaerobic laboratory incubations, Soil Biol. Biochem., 29, 1157–1164, 1997.

Moore, T. R., Bubier, J. L., and Bledzki, L.: Litter Decomposition in Temperate Peatland Ecosystems: The Effect of Substrate and Site, Ecosystems, 10, 949–963, doi:10.1007/s10021-007-9064-5, 2007.

Moriasi, D. N., Arnold G. J., van Liew, M. W., Binger R. L., Harmel, R. D., and Veith, T. L.: Model evaluation guidelines for systematic quantification of accuracy in watershed simulations, Am. Soc. Agri. Biol. Eng., 50, 885–900, 2007.

Mualem, Y.: A new model for predicting the hydraulic conductivity of unsaturated porous media, Water Resour. Res., 12, 513–522, doi:10.1029/WR012i003p00513, 1976.

Murray, K. J., Tenhunen, J. D., and Nowak, R. S.: Photoinhibition as a control on photosynthesis and production of Sphagnum mosses, Oecologia, 96, 200–207, doi:10.1007/BF00317733, 1993.

Nash, J. E. and Sutcliffe, J. V.: River flow forecasting through conceptual models part I – A discussion of principles, J. Hydrol., 10, 282–290, 1970.

Norman, J., Jansson, P.-E., Farahbakhshazad, N., Butterbach-Bahl, K., Li, C., and Klemedtsson, L.: Simulation of NO and N_2O emissions from a spruce forest during a freeze/thaw event using an N-flux submodel from the PnET-N-DNDC model integrated to CoupModel, Ecol. Model., 216, 18–30, doi:10.1016/j.ecolmodel.2008.04.012, 2008.

Nuttonson, M. Y.: The role of bioclimatology in agriculture with special reference to the use of thermal and photo-thermal requirements of pure-line varieties of plants as a biological indicator in ascertaining climatic analogues (Homoclimes), Int. J. Biometeorol., 2, 129–148, 1958.

Paterson, E. and Sim, A.: Effect of nitrogen supply and defoliation on loss of organic compounds from roots of *Festuca rubra*, J. Exp. Bot., 51, 1449–1457, doi:10.1093/jexbot/51.349.1449, 2000.

Paul, K.: Temperature and moisture effects on decomposition, in: Net Ecosystem Exchange Workshop proceedings, Canberra, Australia, 18–20 April 2001, 95–102, 2001.

Poorter, H., Remkes, C., and Lambers, H.: Carbon and Nitrogen Economy of 24 Wild Species Differing in Relative Growth Rate, Plant Physiol., 94, 621–627, 1990.

Poorter, H., van der Werf, A., Atkin, O. K., and Lambers, H.: Respiratory energy requirements of roots vary with the potential growth rate of a plant species, Physiol. Plantarum, 83, 469–475, 1991.

Potter, C., Bubier, J., Crill, P. M., and Lafleur, P.: Ecosystem modeling of methane and carbon dioxide fluxes for boreal forest sites, Can. J. For. Res., 31, 208–223, doi:10.1139/x00-164, 2001.

Proebsting, W. M., Davies, P. J., and Marx, G. A.: Photoperiodic control of apical senescence in a genetic line of peas, Plant Physiol., 58, 800–802, 1976.

Raich, J. W. and Schleisinger, W. H.: The global carbon dioxide flux in soil respiration and its relationship to vegetation and climate, Tellus, 44B, 81–99, 1992.

Ratkowsky, D. A., Olley, J., McMeekin, T. A., and Ball, A.: Relationship between temperature and growth rate of bacterial cultures, J. Bacteriol, 149, 1–5, 1982.

Ravina, M.: Modelling of methane emission from forest ecosystems. Implementation and test of submodel as part of the COUP-Model, Master Thesis, KTH Royal Institute of Technology, Stockholm, 54 pp., 2007.

Reddy, K. R. and Patrick, W. H.: Effect of alternate aerobic and anaerobic conditions on redox potential, organic matter decomposition and nitrogen loss in a flooded soil, Soil Biol. Biochem., 7, 87–94, doi:10.1016/0038-0717(75)90004-8, 1975.

Reich, P. B., Walters, M. B., Ellsworth, D. S., and Uhl, C.: Photosynthesis-nitrogen relations in Amazonian tree species, Oecologia, 97, 62–72, 1994.

Reich, P. B., Ellsworth, D. S., and Walters, M. B.: Leaf structure (specific leaf area) modulates photosynthesis-nitrogen relations: evidence from within and across species and functional groups, Funct. Ecol., 12, 948–958, 1998.

Reich, P., Ellsworth, D., and Walters, M.: Leaf structure (specific leaf area) modulates photosynthesis–nitrogen relations: evidence from within and across species and functional groups, Funct. Ecol., 12, 948–958, 1998.

Reichstein, M., Rey, A., Freibauer, A., Tenhunen, J. D., Valentini, R., Banza, J., Casals, P., Cheng, Y., Grünzweig, J. M., Irvine, J., Joffre, R., Law, B. E., Loustau, D., Miglietta, F., Oechel, W. C., Ourcival, J.-M., Pereira, J. S., Peressotti, A., Ponti, F., Qi, Y., Rambal, S., Rayment, M., Romanya, J., Rossi, F., Tedeschi, V., Giampiero, T., Xu, M., and Yakir, D.: Modeling temporal and large-scale spatial variability of soil respiration from soil water availability, temperature and vegetation productivity indices, Glob. Biogeochem. Cy., 17, 1104, doi:10.1029/2003GB002035, 2003.

Reichstein, M., Falge, E., Baldocchi, D., Papale, D., Aubinet, M., Berbigier, P., Bernhofer, C., Buchmann, N., Gilmanov, T. G., Granier, A., Grünwald, T., Havrankova, K., Ilvesniemi, H., Janous, D., Knohl, A., Laurila, T., Lohila, A., Loustau, D., Matteucci, G., Meyers, T., Miglietta, F., Ourcival, J.-M., Pumpanen, J., Rambal, S., Rotenberg, E., Sanz, M., Tenhunen, J. D., Seufert, G., Vaccari, F., Vesala, T., Yakir, D., and Valentini, R.: On the separation of net ecosystem exchange into assimilation and ecosystem respiration: review and improved algorithm, Glob. Change Biol., 11, 1424–1439, doi:10.1111/j.1365-2486.2005.001002.x, 2005.

Richards, L. A.: Capillary conduction of liquids through porous mediums, J. Appl. Phys., 1, 318–333, doi:10.1063/1.1745010, 1931.

Robson, M. J.: The Growth and Development of Simulated Swards of Perennial Ryegrass, Ann. Bot., 37, 487–500, 1973.

Rovira, A. D.: Plant root exudates, The Botanical Review, 35, 35–57, 1969.

Rudolph, H. and Samland, J.: Occurrence and metabolism of sphagnum acid in the cell walls of bryophytes, Phytochemistry, 24, 745–749, doi:10.1016/S0031-9422(00)84888-8, 1985.

Ryser, P. and Urbas, P.: Ecological significance of leaf life span among Central European grass species, Oikos, 91, 41–50, doi:10.1034/j.1600-0706.2000.910104.x, 2000.

Saarinen, T.: Demography of Carex rostrata in a boreal mesotrophic fen: shoot dynamics and biomass development, Ann. Bot. Fenn., 35, 203–209, 1998.

Sagerfors, J., Lindroth, A., Grelle, A., Klemedtsson, L., Weslien, P., and Nilsson, M.: Annual CO_2 exchange between a nutrient-poor, minerotrophic, boreal mire and the atmosphere, J. Geophys. Res.-Biogeosci., 113, G01001, doi:10.1029/2006JG000306, 2008.

Scheurwater, I., Cornelissen, C., Dictus, F., Welschen, R., and Lambers, H.: Why do fast- and slow-growing grass species differ so little in their rate of root respiration, considering the large differences in rate of growth and ion uptake?, Plant, Cell and Environment, 21, 995–1005, 1998.

Schläpfer, B. and Ryer, P.: Leaf and Root Turnover of Three Ecologically Contrasting Grass Species in Relation to Their Performance along a Productivity Gradient, Oikos, 75, 398–406, 1996.

Schuldt, R. J., Brovkin, V., Kleinen, T., and Winderlich, J.: Modelling Holocene carbon accumulation and methane emissions of boreal wetlands – an Earth system model approach, Biogeosciences, 10, 1659–1674, doi:10.5194/bg-10-1659-2013, 2013.

Sinclair, T. R. and Horie, T.: Leaf Nitrogen, Photosynthesis, and Crop Radiation Use Efficiency: A Review, Crop Science, 29, 90–98, doi:10.2135/cropsci1989.0011183X002900010023x, 1989.

Smith, J., Gottschalk, P., Bellarby, J., Richards, M., Nayak, D., Coleman, K., Hillier, J., Flynn, H., Wattenbach, M., Aitkenhead, M., Yeluripurti, J., Farmer, J., and Smith, P.: Model to Estimate Carbon in Organic Soils – Sequestration and Emissions (ECOSSE), User Manual, available at: http://www.abdn.ac.uk/staffpages/uploads/soi450/ECOSSEUsermanual310810.pdf, (last access: 05 November 2014), 2010.

Smith, P., Smith, J., Powlson, D., McGill, W., Arah, J., Chertov, O., Coleman, K., Franko, U., Frolking, S., and Jenkinson, D.: A comparison of the performance of nine soil organic matter models using datasets from seven long-term experiments, Geoderma, 81, 153–225, 1997.

Steed, J., E., De Wald, L., E., and Kolb, T., E.: Physiological and Growth Responses of Riparian Sedge Transplants to Groundwater Depth, Int. J. Plant Sci., 163, 925–936, doi:10.1086/342634, 2002.

Steele, J. M., Ratliff, R. D., and Ritenour, G. L.: Seasonal variation in total nonstructural carbohydrate levels in Nebraska sedge, J. Range Manage., 37, 465–467, 1984.

St-Hilaire, F., Wu, J., Roulet, N. T., Frolking, S. E., Lafleur, P. M., Humphreys, E. R., and Arora, V.: McGill wetland model: evaluation of a peatland carbon simulator developed for global assessments, Biogeosciences, 7, 3517–3530, doi:10.5194/bg-7-3517-2010, 2010.

Thomas, H. and Stoddart, J. L.: Leaf Senescence, Ann. Rev. Plant Physiol., 31, 83–111, 1980.

Thornton, P. E. and Rosenbloom, N. A.: Ecosystem model spin-up: Estimating steady state conditions in a coupled terrestrial carbon and nitrogen cycle model, Ecol. Model., 189, 25–48, doi:10.1016/j.ecolmodel.2005.04.008, 2005.

Turetsky, M. R.: The Role of Bryophytes in Carbon and Nitrogen Cycling, The Bryologist, 106, 395–409, 2003.

van den Bos, R.: Restoration of former wetlands in the Netherlands; effect on the balance between CO_2 sink and CH_4 source, Neth. J. Geosci., 82, 325–332, 2003.

van der Werf, A., Kooijman, A., Welschen, R., and Lambers, H.: Respiratory energy costs for the maintenance of biomass, for growth and for ion uptake in roots of *Carex diandra* and *Carex acutiformis*, Physiol. Plantarum, 72, 483–491, 1988.

van Huissteden, J., van den Bos, R., and Alvarez, I. M.: Modelling the effect of water-table management on CO_2 and CH_4 fluxes from peat soils, Neth. J. Geosci., 85, 3–18, 2006.

van Huissteden, J., Petrescu, A. M. R., Hendriks, D. M. D., and Rebel, K. T.: Sensitivity analysis of a wetland methane emission model based on temperate and arctic wetland sites, Biogeosciences, 6, 3035–3051, doi:10.5194/bg-6-3035-2009, 2009.

van Oijen, M., Reyer, C., Bohn, F. J., Cameron, D. R., Deckmyn, G., Flechsig, M., Härkönen, S., Hartig, F., Huth, A., Kiviste, A., Lasch, P., Mäkelä, A., Mette, T., Minunno, F., and Rammer, W.: Bayesian calibration, comparison and averaging of six forest models, using data from Scots pine stands across Europe, Forest Ecology and Management, 289, 255–268, doi:10.1016/j.foreco.2012.09.043, 2013.

Verhoeven, J. and Toth, E.: Decomposition of Carex and Sphagnum litter in fens: Effect of litter quality and inhibition by living tissue homogenates, Soil Biol. Biochem., 27, 271–275, doi:10.1016/0038-0717(94)00183-2, 1995.

Ward, S. E., Ostle, N. J., Oakley, S., Quirk, H., Henrys, P. A., Bardgett, R. D., and van der Putten, W.: Warming effects on greenhouse gas fluxes in peatlands are modulated by vegetation composition, Ecol Lett, 16, 1285–1293, doi:10.1111/ele.12167, 2013.

Whalen, J. K., Bottomley, P. J., and Myrold, D. D.: Carbon and nitrogen mineralization from light- and heavy-fraction additions to soil, Soil Biol. Biochem., 32, 1345–1352, doi:10.1016/S0038-0717(00)00040-7, 2000.

White, L. M.: Carbohydrate reserves of grasses: a review, J. Range Manage., 13–18, 1973.

Willmott, C. J.: Some Comments on the Evaluation of Model Performance, Bull. Amer. Meteor. Soc., 63, 1309–1313, doi:10.1175/1520-0477(1982)063<1309:SCOTEO>2.0.CO;2, 1982.

Wingler, A.: The role of sugars in integrating environmental signals during the regulation of leaf senescence, J. Exp. Bot., 57, 391–399, doi:10.1093/jxb/eri279, 2005.

Wohlfahrt, G., Bahn, M., Haubner, E., Horak, I., Michaeler, W., Rottmar, K., Tappeiner, U., and Cernusca, A.: Inter-specific variation of the biochemical limitation to photodynthesis and related leaf traits of 30 species from mountain grassland ecosystems under different land use, Plant Cell Environ., 22, 1281–1296, 1999.

Wu, J., Jansson, P.-E., van der Linden, L., Pilegaard, K., Beier, C., and Ibrom, A.: Modelling the decadal trend of ecosystem carbon fluxes demonstrates the important role of functional changes in a temperate deciduous forest, Ecol. Model., 260, 50–61, doi:10.1016/j.ecolmodel.2013.03.015, 2013a.

Wu, J., Roulet, N. T., Sagerfors, J., and Nilsson, M. B.: Simulation of six years of carbon fluxes for a sedge-dominated oligotrophic minerogenic peatland in Northern Sweden using the McGill Wetland Model (MWM), J. Geophys. Res.-Biogeosci., 118, 795–807, doi:10.1002/jgrg.20045, 2013b.

Wu, S. H. and Jansson, P.-E.: Modelling soil temperature and moisture and corresponding seasonality of photosynthesis and transpiration in a boreal spruce ecosystem, Hydrol. Earth Syst. Sci., 17, 735–749, doi:10.5194/hess-17-735-2013, 2013.

Yeloff, D. and Mauquoy, D.: The influence of vegetation composition on peat humification: implications for palaeoclimatic studies, Boreas, 35, 662–673, doi:10.1080/03009480600690860, 2006.

Yurova, A., Wolf, A., Sagerfors, J., and Nilsson, M.: Variations in net ecosystem exchange of carbon dioxide in a boreal mire: Modeling mechanisms linked to water table position, J. Geophys. Res., 112, doi:10.1029/2006JG000342, 2007.

Zeitz, J. and Velty, S.: Soil properties of drained and rewetted fen soils, Soil Sci. Soc. Am. J., 165, 618–626, 2002.

Zhang, D., Hui, D., Luo, Y., and Zhou, G.: Rates of litter decomposition in terrestrial ecosystems: global patterns and controlling factors, J. Plant Ecol., 1, 85–93, doi:10.1093/jpe/rtn002, 2008.

Zhang, T.: Influence of the seasonal snow cover on the ground thermal regime: An overview, Rev. Geophys., 43, RG4002, doi:10.1029/2004RG000157, 2005.

Zhang, Y., Li, C., Trettin, C. C., Li, H., and Sun, G.: An integrated model of soil, hydrology, and vegetation for carbon dynamics in wetland ecosystems, Glob. Biogeochem. Cy., 16, 9-1–9-17, doi:10.1029/2001GB001838, 2002.

Zimmermann, M., Leifeld, J., Schmidt, M. W. I., Smith, P., and Fuhrer, J.: Measured soil organic matter fractions can be related to pools in the RothC model, Eur. J. Soil Science, 58, 658–667, doi:10.1111/j.1365-2389.2006.00855.x, 2007.

5

Effect of reed canary grass cultivation on greenhouse gas emission from peat soil at controlled rewetting

S. Karki, L. Elsgaard, and P. E. Lærke

Aarhus University, Department of Agroecology, Blichers Allé 20, 8830 Tjele, Denmark

Correspondence to: S. Karki (sandhya.karki@agrsci.dk)

Abstract. Cultivation of bioenergy crops in rewetted peatland (paludiculture) is considered as a possible land use option to mitigate greenhouse gas (GHG) emissions. However, bioenergy crops like reed canary grass (RCG) can have a complex influence on GHG fluxes. Here we determined the effect of RCG cultivation on GHG emission from peatland rewetted to various extents. Mesocosms were manipulated to three different ground water levels (GWLs), i.e. 0, −10 and −20 cm below the soil surface in a controlled semi-field facility. Emissions of CO_2 (ecosystem respiration, ER), CH_4 and N_2O from mesocosms with RCG and bare soil were measured at weekly to fortnightly intervals with static chamber techniques for a period of 1 year. Cultivation of RCG increased both ER and CH_4 emissions, but decreased the N_2O emissions. The presence of RCG gave rise to 69, 75 and 85 % of total ER at −20, −10 and 0 cm GWL, respectively. However, this difference was due to decreased soil respiration at the rising GWL as the plant-derived CO_2 flux was similar at all three GWLs. For methane, 70–95 % of the total emission was due to presence of RCG, with the highest contribution at −20 cm GWL. In contrast, cultivation of RCG decreased N_2O emission by 33–86 % with the major reductions at −10 and −20 cm GWL. In terms of global warming potential, the increase in CH_4 emissions due to RCG cultivation was more than offset by the decrease in N_2O emissions at −10 and −20 cm GWL; at 0 cm GWL the CH_4 emissions was offset only by 23 %. CO_2 emissions from ER were obviously the dominant RCG-derived GHG flux, but aboveground biomass yields, and preliminary measurements of gross photosynthetic production, showed that ER could be more than balanced due to the photosynthetic uptake of CO_2 by RCG. Our results support that RCG cultivation could be a good land use option in terms of mitigating GHG emission from rewetted peatlands, potentially turning these ecosystems into a sink of atmospheric CO_2.

1 Introduction

Peatlands cover 3 % of the world's area but contain 30 % of the soil organic carbon (Parish et al., 2008), signifying an important role in the global carbon cycle. About 15 % of the world's peatlands have been drained for different human purposes, mostly for agriculture and forestry and to a lesser extent for peat extraction (Joosten, 2009). Drained peatlands are major sources of CO_2 emissions and estimated to account for about 6 % of the total anthropogenic CO_2 emission (Joosten, 2009). In order to reduce the large emissions of CO_2 from drained peatlands, extensive rewetting projects have been implemented in Europe and North America (Höper et al., 2008), and rewetted organic soils have been included in the guidelines for national greenhouse gas (GHG) inventories by the Intergovernmental Panel on Climate Change (IPCC, 2014). In addition, agricultural use of wet and rewetted peatlands for crop growth (paludiculture) is considered as a possible land use option that may indirectly reduce the CO_2 emissions by biomass production for energy purposes (Joosten et al., 2012; Günther et al., 2014).

Reed canary grass (RCG) (*Phalaris arundinacea*) is one of the suitable biomass crops for paludiculture (Wichtmann and Tanneberger, 2011). It can be established from seeds as normal agricultural grass (Kandel et al., 2013b), but in some countries it is considered as an invasive species (Maurer et al., 2003). The plants thrive in wet soils due to aerenchyma tissues (Kercher and Zedler, 2004; Askaer et al., 2011) that transport oxygen to the roots in otherwise anaerobic soil compartments. However, cultivating wetland plants like RCG may influence the overall GHG balance by a combination of contrasting effects. First of all, RCG can stimulate the processes of GHG production by increasing the labile soil organic carbon pool, e.g. via root exudates (Ström et al., 2003; Bastviken et al., 2005). Next, the transport of oxygen to anaerobic zones stimulates heterotrophic degradation of organic matter, but at the same time stimulates oxidation of CH_4 (Kao-Kniffin et al., 2010) and suppress CH_4 production due to increase in redox potential (Laanbroek, 2010; Sutton-Grier and Megonigal, 2011). RCG may further increase the emissions of reduced soil gases as the aerenchyma tissues act as a conduit for the direct transport of, for example, CH_4 and N_2O produced in soil (Joabsson et al., 1999; Jørgensen et al., 2012). Also, RCG can decrease N_2O emissions by assimilation of mineral N, which reduces the availability of electron acceptors (nitrate) for denitrifying microorganisms (Roobroeck et al., 2010). In summary, the introduction of RCG at rewetted peatlands may cause a change in the patterns and underlying mechanisms of GHG emission, which are rather complex.

In the natural state, GHG emissions from peatlands are predominantly controlled by the position (depth) of the water table (IPCC, 2014). Basically, due to slow diffusion of oxygen in water (10 000 times slower than in air), ground water level (GWL) has a strong control on the oxic/anoxic soil boundary and thereby on the biogeochemical processes involved in GHG fluxes (Dinsmore et al., 2009; Karki et al., 2014). However, the presence of aerenchymatous plants may strongly interact with GWL in being decisive for the resulting GHG emissions from wet peatlands. The objective of the present study was to quantify the role of RCG cultivation on the resulting GHG emissions of CO_2, N_2O and CH_4 from peat soils rewetted to various extents. Such information is very important for understanding the total GHG balance from paludiculture and improve the basis for modelling future climate. To accomplish this, the GHG emissions of all three gases were measured in an annual study with peat soil mesocosms with RCG and bare soil rewetted to constant GWLs of 0, −10 and −20 cm in a controlled semi-field facility.

2 Materials and methods

2.1 Site description

Soil cores were collected from a fen peatland in the Nørre Å river valley, Denmark (56°44′ N, 9°68′ E). The peatland was drained to a depth of 60–70 cm early in the 20th century and has since then been used for agricultural purposes. RCG experimental plots were established at the site in 2009 (Kandel et al., 2013b). The top soil layer (0–20 cm) at the study site had the following main properties: highly decomposed peat soil corresponding to H9 on the von Post scale; bulk density, 0.27 g cm^{-3}; total organic carbon, 37.8 %; and total nitrogen, 3.2 % (Karki et al., 2014).

2.2 Experimental design

A total of 30 intact soil cores for the mesocosm study were collected in May 2012 by inserting PVC pipes of 60 cm depth and 30 cm diameter into the soil. Half of the soil cores were collected from RCG plots and the other half were collected from a grass field surrounding the RCG plots. The upper 5 cm of the soil and litter layer was removed from the grass field before inserting the PVC pipes, and the soil cores were kept bare during the experiment. The soil cores were retrieved with help of a mini excavator and transported to semi-field facilities at AU Foulum (Karki et al., 2014). The bottom of the PVC pipes was covered with fine-meshed net to allow for free water movement, and the pipes were then installed in plastic cylinders (diameter, 37 cm; height, 70 cm). The plastic cylinders were filled with gravel at the bottom 10 cm and the space between the PVC pipes and the wall of the cylinders (ca. 3 cm) was filled with sand. The whole setup was then installed in a trench at the semi-field facility with the soil surface at ground level.

Mesocosms with bare soil and RCG were randomly divided into three groups and manipulated to three different GWLs of 0, −10 and −20 cm below the soil surface. The water table was adjusted by fitting a rubber tubing (diameter, 1 cm) to the bottom of each plastic cylinder and placing the other end of the rubber tubing at different heights corresponding to the level of GWL treatment. Water was supplied in the space between the PVC pipes and the wall of the cylinders every day for 1 h by a drip irrigation system. Further details on mesocosm incubations and the semi-field facility can be found in Karki et al. (2014).

Due to poor regrowth of RCG (both under mesocosm and field conditions), initial weed biomass was uprooted and new RCG seeds were spread on 21 June 2012. RCG was fertilized with surface application of 0.6 g N, 0.1 g P and 0.5 g K per mesocosm on 23 July 2012 (corresponding to 80 kg N, 13 kg P and 77 kg K ha^{-1}). This fertilization rate corresponded to the rate applied in a previous study at the RCG field site from where the mesocosms were collected (Kandel et al., 2013a), except that the nitrogen rate was slightly increased in the mesocosm study as lower N mineralization was expected at higher GWLs. After the regrowth of RCG in spring 2013, RCG was fertilized with the same amount of fertilizer on 30 April and again in 28 June 2013. RCG plants were harvested twice, first on 29 October 2012 and then on 27 June 2013. In bare soil mesocosms, emerging weeds were uprooted and mosses were eliminated by application of iron sulfate ($FeSO_4$) on 29 August 2012. No fertilizer was added to bare soil mesocosms.

2.3 Gas measurements and flux calculation

Dark PVC chambers (diameter, 30 cm; height, 50 cm) equipped with fans and pressure equilibration vents were used for the measurement of CO_2, CH_4 and N_2O (Karki et al., 2014). Gas measurements were carried out between 10:00 and 13:00 at weekly to fortnightly intervals from July 2012 to July 2013. Four gas samples (10 mL) were drawn from the chamber headspace with polypropylene syringes during 45 min of chamber enclosure and transferred to evacuated 6 mL Exetainers. Gas samples were analysed with an Agilent 7890 gas chromatograph connected to a CTC CombiPAL automatic sample injection system (Agilent, Nærum, Denmark). Fluxes were calculated using the HMR method (Pedersen et al., 2010) in the statistical software R version 3.0.2 (R Core Team, 2013) as non-linear increase in GHG concentration over time was often observed during the non-steady-state chamber measurements (Davidson et al., 2002; Petersen et al., 2012). Thus, according to the statistical HMR analysis, fluxes were calculated either by non-linear or linear models (Pedersen et al., 2010). Out of the total of 435 fluxes for each GHG, the non-linear approach was applied for 41, 40 and 18 % of CO_2, CH_4 and N_2O fluxes from RCG mesocosms, respectively, and 22, 16 and 22 % of CO_2, CH_4 and N_2O fluxes from bare soil mesocosms, respectively. In bare soil at 0 cm GWL, approximately 3 % of the CH_4 fluxes were discarded due to episodic release of CH_4 presumably by ebullition.

2.4 Biomass measurement

Biomass development was monitored through the non-destructive measurement of ratio vegetation index (RVI). RVI was determined for each mesocosm using a SpectroSense 2+ fitted with SKR1800 sensors (Skye Instruments, Powys, UK). The sensors measured the incident and reflected red light (R) at 656 nm and the incident and reflected infrared light (NIR) at 778 nm. RVI was then calculated as $(NIR_r/NIR_i)/(R_r/R_i)$, where the subscripts i and r denote the incident and reflected radiation. RVI has previously been used as a useful predicting factor for modelling ecosystem respiration (ER) and CH_4 fluxes (Kandel et al., 2013a, b; Görres et al., 2014; Karki et al., 2014)

RVI measurements were done on the same days as GHG sampling, except in winter, when the soil was covered with snow or frozen. The total above-ground dry biomass from each mesocosm was also determined after each harvest by oven-drying the plant material at 60 °C to constant weight. After the harvest in 2013, species composition from each mesocosm was determined on dry weight basis to quantify the contribution of volunteer weeds in the total biomass.

2.5 Environmental parameters and pore water analysis

Soil temperature at 5 cm depth and soil moisture was measured by means of temperature and time domain reflectometry (TDR) probes installed permanently in one of the five replicates for each GWL treatment. Soil temperature was measured automatically every hour, while soil moisture measurements with TDR (volumetric water content, VWC) were done on every gas sampling occasion. The instrumented mesocosms also had Pt probes installed at 20 cm depth to measure soil redox potential. Soil redox potential was measured at fortnightly intervals from mid-April to July 2013 with a portable pH meter (PHM220, Radiometer) by gently pushing a double-junction calomel reference electrode (REF251, Hach Lange) into the soil. Measured redox potential were converted to standard hydrogen electrode potential (Eh) by addition of +245 mV (Kjaergaard et al., 2012).

A piezometer (length, 65 cm; diameter, 2 cm) with the screen all the way down was installed in the instrumented mesocosms. Approximately 30 mL of soil water was sampled monthly from these piezometers, except for February to April 2013, when water inside the piezometers was frozen. Water samples were analysed for ammonium, nitrate and sulfate content. Ammonia and nitrate content were measured using an auto-analyser (Bran+Luebbe GmbH, Norderstedt, Germany), and sulfate was determined via ion chromatography on a Dionex ICS-1500 IC system (Dionex Corp., Sunnyvale, CA, USA).

2.6 Cumulative GHG fluxes

For the mesocosms with RCG, CO_2 emissions from ER were modelled as a function of GWL, temperature and biomass (RVI) by model 1 (Karki et al., 2014); for bare soil mesocosms, model 2 excluding RVI was applied:

$$\mathrm{ER} = (b_1 + b_2\mathrm{GWL}) \times \exp\left(b_3\left(\frac{1}{10 - T_0} - \frac{1}{T - T_0}\right)\right) \times (b_4 + \mathrm{RVI}), \tag{1}$$

$$\mathrm{ER} = (b_1 + b_2\mathrm{GWL}) \times \exp\left(b_3\left(\frac{1}{10 - T_0} - \frac{1}{T - T_0}\right)\right), \tag{2}$$

where T_0 is a notional zero respiration temperature, here fixed to −46.02 °C (Lloyd and Taylor, 1994); T is the air or soil temperature (°C); RVI is the ratio vegetation index; GWL is water table depth below the soil surface (cm); and b_1, b_2, b_3 and b_4 are model parameters.

All model parameters were estimated by non-linear regression in SigmaPlot 11 (Systat Software, Chicago, IL, USA). Using the obtained model parameters, continuous temperature data and linearly interpolated RVI data, hourly rates of CO_2 emissions were reconstructed for each GWL. These hourly emissions values were summed to yield the annual flux from 10 July 2012 to 9 July 2013. The uncertainty of

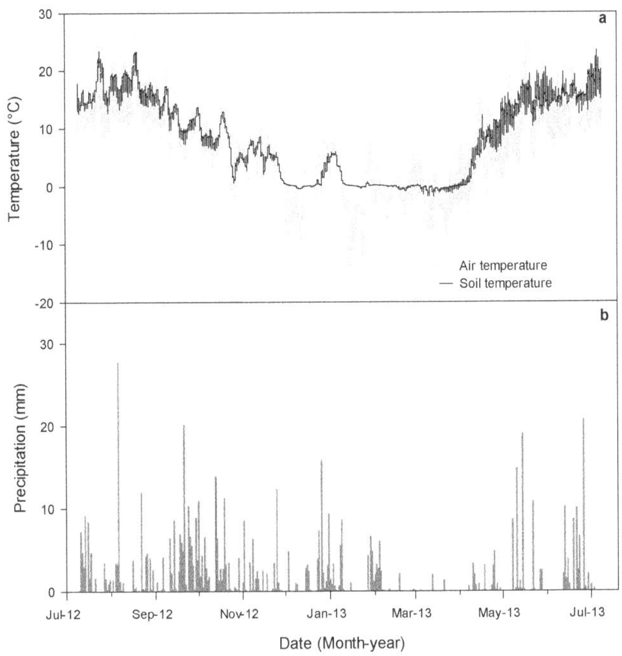

Figure 1. (a) Hourly air temperature at 2 m height at the semi-field facility and hourly average soil temperature at 5 cm depth across all mesocosm treatments, and **(b)** daily precipitation at the semi-field facility during the study period (July 2012 to July 2013).

annual fluxes was addressed by deriving the minimum and maximum cumulative fluxes from upper and lower values of model parameters ± standard errors (SE) (Elsgaard et al., 2012). For model evaluation the Nash–Sutcliffe modelling efficiency (ME) was calculated according to

$$ME = 1 - \frac{\sum_{i=1}^{n} (\mathrm{Mes}_i - \mathrm{Mod}_i)^2}{\sum_{i=1}^{n} \left(\mathrm{Mes}_i - \overline{\mathrm{Mes}}\right)^2}, \qquad (3)$$

where $\mathrm{Mes_i}$ and $\mathrm{Mod_i}$ are measured and modelled values, respectively, and $\overline{\mathrm{Mes}}$ is the mean of measured values (Haefner, 2005).

Cumulative CH_4 and N_2O fluxes were calculated by linear interpolation between the sampling dates using the trapezoidal rule (Petersen et al., 2012). The linear interpolation method was used as there were no common models to predict CH_4 and N_2O fluxes for vegetated and bare soil plots. Cumulative fluxes were calculated for each individual mesocosm and then averaged for each GWL treatment ($n = 5$). Total GHG emissions were calculated by summing annual CO_2, CH_4 and N_2O emissions at each GWL; CH_4 and N_2O emissions were converted to CO_2 equivalents by multiplying by 28 and 265, respectively, according to the revised global warming potential (GWP) of the three GHGs (Myhre et al., 2013). The plant-derived total GHG emission at each GWL was estimated as the difference between the total GHG emissions from RCG mesocosms and bare soil mesocosms. The standard error of annual plant-derived GHG emissions was calculated following the law of error propagation as the square root of the sum of the squared standard error of plant and bare soil emissions.

2.7 Statistical analysis

Statistical analyses were done using R version 3.0.2 (R Core Team, 2013). Data were analysed using a linear mixed model including the fixed effect of vegetation (bare soil/RCG), GWL, date and their two-way interactions. The model also included the random effect of each experimental unit. Prior to analysis, CH_4 and N_2O flux data were log-transformed after addition of a constant (lowest detected fluxes of CH_4 and N_2O) to obtain normal distribution and variance homogeneity. Dates were treated as repeated measurements by applying either compound symmetry structure (each dependent variable has constant covariance independent of time) or autocorrelation structure of order 1 (errors at adjacent time points are correlated) (Maxwell and Delaney, 2004). The best model was selected by use of Akaike's information criterion (AIC). For CH_4 and N_2O, autocorrelation structure was selected, while compound symmetry was selected for CO_2 fluxes.

A similar linear mixed model was run to determine the effect of GWL on RVI development. One-way ANOVA was used to test the difference in mean yield between the treatments. Significance of all tests was accepted at $P < 0.05$.

3 Results

3.1 Environmental conditions

The average air temperature during the study period was 6.9 °C and total precipitation was 667 mm (Fig. 1). Snowfall started in early December 2012 and was observed until the end of March 2013 with intermittent freezing and thawing events. The soil was frozen and covered with ice until mid-April 2013. The annual average soil temperature (5 cm depth) in RCG treatments was 7.4, 7.7 and 7.6 °C at 0, −10 and −20 cm GWL, respectively; for bare soil treatments it was 7.5, 7.4 and 7.9 °C at 0, −10 and −20 cm GWL, respectively. The average volumetric soil water content during the measurement period was 82 ± 5, 67 ± 3, and 58 ± 3 % from RCG treatments at 0, −10 and −20 cm GWL, respectively, and 83 ± 4, 62 ± 6 and 55 ± 7 % from bare soil treatments at 0, −10 and −20 cm GWL, respectively (mean ± standard deviation, $n = 22$). Average soil redox potential was −115, −27 and 40 mV from RCG treatments at 0, −10 and −20 cm GWL, respectively, and −118, −51 and 151 mV from bare soil treatments at 0, −10 and −20 cm GWL, respectively (Fig. 2).

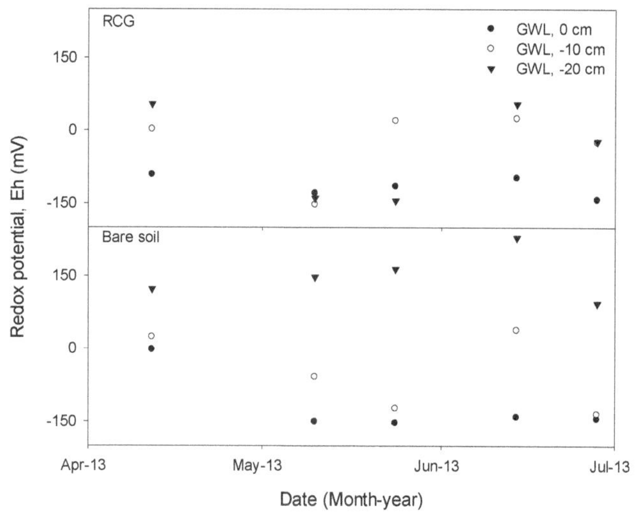

Figure 2. Redox potential (Eh) at different ground water levels (GWL) from reed canary grass (RCG) and bare soil mesocosms. Eh was measured at 20 cm soil depth from April to July 2013.

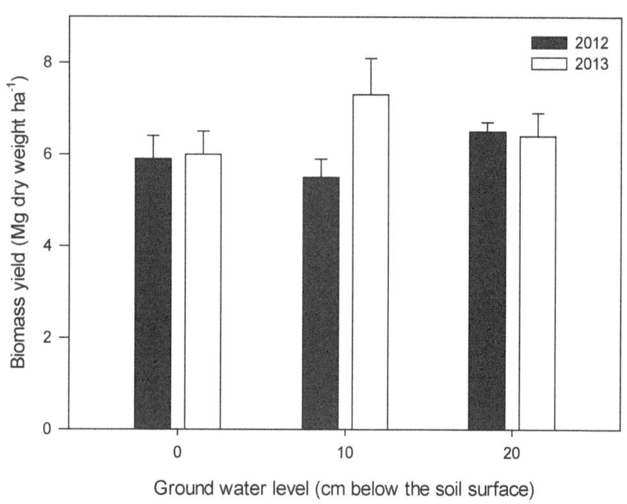

Figure 3. Mean dry biomass yield (Mg ha^{-1}) from mesocosms at different ground water levels in 2012 and 2013. Error bars show standard error ($n = 5$).

3.2 Biomass yield and RVI

The mean biomass yield was 6.0 and 6.6 Mg ha^{-1} across all GWLs in 2012 and 2013, respectively (Fig. 3). During the first year there was a good stand of RCG, but during the second year weed biomass became established, especially at 0 cm GWL; this was notably marsh foxtail (*Alopecurus geniculatus*) and grasses (*Poa* sp.), which made an important contribution to the total biomass at the time of harvest.

The pattern of RVI development was similar among the different GWL treatments; peak values of RVI occurred in late August 2012, whereafter RVI started to decline due to

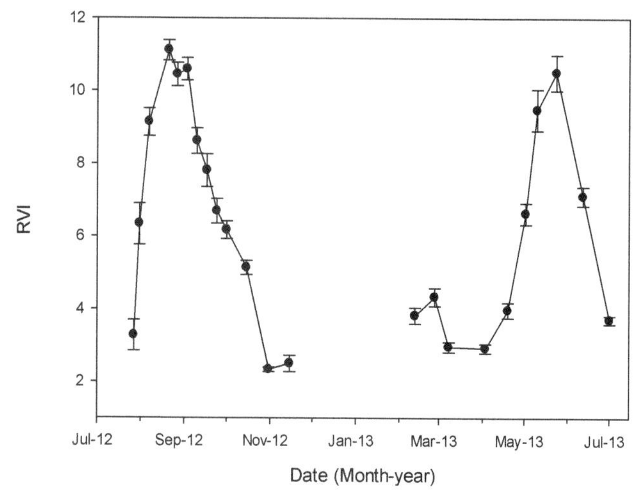

Figure 4. Average ratio vegetation index (RVI) development during the measurement period across all ground water levels. Error bars show standard error ($n = 15$). The dotted line represents the winter period, when RVI was not measured due to ice and snow.

plant senescence. RVI started to increase again during the regrowth of biomass in spring 2013 (Fig. 4).

3.3 Pore water properties

The annual variation in soil water sulfate concentrations ranged from 1.3 to 56.9 mg L^{-1}. Generally, similar SO_4^{2-} concentrations were found in bare soil and RCG mesocosms at 0 and −10 cm GWL, but at −20 cm GWL consistently higher SO_4^{2-} concentrations were found in the bare soil mesocosms (Table 1). For ammonium the concentrations ranged from < 0.1 to 10.2 mg L^{-1}, and higher NH_4^+ concentrations were generally found in bare soil mesocosms than in RCG mesocosms at 0 and −10 cm GWL. In the bare soil treatments the level of NH_4^+ was lower at −20 cm GWL than at 0 and −10 cm GWL, but in RCG treatments NH_4^+ concentrations were similar at all the three GWLs (Table 1). The concentration of nitrate was low (< 3.1 mg L^{-1}) across all treatments; the highest NO_3^- levels were generally seen at bare soil treatments at −20 cm GWL (Table 1).

3.4 Measured GHG fluxes

The emission of CO_2 was measured as ER in RCG and bare soil treatments in order to evaluate the contribution of RCG in the total ER at the different GWLs. The emissions of CO_2 were different between RCG and bare soil mesocosms ($P < 0.001$) and also between the three GWL treatments ($P < 0.001$) (Table 2). CO_2 emissions decreased consistently with higher GWL both from RCG and bare soil mesocosms. The emissions showed expected seasonal variation with highest CO_2 fluxes during summertime ($P < 0.001$) (Fig. 5a, b). CO_2 emissions ranged from 20 to 485 mg m^{-2} h^{-1} across all GWLs in bare soil and from 55

Table 1. Concentration of sulfate, ammonium and nitrate ($mg\,L^{-1}$) in ground water samples collected from piezometers from bare soil and reed canary grass (RGC) mesocosms at different ground water levels (GWL).

Treatment and date	SO_4^{2-} ($mg\,L^{-1}$) at GWL 0 cm	−10 cm	−20 cm	NH_4^+ ($mg\,L^{-1}$) at GWL 0 cm	−10 cm	−20 cm	NO_3^- ($mg\,L^{-1}$) at GWL 0 cm	−10 cm	−20 cm
Bare soil									
26 Jul 2012	20.7	33.4	54.2	0.9	0.7	< 0.1	0.4	0.6	0.7
24 Aug 2012	6.6	10.0	56.9	1.4	2.0	0.1	0.2	0.1	2.0
26 Sep 2012	41.7	13.5	52.9	3.4	4.3	0.3	0.1	< 0.1	1.8
5 Nov 2012	2.7	1.8	46.2	3.6	3.5	0.7	< 0.1	0.1	0.5
30 Nov 2012	2.1	4.7	41.9	4.0	3.2	0.8	< 0.1	< 0.1	0.1
4 Jan 2013	2.3	4.0	29.7	2.9	3.8	0.4	0.1	0.3	1.0
6 May 2013	1.3	1.4	22.7	5.1	4.6	0.7	0.7	0.1	0.6
11 Jun 2013	2.0	2.1	18.1	5.8	3.8	0.2	< 0.1	< 0.1	0.6
16 Jul 2013	1.3	1.4	17.2	10.2	5.9	0.6	< 0.1	< 0.1	2.5
RCG									
26 Jul 2012	10.4	11.4	9.9	2.9	1.2	2.0	2.9	0.4	0.1
24 Aug 2012	3.8	2.4	9.2	0.2	0.1	0.5	< 0.1	< 0.1	0.1
26 Sep 2012	2.0	10.6	3.9	0.6	0.2	0.9	0.1	0.1	0.1
5 Nov 2012	5.6	3.6	2.3	0.8	0.1	1.0	< 0.1	< 0.1	< 0.1
30 Nov 2012	4.1	3.2	3.9	0.6	0.4	0.8	0.5	0.2	0.1
4 Jan 2013	2.3	3.5	5.3	1.5	0.1	0.7	0.1	0.5	0.1
6 May 2013	2.0	1.8	3.5	1.7	0.4	0.2	< 0.1	3.1	< 0.1
11 Jun 2013	3.4	4.5	4.7	0.3	0.8	0.1	< 0.1	0.1	0.3
16 Jul 2013	3.3	1.8	10.6	1.5	1.5	0.1	0.1	< 0.1	0.3

Table 2. Statistical main effects of vegetation (i.e. reed canary grass cultivation or bare soil); ground water level (GWL); and date on fluxes of CO_2, CH_4 and N_2O as explored with linear mixed models. Shown are Df (degrees of freedom) and F and P statistics.

Variables	CO_2			CH_4			N_2O		
	Df	*F* value	*P* value	Df	*F* value	*P* value	Df	*F* value	*P* value
Vegetation	1	956.2	< 0.001	1	165.8	< 0.001	1	0.5	< 0.001
GWL	2	32.2	< 0.001	2	15.4	< 0.001	2	3.1	0.02
Date	28	75.6	< 0.001	28	25.8	< 0.001	28	< 0.1	< 0.001

to 1700 $mg\,m^{-2}\,h^{-1}$ in RCG treatments. Among the air and soil temperature at 5 cm, CO_2 emissions were better correlated with soil temperature in bare soil, but with the air temperature in RCG treatments.

Methane fluxes were significantly affected both by vegetation and GWL (Table 2). CH_4 emissions were highest at 0 cm GWL both from RCG and bare soil treatments (Fig. 5c, d). CH_4 emissions from RCG treatments showed temporal variation ($P < 0.001$) with highest emissions during summertime (Fig. 5c). Peak emissions of CH_4 from RCG treatments were observed in August 2012 across all GWLs, ranging from 4.4 to 8.9 $mg\,CH_4\,m^{-2}\,h^{-1}$. Between November and early April (i.e. winter season), CH_4 emissions from RCG treatments were below 0.1 $mg\,m^{-2}\,h^{-1}$, and even occasional uptake (25 % of total fluxes measured) of CH_4 was recorded. From bare soil treatments, CH_4 fluxes were generally low and fluctuated between apparent net emission and net uptake except for a few episodic peak events, generally from 0 cm GWL. These peak events were considered to represent unsystematic ebullition events.

N_2O fluxes from RCG treatments were generally low, fluctuating in a range between −0.02 and 0.07 $mg\,m^{-2}\,h^{-1}$ except for peak events after fertilizer application (Fig. 5e). Emission peaks of 0.4, 0.7 and 0.4 $mg\,N_2O\,m^{-2}\,h^{-1}$ were observed at 0 cm GWL immediately after the first, second and third fertilization events, respectively. Smaller peak emissions of 0.4 and 0.2 $mg\,N_2O\,m^{-2}\,h^{-1}$ were observed at −10 cm GWL after the first and second fertilization event, but at −20 cm GWL, peak emission after the fertilizer application was absent. N_2O emissions from bare soil treatments generally were higher and ranged from −0.02 to 1.9 $mg\,m^{-2}\,h^{-1}$. Most of the N_2O emission in bare soil mesocosms was measured during the winter period from November 2012 to April 2013, accounting for more than 70 % of the cumulative emission at 0 and −10 cm GWL and more than 50 % at −20 cm GWL.

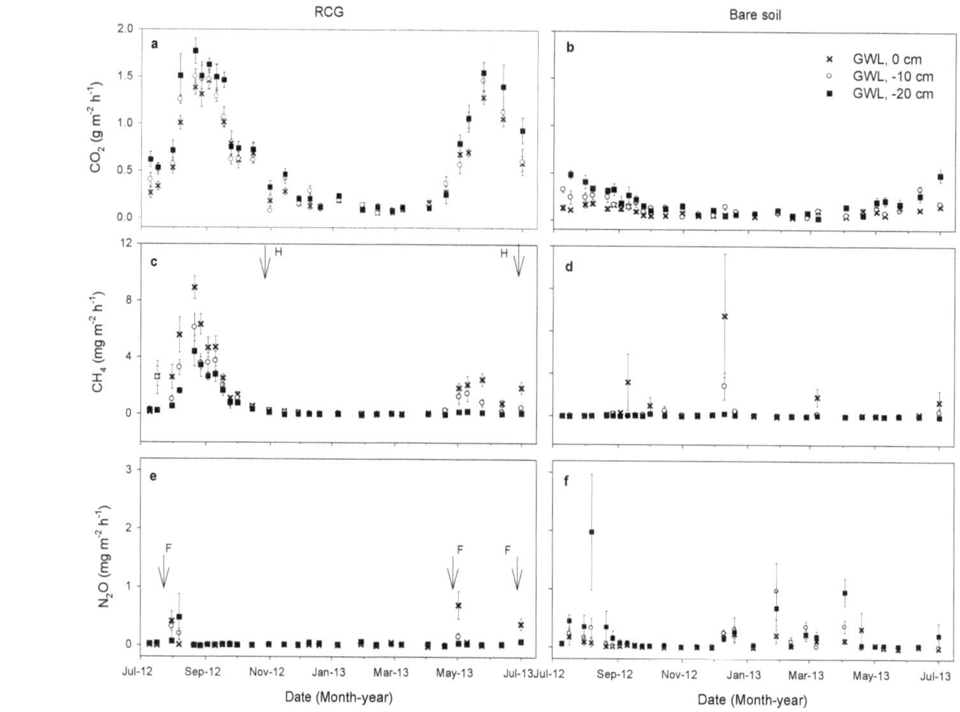

Figure 5. Time series of greenhouse gas fluxes from the rewetted peat soil mesocosms during July 2012 to July 2013 in treatments with RCG cultivation (left panels) and bare soil (right panels). Data are shown for **(a, b)** CO_2 fluxes from ecosystem respiration, **(c, d)** CH_4 fluxes, and **(e, f)** N_2O fluxes. All data are mean and standard error of five replicates from each of the three ground water levels (GWL) at 0, −10 and −20 cm. Arrows marked H indicate the times of harvest, and arrows marked F indicate the times of mineral fertilization.

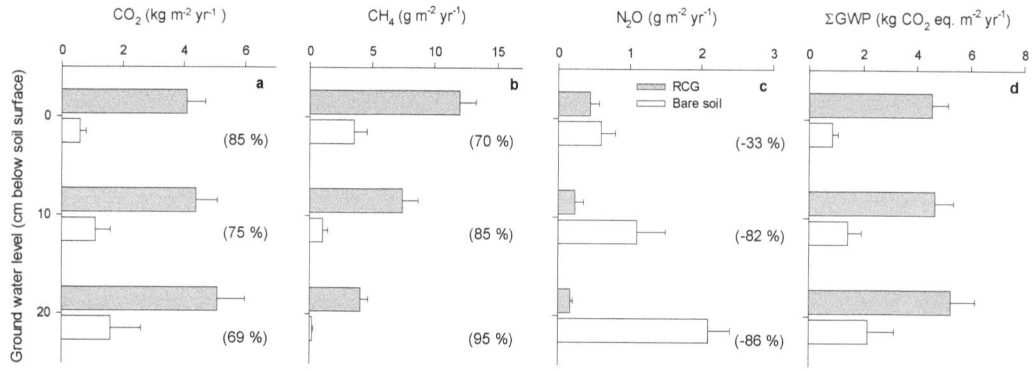

Figure 6. Annual fluxes of **(a)** CO_2 from ecosystem respiration, **(b)** CH_4, **(c)** N_2O and **(d)** total global warming potential ($\sum$GWP) from the rewetted peat soil mesocosms during July 2012 to July 2013 in treatments with RCG cultivation (grey bars) and bare soil (white bars) at ground water levels of 0, −10 and −20 cm. Error bars for CO_2 data show the standard error (SE) derived from SE of model parameters. For CH_4 and N_2O, data are shown as mean and SE of individual mesocosms ($n = 5$). Numbers in parentheses indicate the contribution of RCG in total emission at the different GWLs.

3.5 Annual GHG emissions and contribution of plants to annual GHG emissions

The estimated parameters for CO_2 flux models are presented in Table 3, showing also that the modelling efficiency was considerably higher for the RCG treatments than the bare soil treatments. Annual CO_2 emissions decreased consistently with raising GWL towards the soil surface both in RCG and bare soil treatments (Fig. 6a). In contrast, CH_4 emissions increased systematically both from RCG and bare soil treatments in response to raising GWL (Fig. 6b). The annual N_2O emissions showed a contrasting response to raising GWL in bare soil and RCG treatments; in bare soil treatments, lower N_2O emissions occurred in response to raised GWL, but in RCG treatments there was a tendency of higher N_2O emissions in response to raised GWL (Fig. 6c).

Table 3. Parameter estimates (b_1, b_2, b_3 and b_4) for CO_2 flux models. Uncertainties shown in parentheses are standard error of parameter estimates. Also shown are correlation coefficients (r) between observed and modelled data and modelling efficiencies (ME).

Treatment	CO_2 flux model	b_1 ($mg CO_2 m^{-2} h^{-1}$)	b_2 ($mg CO_2 m^{-2} h^{-1} cm^{-1}$)	b_3 (K)	b_4	r	ME
Reed canary grass	Model 1	49.6 (3.8)	0.4 (0.1)	259.1 (15.5)	5.0 (0.7)	0.90	0.82
Bare soil	Model 2	79.1 (6.9)	5.7 (0.5)	286.4 (24.1)	n/a	0.68	0.46

n/a: not applicable.

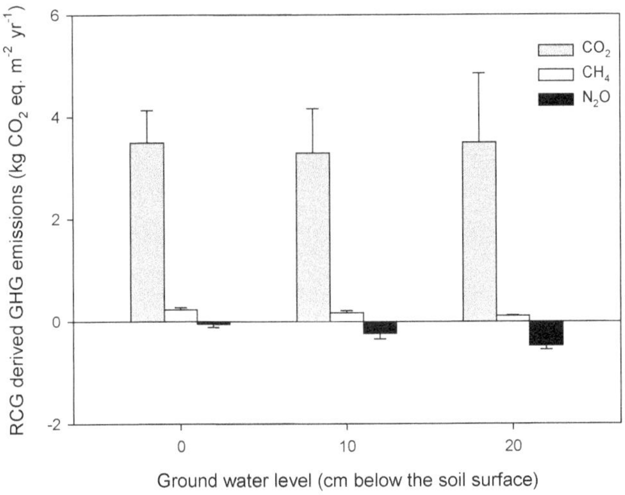

Figure 7. Plant-derived CO_2, CH_4 and N_2O emissions at different ground water levels as compared in terms of CO_2 equivalents (CO_2 eq.). Plant-derived emissions were estimated as the difference between the total emissions from RCG treatments and bare soil treatments.

The presence of plants contributed 69–85 % of the total CO_2 emissions from the RCG mesocosms (Fig. 6a). The highest contribution was observed at 0 cm GWL, and the contribution decreased at lower GWL. RCG likewise accounted for more than 70 % of total CH_4 emissions with the highest contribution of 95 % observed at −20 cm GWL. Thus, at this GWL (−20 cm), CH_4 emission was negligible from bare soil treatments (0.2 g $CH_4 m^{-2} yr^{-1}$), whereas the emissions were substantial from RCG treatments (4.1 g $CH_4 m^{-2} yr^{-1}$). In contrast to CO_2 and CH_4 emissions, cultivation of RCG reduced the annual N_2O emissions despite the application of mineral N fertilizer in RCG mesocosms (Fig. 6c). At −10 and −20 cm GWL, RCG eliminated 82–86 % of the N_2O emissions as compared to bare soil treatments; from 0 cm GWL the reduction corresponded to 33 % of the N_2O emissions. In terms of GWP, the increase in CH_4 emissions due to RCG cultivation was more than offset by the decrease in N_2O emissions at −10 and −20 cm GWL, but apparently not at 0 cm GWL, where CH_4 emissions were offset by only 23 % by the decreased N_2O emission (Fig. 7). CO_2 emissions from ER, though, were the dominant RCG-derived GHG fluxes (Fig. 7).

4 Discussion

During the present study, the effects of RCG cultivation on GHG emission from rewetted peatland were evaluated by comparison of planted and unplanted (bare soil) mesocosms. Mesocosms with RCG and bare soil (rather than, for example, mesocosms with native grasses) were compared in order to tentatively isolate the contribution of RCG in the measured GHG fluxes. One concern with using this plant exclusion method for GHG studies is the difference in soil moisture regime and temperature that may develop between planted and bare soil treatments, which may result in different decomposition rates of soil organic matter (Kuzyakov, 2006). With our experimental setup, we were able to control the GWL throughout the measurement period, and this resulted in soil moisture contents (VWC) that were similar between RCG and bare soil treatments at each GWL; this was generally also seen for the soil redox potential and pore water sulfate concentration at least at 0 and −10 cm GWL. The average soil temperature difference between the RCG and bare soil treatments was found to be less than 1 °C; however during the annual study we observed some seasonal difference in soil temperature, especially during spring days (higher temperature in RCG treatments) and summer days (lower temperature in RCG treatments), which was attributed to the RCG cultivation. However, the differences in moisture and temperature regime between the planted and bare soil mesocosms were considered to be modest and acceptable for an evaluation of the effects of RCG on total GHG emissions.

Monitoring of environmental variables was achieved by instrumentation of one out of five replicate mesocosms at each GWL. We assumed that the measured variables were representative of all replicates and that the instrumentation did not lead to any bias. This was substantiated by the absence of any systematic deviations in measured GHG fluxes from the instrumented and non-instrumented replicates. Thus, the average difference in annual fluxes with and without instrumentation was less than 15 %.

4.1 CO_2 emissions

Plants can enhance CO_2 flux from ER directly by above- and below-ground respiration and indirectly by enhancing the decomposition of soil organic matter by the supply of easily degradable root exudates to the soil (priming effect) (Kuzyakov et al., 2001; Van Huissteden et al., 2006). In vegetated soils, ER is essentially balanced by photosynthetic CO_2 uptake, and therefore CO_2 emissions from ER do not represent the net ecosystem exchange (NEE) of CO_2. Rather than quantifying NEE, an important result of the present study was that plant-derived ER from RCG mesocosms (the major part of total CO_2 emissions) was similar at all three GWLs (Fig. 7), substantiating the results of Lafleur et al. (2005) and Riutta et al. (2007), who reported autotrophic respiration to be independent of water table depth. Thus, the observed increase (from 69 to 85 %) in total ER with rising GWL was promoted mainly by decreasing soil respiration at the higher GWL (Fig. 6). The observed contribution of RCG to total CO_2 emissions was higher than the values of 55 % previously reported by Shurpali et al. (2008). However, the results of Shurpali et al. (2008) were obtained for a drained peatland with an average GWL of −65 cm, which would favour aerobic soil respiration to a larger extent than in our soils, which had a GWL no deeper than −20 cm. In accordance with this we also observed a larger soil respiration at −20 cm than at 0 cm GWL.

4.2 Methane emissions

Methane fluxes from soil is the result of CH_4 production, consumption and transport (Lai, 2009). Plants play a key role in CH_4 fluxes as they have the potential to influence all three processes (Joabsson et al., 1999). CH_4 emissions were higher from RCG than bare soil treatments even though the GWL was raised to the soil surface. Plant roots release organic compounds to soil, which are easily available carbon sources to anaerobic microbial consortia eventually producing the precursors (acetate or H_2 / CO_2) for methanogenesis (Ström et al., 2003). Such fresh organic carbon is suggested to be an important substrate for methanogenesis as peat carbon is shown to be more recalcitrant to anaerobic decomposition (Tuittila et al., 2000; Hahn-Schöfl et al., 2011).

Methane produced in soil can be emitted to the atmosphere by diffusion, ebullition (release of gas bubbles) and plant-mediated transport (Whalen, 2005; Lai, 2009). Indeed, RCG can transport CH_4 from soil to the atmosphere directly through its aerenchyma tissue, thereby bypassing the microbial methane oxidation layer in the soil. On an annual basis it has been estimated that RCG may actually transport 70 % of the total CH_4 emissions from a natural wetland in Denmark (Askaer et al., 2011). In the absence of plant-mediated transport, diffusion would expectedly be the dominant pathway of CH_4 emissions in bare soil treatments. CH_4 transport through diffusion is a slow but important process for bringing CH_4 in contact with the CH_4 oxidizing microbial community (Whalen 2005; Lai 2009). In our study there were negligible CH_4 emissions from bare soil at −20 cm GWL, aligning with the results of Schäfer et al. (2012), who reported this drainage depth to be sufficient to suppress diffusive CH_4 emissions due to methane oxidation and reduced methanogenesis.

The transport of oxygen by aerenchyma plants to anoxic soil compartments has been reported to increase the redox potential, which could suppress CH_4 emission (Sutton-Grier and Megonigal, 2011). However, in our study neither the redox potential nor the sulfate content was consistently increased by the presence of plants, suggesting the role of substrate availability and transport of CH_4 through RCG to be the more important factors for controlling CH_4 emissions from the RCG treatments.

It is possible that we could have underestimated the total CH_4 emission from bare soil treatments at 0 cm GWL as episodic CH_4 release through ebullition was not taken into account in the annual balance. Ebullition events were identified by occasional erratic time courses of CH_4 concentrations during the flux measurements; however as these events were generally associated with the initial (time 0 and 15 min) chamber gas samplings it was believed to represent artifacts created during chamber deployment. However, episodic release of CH_4 may be more important in bare soil than in vegetated soil, as plants may reduce the soil concentration of CH_4 by mediating CH_4 transport and also by rhizospheric oxidation of CH_4; these processes reduce the potential formation of CH_4 bubbles (Chanton, 2005). With the observed episodic CH_4 release tentatively accounted for, a total of 0.04 g m^{-2} of CH_4 was released during the study; this was a negligible contribution (< 1 %) to the annual CH_4 flux from bare soil at 0 cm GWL. However, as ebullition events are short-lived and unsystematic, they could easily be missed with the chamber measurements (Coulthard et al., 2009).

4.3 N_2O emissions

Annual fluxes of N_2O (0.2 to 0.4 g $N_2O\,m^{-2}\,yr^{-1}$) from RCG mesocosms were within the range (−0.4 to 0.8 g $N_2O\,m^{-2}\,yr^{-1}$) reported for undisturbed Danish riparian wetland (Audet et al., 2014). However, annual fluxes were higher in bare soil (0.6 to 2.1 g $N_2O\,m^{-2}\,yr^{-1}$) as compared to RCG treatments. Thus, RCG decreased the annual N_2O emissions, contradictory to the finding of Hyvönen et al. (2009), where fertilization in RCG increased the N_2O emissions by 90 % as compared to bare soil. However, in the study by Hyvönen et al. (2009), N_2O emissions were quite low (0.01 g $N_2O\,m^{-2}\,yr^{-1}$) from the soil without vegetation. Their site was an abandoned peatland (Hyvönen et al., 2009) probably with limited nitrification because of a high C / N ratio (42.3) (Klemedtsson et al., 2005) compared to our peat soil with rich N content (3.2 %) and a low C / N ratio (11.6). Thus, the ecosystem studied by Hyvönen et al. (2009) might have been more N-limited at the unfertilized sites than was the case for our study site.

The effect of RCG cultivation on N_2O emissions was highly dependent upon the GWL. The least effect of RCG cultivation on N_2O emissions was observed at 0 cm GWL, due to peak emissions observed after fertilization. Peak emission observed after fertilization events suggests that N_2O

emission was limited by mineral N content at 0 cm GWL. Saari et al. (2013) and Silvan et al. (2002) also reported a significant increase in N_2O emission after addition of inorganic nitrogen in riparian wetland due to favourable conditions for denitrification.

Previous studies have reported that winter emissions significantly contributed to annual N_2O emissions (Maljanen et al., 2004; Regina et al., 2004). Such emissions in winter have been related to the physical release of N_2O that is produced and trapped under frozen surface layers as well as the emissions of newly produced N_2O (de novo emissions) at the onset of thaw stimulated by increased biological activity and changes in physical and chemical soil conditions (Risk et al., 2013). Significant emissions at all GWLs were observed in winter from bare soil treatments, but not from RCG treatments. After harvesting, there was regrowth of RCG and also other volunteer grasses which survived throughout the winter and which may have competed with microorganisms for available N. Maljanen et al. (2004) also observed higher N_2O emissions from bare soil as compared to vegetated plots during winter and likewise related the low emission in vegetated plots to low mineral N content due to uptake of nitrate by plants. Bare soil treatments indeed had a higher availability of mineral N (Table 1), and could be more prone to physical damage by freeze and thaw cycles due to lack of plant cover; both these factors stimulate the biological activities related to N_2O emissions as also substantiated by the observed slight increase in CO_2 emissions coinciding with increased N_2O emission, especially at 0 and −10 cm GWL.

4.4 Effect of RCG cultivation on GHG balance from rewetted peatland

Two of the major concerns of growing wetland plants like RCG in rewetted peatland are the possible increase in CH_4 emissions due to supply of fresh plant material and transport of CH_4 by aerenchyma tissue (Ström et al., 2003; Askaer et al., 2011) and the possible increase in N_2O due to application of N fertilizers (Maljanen et al., 2010). However, in the present experiment, cultivation of RCG decreased N_2O emission to an extent that could offset the increase in CH_4 emission at −10 and −20 cm GWL, but apparently not at 0 cm GWL – the latter case being due to peak emissions in N_2O after fertilization events in RCG. This result suggests that emissions at 0 cm GWL can be reduced by reducing the N fertilization rate. Further studies are needed to assess the optimum amount and timing of fertilization required for optimum growth of RCG with acceptable N_2O emissions. Emissions of N_2O caused by N fertilization should not offset the benefit of fossil fuel substitution obtained by the fertilizer-induced increase of biomass production (Kandel et al., 2013a). Regarding the overall GHG emission, the CO_2 emissions from ER was clearly the dominant RCG-derived GHG flux. However, CO_2 flux from ER would to a large extent be counterbalanced by gross photosynthesis, which, as expected, was similar at all GWL treatments (based on the similar biomass yields), though CO_2 flux from photosynthesis was not measured in this annual study. However, a photosynthetic uptake of 6.2 kg CO_2 m^{-2} was measured from RCG mesocosms at 0 cm GWL during the growing season from May to September 2013 (S. Karki, unpublished data), reflecting that RCG potentially can turn the rewetted ecosystem into a sink of CO_2 from an atmospheric perspective. Adaptation or selection of RCG varieties that thrive especially well under distinct climate and shallow GWL conditions could further help to improve the GHG balance of paludiculture with RCG.

5 Conclusions

The present study is, to our knowledge, the first to compare the annual GHG emission from RCG and bare soil treatments of rewetted peatland at controlled GWL. The following conclusions were derived: (i) soil respiration decreased with increasing GWL from −20 to −10 to 0 cm, but RCG-derived ER was similar at all three GWLs, resulting in the highest contribution of RCG to total ER (85 %) at 0 cm GWL; (ii) cultivation of RCG increased CH_4 emission at all GWLs, but relatively most at −20 cm GWL; (iii) N_2O emissions decreased due to RCG cultivation, especially during winter – winter emissions were a more important component of annual emission from bare soil than from RCG treatments; (iv) in terms of GWP, the increase in CH_4 emissions due to RCG cultivation was more than offset by the decrease in N_2O emissions at −10 and −20 cm GWL; and (v) CO_2 emissions from ER (the dominant RCG-derived GHG flux) could be balanced by photosynthetic CO_2 uptake at all three GWLs, as indicated by the large and similar above-ground biomass yields at all GWLs, signifying a potential of RCG cultivation to turn the rewetted peatland into a sink of atmospheric CO_2.

Acknowledgements. The study was partly supported by the Danish Centre for Food and Agriculture (DCA) as a part of the BioBase programme. The authors like to thank Sanmohan Baby for statistical advice. We are also grateful for the technical support we received from Bodil Stensgaard, Finn Henning Christensen, Holger Bak, Jørgen M. Nielsen and Stig T. Rasmussen.

Edited by: P. Stoy

References

Askaer, L., Elberling, B., Friborg, T., Jørgensen, C. J., and Hansen, B. U.: Plant-mediated CH_4 transport and C gas dynamics quantified in-situ in a *Phalaris arundinacea*-dominant wetland, Plant Soil, 343, 287–301, 2011.

Audet, J., Hoffmann, C. C., Andersen, P. M., Baattrup-Pedersen, A., Johansen, J. R., Larsen, S. E., Kjaergaard, C., and Elsgaard,

L.: Nitrous oxide fluxes in undisturbed riparian wetlands located in agricultural catchments: Emission, uptake and controlling factors, Soil Biol. Biochem., 68, 291–299, 2014.

Bastviken, S. K., Eriksson, P. G., Premrov, A., and Tonderski, K.: Potential denitrification in wetland sediments with different plant species detritus, Ecol. Eng., 25, 183–190, 2005.

Chanton, J. P.: The effect of gas transport on the isotope signature of methane in wetlands, Org. Geochem., 36, 753–768, 2005.

Coulthard, T. J., Baird, A. J., Ramirez, J., and Waddington, J. M.: Methane dynamics in peat: Importance of shallow peats and a novel reduced-complexity approach for modeling ebullition, in: Carbon Cycling in Northern Peatlands, edited by: Baird, A. J., Belyea, L. R., Comas, X., Reeve, A. S., and Slater, L. D., American Geophysical Union, Washington, DC, 173–185, 2009.

Davidson, E. A., Savage, K., Verchot, L. V., and Navarro, R.: Minimizing artifacts and biases in chamber-based measurements of soil respiration, Agric. For. Meteorol., 113, 21–37, 2002.

Dinsmore, K. J., Skiba, U. M., Billett, M. F., and Rees, R. M.: Effect of water table on greenhouse gas emissions from peatland mesocosms, Plant Soil, 318, 229–242, 2009.

Elsgaard, L., Görres, C.-M., Hoffmann, C. C., Blicher-Mathiesen, G., Schelde, K., and Petersen, S. O.: Net ecosystem exchange of CO_2 and carbon balance for eight temperate organic soils under agricultural management, Agr. Ecosyst. Environ., 162, 52–67, 2012.

Görres, C. M., Kutzbach, L., and Elsgaard, L.: Comparative modeling of annual CO_2 flux of temperate peat soils under permanent grassland management, Agr. Ecosyst. Environ., 186, 64–76, 2014.

Günther, A., Huth, V., Jurasinski, G., and Glatzel, S.: The effect of biomass harvesting on greenhouse gas emissions from a rewetted temperate fen, GCB Bioenergy, doi:10.1111/gcbb.12214, 2014.

Haefner, J. W.: Modelling Biological Systems: Principles and Applications, 2nd Edn., Springer, New York, NY, 2005.

Hahn-Schöfl, M., Zak, D., Minke, M., Gelbrecht, J., Augustin, J., and Freibauer, A.: Organic sediment formed during inundation of a degraded fen grassland emits large fluxes of CH_4 and CO_2, Biogeosciences, 8, 1539–1550, doi:10.5194/bg-8-1539-2011, 2011.

Höper, H., Augustin, J., Cagampan, J. P., Drösler, M., Lundin, L., Moors, E., Vasander, H., Waddington, J. M., and Wilson, D.: Restoration of peatlands and greenhouse gas balances, in: Peatlands and Climate Change, edited by: Starck, M., International Peat Society, Saarijärvi, Finland, 182–210, 2008.

Hyvönen, N. P., Huttunen, J. T., Shurpali, N. J., Tavi, N. M., Repo, M. E., and Martikainen, P. J.: Fluxes of nitrous oxide and methane on an abandoned peat extraction site: effect of reed canary grass cultivation, Bioresoure Technol., 100, 4723–4730, 2009.

IPCC: 2013 Supplement to the 2006 IPCC Guidelines for National Greenhouse Gas Inventories: Wetlands in, edited by: Hiraishi, T., Krug, T., Tanabe, K., Srivastava, N., Baasansuren, J., Fukuda, M., and Troxler, T. G., IPCC, Switzerland, 2014.

Joabsson, A., Christensen, T. R., and Wallen, B.: Vascular plant controls on methane emissions from northern peatforming wetlands, Trends Ecol. Evol., 14, 385–388, 1999.

Joosten, H.: The Global Peatland CO_2 Picture: peatland status and drainage related emissions in all countries of the world, Wetland International, Ede., 2009.

Joosten, H., Tapio-Biström, M.-L., and Tol, S.: Peatlands-guidance for climate change mitigation through conservation, rehabilitation and sustainable use, Food and Argiculture Organisation of the United Nations and Wetlands International, Rome, Italy, 2012.

Jørgensen, C. J., Struwe, S., and Elberling, B.: Temporal trends in N_2O flux dynamics in a Danish wetland – effects of plant-mediated gas transport of N_2O and O_2 following changes in water level and soil mineral-N availability, Glob. Change Biol., 18, 210–222, 2012.

Kandel, T. P., Elsgaard, L., Karki, S., and Lærke, P. E.: Biomass yield and greenhouse gas emissions from a drained fen peatland cultivated with reed canary grass under different harvest and fertilizer regimes, Bioenergy Res., 6, 883–895, 2013a.

Kandel, T. P., Elsgaard, L., and Lærke, P. E.: Measurement and modelling of CO_2 flux from a drained fen peatland cultivated with reed canary grass and spring barley, GCB Bioenergy, 5, 548–561, 2013b.

Kao-Kniffin, J., Freyre, D. S., and Balser, T. C.: Methane dynamics across wetland plant species, Aquat. Bot., 93, 107–113, 2010.

Karki, S., Elsgaard, L., Audet, J., and Lærke, P. E.: Mitigation of greenhouse gas emissions from reed canary grass in paludiculture: effect of groundwater level, Plant Soil, 383, 217–230, 2014.

Kercher, S. M. and Zedler, J. B.: Flood tolerance in wetland angiosperms: a comparison of invasive and noninvasive species, Aquat. Bot., 80, 89–102, 2004.

Kjaergaard, C., Heiberg, L., Jensen, H. S., and Hansen, H. C. B.: Phosphorus mobilization in rewetted peat and sand at variable flow rate and redox regimes, Geoderma, 173, 311-321, 2012.

Klemedtsson, L., Von Arnold, K., Weslien, P., and Gundersen, P.: Soil CN ratio as a scalar parameter to predict nitrous oxide emissions, Glob. Change Biol., 11, 1142–1147, 2005.

Kuzyakov, Y.: Sources of CO_2 efflux from soil and review of partitioning methods, Soil Biol. Biochem., 38, 425–448, 2006.

Kuzyakov, Y., Ehrensberger, H., and Stahr, K.: Carbon partitioning and below-ground translocation by *Lolium perenne*, Soil Biol. Biochem., 33, 61–74, 2001.

Laanbroek, H. J.: Methane emission from natural wetlands: interplay between emergent macrophytes and soil microbial processes. A mini-review, Ann. Bot., 105, 141–153, 2010.

Lafleur, P. M., Moore, T. R., Roulet, N. T., and Frolking, S.: Ecosystem respiration in a cool temperate bog depends on peat temperature but not water table, Ecosystems, 8, 619–629, 2005.

Lai, D. Y. F.: Methane dynamics in northern peatlands: A review, Pedosphere, 19, 409–421, 2009.

Lloyd, J. and Taylor, J. A.: On the temperature dependence of soil respiration, Funct. Ecol., 8, 315–323, 1994.

Maljanen, M., Komulainen, V.-M., Hytönen, J., Martikainen, P. J., and Laine, J.: Carbon dioxide, nitrous oxide and methane dynamics in boreal organic agricultural soils with different soil characteristics, Soil Biol. Biochem., 36, 1801–1808, 2004.

Maljanen, M., Sigurdsson, B. D., Guðmundsson, J., Óskarsson, H., Huttunen, J. T., and Martikainen, P. J.: Greenhouse gas balances of managed peatlands in the Nordic countries – present knowledge and gaps, Biogeosciences, 7, 2711–2738, doi:10.5194/bg-7-2711-2010, 2010.

Maurer, D. A., Lindig-Cisneros, R., Werner, K. J., Kercher, S., Miller, R., and Zedler, J. B.: The Replacement of Wetland

Vegetation by Reed Canarygrass (*Phalaris arundinacea*), Ecol. Restor., 21, 116–119, 2003.

Maxwell, S. E. and Delaney, H. D.: Designing experiments and analyzing data: A model comparison perspective, Psychology Press, 2004.

Myhre, G., Shindell, D., Bréon, F.-M., Collins, W., Fuglestvedt, J., Huang, J., Koch, D., Lamarque, J.-F., Lee, D., Mendoza, B., Nakajima, T., Robock, A., Stephens, G., Takemura, T., and Zhang, H.: Anthropogenic and natural radiative forcing, in: Climate change 2013: The physical science basis. Contribution of working group I to the fifth assessment report of the Intergovernmental Panel on Climate Change, edited by: Stocker, T. F., Qin, D., Plattne, G.-K., Tignor, M., Allen, S. K., Boschung, J., Nauels, A., Xia, Y., Bex, V., and Midgley, P. M., Cambridge University Press, Cambridge, United Kingdom and New York, NY, USA, 658–740, 2013.

Parish, F., Sirin, A., Charman, D., Joosten, H., Minayeva, T., Silvius, M., and Stringer, L.: Assessment on peatlands, biodiversity and climate change: main report, Global Environment Centre, Kuala Lumpur and Wetlands International, Wageningen, 2008.

Pedersen, A. R., Petersen, S. O., and Schelde, K.: A comprehensive approach to soil-atmosphere trace-gas flux estimation with static chambers, Eur. J. Soil Sci., 61, 888–902, 2010.

Petersen, S. O., Hoffmann, C. C., Schäfer, C.-M., Blicher-Mathiesen, G., Elsgaard, L., Kristensen, K., Larsen, S. E., Torp, S. B., and Greve, M. H.: Annual emissions of CH_4 and N_2O, and ecosystem respiration, from eight organic soils in Western Denmark managed by agriculture, Biogeosciences, 9, 403–422, doi:10.5194/bg-9-403-2012, 2012.

R Core Team: R: A language and environment for statistical computing. R foundation for statistical computing, Vienna, Austria, 2013.

Regina, K., Syväsalo, E., Hannukkala, A., and Esala, M.: Fluxes of N_2O from farmed peat soils in Finland, Eur. J. Soil Sci., 55, 591–599, 2004.

Risk, N., Snider, D., and Wagner-Riddle, C.: Mechanisms leading to enhanced soil nitrous oxide fluxes induced by freeze-thaw cycles, Can. J. Soil Sci., 93, 401–414, 2013.

Riutta, T., Laine, J., and Tuittila, E.-S.: Sensitivity of CO_2 exchange of fen ecosystem components to water level variation, Ecosystems, 10, 718–733, 2007.

Roobroeck, D., Butterbach-Bahl, K., Brüggemann, N., and Boeckx, P.: Dinitrogen and nitrous oxide exchanges from an undrained monolith fen: short-term responses following nitrate addition, Eur. J. Soil Sci., 61, 662–670, 2010.

Saari, P., Saarnio, S., Heinonen, J., and Alm, J.: Emissions and dynamics of N_2O in a buffer wetland receiving water flows from a forested peatland, Boreal Environ. Res., 18, 164–180, 2013.

Schäfer, C. M., Elsgaard, L., Hoffmann, C. C., and Petersen, S. O.: Seasonal methane dynamics in three temperate grasslands on peat, Plant Soil, 357, 339–353, 2012.

Shurpali, N. J., Hyvönen, N. P., Huttunen, J. T., Biasi, C., Nykänen, H., Pekkarinen, N., and Martikainen, P. J.: Bare soil and reed canary grass ecosystem respiration in peat extraction sites in Eastern Finland, Tellus B, 60, 200–209, 2008.

Silvan, N., Regina, K., Kitunen, V., Vasander, H., and Laine, J.: Gaseous nitrogen loss from a restored peatland buffer zone, Soil Biol. Biochem., 34, 721–728, 2002.

Ström, L., Ekberg, A., Mastepanov, M., and Røjle Christensen, T.: The effect of vascular plants on carbon turnover and methane emissions from a tundra wetland, Glob. Change Biol., 9, 1185–1192, 2003.

Sutton-Grier, A. E. and Megonigal, J. P.: Plant species traits regulate methane production in freshwater wetland soils, Soil Biol. Biochem., 43, 413–420, 2011.

Tuittila, E.-S., Komulainen, V.-M., Vasander, H., Nykänen, H., Martikainen, P. J., and Laine, J.: Methane dynamics of a restored cutaway peatland, Glob. Change Biol., 6, 569–581, 2000.

Van Huissteden, J., van den Bos, R., and Alvarez, I. M.: Modelling the effect of water-table management on CO_2 and CH_4 fluxes from peat soils, Neth. J. Geosci, 85, 3–18, 2006.

Whalen, S. C.: Biogeochemistry of methane exchange between natural wetlands and the atmosphere, Environ. Eng. Sci, 22, 73–94, 2005.

Wichtmann, W. and Tanneberger, F.: Landuse options for rewetted peatlands, in: Carbon credits from peatland rewetting, Climate, biodiversity, land use, edited by: Tanneberger, F., and Wichtmann, W., Stuttgart, Schweizerbart Science Publishers, 107–132, 2011.

6

Contrasting responses of terrestrial ecosystem production to hot temperature extreme regimes between grassland and forest

Y. Zhang[1], M. Voigt[1], and H. Liu[2]

[1]Institute for Space Sciences, Freie Universität Berlin, 12165 Berlin, Germany
[2]Changjiang River Scientific Research institute, Changjiang Water Resources Commission, 430010 Wuhan, China

Correspondence to: Y. Zhang (yongguang.zhang@wew.fu-berlin.de)

Abstract. During the past several decades, observational data have shown a faster increase in hot temperature extremes than the change in mean temperature. Increasingly high extreme temperatures are expected to affect terrestrial ecosystem function. The ecological impact of hot extremes on vegetation production, however, remains uncertain across biomes in natural climatic conditions. In this study, we investigated the effects of hot temperature extremes on vegetation production by combining the MODIS enhanced vegetation index (EVI) data set and in situ climatic records during the period 2000 to 2009 from 12 long-term experimental sites across biomes and climate. Our results show that higher mean annual maximum temperatures (T_{max}) greatly reduced grassland production, and yet enhanced forest production after removing the effect of precipitation. The relative decrease in vegetation production was 16 % for arid grassland and 7 % for mesic grassland, and the increase was 5 % for forest. We also observed a significantly positive relationship between interannual aboveground net primary production (ANPP) and T_{max} for the forest biome ($R^2 = 0.79$, $P < 0.001$). This line of evidence suggests that hot temperature extremes lead to contrasting ecosystem-level responses of vegetation production between grassland and forest biomes. Given that many terrestrial ecosystem models use average daily temperature as input, predictions of ecosystem production should consider such contrasting responses to increasingly hot temperature extreme regimes associated with climate change.

1 Introduction

The observed global temperature showed a warming of 0.85°C (0.65 to 1.06°C) over the period of 1880 to 2012, and the number of warm days and nights has increased at the global scale (IPCC, 2013). Future temperature is expected to continue to warm more rapidly over land than ocean, and there will be more frequent hot and fewer cold temperature extremes over most land area (IPCC, 2013). Terrestrial ecosystems are strongly impacted by climate and climate change (Nemani et al., 2003), and hence this trend towards hot temperature extremes would have important consequences on terrestrial ecosystems (IPCC, 2012). Numerous modeling and observational climate warming studies have shown the general enhancement of vegetation growth or increases in vegetation greenness in northern terrestrial ecosystems (e.g., Keeling, et al., 1996; Myneni, et al., 1997; Zhou et al., 2001; Neigh et al, 2008; Wu et al., 2011). However, knowing the general response of ecosystems tells us little about how the ecosystem in a particular location will respond or how different ecosystem responds to hot temperature extremes. For example, Peng et al. (2013) recently reported that the growing-season greenness was positively correlated with the maximum daily temperature (T_{max}) in northwestern North America and Siberia but negatively correlated in drier temperate regions such as western China, central Eurasia, and central and southwestern North America.

Usually, field manipulated experiments have been conducted to investigate the effects of climatic warming on ecosystem (Alward et al., 1999; Shaver et al., 2000; Wu et al., 2011). Such experimental manipulations are important to understand and quantify the individual contribution of climatic

Table 1. Descriptions of the sites in this study[a].

Site and location	Latitude (degree)	Longitude (degree)	Land cover	MAP (mm)[b]	Max. temperature (°C)	Code
Desert Exp. Range, UT	38.547	−113.712	Arid grassland	216 (65)	19 (1.1)	DE
Jornada Exp. Range, NM	32.589	−106.844	Arid grassland	242 (78)	25 (0.7)	JE
Walnut Gulch Exp. Watershed, AZ	31.736	−109.938	Arid grassland	311 (85)	25 (1.0)	WG
Santa Rita Exp. Range, AZ	31.846	−110.839	Arid grassland	447 (129)	29 (0.7)	SR
Central Plains Exp. Range, CO	40.819	−104.748	Arid grassland	381 (91)	16 (1.4)	CP
Southern Plains Exp. Range, OK	36.614	−99.576	Mesic grassland	586 (153)	22 (0.9)	SP
Little Washita Creek, OK	34.918	−97.956	Mesic grassland	796 (195)	24 (1.2)	LW
Little River Watershed, GA	31.537	−83.626	Temperate conifer forest	1148 (257)	25 (0.6)	LR
Mahatango Creek, PA	40.731	−76.592	Temperate broadleaf forest	1058 (179)	16 (0.9)	MC
Cutfoot Experimental Forest, MN	47.4264	−94.0141	Temperate broadleaf forest	665 (101)	11 (1.1)	CF
Bent Creek Exp. Forest, NC	35.500	−82.624	Temperate mixed forest	1227 (239)	19 (0.6)	BC
Caspar Creek, CA	39.337	−123.748	Mediterranean forest	1054 (301)	16 (0.7)	CC

[a] Precipitation and temperature for the 40-year period 1970–2009 were available for all sites except JE, for which data were available for a 32-year period 1978–2009.
[b] Average annual sum of precipitation (MAP) and average annual mean max temperature with standard deviation in parentheses.

warming on vegetation growth by controlling other global change factors. However, these studies have usually been conducted either on an individual ecosystem or over short-term periods, which render the comparisons across biomes that may differ between regions and ecosystem difficult. A main problem with these experiments is that they do not incorporate the entire micro- and macro-environmental aspects of variable weather. In addition, the long-term responses of ecosystem function are difficult to capture in warming experiments most of which were short term (< 5 years) (Wu et al., 2011). In other words, manipulated experiments are spatially and temporally restricted. The results from these manipulated studies are needed to understand in the context of long-term experiments in the natural field settings. An alternative to manipulated experiments is to analyze these effects on ecosystem processes in natural field settings with long-term measurements across biomes (Huxman et al., 2004).

Although individual grassland or forest sites have been conducted for their response to climatic warming or extremes (Wu et al., 2011; Smith, 2011), there is still no consensus on how ecosystem production will respond to temperature extremes across biomes from arid grassland to forest. Furthermore,many comparison studies used ecosystem production from a compilation of in situ measurements from long-term experimental sites (Huxman et al., 2004). However, the measuring procedures of vegetation production are not consistent across sites, and in some cases, not consistent over time at a given site (Sala et al., 1988), and this will result in some uncertainties. As a result, we are lacking generalizations about the regional behavior of terrestrial ecosystem with more hot temperature extreme regimes.

The last decade has witnessed dramatic global warming: 9 of the 10 warmest years on record have occurred during the 21st century (NOAA, 2013). These conditions are similar to those expected due to climate change (IPCC, 2013). In particular, the United States has warmed faster than the global rate since the late 1970s, and heat waves in 2005, 2006, and 2007 broke all-time records for high maximum and minimum temperatures (NOAA, 2013). Therefore, this recent climatic condition provides an opportunity to study the functional response of biomes to hot temperature extremes with respect to future climate change.

In this study, we used a 10-year data set of MODerate resolution Imaging Spectroradiometer (MODIS) enhanced vegetation index (EVI) (Huete et al., 2002) as an indicator of aboveground net primary production (ANPP), in combination with field observations from 12 long-term experimental sites in the conterminous United States. The primary objective of this study was to examine the response of vegetation production to hot temperature extremes, with particular focus on quantifying the direction and magnitude of ANPP long-term response across biomes from semi-arid grassland to temperate forest. We first assessed the changes in annual ANPP due to higher temperature extremes across biomes. Then, we examined the link between temperature extremes and annual ANPP after controlling the effects of precipitation and low temperature.

2 Materials and methods

2.1 Study sites and meteorological data

Twelve USDA experimental sites across the conterminous United States were used. These sites included different precipitation regimes and biomes representative of ecosystems ranging from arid grassland to temperate forest. They represent a broad range of production, climatic and soil conditions, and life history characteristics of the dominant species. At each site, a location was selected in an undisturbed vegetated area of at least 2.25 km × 2.25 km in size (Table 1).

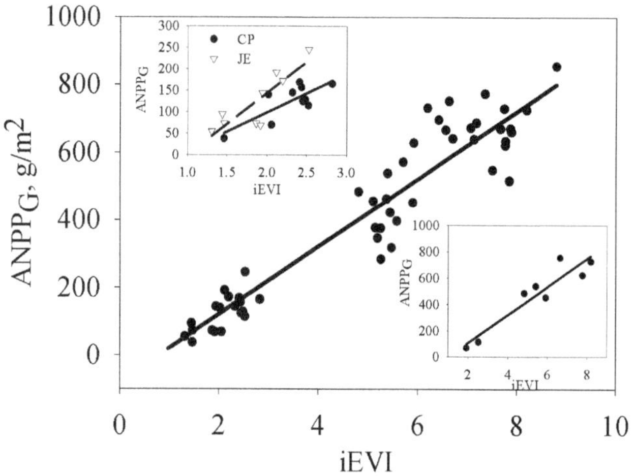

Figure 1. Relationship between annual ground measurements of ANPP ($ANPP_G$) and the corresponding iEVI derived from MODIS data during the period between 2000 and 2009 for nine selected sites across biomes ($R^2 = 0.90$, $P < 0.0001$; Table 2). The upper inset shows the relationship at two sites of CP and JE ($R^2 = 0.63$ and 0.74, respectively, $P < 0.01$). The lower inset shows the relationship between $ANPP_G$ and iEVI for the year 2001 across the 9 sites ($R^2 = 0.88$, $P < 0.001$).

According to Köppen–Geiger climate classification (Peel et al., 2007), arid grassland (DE, JE, WG, and SR) and Mediterranean forest (CC) sites experience a climate with a dry season and are seasonally water-limited, whereas mesic grassland (CP, SP, and LW) and temperate forest (LR, MC, BC, and CF) sites experience humid climate and can be temperature-limited.

The climate data set used in this study was from in situ daily precipitation and maximum and minimum air temperature measured at the local weather station representative of each site from 1970 to 2009 except for JE, for which data were available from 1978 to 2009. Long-term (40-year) in situ temperature data sets were used to identify climate extremes within the past decade. In this study, we considered two extreme temperature indices. Maximum temperature index (T_{max}) represents annual mean daily maximum temperature, and minimum temperature index (T_{min}) represents annual mean daily minimum temperature. Annual values were based on the hydrologic year extending from 1 October to 30 September. The interannual variability of temperature extremes was represented by the anomaly, which was calculated as the departure of a given year from the mean of the period between 1970 and 2009 divided by the standard deviation. Positive anomaly means higher T_{max} above the long-term average, and vice versa for negative anomaly.

2.2 Satellite data

We used satellite observations of the EVI from the MODIS as a proxy for annual ANPP. The EVI data set was derived from the MODIS land product subset (MOD13Q1) with 16-day and 250 m resolutions for the period of 2000–2009. To compare EVI with in situ climatic measurements, we averaged the EVI data over an area of ~2.25 km × 2.25 km (9 × 9 pixels) based on the coordinates for each site in Table 1. A total of 230 scenes (23 scenes × 10 years) was obtained for each of the 12 sites. In order to eliminate the noise of low-quality, cloud- and aerosol-contaminated pixels, a pixel-based quality assurance (QA) control was applied to generate a less noisy time series data set based on the quality flag in MOD13Q1 product (Ponce-Campos et al., 2013). Then the software TIMESAT was used to smooth the QA-filtered time series of EVI as well as to estimate the vegetation parameters such as EVI integrals of the growing season (Jönsson and Eklundlh, 2004). The large integral of MODIS EVI measurements (referred to as iEVI hereafter) over the whole year was used as our surrogate measure of ANPP (Fig. 1). The MODIS iEVI has been used to quantify the dynamics of ANPP across biomes ranging from arid grassland to forest (Zhang et al., 2013; Ponce-Campos et al., 2013). For this study, to validate the relation between iEVI and annual ANPP for the data set in this study, ground measurements of ANPP ($ANPP_G$) during the period 2000–2009 were compiled for nine sites (53 years in total) across the United States (Table 2). A strong relationship (Eq. 1) between $ANPP_G$ and the corresponding iEVI was derived across biomes for these long-term experimental sites (Fig. 1):

$$ANPP_G = 100.97 \times iEVI - 85.28 \quad (1)$$

$$R^2 = 0.90 \quad P < 0.001 \quad (n = 53).$$

It should be noted that Eq. (1) is a spatiotemporal relationship between $ANPP_G$ and iEVI across biomes. At the site scale, the temporal relationship between $ANPP_G$ and iEVI is not as strong as Eq. (1), but the site-specific comparison with $ANPP_G$ from the plot-scale ground measurements at two sites in Table 1 (CP and JN) showed generally good agreement during the period between 2000 and 2009 (Fig. 1 inset, $R^2 = 0.70$, $P < 0.01$) in spite of the scale differences between measurements of iEVI and $ANPP_G$. Hence iEVI could show the interannual variability of vegetation growth. On the other hand, the spatial correlation is also significantly positive between $ANPP_G$ and the corresponding iEVI, as shown in Fig. 1 (inset) for the year of 2001 across biomes ($R^2 = 0.88$, $P < 0.001$). Therefore, we assumed that the iEVI can be a reasonable surrogate for ANPP interannual variability and provide consistent sensitivity across biomes ranging from arid grassland to forest in our analyses.

2.3 Data analysis

To investigate the sensitivity of ANPP to temperature extreme (T_{max}) across biomes, we compared the iEVI measured during years with extremely high temperatures with the mean iEVI of all other years during 2000–2009 for each site. Years

Table 2. Sites with "in-situ" ANPP measurements within the period of 2000–2009 for validation with iEVI.

Site	Biome and Location	Period	Source
Jornada LTER	Arid grassland, New Mexico	2000–2009	http://www.lternet.edu/sites/
Sevilleta LTER	Arid grassland, New Mexico	2002–2003	http://www.lternet.edu/sites/
Shortgrass Steppe LTER	Grassland, Colorado	2000–2009	http://www.lternet.edu/sites/
Cedar Creek LTER	Grassland, Minnesota	2000–2007	http://www.lternet.edu/sites/
Konza Prairie LTER	Grassland, Kansas	2000–2002	http://www.lternet.edu/sites/
Harvard Forest	Mixed forest, Massachusetts	2000–2006	http://www.lternet.edu/sites/
Metolius Intermediate Pine	Evergreen needleleaf forest, Oregon	2001	http://public.ornl.gov/ameriflux/
Park Falls	Deciduous broadleaf forest, Wisconsin	2000, 2004	http://public.ornl.gov/ameriflux/
Ohio Hills FFs	Mixed forest, Ohio	2001–2002	Chiang et al. (2008)
University of Michigan Biological Station	Deciduous broadleaf forest, Michigan	2000–2006	Gough et al. (2008)

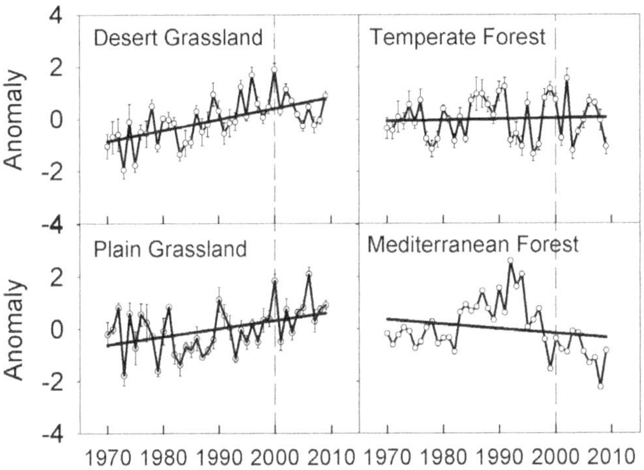

Figure 2. Long-term trends of the anomaly of T_{max} during 1970–2009 for different biome types. DG, arid grassland sites (DE, JE, WG, SR, and CP in Table 1); MG, mesic grassland sites (SP and LW in Table 1); TF, temperate forested sites (LR, MC, BC, and CF in Table 1); MF, Mediterranean forested site (CC). The dotted line shows the year of 2000 for the starting year of the MODIS EVI data set.

with extremely high temperatures were defined as those for which the T_{max} anomaly was ≥ 1 or the maximum anomaly year when there is no anomaly > 1 during 2000–2009. Since both precipitation and temperature (T_{max} and T_{min}) have limitations on vegetation production (iEVI) and they covary with one another, we also used partial correlation analysis to assess the relationship between iEVI and T_{max} by removing the effects of precipitation and T_{min}. Partial correlation analysis is widely used to isolate the relationship between two variables by removing the effects of many correlated variables. A Duncan multiple range test was used to determine significant differences in temperature and iEVI among groups.

3 Results and discussion

3.1 Long-term trends of the anomaly of T_{max}

Figure 2 shows the long-term trends of T_{max} for the four biome types (Table 1). For desert grassland, annual mean maximum temperature increased by 1.66 °C ($P < 0.0001$) during the 40-year period from 1970 and 2009 (Fig. 2). For mesic grassland, T_{max} increased by 1.21 °C ($P < 0.0001$) during 1970–2009 (Fig. 2). There was no significant trend for T_{max} for temperate forest sites. In contrast, T_{max} decreased slightly for Mediterranean forest even though not statistically significant for the whole 40-year period (Fig. 2, $P > 0.1$). However, Fig. 2 shows that there were two different periods for T_{max} at the Mediterranean forest site (CC in Table 1). T_{max} increased by 1.86 °C ($P < 0.0001$) before the earlier 1990s but then dropped dramatically by −3.46 °C ($P < 0.0001$) after 1992 (Fig. 2). The temperature rises observed in desert and mesic grassland are consistent with the observation in the southwestern US and the Great Plains (USGCRP, 2009; MacDonald, 2010). However, the unchanged annual mean T_{max} in the temperate forest sites is not consistent with the regional temperature rise in the eastern US (USGCRP, 2009).

3.2 Contrasting responses to T_{max} between grassland and forest biomes

Figure 3 shows that annual iEVI was significantly correlated with T_{max} ($R^2 = 0.79$, $P < 0.001$) across the temperature gradient of forested sites, and a stronger relation was identified between the decadal maximum T_{max} and corresponding iEVI ($R^2 = 0.95$, $P < 0.005$; Fig. 3). Because the slopes of these two relations are not significantly different (F test, $P > 0.05$; Fig. 3), this confirms that forest production increases with elevated temperature across the temperature gradient (Magnani et al., 2001; Wullschleger et al., 2003; Huxman et al., 2004). This also suggests that the decadal maximum T_{max} may not affect the overall sensitivity of interannual ANPP to mean annual temperature. Figure 3 also

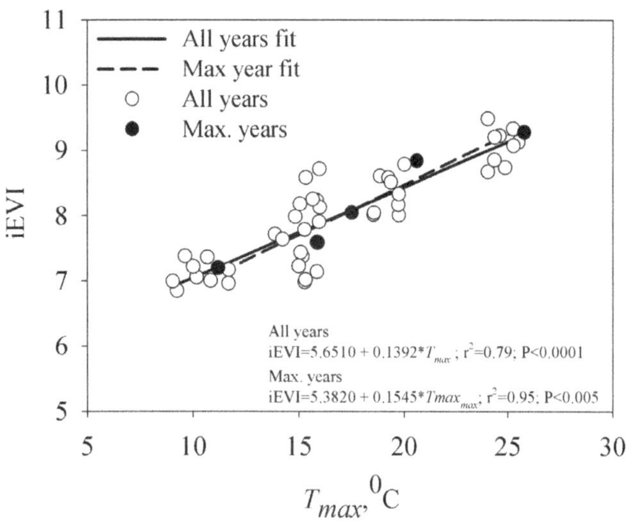

Figure 3. Relationship between iEVI and T_{max} and their maximum-year T_{max}–iEVI relationship for five forested sites. The dotted line shows the linear relationship between maximum-year T_{max} and the relevant iEVI for all the sites.

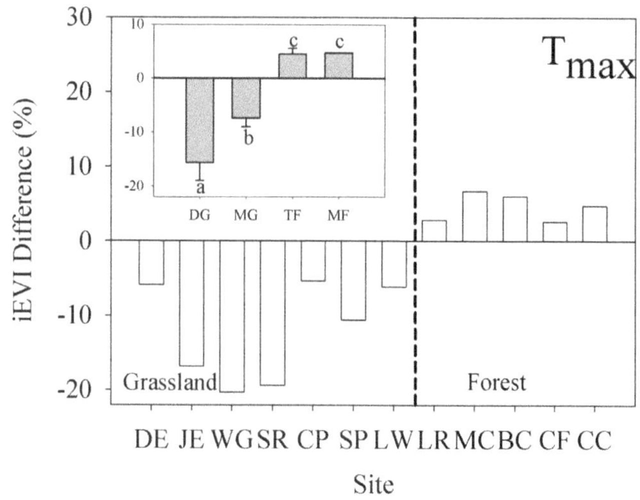

Figure 4. Comparison of iEVI differences between the extreme years and average of all other years of T_{max} across sites. Extreme years mean that the years of T_{max} anomaly are ≥ 1. The inset denotes the average differences by biome type. DG, arid grassland sites (DE, JE, WG, and SR in Table 1); MG, mesic grassland sites (CP, SP, and LW in Table 1); TF, temperate forested sites (LR, MC, BC, and CF in Table 1); MF, Mediterranean forested site (CC). Different letters indicate significant differences at $P < 0.05$.

suggests that maximum temperature can explain 80 % of the variability of vegetation production across these forest sites. For the grassland sites, however, there is no significant correlation between mean annual iEVI and T_{max} ($R^2 = 0.05$, $P = 0.64$). This is consistent with the fact that vegetation production is more controlled by water availability for grasslands in arid and semi-arid regions and interannual ANPP being related to soil moisture variability (Knapp and Smith, 2001; Knapp et al., 2002; Cherwin and Knapp, 2012). On the other hand, the forest biome is more temperature-limited with regard to productivity, particularly in regions that are not constrained by water, and climatic warming can stimulate vegetation growth through enhancing summer photosynthesis (Churkina and Running, 1998; Nemani et al., 2003; Piao et al., 2007).

Within sites, however, the interannual iEVI was not correlated with interannual variations in T_{max} at any forest site ($P > 0.05$). The differences between spatial and temporal patterns of forest ANPP responses to T_{max} reflect different underlying mechanisms at regional and local scale. The regional pattern of forest ANPP is determined primarily by temperature, while the temporal pattern for a given ecosystem is most likely affected by interactions between temperature and nutrient availability. Several studies have found limited forest production response to warming alone but significant response to warming with fertilization (Parsons et al., 1994; Press et al., 1998; Rustad et al. 2001; Strömgren and Linder 2002; Xu et al. 2011).

Among biomes, higher T_{max} with an anomaly > 1 had a direct negative effect on vegetation production in grassland ecosystems, especially for arid grassland, but a positive effect on forest ecosystems (Fig. 4, $P < 0.05$). On average, the decreases in iEVI were up to 7 % for mesic grassland, and 16 % for arid grassland (Fig. 4, inset). This may be attributed to the negative effects of warming temperatures on water availability through enhanced evapotranspiration (Seager and Vecchi, 2010). In contrast, higher T_{max} enhanced mean annual iEVI by 5 % for both temperate and Mediterranean forest sites (Fig. 4). There were larger, positive responses of ANPP to higher temperatures for forested sites in colder environments, which are the sites of BC and MC (Fig. 4).

The results stated above demonstrate the effects of hot temperature extreme on vegetation production without considering the confounding effects of other variables such as precipitation and T_{min}. There is a highly positive correlation between T_{max} and T_{min}. To isolate the role of T_{max} from precipitation and T_{min}, we alternately investigated the apparent responses of iEVI to T_{max} with partial correlation analyses to remove the confounding effects. Figure 5 shows how interannual iEVIs respond to variations in interannual T_{max} across sites. After removing the effects of T_{min} and precipitation in the partial correlation analyses, the individual T_{max} interannual changes again show the contrasting effects on the interannual iEVI between grassland and forest sites (Fig. 5). For desert grassland sites, interannual iEVI is negatively correlated with interannual T_{max} with statistical significance at the 0.05 level ($R = 0.35$). There is no significantly partial correlation between T_{max} and annual iEVI for mesic grassland sites (Fig. 5), implying little or no response of ecosystem production to T_{max} after removing the effects of T_{min} and precipitation. In contrast, interannual

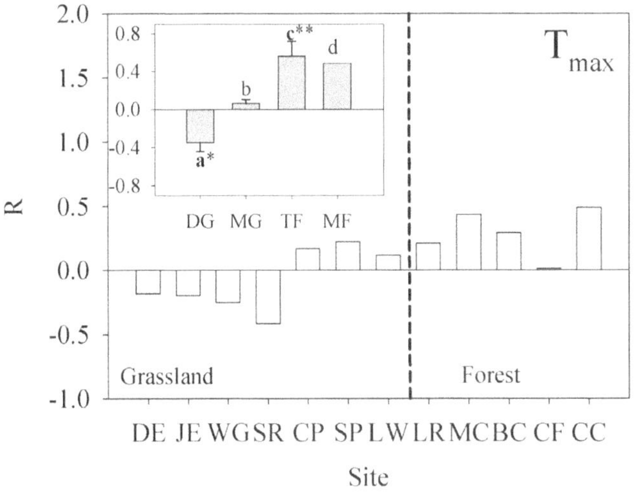

Figure 5. Partial correlations between iEVI and T_{max} after controlling the effects of T_{min} and precipitation across sites. (1) Statistically significant at the 95 % ($P < 0.05$) level; (2) statistically significant at the 99.9 % ($P < 0.001$) level.

T_{max} variations exhibit significantly positive partial correlations with interannual iEVI changes for temperate forest sites ($R = 0.57$; $P < 0.001$). For the Mediterranean forest site of Caspar Creek, it also shows positive partial correlations between interannual T_{max} and iEVI but without statistical significance ($R = 0.49$; $P = 0.22$) due to fewer data sets from only site. This opposite response of interannual iEVI to T_{max} between wet and dry temperate regions of the North America agrees well with a recent global study by Peng et al., 2013, in which remarkable spatial patterns of the partial correlations between growing-season greenness and T_{max} over the Northern Hemisphere were shown.

In all, the two approaches in the present study suggest that hot temperature extremes imposed a negative effect on vegetation production for grassland, especially desert grassland in the southwestern US, while it has a positive effect on forest (Figs. 4 and 5). This difference in response between grassland versus forest may be related to the adaptations of dominant species in terms of their response to warming temperature. Higher T_{max} and warming climate would imply drier soils through increased evaporative demand (Manabe and Wetherald, 1986) and decreased production due to decreases in stomatal conductivity, down-regulation of the photosynthetic processes, and increased allocation to roots in arid and semi-arid regions (Chaves et al., 2002). Our results agree well with the results of previous studies (Braswell et al., 1997; Piao et al., 2006; Munson et al., 2012) that higher temperatures may have directly negative effects on vegetation growth in arid and semi-arid grassland. With more atmospheric carbon dioxide in the future, however, such warming desiccation effects would be likely modified at least for arid grasslands as shown by Morgan et al. (2011). For forest, the positive effect is consistent with the results reported by Rustad et al. (2001) and McMahon et al. (2010) for ANPP in ecosystem warming experiments that higher T_{max} would have a positive impact on forest production (Boisvenue and Running, 2006). Previous studies have also shown that higher temperatures favor tree growth by enhancing photosynthesis (Lukac et al., 2010) and nutrient uptake (Weih and Karlsson, 2002), especially in sites where trees were not typically constrained by moisture stress. Thus, these contrasting responses to T_{max} in different ecosystems could have different effects on regional vegetation carbon uptake (Braswell et al., 1997). However, it should be noted that, due to data limitations, only 12 sites were used in this study. We need more data from all terrestrial ecosystems to test whether the contrasting effect of temperature extremes is a general behavior in forest and grassland ecosystems. Furthermore, the results reported here may be temporal, and the long-term impact of temperature extremes on ecosystem functional integrity across biomes is yet unknown. The ongoing field measurements of carbon flux and meteorological data from eddy covariance flux method for different ecosystem may provide an opportunity to validate the assumption in this study.

4 Conclusions

Understanding how vegetation production responds to extreme warm temperature regimes is crucial for assessing the impacts of climate change on terrestrial ecosystem. Recent record-breaking high temperatures in the contiguous US provide the opportunity to study such effects. Our study offers a generalization of the functional response of ecosystem to hot extreme conditions predicted with climate change across biomes in the natural climatic conditions. By using satellite and in situ meteorological data, we found a contrasting response of terrestrial ecosystem to extreme warm temperature anomalies between grassland and forest biomes in natural settings. The opposite direction and magnitude of response indicates different sensitivities across ecosystems to hot temperature extremes. A recent study has shown that there is a continuous increase in hot extremes over land, despite the slowed rate of increase in annual global mean temperature (Seneviratne et al., 2014). Hence, the sensitivity of ecosystem production in response to hot extremes across biomes found here has important implications. Current terrestrial ecosystem models usually utilize daily mean or monthly temperature data as input, and hence they may neglect the response of vegetation to extreme warm temperature (T_{max}). To some extent, the effects of hot extremes are more relevant for climate change impacts than global mean temperature on ecosystems (IPCC, 2012, 2013). Hence, this work further strengthens our understandings of the ecosystem-level response to extreme warm temperatures across biomes. This study also clarifies the value of long-term experimental sites together with continuing satellite-based observations such as EVI in future studies. This compelling result in a natural setting at the ecosystem level should play a role in future climate change impact studies.

Acknowledgements. The work by Y. Zhang and M. Voigt was funded by the Emmy Noether Programme (GlobFluo project) of the German Research Foundation. Thanks to M. Susan Moran, Anthony R. Buda, David D. Bosch, Stacey A. Gunter, Stanley G. Kitchen, W. Henry McNab, Jack A. Morgan, Mitchel P. McClaran, Diane S. Montoya, Debra P. C. Peters, and Patrick J. Starks for providing the climatic data. We are also grateful for access to multi-site, long-term data provided through the EcoTrends Project (www.ecotrends.info). This work used biological data acquired by the LTER network. We acknowledge the Oak Ridge National Laboratory (ORNL) Distributed Active Archive Center (DAAC) for making these data available. We thank Bai Yang, Jyh-Min Chiang, Beverly Law, and Arlyn Andrews for providing suggestions about the ANPP data. We also thank the principal investigators, information managers, scientists, and technicians responsible for measuring ANPP at the LTER sites. The authors also gratefully acknowledge the valuable suggestions made by Christopher Kyba at the Freie Universität Berlin. The authors are also are grateful to the three anonymous reviewers for their valuable comments.

Edited by: E. Pendall

References

Alward, R. D., Detling, J. K., and Milchunas, D. G.: Grassland vegetation changes and nocturnal global warming, Science, 283, 229–231, 1999.

Braswell, B. H., Schimel, D. S., Linder, E., Moore, B.: The response of global terrestrial ecosystems to interannual temperature variability, Science, 238, 870–872, 1997.

Boisvennue, C. and Running, S. W.: Impacts of climate change on natural forest production- evidence since the middle of the 20th century, Glob. Change Biol., 12, 862–882, 2006.

Chaves, M. M., Pereira, J. S., Maroco, J. P.: How plants cope with water stress in the field: photosynthesis and growth, Ann. Bot., 89, 907–916, 2002.

Cherwin, K. and Knapp, A. K.: Unexpected patterns of sensitivity to drought in three semi-arid grasslands, Oecologia, 169, 845–852, 2012.

Chiang, J. M., McEwan, R. W., Yaussy, D. A., and Brown, K. J.: The effects of prescribed fire and silvicultural thinning on the aboveground carbon stocks and net preimary production of overstory trees in an oak-hickory ecosystem in southern Ohio, Forest. Ecol. Manage., 255, 1584–1594, 2008.

Churkina, G. and Running, S. W.: Contrasting climate controls on the estimated production of global terrestrial biomes. Ecosystems, 1 pp., 206–215, 1998.

Gough, C. M., Vogel, C. S., Schmid, H. P., Su, H. B., and Curtis, P. S.: Multi-year convergence of biometric and meteorological estimates of forest carbon storage, Agr. Forest. Meteorol., 148, 158–170, 2008.

Huete, A. R., Didan, K., Miura, T., Rodriguez, E. P., Gao, X., and Ferreira, L. G.: Overview of the radiometric and biophysical performance of the MODIS vegetation indices, Remote Sens. Environ., 83, 195–213, 2002.

Huxman, T. E., Smith, M. D., Fay, P. A., Knapp, A. K., Shaw, M. R., Loik, M. E., Smith, S. D., Tissue, D. T., Zak, J. C., Weltzin, J. F., Pockman, W. T., Sala, O. E., Haddad, B. M., Harte, J., Koch, G. W., Schwinning, S., Small, E. E., and Williams D. G.: Convergence across biomes to a common rain-use efficiency, Nature, 429, 651–654, 2004.

IPCC: Managing the Risks of Extreme Events and Disasters to Advance Climate Change Adaptation, in: A Special Report of Working Groups I and II of the Intergovernmental Panel on Climate Change, edited by: Field, C. et al., Cambridge University Press, Cambridge, UK, and New York, NY, USA, 582 pp., 2012.

IPCC: Climate Change 2013: The Physical Science Basis, in: Contribution of Working Group I to the Fifth Assessment Report of the Intergovernmental Panel on Climate Change, edited by: Stocker, et al., Cambridge University Press, Cambridge, United Kingdom and New York, NY, USA, 1535 pp., 2013.

Jonsson, P. and Eklundh, L.: TIMESAT–a program for analyzing time-series of satellite sensor data, Comput. Geosci., 30, 833–845, 2004.

Knapp, A. K. and Smith, M. D.: Variation among biomes in temporal dynamics of aboveground primary production, Science, 291, 481–484, 2001.

Knapp A. K., Fay, P. A., Blair, J. M., Collins, S. L., Smith, M. D., Carlisle, J. D., Harper, C. W., Danner, B. T., Lett, M. S., and McCarron, J. K.: Rainfall variability, carbon cycling, and plant species diversity in a mesic grassland, Science, 298, 2202–2205, 2002.

Keeling, R. F., Piper, S. C., and Heimann, M.: Global and hemispheric CO2 sinks deduced from changes in atmospheric O2 concentration., Nature, 381, 218–221, 1996.

Lukac, M., Calfapietra, C., Lagomarsino, A., and Loreto, F.: Global climate change and tree nutrition: effects of elevated CO_2 and temperature, Tree Physiology, 30, 1209–1220, 2010.

MacDonald, G. M.: Water, climate change, and sustainability in the southwest, Proc. Natl. Acad,Sci. USA, 107, 21256–21262, 2010.

Manabe, S. and Wetherald, R. T.: Reduction in summer soil wetness induced by an increase in atmospheric carbon dioxide, Science, 232, 626–628,1986.

McMahon, S. M., Parker, G. G., and Miller, D. R.: Evidence for a recent increase in forest growth, Proc. Natl. Acad. Sci. USA, 107, 3611–3615, 2010.

Myneni, R. B., Keeling, C. D., Tucker, C. J., Asrar, G., and Nemani, R. R.: Increased plant growth in the northern high latitudes from 1981 to 1991, Nature, 386, 698–702, 1997.

Myneni, R. B., Dong, J., Tucker, C. J., Kaufmann, R. K., Kauppi, P. E., Lisk, J., Zhou, L., Alexeyev, V., and Hughes, M. K.: A large carbon sink in the woody biomass of northern forests, Proc. Natl. Acad. Sci. USA, 98, 14784–14789, 2001.

Nemani, R. R., Keeling, C. D., Hashimoto, H., Jolly, W. M., Piper, S. C., Tucker, C. J., Myneni, R. B., and Running, S. W.: Climate-driven increases in global terrestrial net primary production from 1982 to 1999, Science, 300, 1560–1563, 2003.

NOAA National Climatic Data Center, State of the Climate: Global Analysis for Annual 2013, published online December 2013, retrieved on February 21, 2014 from http://www.ncdc.noaa.gov/sotc/global/2013/13, 2014.

Parsons, A. N., Welker, J. M., Wookey, P. A., Press, M. C., Callaghan, T. V., and Lee, J. A.: Growth responses of four dwarf shrubs to simulated climate change, J. Ecol., 82, 307–318, 1994.

Peel, M. C., Finlayson, B. L., and McMahonl, T. A.: Updated world map of the Köppen-Geiger climate classification, Hydrol. Earth Syst. Sci., 11, 1633–1644, 2007.

Peng, S., Piao, S. L., Ciais, P., Myneni, R., Chen, A., Chevallier, F., Dolman, A. J., Janssens, I.A., Peñuelas, J., Zhang, G., Vicca, S., Wan, S., Wang, S., and Zeng, H.: Asymmetric effects of day-time and night-time warming on Northern Hemisphere vegetation, Nature, 501, 88–92, 2013.

Piao, S. L., Mohammat, A., Fang, J. Y., Cai, Q., and Feng, J. M.: NDVI-based increase in growth of temperate grasslands and its responses to climate change in China, Glob. Environ. Change, 16, 340–348, 2006.

Piao, S. L., Friedlingstein, P., Ciais, P., Viovy, N., and Demarty, J.: Growing season extension and its effects on terrestrial carbon flux over the last two decades, Global Biogeochem. Cy. ,21, GB3018, doi:10.1029/2006GB00288, 2007.

Ponce-Campos, C. G. E., Moran, M. S., Huete, A., Zhang, Y., Bresloff, C., Huxman, T., Bosch, D., McClaran, M. P., Kitchen, S .G., Buda, A., Gunter, S., McNab, H., Peters, D., Starks, P., Montoya, D. S., Heartsill, T., McVicar, T., Eamus, D., and Donohue, R.: Ecosystem resilience despite large-scale altered hydroclimatic conditions, Nature, 494, 349–352, 2013.

Press, M. C., Potter, J. A., Burke, M. J. W., Callaghan, T. V., and Lee, J. A.: Responses of a sub-arctic dwarf shrub community to simulated environmental change, J. Ecol., 86, 315–327, 1998.

Rustad, L. E., Campbell, J. L., Marion, G. M., et al.: A meta-analysis of the response of soil respiration, net nitrogen mineralization, and aboveground plant growth to experimental ecosystem warming, Oecologia, 126, 543–562, 2001.

Sala, O. E., Parton, W. J., Joyce, L. A., and Lauenroth, W. K.: Primary production of the central grassland region of the United States: spatial pattern and major controls, Ecology, 69, 40–45, 1988.

Seager, R., and Vecchi, G. A.: Greenhouse warming and the 21st century hydroclimate of southwestern North America, Proc. Natl. Acad. Sci. USA, 107, 21277–21282, 2010.

Seneviratne, S. I., Donat, M. G., Mueller, B., and Alexander, L. V.: No pause in the increase of hot temperature extremes, Nat. Clim. Change, 4, 161–163, 2014.

Shaver, G. R., Canadell, J., Chapin III, F. S., et al.: Global warming and terrestrial ecosystems: a conceptual framework for analysis, Bioscience, 50, 871–882, 2000.

Smith, M. D.: The ecological role of climate extremes: current understanding and future prospects, J. Ecol., 99, 651–655, 2011.

Strömgren, M. and Linder, S.: Effects of nutrition and soil warming on stemwood production in a boreal Norway spruce stand, Glob. Change Biol., 8, 1195–1204, 2002.

USGCRP: Global Climate Change Impacts in the United States, edited by: Karl, T. R., Melillo, J. M., and Peterson, T. C., United States Global Change Research Program. Cambridge University Press, New York, NY, USA, 2009.

Weih, M. and Karlsson, P. S.: Low winter soil temperature affects summertime nutrient uptake capacity and growth rate of mountain birch seedlings in the subarctic, Swedish lapland,Arctic, Antarctic, and Alpine Res., 34, 434–439, 2002.

Wu, Z., Dijkstra, P., Koch, G. W., Peñuelas, J., and Hungate, B. A.: Responses of terrestrial ecosystems to temperature and precipitation change: a meta-analysis of experimental manipulation, Glob. Change Biol., doi:10.1111/j.1365-2486.2010.02302.x, 2010.

Wullschleger, S. D., Hanson, P. J., and Todd, D. E.: Forest water use and the Influence of precipitation change, in: North American temperate deciduous forest responses to changing precipitation regimes, edited by: Hanson, P. J. and Wullschleger, S. D., Springer, New York, USA, 2003.

Xu, C. G., Liang, C., Wullschleger, S., Wilson, C., and McDowell, N.: Importance of feedback loops between soil inorganic N and microbial communities in the heterotrophic soil respiration response to global warming, Nat. Rev. Microbiol., 9, 222, doi:10.1038/nrmicro2439-c1, 2011.

Zhang, Y., Moran, M. S., Nearing, M. A., Ponce, C. G. E., et al.: Extreme precipitation patterns and reductions of terrestrial ecosystem production across biomes, J. Geophys. Res.-Biogeosci., 118, 1–10, 2013.

Zhou, L. M., Tucker, C. J., Kaufmann, R. K., Slayback, D., Shabanov, N. V., and Myneni, R. B.: Variations in northern vegetation activity inferred from satellite data of vegetation indexduring 1981 to 1999, J. Geophys. Res., 106, 20069–20083, 2001.

7

Moderate forest disturbance as a stringent test for gap and big-leaf models

B. Bond-Lamberty[1], J. P. Fisk[2], J. A. Holm[3], V. Bailey[4], G. Bohrer[5], and C. M. Gough[6]

[1]Pacific Northwest National Laboratory, Joint Global Change Research Institute at the University of Maryland–College Park, 5825 University Research Court, Suite 3500, College Park, Maryland, MA 20740, USA
[2]Department of Geographical Sciences, 1150 LeFrak, University of Maryland, College Park, Maryland, MA 20742, USA
[3]Climate Sciences Department, Lawrence Berkeley National Laboratory, 1 Cyclotron Rd., MS 74-0171, Berkeley, CA 94720, USA
[4]Pacific Northwest National Laboratory, 902 Battelle Boulevard, Richland, WA 99352, USA
[5]Department of Civil, Environmental and Geodetic Engineering, The Ohio State University, 470 Hitchcock Hall, 2070 Neil Avenue, Columbus, Ohio, OH 43210, USA
[6]Virginia Commonwealth University, Department of Biology, P.O. Box 842012, 1000 West Cary Street, Richmond, VA 23284-2012, USA

Correspondence to: B. Bond-Lamberty (bondlamberty@pnnl.gov)

Abstract. Disturbance-induced tree mortality is a key factor regulating the carbon balance of a forest, but tree mortality and its subsequent effects are poorly represented processes in terrestrial ecosystem models. It is thus unclear whether models can robustly simulate moderate (non-catastrophic) disturbances, which tend to increase biological and structural complexity and are increasingly common in aging US forests. We tested whether three forest ecosystem models – Biome-BGC (BioGeochemical Cycles), a classic big-leaf model, and the ZELIG and ED (Ecosystem Demography) gap-oriented models – could reproduce the resilience to moderate disturbance observed in an experimentally manipulated forest (the Forest Accelerated Succession Experiment in northern Michigan, USA, in which 38 % of canopy dominants were stem girdled and compared to control plots). Each model was parameterized, spun up, and disturbed following similar protocols and run for 5 years post-disturbance. The models replicated observed declines in aboveground biomass well. Biome-BGC captured the timing and rebound of observed leaf area index (LAI), while ZELIG and ED correctly estimated the magnitude of LAI decline. None of the models fully captured the observed post-disturbance C fluxes, in particular gross primary production or net primary production (NPP). Biome-BGC NPP was correctly resilient but for the wrong reasons, and could not match the absolute observational values. ZELIG and ED, in contrast, exhibited large, unobserved drops in NPP and net ecosystem production. The biological mechanisms proposed to explain the observed rapid resilience of the C cycle are typically not incorporated by these or other models. It is thus an open question whether most ecosystem models will simulate correctly the gradual and less extensive tree mortality characteristic of moderate disturbances.

1 Introduction

Natural and anthropogenic disturbances have numerous effects on the carbon (C) and energy dynamics in forested ecosystems and result in a variety of feedbacks between terrestrial ecosystems and climate (Goetz et al., 2012). In particular, disturbance-induced tree mortality is a key factor regulating the forest C balance but a complicated one due to high temporal and spatial heterogeneity (Vanderwel et al., 2013). Partly as a result, mortality and disturbance are poorly represented processes in terrestrial ecosystem models (Medvigy and Moorcroft, 2012; Peters et al., 2013; Dietze and Matthes, 2014).

Most North American forests are at some stage of recovery from either natural or anthropogenic disturbance (Pan et al., 2011). In the US upper Midwest and northeast, low-severity disturbance is increasing in frequency and extent in regional forests, which have regrown following stand-replacing disturbances over a century ago (Frelich and Reich, 1995). The resulting cohort of fast-growing, deciduous trees is now past maturity and beginning to decline, while longer-lived species representation is increasing (Gough et al., 2010b). At the same time, forest disturbances in the region are transitioning away from severe events that historically caused complete stand replacement towards more subtle disturbances that result in only partial canopy defoliation or loss of selected species (Pregitzer and Euskirchen, 2004; Williams et al., 2012; Birdsey et al., 2006). These subtler disturbances include partial harvests, wind, pathogenic insects, diseases, and age-related senescence (e.g., Caspersen et al., 2000), which contribute to a gradient of disturbance intensities across the landscape. Unlike stand-replacing disturbance, moderate disturbances tend to increase biological and structural complexity and, consequently, are expected to have entirely different functional consequences for ecosystems (Nave et al., 2011; Peters et al., 2013).

Moderate disturbances have mixed effects on successional trajectories of forest C production and storage (Birdsey et al., 2006; Knohl et al., 2002; Vanderwel et al., 2013). In many forests, C storage shows unexpected resilience or even resistance to partial canopy defoliation (Hicke et al., 2011; Gough et al., 2013; Mathys et al., 2013) or thinning (Granier et al., 2008). The reasons and mechanisms for different functional responses to moderate disturbance are not clear, but these results have large potential implications, as the long-assumed future decline of production in aging stands is expected to reduce continental C sink strength (Birdsey et al., 2006). Recent empirical evidence indicates, however, that net ecosystem production (NEP, the ecosystem carbon balance) may be sustained or even increase in older forests that experience moderate disturbance (Luyssaert et al., 2008). For example, NEP in the ~ 100 yr old Harvard Forest has more than doubled in the last 18 years (Keenan et al., 2012). More broadly, recent syntheses of North America's mixed temperate forests found no evidence for a substantial decline in NEP or net primary production (NPP) with age (He et al., 2012; Amiro et al., 2010).

Many ecosystem-scale models, designed for and tested in early to mid-successional forests with low biological and structural complexity, can be expected to have trouble reproducing these results (Landsberg and Waring, 1997; Raulier, 1999; Law et al., 2003; Li et al., 2003; Zhao et al., 2009). Such models are typically developed from, and tested most thoroughly against, classic primary- and secondary-succession scenarios featuring stand-replacing or at least gap-size disturbances (Peters et al., 2013; Weng et al., 2012; Wang et al., 2014). Most model experiments using moderate (non-catastrophic) disturbance intensities have been performed in the context of timber management, e.g., assessing the sustainability of harvesting for a particular ecosystem or region (e.g., Peng et al., 2002; Rolff and Ågren, 1999). As a result, it is unclear whether most ecosystem models will be able to correctly simulate naturally occurring disturbances in mature forests, which may be spatially more heterogeneous and generally do not involve biomass removals. This is particularly important given the rapidly aging distribution of eastern US forests (USDA, 2013; Radeloff et al., 2012).

With moderate disturbances increasing in aging North American forests, and only an emerging understanding of the mechanisms underpinning such forests' resilience to disturbance, it is clearly important to understand how, and how well, forest models simulate these events. Doing so not only provides a quantitative assessment of model performance, but also may help identify knowledge gaps and processes missing or not properly implemented in ecosystem models more generally. This study tested three forest ecosystem models – a classic big-leaf model and two gap models – to understand how well they reproduce observed resilience to moderate disturbance in an experimentally manipulated forest and to explore specific mechanisms limiting model skill.

2 Methods

2.1 Site description

The study site is the University of Michigan Biological Station (UMBS, 45°35.5′ N, 84°43′ W), nested within a secondary successional forest that is comprised of bigtooth aspen (*Populus grandidentata*), northern red oak (*Quercus rubra*), red maple (*Acer rubrum*), paper birch (*Betula papyrifera*), and eastern white pine (*Pinus strobus*). Average overstory tree age in 2013 was 95 years. NEP in the unmanipulated footprint of the UMBS control tower (US-UMBS) was 0.80–1.98 $\mathrm{Mg\,C\,ha^{-1}\,yr^{-1}}$ from 1999 to 2006, averaging 1.58 $\mathrm{Mg\,C\,ha^{-1}\,yr^{-1}}$ with substantial landscape variation (Gough et al., 2009). The forest was heavily logged in the late 1800s and early 1900s and disturbed by fire until 1923; its present-day plant composition is typical of many forests in the upper Great Lakes region (Gough et al., 2007).

2.2 The Forest Accelerated Succession Experiment

The Forest Accelerated Succession Experiment (FASET) is an ongoing experiment, in which >6.700 aspen and birch trees (equivalent to 38 % of stand basal area) were stem girdled in 2008 within a 39 ha area. FASET is investigating how C storage and fluxes change following moderate disturbance as Great Lakes forests transition from an assemblage of early successional canopy trees to later successional canopy dominants. The experiment's overarching hypothesis is that forest NEP will be resilient following partial canopy defoliation and subsequently increase as canopies become more biologically and structurally complex and as nitrogen (N) not taken up by senescing aspen and birch trees is redistributed to other, longer-lived species assuming canopy dominance. The experiment employs a suite of paired C cycling measurements within separate treatment and control meteorological flux tower footprints. The C cycling parameters reported here for the control and treatment forests are aboveground biomass (AGB), gross primary production (GPP), ecosystem respiration (ER), leaf area index (LAI), total (above- and belowground) NPP, and NEP. Site methodological approaches for the derivation of each are described by Gough et al. (2013, 2008), but, briefly, AGB was estimated biometrically, using dendrometers and site-specific allometry; LAI from litter traps; NPP from biometry and fine root cores; and ER, GPP, and NEP (here treated as equivalent to net ecosystem exchange) from eddy covariance (Gough et al., 2013).

FASET results were most recently summarized by Gough et al. (2013). Briefly, the girdling treatment successfully expedited the mortality of early successional aspen and birch, promoting an emerging canopy that approximates projected regional changes in forest composition and structure (e.g., Wolter and White, 2002). In the first 4 years following disturbance, GPP and ER both initially rose in the treatment plots relative to the controls, while NPP and NEP were not significantly different in the control and treatment forests even though LAI in the latter declined by up to 44 % (summarized in Fig. 1). This high resilience of the C cycle was attributed to high N retention and rapid reallocation of this limiting resource in support of new leaf area production as aspen and birch declined (Nave et al., 2011). Decadal records of tree growth indicate that resilience to age-related declines in NPP is highest where a diversity of canopy tree species is present because later successional species rapidly compensate for the declining growth of early successional species (Gough et al., 2010b). Investigators are also finding that resilience of forest production to disturbance is dependent upon canopy structural reorganizations that enhance C uptake by increasing light use efficiency (Hardiman et al., 2011; Gough et al., 2013) and by hydrodynamic responses that increase post-disturbance water use efficiency in some species (Matheny et al., 2015).

2.3 Model descriptions

We tested three complementary models for their ability to replicate disturbance-related changes in production and LAI observed in FASET; model attributes and differences are summarized in Table 1. The first was a version of Biome-BGC (BioGeochemical Cycles) (Running and Hunt, 1993; Thornton et al., 2002). This model has coupled water, carbon, and nitrogen cycles (Thornton et al., 2007), uses a Farquhar photosynthesis submodel linked to prognostic leaf area, and runs on a daily timestep. The model partitions NPP into the leaves, roots and stems using dynamic allocation patterns, accounting for nitrogen and water limitations. It has been widely used for simulating carbon flows in forest ecosystems (Kimball et al., 1997; Pietsch et al., 2003; Tatarinov and Cienciala, 2009; Warren et al., 2011). We used a version of the model that incorporates an explicit disturbance mechanism (Bond-Lamberty et al., 2007).

The second model tested was ZELIG, a gap model based on the original principles of the JABOWA (Botkin et al., 1972) and FORET (Shugart and West, 1977) models. ZELIG simulates the growth, death, and regeneration of individual trees (Urban, 1990; Urban et al., 1991) in a two-dimensional grid of 400 m^2 cells (i.e., gaps) representing the forest canopy. Trees in each cell influence the availability of resources in adjacent cells, although direct tree-to-tree interactions are not represented (Taylor et al., 2009). ZELIG's main routines include growth, mortality, regeneration, and tracking environmental conditions. In each model timestep, forest processes (e.g., seedling establishment rate, diameter increment, survival rate) are reduced from their maximum potential rates based on available resources. Potential tree regeneration, growth, and survival are functions of light conditions, soil moisture, level of soil fertility resources, and temperature. The model runs on a monthly timestep. Specific details on the methodical approaches used in the model can be found in Urban (1990), Urban et al. (1991) and Larocque et al. (2006). ZELIG has been applied over many large-scale and diverse landscapes (see list and further references in Holm et al., 2012).

The third model was ED (Ecosystem Demography), a terrestrial biosphere model that uses size- and age-structure partial differential equations (PDEs) (Moorcroft et al., 2001) to approximate the behavior of a stochastic gap model on medium to large scales. It combines an individual-based gap model, describing a particular plant community, with a biogeochemical simulation of carbon, water, and nitrogen fluxes. Modeled processes include leaf-level photosynthesis, explicit competition for water and mortality, and C and N allocation above- and belowground (Moorcroft et al., 2001). Much of the soil model is based on that of CENTURY (Parton et al., 1987). ED then models subgrid ($\sim$ 10 ha) disturbance heterogeneity using its PDEs to approximate the behavior of a spatially distributed ensemble of individual plants, and it has been used for a variety of optimization and data assimilation exercises (Medvigy et al., 2009).

It is important to note the complementary nature of these models: one is a classic "big-leaf" biogeochemical model focusing on process representation in a nonspatial framework, another a gap model representative of its class, and the third emphasizes mathematical scaling of a gap model across time and space. In addition, Biome-BGC's algorithms underlie the current version of the Community Land Model (CLM) (e.g., Bonan and Levis, 2010), while work is underway to make ED's algorithms an optional component in the next version of CLM. This provides a strong framework and motivation for examining whether the high C cycling resilience observed following FASET's moderate disturbance can be reproduced in modeling experiments.

Table 1. Comparison of the models used in this study. LUE stands for light use efficiency, APAR stands for absorbed photosynthetically active radiation, and PFT stands for plant functional type.

	Model		
	Biome-BGC	ZELIG	ED
Category	Big-leaf	Gap	Gap hybrid
Time step	Daily	Monthly	Hourly
Spatial scale	Indeterminate	400 m^2 cells (gaps)	Variable
Nitrogen cycle?	Yes	No	No
Soil model	Four pools	One pool; pseudo bucket model	Eight pools
Phenology	Calculated based on soil temperature	Seasonal heat sum, growing degree days	Calculated based on monthly air temperature
Allocation	Fixed ratios	Fixed ratios	Allometric
Canopy	Two layers, sun and shade	Species-specific	Both PFT-specific (individual) and distributional (site)
GPP	Enzyme kinetic: Farquhar, Ball–Berry	APAR and LUE	Enzyme kinetic: Farquhar, Ball-Berry
Respiration	Q_{10}, modified by temperature and moisture	Modified by temperature	Arrhenius, modified by temperature and moisture
Succession	None	Species-specific	PFT-specific
Mortality	Fixed rate	Competition-driven, stochastic	From size- and age-structure PDEs

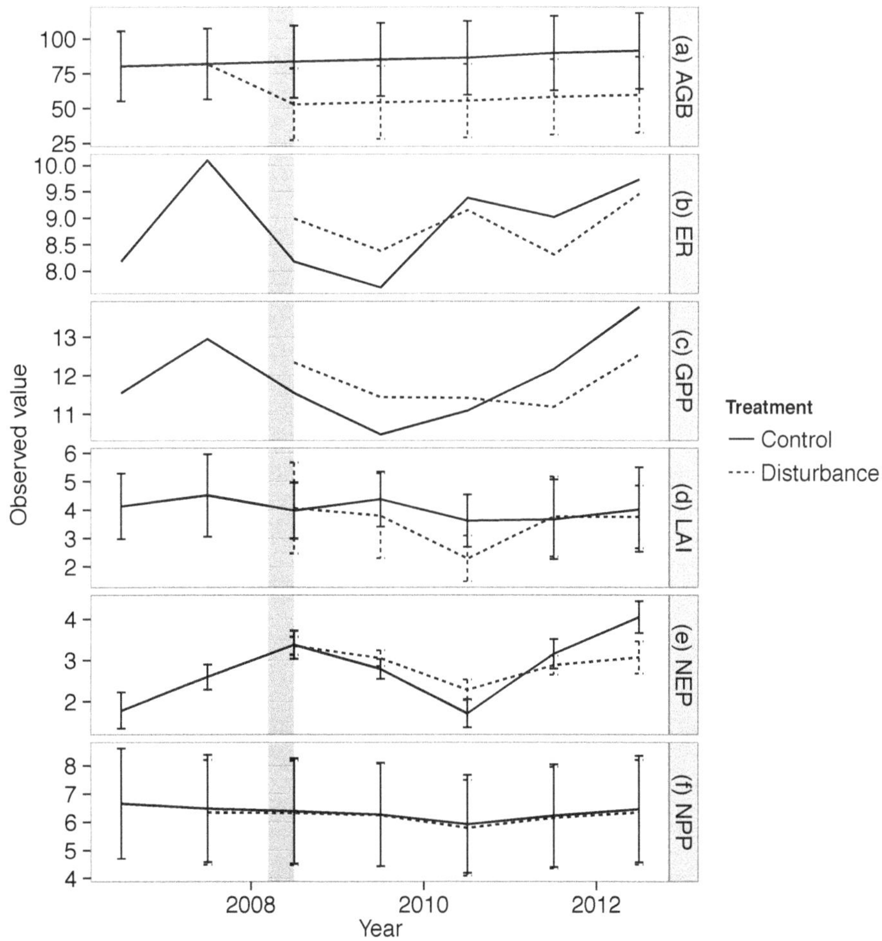

Figure 1. Observed data from FASET treatment and control forests. Panels include **(a)** aboveground biomass (AGB, in $MgC ha^{-1}$), **(b)** ecosystem respiration (ER, $MgC ha^{-1}$), **(c)** GPP ($MgC ha^{-1}$), **(d)** leaf area index (LAI, unitless), **(e)** net ecosystem production (NEP, $MgC ha^{-1}$), and **(f)** net primary production (NPP, $MgC ha^{-1}$). Vertical shaded area shows approximate time of the girdling treatment described in the text. Error bars indicate ±1 SD based on eight measurement plots (Gough et al., 2013). Control and treatment sites had near-identical data in 2006 and 2007, and thus the latter (dashed) line is not visible in panels **(a)**, **(d)**, and **(f)** in those years.

2.4 Parameters and optimization

Biome-BGC was subjected to a pre-experiment optimization exercise, with the goal of algorithmically adjusting its parameters, within observational ranges, such that model output best matched the pre-experiment carbon stocks and pools of the UMBS forest. The choice of parameters to include was based on three factors: the known sensitivities of Biome-BGC (White et al., 2000); our a priori knowledge of the FASET research site and possible physiological mechanisms underlying forest resilience to disturbance (Gough et al., 2013); and known uncertainties in measured data (C. M. Gough et al., unpublished data). The final set of optimized parameters is shown in Table 2. Constraining against observed C stocks can provide significant improvements in model performance (Carvalhais et al., 2010); in this study, slow-turnover soil C, tree stem C, and NPP were used as constraining variables. For the parameter-space search itself we used a variant of the simplex algorithm (Nelder and Mead, 1965) that uses a randomly oriented set of basis vectors instead of fixed coordinate axes, as implemented (*gsl_multimin_fminimizer_nmsimplex2rand*) in Gnu Scientific Library version GSL-1.16 (Gough, 2009). For each combination of parameter values selected by the algorithm, Biome-BGC was "spun up", i.e., its slow soil pools were brought to equilibrium, and the C pools noted above compared to observed soil C values. A linear cost function ranked model performance, imposing a large penalty if a parameter varied more than 2σ (based on expert judgment) from its observed mean.

ZELIG was parameterized with species-specific and site-specific parameters representative of the UMBS study site. The silvicultural and biological parameters for each of the eight temperate tree species required for ZELIG are listed in Table 3, with species data collected in previous studies (Larocque et al., 2006; Leemans and Prentice, 1989; Holm et al., 2013). Soil field capacity (cm) and wilting point (cm) were determined from measurements at the study site (unpublished data). We used allometric equations to estimate aboveground biomass (AGB, $\mathrm{Mg\,C\,ha^{-1}}$), these were generated from on-site harvests at the UMBS site or from general allometric equations typical of northeastern trees (Gough et al., 2008).

ED's parameters were used from the versions developed for studying both anthropogenic and natural disturbance across US forests (Hurtt et al., 2002; Fisk et al., 2013). This configuration uses two tree functional types: a cold deciduous and an evergreen type. Allometric equations, leaf characteristics, and phenology parameters are described in Hurtt et al. (2002) and summarized in Table 4.

For the main modeling experiment, Biome-BGC and ZELIG were driven by identical reanalysis daily climate (NCEP (National Centers for Environmental Prediction); Kanamitsu et al., 2002) from 1970–2012 data, with mean values of air temperature (5.1 °C) and precipitation ($575\,\mathrm{mm\,yr^{-1}}$). In contrast, ED used a climatology (i.e., with no year-to-year variation) comprised of the average monthly diurnal cycle for light, temperature and humidity, and mean monthly precipitation from the slightly warmer (mean 6.5 °C) North American Regional Reanalysis for 1979–2010 (NARR, 2013). We recognize that using different climatic inputs is not ideal, but Biome-BGC and ZELIG both took steps that made their results comparable to those of ED. For Biome-BGC, we used ensembling (Thornton et al., 2002) to characterize the mean climate and effect of interannual climate variability on model outputs, reporting model outputs as means $\pm$ standard deviation computed by running the model starting with each successive year in the climate data. For ZELIG, each year the model stochastically generated new monthly temperature and precipitation, based on the range provided by the NCEP data, thus also diminishing the effect of year-to-year variability in the input data. In summary, all model results are reported based on mean climatic conditions, not exact year-to-year changes.

2.5 Modeling experiment

As far as possible, we used the same experimental protocol with each of the three models. The models were spun up, i.e., brought to a steady state with a mature forest, and then the entire site was clear-cut, with all trees removed, i.e., harvested and the biomass taken away. This approximates the known stand-replacing disturbances of the early 20th century (Gough et al., 2007) in the UMBS forest. The models then allowed the forest to recover over 90 years before imposing 13–14 % harvests of basal area (ED and ZELIG) and biomass (Biome-BGC) in 2008, 2009, and 2010. This approach was used, as opposed to a single $\sim$40 % cut in 2008, to better mimic the slow death of girdled trees observed over 2–3 years in the FASET study, as lagged mortality has been shown to exert a strong influence on the modeling of forest disturbances (Dietze and Matthes, 2014). None of the models allowed for tree girdling, and we used harvests as a second-best alternative; under this protocol, the models remove tree stems while allowing leaves and fine litter to decay on-site. This was consistent with our observations that girdled trees in FASET did not senesce at once and remained standing for multiple years without significantly decaying (Gough et al., 2013).

As ZELIG is an individual-based, species-specific forest demographic model, we had the ability to more precisely replicate the FASET experiment by only harvesting aspen and birch trees in the forest simulator. This allowed the remaining species to continue growing, starting from their trajectories prior to the harvest, but to be subject to less competition due to the removal of aspen and birch trees. Prior to beginning the girdling experiment, early successional aspen and birch accounted for 49 % of the basal area in ZELIG (versus 38 % in the FASET study site), and these species were preferentially removed to match the 13–14 % annual har-

Table 2. Selected site-specific parameters used by Biome-BGC. Model inputs differ from observed values because of the optimization procedure used (see Methods section).

Parameter	Observed value (±se)	Model value	Units
Fine root C : N ratio	77	77.0	$kg\,C\,kg\,N^{-1}$
Fine root : leaf C allocation	1.18	1.14	Ratio
Fraction of leaf N in Rubisco		0.12	Fraction
Leaf C : N ratio	25 ± 3.4	25.0	$kg\,C\,kg\,N^{-1}$
Maximum stomatal conductance	0.03	0.0065	$m\,s^{-1}$
Nitrogen deposition	0.00085	0.001	$kg\,N\,m^{-2}\,yr^{-1}$
Specific leaf area		19.42	$m^2\,kg\,C^{-1}$
Stem : leaf C allocation	1.16	1.16	Ratio
Whole plant mortality fraction	0.014	0.015	$1\,yr^{-1}$

Table 3. Species-specific allometric and ecological parameters for the eight tree species used in ZELIG, representing species found in the upper Great Lakes. Species shown include *Populus grandidentata* (POGR), *Betula papyrifera* (BEPA), *Quercus rubra* (QURU), *Pinus strobus* (PIST), *Acer saccharum* (ACSA), *Acer rubrum* (ACRU), *Populus tremuloides* (POTR), and *Fagus grandifolia* (FAGR). All species were assigned a probability factor of stress mortality of 0.369, a probability factor of natural mortality of 2.408, and a zone of seed influence of 200. Full explanations of all parameters can be found in the original ZELIG paper (Urban, 1990).

Species	Age_{max}	DBH_{max}	HT_{max}	G	$DegD_{min}$	$DegD_{max}$	L	D	N	RSER	Stock
POGR	150	70	30	42	800	3169	4	5	2	0.82	0.8
PIST	450	150	37	68	800	3183	3	2	3	0.90	0.7
QURU	400	100	30	92	800	4903	2	3	2	0.44	0.7
ACRU	150	100	30	244	800	6986	2	2	1	0.56	0.8
BEPA	140	100	25	160	800	2500	4	3	3	0.33	0.2
FAGR	366	80	30	100	800	5894	2	2	2	0.44	0.5
ACSA	400	150	40.1	89	800	3200	1	2	2	0.30	0.4
POTR	150	75	37	158	889	5556	4	3	2	0.50	0.4

Key: Age_{max} – maximum age for the species (yr); DBH_{max} – maximum diameter at breast height (cm); HT_{max} – maximum height (m); G – growth rate scaling coefficient; $DegD_{min}$ – minimum growing degree day; $DegD_{max}$ – maximum growing degree day; Light (L), Drought (D), Nutrient (N) – light/shade tolerance class, maximum drought tolerance class, and soil nutrient tolerance class; RSER – relative seedling establishment rate; Stock – regeneration stocking.

vests used by the two other models. Although ED also tracks the dynamics of individual trees, the configuration used here was limited to two tree functional types (cf. Table 4). This precluded species-specific girdling; instead, 13–14 % of the basal area across all individuals was harvested annually for the 3-year period.

The disturbances occurred on 1 May in all models, replicating the timing of the girdling treatment just prior to spring leaf-out (Gough et al., 2013). We examined six primary model outputs at an annual resolution – GPP, ER, NPP, NEP (all these fluxes in $Mg\,C\,ha^{-1}\,yr^{-1}$), maximum LAI (unitless), and aboveground biomass ($Mg\,C\,ha^{-1}$) – comparing them to observed data for 0 to 4 years after disturbance. We particularly focused on the models' structure and flux dynamics, i.e., whether they could replicate the relative changes observed in FASET.

3 Results

Summarizing the models' absolute performance provides a useful context for evaluating their relative changes discussed below. Pretreatment (i.e., control plots in 2007–8) aboveground biomass and LAI were 81.2 ± 25.4 $Mg\,C\,ha^{-1}$ and 4.3 ± 1.3, respectively (Fig. 1). The models' comparable values ranged from 51 (Biome-BGC) to 101 (ED) to 109 (ZELIG) $Mg\,C\,ha^{-1}$ for biomass and 1.5 to 4.9 to 6.4 for LAI, respectively. Biome-BGC's forest, in other words, was significantly smaller than the observed data; ZELIG's slightly larger; and that simulated by ED roughly comparable. Observed pretreatment gross primary production (GPP) was 12.2 $Mg\,C\,ha^{-1}\,yr^{-1}$, and ecosystem respiration (ER) was 9.1 $Mg\,C\,ha^{-1}\,yr^{-1}$, resulting in net C fluxes of 6.6 and 2.2 $Mg\,C\,ha^{-1}\,yr^{-1}$ for NPP and NEP, respectively (Fig. 1e, f). The models' pretreatment GPP values ranged from 2.2 (ED) to 6.8 (ZELIG) $Mg\,C\,ha^{-1}\,yr^{-1}$, with Biome-BGC roughly halfway between these two; all were thus much lower than observations. Con-

Table 4. Allometric and ecological parameters used in the ED model. The two plant functional types represent generic cold deciduous hardwood and evergreen needleleaf trees, respectively.

Parameter	Cold Deciduous	Evergreen	Units
V_{max}	12.5	12.5	μmol m^{-2} s^{-1}
Height computation*	$H = 2.34\,D^{0.64}$	$H = 1.04\,D^{0.94}$	m (H) and cm (D)
Max height	35	35	m
Specific leaf area	18.2	5.5	m^2 kg C^{-1}
Phenology temperature	10	–	°C
Density-independent mortality	0.014	0.014	1 yr^{-1}

* Height (H, m) is computed based on DBH (D, cm).

trol forest NPP values of both Biome-BGC and ZELIG were low (2.6 and 3.7 Mg C ha^{-1} yr^{-1}, respectively), while ED was 8.2 Mg C ha^{-1} yr^{-1}. ZELIG was very close (2.1 Mg C ha^{-1} yr^{-1}) to the observed NEP value, with ED and Biome-BGC much smaller (1.4 and 0.3 Mg C ha^{-1} yr^{-1}, respectively). In summary, pretreatment carbon stocks and fluxes varied significantly among the models, with Biome-BGC consistently low – a smaller forest producing and sequestering less C. The other two models varied in their fidelity to observations, with only ED able to achieve observed NPP, while ZELIG was closest to overall C balance, but neither could achieve the high observed GPP values.

Aboveground biomass declined by 35–36 % between 2006 and 2010 in the FASET experiment. The models tracked this well (Fig. 2a), although the decline occurred more slowly because of the protocol used in this modeling experiment (i.e., three successive years of 13–14 % cut instead of a single large girdling event). Leaf area index was less well reproduced: ED and ZELIG came close to capturing the magnitude of the observed decline (−30 and 33 %, respectively, compared to −37 to −44 % observed) but not the observed rebound of LAI by 2011 (Fig. 2d). Leaf area in Biome-BGC, in contrast, captured the timing and rebound of observed LAI but not its magnitude, as LAI only declined by 13 % in the model.

None of the models fully captured the main C flux dynamics observed in FASET. GPP initially rose in the treatment plots relative to the observed plots, but the models all simulated GPP declines (Fig. 1c) of up to 5 % (Biome-BGC), 10 % (ED), and 14 % (ZELIG). The models also all produced modest ER declines for 2008–2010, whereas observed ER rose by 10 % relative to control values; this is perhaps not surprising, given that our modeling protocol removed "girdled" trees from the ecosystem. Observed net primary production did not significantly differ between treatment and control plots (Fig. 1f), but the models all exhibited NPP declines, by up to 3 % (Biome-BGC), 10 % (ED), and 14 % (ZELIG). All models' treatment NPP had, however, recovered to control levels by 2012 (Fig. 2f). Net ecosystem production was also unchanged in the observations, while Biome-BGC NEP declined by 23–27 % (Fig. 2e). ED and ZELIG recorded even larger drops – of 79 and 43 %, respectively – although NEP had, like NPP, recovered to control levels 4 years following disturbance in all models.

The models' skill – i.e., how well they replicated both the magnitude and timing of all observed variables – is summarized in Fig. 3, a Taylor plot (Taylor, 2001) that is useful for summarizing both multiple aspects of complex models and relative skill. Here, all models exhibited low correlation (0.08–0.29) with observations, high root-mean-square difference (9–18 %) between simulated and observed values, and high standard deviation, implying overall low model skill.

In ZELIG, aspen and birch exhibited low to moderate resilience (i.e., full recovery to pretreatment basal area was not achieved) following moderate forest disturbance. The model also predicted which species thrived or declined post-disturbance (Fig. 4). Of the two treatment species that were girdled, aspen showed a stronger resilience and recovered to 71 % of pretreatment basal area after 4 years, increasing by 3.1 m^2 ha^{-1}. In contrast, birch remained at post-treatment basal area over the next 60 years, increasing by only 0.2 m^2 ha^{-1}. The ZELIG forest became dominated by red oak (Fig. 4), with that species' basal area increasing nearly 2-fold, followed by sugar maple and white pine, which increased by 72 and 6 %, respectively. Thirty years after disturbance, the total basal area as predicted by ZELIG was 33.6 vs. 32.7 m^2 ha^{-1} pretreatment, and recovery of basal area (a proxy for recovery of biomass) was achieved, even though ZELIG failed to capture the observed high resilience in C fluxes during the first 4 years after disturbance.

Similarly, the reduction in the number of individuals in ED resulted in a direct reduction in LAI, due to the strict allometric relationships used. Because NPP and NEP are so closely tied to LAI in ED, this resulted in low resistance to the disturbance event.

4 Discussion

Relatively few previous studies have examined how well models can simulate non-catastrophic forest disturbance. Peters et al. (2013) used the PnET-CN model to examine how disturbance type, intensity, and frequency influenced forest

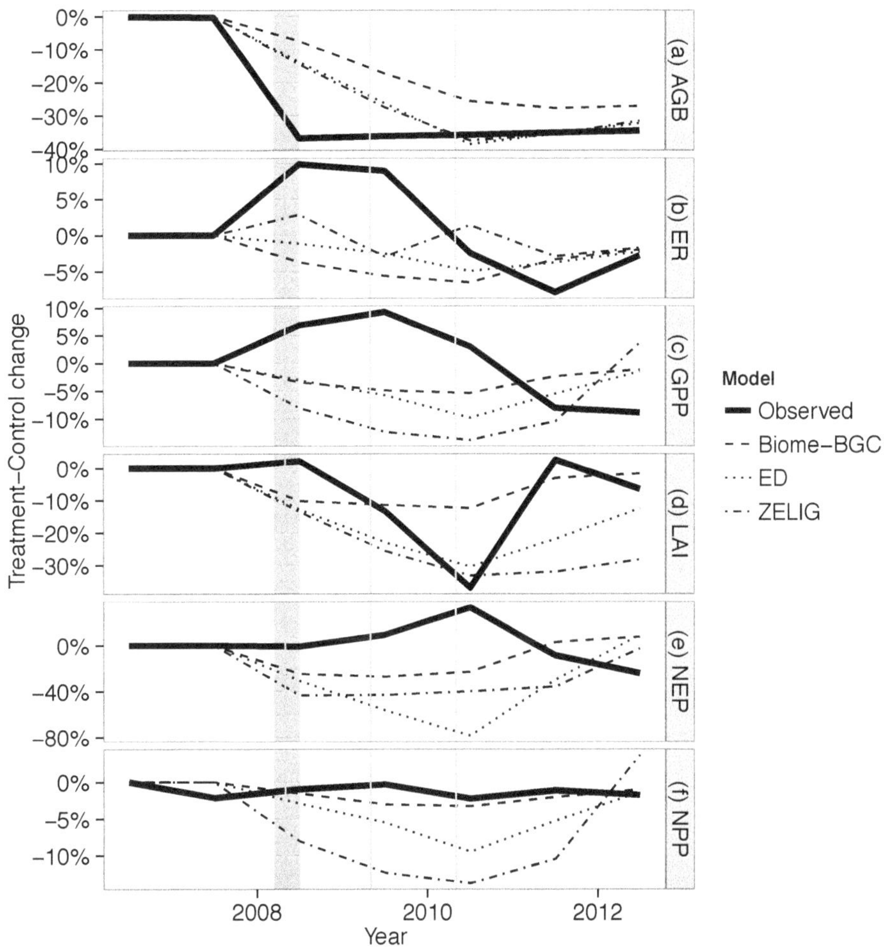

Figure 2. Model performance in replicating the FASET experiment. Panels include **(a)** aboveground biomass (AGB, in $Mg\,C\,ha^{-1}$), **(b)** ecosystem respiration (ER, $Mg\,C\,ha^{-1}$), **(c)** GPP ($Mg\,C\,ha^{-1}$), **(d)** leaf area index (LAI, unitless), **(e)** net ecosystem production (NEP, $Mg\,C\,ha^{-1}$), and **(f)** net primary production (NPP, $Mg\,C\,ha^{-1}$), all expressed on a common normalized scale (relative change between treatment and control). Vertical shaded area shows approximate time of the girdling treatment described in the text. Vertical lines show May 1 forest harvests imposed in the Biome-BGC, ED, and ZELIG models.

NPP for forest stands across the upper Midwest, and found that increasing intensity had no effect on deciduous species but decreased evergreen NPP. Wang et al. (2014) also used PnET-CN and reported that measured and modeled evergreen needleleaf forests had lower resilience to disturbance than deciduous forests. This agrees with Biome-BGC's behavior, in which broadleaf deciduous trees (such as simulated here) are less sensitive to moderate disturbance than are evergreen conifers (Thornton et al., 2002). The interaction of disturbance intensity and forest resilience thus has both short- and long-term effects, presenting significant challenges to models (Seidl et al., 2014; Dietze and Matthes, 2014).

4.1 Model mechanisms and behaviors

Gough et al. (2013) proposed several mechanisms supporting sustained C uptake and storage (in particular the fluxes in NPP and NEP) after the FASET disturbance: enhancement of canopy light use efficiency, maintenance of light absorption as later successional species take advantage of increased light availability, and redistribution of N from senescent to early successional trees (Nave et al., 2011). The three models used in this study are highly variable in their assumptions, parameters, and processes, and it is instructive to understand how and why each had difficulty reproducing the FASET results with respect to these proposed mechanisms.

All the models here, along with most others (e.g., Potter et al., 2003), assume a fixed light use efficiency (LUE): trees in the model can produce more or less leaf area, intercepting more or less radiation, but that area will produce a fixed amount of photosynthate under particular environmental conditions of light, temperature, etc. In reality trees can produce leaves with different structural, chemical, and photosynthetic characteristics (e.g., Sardans et al., 2012). These changes, integrated across leaves within a forest canopy, would likely result in different post-disturbance biotic and

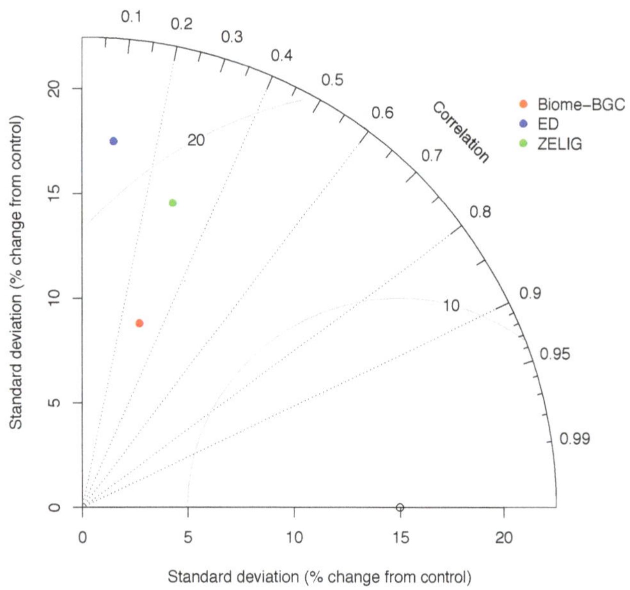

Figure 3. Taylor diagram (Taylor, 2001) summarizing model skill at predicting all (AGB, ER, GPP, LAI, NEP, NPP; cf. Fig. 1) observed data, normalized relative to the control forest. The standard deviation of the simulated data (colored by model) is gauged by the radial distance from the origin and can be compared to the observed data (circle on horizontal axis); model correlation to observations is found by azimuthal position; the curve contours show root mean square error (%).

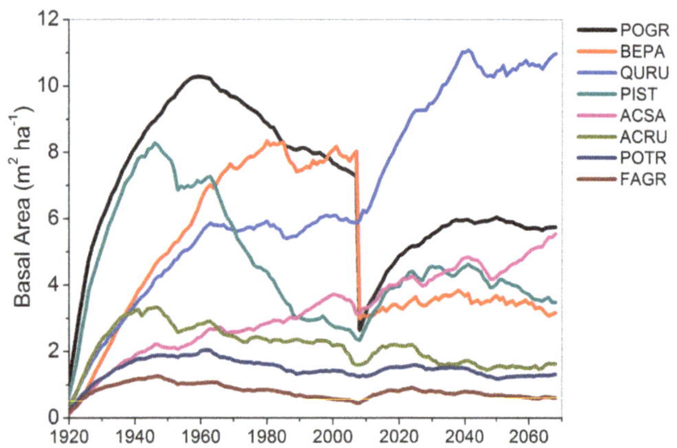

Figure 4. Species-specific basal area trajectories simulated by ZELIG, before and after the 2008–2010 tree removals mimicking the FASET experiment. Species codes are as in Table 1.

abiotic dynamics; FASET has already shown the assumption of a fixed LUE not to be true at the stand level (Gough et al., 2013). Recent work has also shown that the use of a spatially variable LUE parameterization, using C flux measurements from the Fluxnet data set, can significantly improve the accuracy of modeled GPP (Madani et al., 2014).

Maintenance of canopy light absorption in the FASET forest depends on a structurally heterogeneous canopy so that subdominant trees quickly increase their absorption following the girdling of canopy dominants (Gough et al., 2013). We would have expected, a priori, that ZELIG would be best able to simulate this dynamic, as it models a wide range of competing tree species, both early and late successional, competing in the same forest (Fig. 4). All models simulated small to moderate declines in both GPP and ER with disturbance, in contrast to the small observational increases. The differences were generally small, however, and both fluxes are not direct observations but rather derived from tower measurements of ecosystem exchange, and they thus use less well constrained than NPP and NEP. For this reason we consider the models' inability to replicate the *absolute* GPP and ER (which were 2–3 times higher than simulations) more troubling than their failure to exactly match the relative patterns shown in Fig. 2. NPP and NEP are better constrained observationally than are derived fluxes. Biome-BGC best maintained these fluxes with disturbance but for the wrong reason, i.e., too resilient a leaf area (Fig. 2b), rather than by increasing LUE when LAI declined in the FASET study. We note, however, that the Biome-BGC phenology submodel was quite accurate (cf. Gough et al., 2010a), a critical first step to accurately simulate stand C dynamics (Richardson et al., 2012).

The proximal reason for Biome-BGC's too strong resilience is that the fraction of photosynthetically active radiation absorbed by the canopy, FPAR, does not diminish change linearly with LAI changes. Radiation transmission and absorption through canopies is a complex, computationally expensive process, and the three models studied here all use a common simplification: Beer's law (Campbell and Norman, 1998), which models this process as an exponential decrease downwards through the canopy. Biome-BGC, ZELIG, and ED also all assume a (mostly) equal extinction coefficient, and this implies that the models' FPAR declines theoretically peaked at 3, 12, and 8 %, respectively (Fig. 5), compared to 6 % as measured in the field (Gough et al., 2013). The mathematical form of Beer's law means that FPAR declines are smallest at low and high LAI values. For Biome-BGC, with its low-biomass forest, this meant relatively small FPAR declines with disturbance, small to moderate quantities of stored C and N lost to disturbance, and enough stored C resources to fully leaf out the canopy and support photosynthesis over the growing season.

ZELIG and ED both matched the observed LAI decline, and reasonably approximated FPAR as well, but exhibited large declines in NPP and NEP for both models. In ZELIG, even with the post-disturbance increase in available light, the remaining subdominant species were not able to quickly increase their growth to make up the difference in NPP loss. This may be due to the inherent growth and life history strategies of these subdominant species, which is accounted for in the species parameterization and initialization of ZELIG (Table 3). Only one species, red oak, recovered quickly (Fig. 4), while the remaining dominant species and subdominant species could not contribute to an increase in NPP and NEP. Based on the current model structure of ZELIG, leaf production and leaf loss are tightly linked with NPP and NEP; therefore, the decline in LAI corresponded to a resulting decline in C fluxes.

In a separate study, ZELIG-TROP, a modified version of ZELIG that simulates tropical forests, was successful at replicating a nonsignificant change in NPP as a result of gradual, less extensive tree mortality (Holm et al., 2014). That study used a continual low-level elevated mortality rate as a treatment, i.e., doubling annual background mortality rate, and ZELIG-TROP predicted highly resilient NPP. However, following a one-time dramatic disturbance event (removing 20 % of basal area) NPP also declined, matching the modeled results seen here. Thus, the ZELIG results are characteristic of the model and not dependent on the particular forest type, soils, or climate of the FASET experiment.

In ED, despite the increase in light availability following disturbance, the remaining undisturbed trees were not able to respond sufficiently to offset NPP loss. This may be due in part to the limited number of plant functional types used here not representing the competition of early and late successional species. Additionally, ED's scaling of individual trees to stand dynamics does not maintain the full level of canopy complexity, which may be required for resilience to a disturbance of this type.

Among the models tested here, nitrogen redistribution and limitation was only possible in Biome-BGC, as ZELIG lacks an N cycle, and ED's integrated N cycle was not parameterized or enabled in this study. Biome-BGC's integrated N cycle encompasses N fixation, its deposition and leaching, plant growth, and microbial decomposition, and should, in theory, constrain C uptake in many circumstances (Thornton et al., 2007). Such an effect was not noticeable here, however, as equal percentages of C and N were removed in the Biome-BGC disturbances (data not shown); this implies leaching/loss, i.e., a lack of N conservation as opposed to what was observed in FASET (Nave et al., 2011). This may also partly be an artifact, as all models used stem biomass removals to simulate the real-world girdling (although in Biome-BGC leaves were transferred to the litter pool, providing some N reallocation). We speculate, however, that excessive N limitation was a factor in the model's inability to match the C stock and flux values of the UMBS forest.

In summary, the biological mechanisms proposed (Gough et al., 2013) to explain the carbon-cycle resilience of a mid-successional forest to disturbance are ones that most models either do not simulate (integrated C and N cycles, changing light use efficiency) or do so only crudely (canopy structure, heterotrophic respiration). At fine spatial scales, factors such as canopy structure can be simulated, but the computational demands are large and thus impractical for larger-scale models (Caspersen et al., 2011), a consideration that inspired the development of models such as ED (Moorcroft et al., 2001). Similarly, how to translate the N recycling microbial dynamics into ecosystem- to global-scale models is an area of intense research (Wieder et al., 2013), as most models (including those tested here) use a few conceptual soil pools following simple first-order kinetics. C–N integration inside such models is increasingly common (Zaehle et al., 2014; Thornton et al., 2007), enabling N redistribution and limitation dynamics, and should improve future simulations of moderate disturbances.

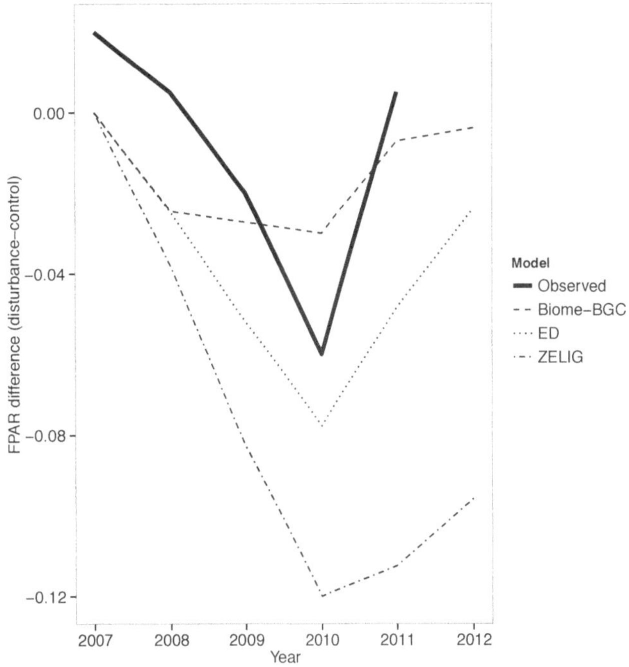

Figure 5. Effect of disturbance on fraction of photosynthetically active radiation absorbed by the canopy (FPAR). Observed line is based on data from Fig. 4 in Gough et al. (2013). Model lines show implied (i.e., theoretical, based on Beer's law) FPAR based on the observed and modeled leaf area index values and a common extinction coefficient of $k = -0.45$, the model mean.

5 Conclusions

The FASET results were unexpected and intriguing (Nave et al., 2011; Gough et al., 2013; Hardiman et al., 2013). How well can current forest models simulate such moderate, i.e., not stand-replacing, disturbances? Not all disturbances, even of the same severity, equally affect biogeochemical processes that support recovery – for example, slow versus immediate tree death have very different consequences (Franklin et al., 1987). Our results suggest that some ecosystem models, developed to simulate processes following stand-replacing disturbances, may not simulate gradual death scenarios well (McDowell et al., 2013), specifically nonlinear or threshold responses of the carbon cycle in disturbance intensity (Goodrich-Stuart et al., 2015) over short timescales. Their skill over longer (decadal) periods remains an open question. This is particularly important as the moderate disturbances associated with slow tree death (insect outbreaks, fungal pathogens) are on the rise worldwide (Allen et al., 2010) and in aging US forests. It is thus increasingly important to confront models with non-catastrophic disturbance scenarios.

Acknowledgements. B. Bond-Lamberty and V. L. Bailey were supported by the Office of Science of the US Department of Energy, as part of the Terrestrial Ecosystem Sciences Program. J. A. Holm was supported by Jeffrey Chambers for part of this work. C. M. Gough and FASET were supported by the Climate and Environmental Sciences Division, Office of Science, US Department of Energy (DOE) (Award No. DE-SC0006708) and by the Ameriflux Management Project (Flux 506 Core Site agreement No. 7096915) through Lawrence Berkeley National Laboratory. We acknowledge the University of Michigan Biological Station for infrastructure and logistics support and are grateful for the insightful comments of two anonymous referees.

Edited by: J. Xiao

References

Allen, C. D., Macalady, A. K., Chenchouni, H., Bachelet, D., McDowell, N. G., Vennetier, M., Kitzberger, T., Ringling, A., Breshears, D. D., Hogg, E. H., Gonzalez, P., Fensham, R., Zhang, Z., Castro, J., Demidova, N., Lim, J.-H., Allard, G., Running, S. W., Semerci, A., and Cobb, N.: A global overview of drought and heat-induced tree mortality reveals emerging climate change risks for forests, Forest Ecol. Manage., 259, 660–684, doi:10.1016/j.foreco.2009.09.001, 2010.

Amiro, B. D., Barr, A. G., Barr, J. G., Black, T. A., Bracho, R., Brown, M., Chen, J. M., Clark, K. L., Davis, K. J., Desai, A. R., Dore, S., Engel, V., Fuentes, J. D., Goldstein, A. H., Goulden, M. L., Kolb, T. E., Lavigne, M. B., Law, B. E., Margolis, H. A., Martin, T. A., McCaughey, J. H., Misson, L., Montes-Helu, M. C., Noormets, A., Randerson, J. T., Starr, G., and Xiao, J.: Ecosystem carbon dioxide fluxes after disturbance in forests of North America, J. Geophys. Res.-Biogeosci., 115, G00K02, doi:10.1029/2010JG001390, 2010.

Birdsey, R. A., Pregitzer, K. S., and Lucier, A.: Forest carbon management in the United States: 1600-2100, J. Environ. Qual., 35, 1461–1469, doi:10.2134/jeq2005.0162, 2006.

Bonan, G. B. and Levis, S.: Quantifying carbon-nitrogen feedbacks in the Community Land Model (CLM4), Geophys. Res. Lett., 37, L07401, doi:10.1029/2010GL042430, 2010.

Bond-Lamberty, B., Peckham, S. D., Ahl, D. E., and Gower, S. T.: The dominance of fire in determining carbon balance of the central Canadian boreal forest, Nature, 450, 89–92, doi:10.1038/nature06272, 2007.

Botkin, D. B., Janak, J. F., and Wallis, J. R.: Rationale, limitations and assumptions of a northeastern forest growth simulator, IBM J. Res. Dev., 16, 101–116, 1972.

Campbell, G. S. and Norman, J. M.: An Introduction to Environmental Biophysics, Springer-Verlag, New York, 286 pp., 1998.

Carvalhais, N., Reichstein, M., Ciais, P., Collatz, G. J., Mahecha, M. D., Montagnani, L., Papale, D., Rambal, S., and Seixas, J.: Identification of vegetation and soil carbon pools out of equilibrium in a process model via eddy covariance and biometric constraints, Global Change Biol., 16, 2813–2829, doi:10.1111/j.1365-2486.2010.02173.x, 2010.

Caspersen, J. P., Pacala, S. W., Jenkins, J. C., Hurtt, G. C., Moorcroft, P. R., and Birdsey, R. A.: Contributions of land-use history to carbon accumulation in US forests, Science, 290, 1148–1151, doi:10.1126/science.290.5494.1148, 2000.

Caspersen, J. P., Vanderwel, M. C., Cole, W. G., and Purves, D.: How stand productivity results from size- and competition-dependent growth and mortality, PLOS ONE, 6, e28660, doi:10.1371/journal.pone.0028660, 2011.

Dietze, M. C. and Matthes, J. H.: A general ecophysiological framework for modelling the impact of pests and pathogens on forest ecosystems, Ecol. Lett., 17, 1418–1426, doi:10.1111/ele.12345, 2014.

Fisk, J. P., Hurtt, G. C., Chambers, J. Q., Zeng, H., Dolan, K. A., and Negrón-Juárez, R. I.: The impacts of tropical cyclones on the net carbon balance of eastern US forests (1851–2000), Environ. Res. Lett., 8, 045017, doi:10.1088/1748-9326/8/4/045017, 2013.

Franklin, J. F., Shugart, H. H., and Harmon, M. E.: Tree death as an ecological process: the causes, consequences and variability of tree mortality, BioScience, 37, 55-0-556, doi:10.2307/1310665, 1987.

Frelich, L. E. and Reich, P. B.: Spatial patterns and succession in a Minnesota southern-boreal forest, Ecol. Monogr., 65, 325–346, 1995.

Goetz, S. J., Bond-Lamberty, B., Harmon, M. E., Hicke, J. A., Houghton, R. A., Kasischke, E. S., Law, B. E., McNulty, S. G., Meddens, A. J. H., Mildrexler, D., O'Halloran, T. L., and Pfeifer, E. M.: Observations and assessment of forest carbon recovery following disturbance in North America, J. Geophys. Res.-Biogeosci., 117, G02022, doi:10.1029/2011JG001733, 2012.

Goodrich-Stuart, E. J., Curtis, P. S., Fahey, R. T., Vogel, C. S., and Gough, C. M.: Forest net primary production exhibits a threshold response to increasing disturbance severity Ecosystems, submitted, 2015.

Gough, C. M., Vogel, C. S., Kazanski, C., Nagel, L., Flower, C. E., and Curtis, P. S.: Coarse woody debris and the carbon balance of a north temperate forest, Forest Ecol. Manage., 244, 60–67, doi:10.1016/j.foreco.2007.03.039, 2007.

Gough, C. M., Vogel, C. S., Schmid, H. P., and Curtis, P. S.: Multi-year convergence of biometric and meteorological estimates of forest carbon storage, Agric. Forest Meteorol., 148, 158–170, doi:10.1016/j.agrformet.2007.08.004, 2008.

Gough, C. M., Flower, C. E., Vogel, C. S., Dragoni, D., and Curtis, P. S.: Whole-ecosystem labile carbon production in a north temperate deciduous forest, Agric. Forest Meteorol., 149, 1531–1540, doi:10.1016/j.agrformet.2009.04.006, 2009.

Gough, C. M., Flower, C. E., Vogel, C. S., and Curtis, P. S.: Phenological and Temperature Controls on the Temporal Non-Structural Carbohydrate Dynamics of Populus grandidentata and Quercus rubra, Forests, 1, 65–81, doi:10.3390/f1010065, 2010a.

Gough, C. M., Vogel, C. S., Hardiman, B., and Curtis, P. S.: Wood net primary production resilience in an unmanaged forest transitioning from early to middle succession, Forest Ecol. Manage., 260, 36–41, doi:10.1016/j.foreco.2010.03.027, 2010b.

Gough, C. M., Hardiman, B., Nave, L. E., Bohrer, G., Maurer, K. D., Vogel, C. S., Nadelhoffer, K. J., and Curtis, P. S.: Sustained carbon uptake and storage following moderate disturbance in a Great Lakes forest, Ecol. Appl., 23, 1202–1215, doi:10.1890/12-1554.1, 2013.

Granier, A., Breda, N., Longdoz, B., Gross, P., and Ngao, J.: Ten years of fluxes and stand growth in a young beech forest at Hesse, north-eastern France, Ann. For. Sci., 65, 704, doi:10.1051/forest:2008052, 2008.

Hardiman, B., Bohrer, G., Gough, C. M., Vogel, C. S., and Curtis, P. S.: The role of canopy structural complexity in wood net primary production of a maturing northern deciduous forest Ecology, 92, 1818–1827, doi:10.1890/10-2192.1, 2011.

Hardiman, B., Gough, C. M., Halperin, A., Hofmeister, K. L., Nave, L. E., Bohrer, G., and Curtis, P. S.: Maintaining high rates of carbon storage in old forests: A mechanism linking canopy structure to forest function, Forest Ecol. Manage., 298, 111–119, doi:10.1016/j.foreco.2013.02.031, 2013.

He, L., Chen, J. M., Pan, Y., Birdsey, R. A., and Kattge, J.: Relationships between net primary productivity and forest stand age in US forests, Global Biogeochem. Cy., 26, GB3009, doi:10.1029/2010GB003942, 2012.

Hicke, J. A., Allen, C. D., Desai, A. R., Dietze, M. C., Hall, R. J., Hogg, E. H., Kashian, D. M., Moore, D. J. P., Raffa, K. F., Sturrock, R. N., and Vogelmann, J. E.: Effects of biotic disturbances on forest carbon cycling in the United States and Canada, Global Change Biol., 18, 7–34, doi:10.1111/j.1365-2486.2011.02543.x, 2011.

Holm, J. A., Shugart, H. H., Van Bloem, S. J., and Larocque, G. R.: Gap model development, validation, and application to succession of secondary subtropical dry forests of Puerto Rico, Ecol. Model., 233, 70–82, doi:10.1016/j.ecolmodel.2012.03.014, 2012.

Holm, J. A., Thompson, J. R., McShea, W. J., and Bourg, N. A.: Interactive effects of chronic deer browsing and canopy gap disturbance on forest successional dynamics, Ecosphere, 4, 44, doi:10.1890/es13-00223.1, 2013.

Holm, J. A., Chambers, J. Q., Collins, W. D., and Higuchi, N.: Forest response to increased disturbance in the central Amazon and comparison to western Amazonian forests, Biogeosciences, 11, 5773–5794, doi:10.5194/bg-11-5773-2014, 2014.

Hurtt, G. C., Pacala, S. W., Moorcroft, P. R., Caspersen, J. P., Shevliakova, E., Houghton, R. A., and Moore III, B.: Projecting the future of the US carbon sink, P. Natl. Acad. Sci. USA, 99, 1389–1394, doi:10.1073/pnas.012249999, 2002.

Kanamitsu, M., Ebisuzaki, W., Woollen, J., Yang, S.-K., Hnilo, J. J., Fiorino, M., and Potter, G. L.: NCEP–DOE AMIP-II Reanalysis (R-2), B. of the Am. Meteorol. Soc., 83, 1631–1643, doi:10.1175/BAMS-83-11-1631, 2002.

Keenan, T. F., Davidson, E. A., Moffat, A. M., Munger, J. W., and Richardson, A. D.: Using model-data fusion to interpret past trends, and quantify uncertainties in future projections, of terrestrial ecosystem carbon cycling, Global Change Biol., 18, 2555–2569, doi:10.1111/j.1365-2486.2012.02684.x, 2012.

Kimball, J. S., Thornton, P. E., White, M. A., and Running, S. W.: Simulating forest productivity and surface-atmosphere carbon exchange in the BOREAS study region, Tree Physiol., 17, 589–599, 1997.

Knohl, A., Kolle, O., Minayeva, T. Y., Milyukova, I. M., Vygodskaya, N. N., Foken, T., and Schulze, E.-D.: Carbon dioxide exchange of a Russian boreal forest after disturbance by wind throw, Global Change Biol., 8, 231–246, 2002.

Landsberg, J. J. and Waring, R. H.: A generalised model of forest productivity using simplified concepts of radiation-use efficiency, carbon balance and partitioning, Forest Ecol. Manage., 95, 209–228, doi:10.1016/S0378-1127(97)00026-1, 1997.

Larocque, G. R., Archambault, L., and Delisle, C.: Modelling forest succession in two southeastern Canadian mixedwood ecosystem types using the ZELIG model, Ecol. Model., 199, 350–362, doi:10.1016/j.ecolmodel.2006.05.010, 2006.

Law, B. E., Sun, O. J., Campbell, J. S., Van Tuyl, S., and Thornton, P. E.: Changes in carbon storage and fluxes in a chronosequence of ponderosa pine, Global Change Biol., 9, 510–524, 2003.

Leemans, R. and Prentice, I. C.: FORSKA: A general forest succession model, Meddelanden, Vaxtbiologiska institutionen, Vaxtbiologiska institutionen, Uppsala, Sweden, 2, 1–145, 1989.

Li, Z., Apps, M. J., Kurz, W. A., and Banfield, E.: Temporal changes of forest net primary production and net ecosystem production in west central Canada associated with natural and anthropogenic disturbances, Can. J. For. Res., 33, 2340–2351, doi:10.1139/x03-168, 2003.

Luyssaert, S., Schulze, E.-D., Börner, A., Knohl, A., Hessenmöller, D., Law, B. E., Ciais, P., and Grace, J.: Old-growth forests as global carbon sinks, Nature, 455, 213–215, doi:10.1038/nature07276, 2008.

Madani, N., Kimball, J. S., Affleck, D. L. R., Kattge, J., van Bodegom, P. M., Reich, P. B., and Running, S. W.: Improving ecosystem productivity modeling through spatially explicit estimation of optimal light use efficiency, J. Geophys. Res.-Biogeosci., 119, 1755–1769, doi:10.1002/2014JG002709, 2014.

Matheny, A. M., Bohrer, G., Vogel, C. S., Morin, T. H., He, L., Frasson, R. P. d. M., Mirfenderesgi, G., Schäfer, K. V. R., Gough, C. M., Ivanov, V. Y., and Curtis, P. S.: Species-specific transpiration responses to intermediate disturbance in a northern hardwood forest, J. Geophys. Res.-Biogeosci., in press, doi:10.1002/2014JG002804, 2015.

Mathys, A., Black, T. A., Nesic, Z., Nishio, G., Brown, M., Spittlehouse, D. L., Fredeen, A. L., Bowler, R., Jassal, R. S., Grant, N. J., Burton, P. J., Trofymow, J. A., and Meyer, G.: Carbon balance of a partially harvested mixed conifer forest following mountain pine beetle attack and its comparison to a clear-cut, Biogeosciences, 10, 5451–5463, doi:10.5194/bg-10-5451-2013, 2013.

McDowell, N. G., Fisher, R. A., Xu, C., Domec, J.-C., Höltta, T., Mackay, D. S., Sperry, J. S., Boutz, A., Dickman, L. T., Gehres, N., Limousin, J. M., Macalady, A. K., Martínez-Vilalta, J., Mencuccini, M., Plaut, J. A., Ogée, J., Pangle, R. E., Rasse, D. P., Ryan, M. G., Sevanto, S., Waring, R. H., Williams, A. P., Yepez, E. A., and Pockman, W. T.: Evaluating theories of drought-induced vegetation mortality using a multimodel–experiment framework, New Phytol., 200, 304–321, doi:10.1111/nph.12465, 2013.

Medvigy, D. and Moorcroft, P. R.: Predicting ecosystem dynamics at regional scales: an evaluation of a terrestrial biosphere model for the forests of northeastern North America, Philos. T. Roy. Soc. B, 367, 222–235, doi:10.1098/rstb.2011.0253, 2012.

Medvigy, D., Wofsy, S. C., Munger, J. W., Hollinger, D. Y., and Moorcroft, P. R.: Mechanistic scaling of ecosystem function and dynamics in space and time: Ecosystem Demography model version 2, J. Geophys. Res.-Biogeosci., 114, G01002, doi:10.1029/2008JG000812, 2009.

Moorcroft, P. R., Hurtt, G. C., and Pacala, S. W.: A method for scaling vegetation dynamics: the Ecosystem Demography model (ED), Ecol. Monogr., 71, 557–586, doi:10.1890/0012-9615(2001)071[0557:AMFSVD]2.0.CO;2, 2001.

Nave, L. E., Gough, C. M., Maurer, K. D., Bohrer, G., Le Moine, J., Munoz, A. B., Nadelhoffer, K. J., Sparks, J. P., Strahm, B. D., Vogel, C. S., and Curtis, P. S.: Disturbance and the resilience of coupled carbon and nitrogen cycling in a north temperate forest, J. Geophys. Res.-Biogeosci., 116, G04016, doi:10.1029/2011JG001758, 2011.

Nelder, J. A. and Mead, R.: A simplex method for function minimization, Computer Journal, 7, 308–313, doi:10.1093/comjnl/7.4.308, 1965.

Pan, Y., Chen, J. M., Birdsey, R., McCullough, K., He, L., and Deng, F.: Age structure and disturbance legacy of North American forests, Biogeosciences, 8, 715–732, doi:10.5194/bg-8-715-2011, 2011.

Parton, W. J., Schimel, D. S., Cole, C. V., and Ojima, D. S.: Analysis of factors controlling soil organic matter levels in Great Plains grasslands, Soil Sci. Soc. Am. J., 51, 1173–1179, 1987.

Peng, C., Jiang, H., Apps, M. J., and Zhang, Y.: Effects of harvesting regimes on carbon and nitrogen dynamics of boreal forests in central Canada: a process model simulation, Ecol. Model., 155, 177–189, doi:10.1016/S0304-3800(02)00134-5, 2002.

Peters, E. B., Wythers, K. R., Bradford, J. B., and Reich, P. B.: Influence of disturbance on temperate forest productivity, Ecosystems, 16, 95–110, doi:10.1007/s10021-012-9599-y, 2013.

Pietsch, S. A., Hasenauer, H., Kucera, J., and Cermák, J.: Modeling effects of hydrological changes on the carbon and nitrogen balance of oaks in floodplains, Tree Physiol., 23, 735–746, 2003.

Potter, C. S., Klooster, S., Myneni, R. B., Genovese, V., Tan, P.-N., and Kumar, V.: Continental-scale comparisons of terrestrial carbon sinks estimated from satellite data and ecosystem modeling 1982–1998, Global Planet. Change, 39, 201–213, 2003.

Pregitzer, K. S. and Euskirchen, E. S.: Carbon cycling and storage in world forests: biome patterns relating to forest age, Global Change Biol., 10, 2052–2077, doi:10.1111/j.1365-2486.2004.00866.x, 2004.

Radeloff, V. C., Nelson, E., Plantinga, A. J., Lewis, D. J., Helmers, D., Lawler, J. J., Withey, J. C., Beaudry, F., Martinuzzi, S., Butsic, V., Lonsdorf, E., White, D. C., and Polasky, S.: Economic-based projections of future land use in the conterminous United States under alternative policy scenarios, Ecol. Appl., 22, 1036–1049, 2012.

Raulier, F.: Canopy photosynthesis of sugar maple (*Acer saccharum*): comparing big-leaf and multilayer extrapolations of leaf-level measurements., Tree Physiol., 19, 407–420, 1999.

Richardson, A. D., Anderson, R. S., Arain, M. A., Barr, A. G., Bohrer, G., Chen, G., Chen, J. M., Ciais, P., Davis, K. J., Desai, A. R., Dietze, M. C., Dragoni, D., Garrity, S. R., Gough, C. M., Grant, R. F., Hollinger, D. Y., Margolis, H. A., McCaughey, J. H., Migliavacca, M., Monson, R. K., Munger, J. W., Poulter, B., Raczka, B. M., Ricciuto, D. M., Sahoo, A. K., Schaefer, K., Tian, H.-Q., Vargas, R., Verbeeck, H., Xiao, J., and Xue, Y.: Terrestrial biosphere models need better representation of vegetation phenology: results from the North American Carbon Program Site Synthesis, Global Change Biol., 18, 566–584, doi:10.1111/j.1365-2486.2011.02562.x, 2012.

Rolff, C. and Ågren, G. I.: Predicting effects of different harvesting intensities with a model of nitrogen limited forest growth, Ecol. Model., 118, 193–211, doi:10.1016/S0304-3800(99)00043-5, 1999.

Running, S. W. and Hunt, R. E.: Generalization of a forest ecosystem process model for other biomes, BIOME-BGC, and an application for global-scale models, in: Scaling Physiologic Processes: Leaf to Globe, edited by: Ehleringer, J. R. and Field, C. B., Academic Press, San Diego, CA, 141–158, 1993.

Sardans, J., Rivas-Ubach, A., and Peñuelas, J.: The C:N:P stoichiometry of organisms and ecosystems in a changing world: A review and perspectives, Perspectives in Plant Ecology and Evolution, 14, 33–47, doi:10.1016/j.ppees.2011.08.002, 2012.

Seidl, R., Rammer, W., and Spies, T. A.: Disturbance legacies increase the resilience of forest ecosystem structure, composition, and functioning, Ecol. Appl., 24, 2063–2077, doi:10.1890/14-0255.1, 2014.

Shugart, H. H. and West, D. C.: Development of an Appalachian deciduous forest succession model and its application to assessment of the impact of the chestnut blight, J. Environ. Manage., 5, 161–179, 1977.

Tatarinov, F. A. and Cienciala, E.: Long-term simulation of the effect of climate changes on the growth of main Central-European forest tree species, Ecol. Model., 220, 3081–3088, doi:10.1016/j.ecolmodel.2009.01.029, 2009.

Taylor, A. R., Chen, H. Y. H., and VanDamme, L.: A review of forest succession models and their suitability for forest management planning, Forest Science, 55, 23–36, 2009.

Taylor, K. E.: Summarizing multiple aspects of model performance in a single diagram, J. Geophys. Res., 106, 7183–7192, 2001.

Thornton, P. E., Law, B. E., Gholz, H. L., Clark, K. L., Falge, E., Ellsworth, D. S., Goldstein, A. H., Monson, R. K., Hollinger, D. Y., Falk, M., Chen, J., and Sparks, J. P.: Modeling and measuring the effects of disturbance history and climate on carbon and water budgets in evergreen needleleaf forests, Agric. Forest Meteorol., 113, 185–222, doi:10.1016/S0168-1923(02)00108-9, 2002.

Thornton, P. E., Lamarque, J.-F., Rosenbloom, N. A., and Mahowald, N. M.: Influence of carbon-nitrogen cycle coupling on land model response to CO2 fertilization and climate variability, Global Biogeochem. Cy., 21, GB4018, doi:10.1029/2006GB002868, 2007.

Urban, D. L.: A Versatile Model to Simulate Forest Pattern: A User's Guide to ZELIG Version 1.0., University of Virginia, Charlottesville, Virginia, 108 pp., 1990.

Urban, D. L., Bonan, G. B., Smith, T. M., and Shugart, H. H.: Spatial applications of gap models, Forest Ecol. Manage., 42, 95–110, doi:10.1016/0378-1127(91)90067-6, 1991.

Vanderwel, M. C., Coomes, D. A., and Purves, D.: Quantifying variation in forest disturbance, and its effects on aboveground biomass dynamics, across the eastern United States, Global Change Biol., 19, 1504–1517, doi:10.1111/gcb.12152, 2013.

Wang, W., Xiao, J., Ollinger, S. V., Desai, A. R., Chen, J., and Noormets, A.: Quantifying the effects of harvesting on carbon fluxes and stocks in northern temperate forests, Biogeosciences, 11, 6667–6682, doi:10.5194/bg-11-6667-2014, 2014.

Warren, J. M., Pötzelsberger, E., Wullschleger, S. D., Thornton, P. E., Hasenauer, H., and Norby, R. J.: Ecohydrologic impact of reduced stomatal conductance in forests exposed to elevated CO_2, Ecohydrology, 4, 196–210, doi:10.1002/eco.173, 2011.

Weng, E., Luo, Y., Wang, W., Wang, H., Hayes, D. J., McGuire, A. D., Hastings, A., and Schimel, D. S.: Ecosystem carbon storage capacity as affected by disturbance regimes: A general theoretical model, J. Geophys. Res.-Biogeosci., 117, G03014, doi:10.1029/2012JG002040, 2012.

White, M. A., Thornton, P. E., Running, S. W., and Nemani, R. R.: Parameterization and sensitivity analysis of the BIOME-BGC terrestrial ecosystem model: net primary production controls, Earth Interact., 4, 3, 1–85, 2000.

Wieder, W. R., Bonan, G. B., and Allison, S. D.: Global soil carbon projections are improved by modelling microbial processes, Nature Climate Change, 3, 909–912, doi:10.1038/NCLIMATE1951, 2013.

Williams, C. A., Collatz, G. J., Masek, J. G., and Goward, S. N.: Carbon consequences of forest disturbance and recovery across the conterminous United States, Global Biogeochem. Cy., 26, GB1005, doi:10.1029/2010GB003947, 2012.

Wolter, P. T. and White, M. A.: Recent forest cover type transitions and landscape structural changes in northeast Minnesota, USA, Landscape Ecol., 17, 133–155, 2002.

Zaehle, S., Medlyn, B. E., De Kauwe, M., Walker, A. P., Dietze, M. C., Hickler, T., Luo, Y., Wang, Y.-P., El-Masri, B., Thornton, P. E., Jain, A. K., Wang, S., Warlind, D., Weng, E., Parton, W. J., Iversen, C. M., Gallet-Budynek, A., McCarthy, H. R., Finzi, A. C., Hanson, P. J., Prentice, I. C., Oren, R., and Norby, R. J.: Evaluation of 11 terrestrial carbon–nitrogen cycle models against observations from two temperate Free-Air CO2 Enrichment studies, New Phytol., 202, 803–822, doi:10.1111/nph.12697, 2014.

Zhao, M., Xiang, W., Peng, C., and Tian, D.: Simulating age-related changes in carbon storage and allocation in a Chinese fir plantation growing in southern China using the 3-PG model, Forest Ecol. Manage., 257, 1520–1531, doi:10.1016/j.foreco.2008.12.025, 2009.

8

Divergence of above- and belowground C and N pool within predominant plant species along two precipitation gradients in North China

X. H. Ye[1], X. Pan[1,5], W. K. Cornwell[2,4], S. Q. Gao[1], M. Dong[1,3], and J. H. C. Cornelissen[2]

[1]State Key Laboratory of Vegetation and Environmental Change, Institute of Botany, Chinese Academy of Sciences, Beijing 100093, China
[2]Systems Ecology, Department of Ecological Science, VU University, Amsterdam, De Boelelaan 1085, 1081 HV Amsterdam, the Netherlands
[3]Key Laboratory of Hangzhou City for Ecosystem Protection and Restoration, College of Life and Environmental Sciences, Hangzhou Normal University, Hangzhou 310036, China
[4]School of Biological, Earth and Environmental Sciences, University of New South Wales, Sydney, Australia
[5]College of Life Sciences, University of Chinese Academy of Sciences, Beijing 100049, China

Correspondence to: M. Dong (dongming@ibcas.ac.cn)

Abstract. The coupling of carbon cycle and nitrogen cycle drives the food web structure and biogeochemistry of an ecosystem. However, across precipitation gradients, there may be a shift in C pool and N pool from above- to belowground because of shifting plant stoichiometry and allocation. Based on previous evidence, biomass allocation to roots should increase with aridity, while leaf [N] should increase. If their effect sizes are equal, they should cancel each other out, and the above- and belowground proportions of the N would remain constant. Here, we present the first study to explicitly compare above- and belowground pool sizes of N and C within predominant plant species along precipitation gradients. Biomass and nutrient concentrations of leaves, stems and roots of three predominant species were measured along two major precipitation gradients in Inner Mongolia, China. Along the two gradients, the effect sizes of the biomass shifts were remarkably consistent among three predominant species. However, the size of the shift in aboveground [N] was not, leading to a species-specific pattern in above- and belowground pool size. In two species (*Stipa grandis* and *Artemisia ordosica*) the effect sizes of biomass allocation and [N] were equal and the proportion of N of above- and belowground did not change with aridity, but in *S. bungeana* the increase in leaf [N] with aridity was much weaker than the biomass shift, leading to a decrease in the proportion of N aboveground at dry sites. We have found examples of consistent N pool sizes above- and belowground and a shift to a greater proportion of belowground N in drier sites depending on the species. We suggest that precipitation gradients do potentially decouple the C and N pool, but the exact nature of the decoupling depends on the dominant species' capacity for intraspecific variation.

1 Introduction

Plant carbon economy and nitrogen economy are closely coupled (Shaver and Chapin, 1991; Aerts and Chapin, 2000), and this coupling has important feedbacks to terrestrial carbon budgets and climate (IPCC, 2007; Chapin et al., 2009). To capture and utilize light, nutrients, carbon and water, plants allocate resources both to aboveground shoots and belowground roots. Nitrogen is obtained belowground via roots and partly transported to the aboveground shoots to support photosynthesis principally through the enzyme Rubisco and other parts of the photosynthetic systems, while carbon is obtained aboveground via photosynthesis and partly transported to belowground roots to support both acquisition of water and mineral nutrients. Important differences in allocation strategy occur both among species and among abiotic environments within species (Lambers et al., 1998; Poorter and Nagel, 2000). When a given resource is in short supply, plants typically allocate resources differently to compensate. If the resource in short supply is belowground for example, limited water, species and populations within species typically will allocate more resources belowground to compensate (Brouwer, 1963; Tilman, 1991; Gleeson and Tilman, 1992; Reynolds and Chen, 1996; Craine et al., 2002). So comparing dry versus mesic sites, it is well documented that the allocation of biomass shifts to a higher proportion of roots versus leaves (Orians and Solbrig, 1977; Mooney et al., 1978).

There is a key shift in the use of nitrogen when water is in short supply. Increased N allocation to leaves can support greater Rubisco-N investment per unit leaf area, which in turn can drive a greater photosynthetic capacity per unit transpiring area (Field and Mooney, 1986). Faster photosynthetic production at a given stomatal conductance increases intrinsic water use efficiency (Field et al., 1983; Buckley et al., 2002; Wright et al., 2003), which may compensate for prolonged periods of stomatal closure. Going from mesic to dry sites, species commonly allocate more N per leaf area partially by increasing N per unit leaf mass and partially by increasing the leaf mass per area (Wright et al., 2002). This pattern has been documented both among species in Australia (Wright et al., 2001) and across China (Han et al., 2011), and within species in Hawaii (Cornwell et al., 2007).

Interestingly, aridity creates a condition under which biomass may be allocated in greater quantities belowground, but N is needed aboveground to facilitate greater water use efficiency. This suggests a potential decoupling of pool sizes for C and N, above- and belowground, with the pattern in pool size determined by the relative effect sizes of the trends in biomass allocation and [N]. These shifts could have implications for biogeochemical functions – especially the residence time of C in the biosphere, both via leaf quality to herbivores and litter quality to decomposers. Both herbivory and decomposition are known to be at least in part a function of tissue C : N ratio (Swift et al., 1979; Austin and Vitousek, 2000; Perez-Harguindeguy et al., 2000). Here we argue that a particular consequence of increasing divergence of above- versus belowground N and C pool with increasing aridity could be accelerated by land degradation; as many semi-arid and arid lands are grazed by livestock, the proportional removal of a plant's aboveground N would increase with aridity, thereby potentially leading to faster exhaustion of whole-plant N to support productivity. Such positive feedback to land degradation by livestock would be a significant and previously unrecognized consequence of increased allocation of N to leaves to increase water use efficiency.

We investigate the above- and belowground pool sizes for C and N in the predominant plant species across two precipitation gradients in the semiarid to arid region of northern China. We then decompose the pool size patterns into the respective drivers – shifts in both the allocation of biomass and in the concentration of macronutrients. Our approach is to use three predominant species, which together represent much of the biomass along these gradients. Our hypothesis is that, from mesic towards dry sites, above- and belowground C pool and N pool will diverge as a consequence of increasing biomass allocation belowground combined with greater leaf N concentration (Fig. 1). To our knowledge, this is the first large-scale field study to explicitly dissect carbon pool and nitrogen pool both between and within organs of whole plants.

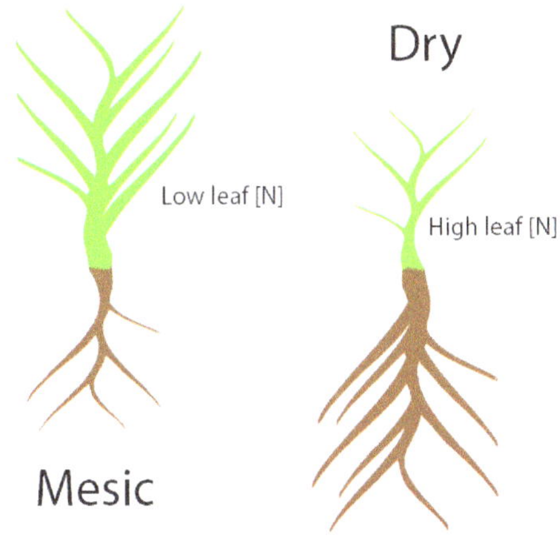

Figure 1. The allocation of biomass under different water regimes. Going from mesic to dry sites, the allocation of biomass shifts to a higher proportion of roots versus leaves and higher leaf N concentrations.

2 Methods

2.1 Study area and species

This study was conducted along two transects, each following a major precipitation gradient, in the Inner Mongolia Autonomous Region (hereinafter, Inner Mongolia), N China (Fig. 2). The North-East China Transect (NECT) ranges from Erenhot to Chifeng city (longitude 112.12 to 117.42° E, latitude 43 to 44° N, altitudes 920 to 1300 m), with mean annual precipitation (MAP) ranging from approx. 200 to 400 mm; The Ordos Transect (OT) stretches also from the west to east in Ordos City (longitude 107.00 to 111.31° E, latitude 38 to 39° N, altitudes 1200 to 1500 m), with MAP ranging from 160 to 440 mm. Along the west of NECT, temperature only shows a modest trend, with mean annual temperature (MAT) from 1 to 3 °C, mean temperature of the coldest month (CMT) from −20 to −12 °C (January), and mean temperature of the warmest month (WMT) from 19 to 24 °C (July). The precipitation from May through September accounts for almost 80 % of MAP (Zhang et al., 1997). The warmest sites have the lowest relative inner-annual

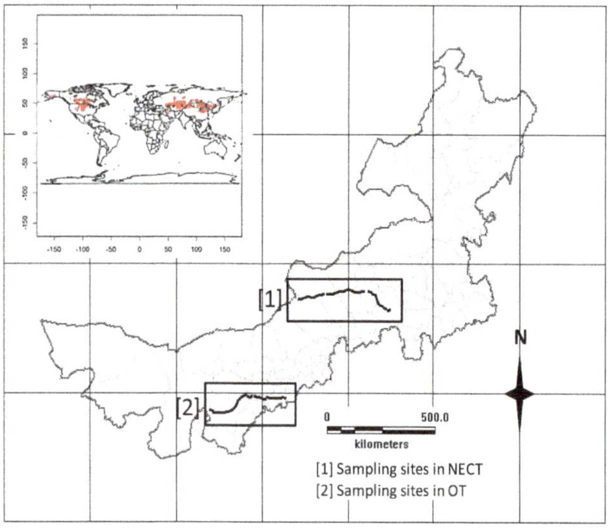

Figure 2. The sampling sites in Inner Mongolia, China. The 46 northern sites (1) were along the North-East China Transect (NECT), and the 41 southern sites (2) were along the Ordos Transect (OT). The map in the upper left corner shows places in the world with a similar climatic regime to the study area – with similar annual and seasonal pattern of precipitation and temperature; that is, with mean annual precipitation ranging from 160 to 440 mm, and most of MAP occurring from May through September, and with mean annual temperature from 1 to 9°, mean temperature of the coldest month from −20 to −10° (January), and mean temperature of the warmest month from 19 to 24° (July).

variability in temperature and the coldest sites have the highest variability, while the driest sites have the highest seasonality in precipitation and the wettest sites have the lowest variability along NECT (Ni and Zhang, 2000). The Ordos Transect, about 450 km south of NECT, features a weak east-to-west temperature gradient (MAT from 6 to 9 °C, CMT from −13 to −10 °C (January), and WMT from 20 to 23 °C (July)) and a strong east-to-west precipitation gradient, with June through September accounting for 60–80 % of MAP (Niu and Li, 1992).

Artemisia ordosica Krasch (Asteraceae) is a deciduous semi-shrub with plumose, linearly lobate leaves, occurring in the Inner Mongolia Autonomous Region, Ningxia Autonomous Region and Shanxi Province of N China. Its root system is mainly distributed in the upper 30 cm of sandy soils. This species is overwhelmingly dominant in the semi-fixed and fixed dune landscape of the Mu Us Sandland, which is the main body of the Ordos Plateau (Zhang, 1994). *Stipa grandis* P. Smim is a tall bunchgrass and can grow to almost 1 m at the peak of the growing season (late August). Its aboveground parts die off in autumn and new growth is in spring. It often forms nearly pure stands in the semi-arid region of N China. *Stipa bungeana* Trin. is smaller than *S. grandis*, with height ranging from 20 to 60 cm. It is distributed in N, SW and SE China as well as in Mongolia and Japan. Both *Stipa* species predominate on the loamier soils in the region, as opposed to *A. ordosica* on the sandy substrates.

2.2 Field sampling and laboratory analysis

The field investigation was conducted from 6 August to 18 September 2006, i.e. the first 3 weeks along NECT and the latter 3 weeks along OT. Along each transect this was during the peak biomass period and well before the onset of senescence (which is in September at NECT; end of September/early October at OT). In total 46 sites were chosen along NECT and 41 sites along OT (Fig. 2), each two adjacent sites at intervals of more than 10 km. MAP at each site was extracted from the interpolated climate grids (New et al., 2002), using the Raster package in *R* (Hijmans and van Etten, 2011). The site values closely matched previous estimates based on regional regressions with altitude, latitude and longitude for NECT (Zhang et al., 1997) and OT (Niu and Li, 1992), respectively.

Three 1 m × 1 m quadrats for grass and/or 2 m × 2 m quadrats for semi-shrub were measured at each site. Within each quadrat, all living biomass was harvested, separated into species and oven-dried at 70 °C for ≥ 24 h to constant mass before they were weighed. For each species, cover %, height and number of individuals were also measured within each quadrat at the same time. Leaves of *S. grandis* in NECT and of *S. bungeana* and *A. ordosica* in OT were collected for lab analyses within each quadrat, as well as the stems of *A. ordosica* in OT. For each species roots were collected by taking three 10 cm diameter soil cores from 0–40 cm depths, mixed in situ into one composite sample. Roots of *S. grandis* in NECT and roots of *S. bungeana* and *A. ordosica* in OT were hand-sorted and washed clean for lab analyses. It is often difficult to differentiate between the roots of different species, but it was perfectly doable to differentiate between the roots of the three target species in this study. Both *S. grandis* and *S. bungeana* are bunch-grasses and could be identified visually; their roots are fibrous and shallow. Confusion with *Leymus* can be excluded, as this grass is rhizomatous and has its root distribution much deeper in the soil profile than *Stipa*. The roots of the shrub *A. ordosica* were easily recognized by their strong and typical aroma as well as by their morphology. Most other species in NECT and OT were non-grass herbaceous species, such as *Allium*, *Carex*, *Euphorbia*, *Iris*, *Potentilla* and *Taraxacum*. Only roots of the three target species were hand-sorted and washed clean for lab analyses, and all other roots in NECT were mixed into one composite sample.

Leaf, stem and root biomass of the three species were measured after ≥ 24 h oven-drying at 70 °C. Leaf, stem and root potassium (K) concentration was determined using atomic absorption spectrophotometry (Model-932, GBC Scientific Equipment, Melbourne, Australia), while nitrogen (N) and phosphorus (P) concentrations were analysed according to the micro-Kjeldahl method (Kjeltec 2300 Analyser Unit,

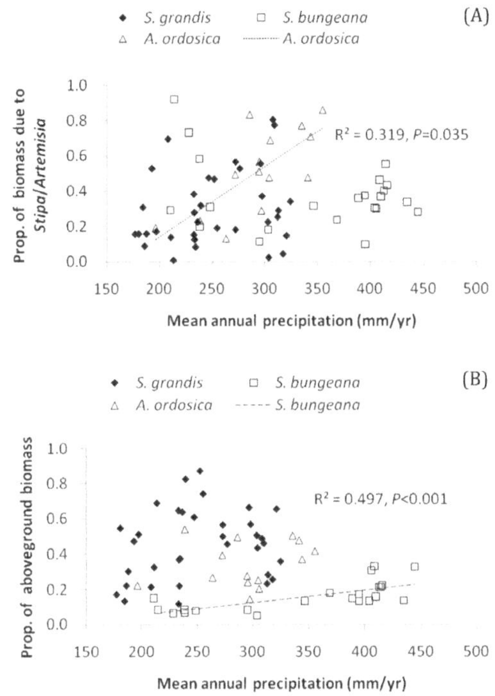

Figure 3. Relationship between percentage of vegetation biomass due to *Stipa*/*Artemisia* and mean annual precipitation **(a)**, percentage of biomass aboveground in *Stipa*/*Artemisia* and mean annual precipitation **(b)**, both along the two transects.

Sweden), and total carbon (C) concentration was determined using the method of $K_2Cr_2O_7 - H_2SO_4$ solution digestion with oil-bath heating (Dong et al., 1996; Zheng et al., 2010). Also, standard reference materials were used throughout to guarantee the measurement accuracy and repeatability.

In total, 15 of the sites provided data for plant aboveground and underground biomass and element contents for *A. ordosica* in OT, 22 sites for *S. bungeana* in OT, and 35 sites for *S. grandis* in NECT. Unfortunately, it was not logistically possible to measure soil nutrient availabilities simultaneously and repeatedly at so many points along the two gradients, given that nutrient availability is notoriously variable in time. However, we expect nutrient availabilities to have varied little over each of the transects for given species, as both *Stipa* species were generally sampled from unfertilized loamy soils and *A. ordosica* generally from unfertilized sandy soils, The remarkably flat pattern for root N content with MAP for all three species (see Results, Fig. 4b) strongly suggests that there was no correspondence between soil nutrient availability and precipitation regime, although we cannot exclude effects of nutrient availability on intra-specific variability in nutrient-related traits at finer spatial scale.

2.3 Data analysis

Statistical analyses were performed using SPSS13.0 software (2004, SPSS Inc., USA). We analysed variation in each trait for the effect of species and precipitation using ANCOVA, with species as an independent variable and precipitation as a covariate. And we analysed relationship between each trait and precipitation for each species using linear regression. Data for leaf [P], leaf [K], root [P], root [K], and proportion of aboveground N, C, P and K content were log-transformed before analyses in order to meet the assumptions of normality.

3 Results

The relative contribution of *A. ordosica* to the plant community biomass ranged from 13 to 86 % (mean 52.6 %) and increased with higher precipitation within the sandy region in OT ($r^2 = 0.319$, $P = 0.035$) (Fig. 3a). *S. bungeana* made up a higher proportion of the total biomass (range 10 to 91 %, mean 36.8 %) on the loamier OT sites where *A. ordosica* did not dominate. *S. bungeana* on the OT achieved a greater dominance compared to *S. grandis* in NECT (range 1 to 80 %, mean 29.8 %). For neither *Stipa* species was the proportion correlated with precipitation ($P > 0.1$).

3.1 Plant biomass and its allocation along transects

The ANCOVA results showed that the proportion of aboveground biomass increased linearly with MAP in all three species (df = 1, $F = 8.135$, $P = 0.006$); the slope of proportion of aboveground biomass as a function of MAP was not heterogeneous across species. The size of the trend, the crucial element affecting pool sizes, was a mean of 7.0 % additional allocation aboveground for every 100 mm yr^{-1} of additional rainfall. Also, there was a significant difference in elevation of these regressions among species (df = 2, $F = 6.521$, $P = 0.003$); *S. grandis* in NECT had the largest proportion aboveground biomass allocation, followed by *A. ordosica* and *S. bungeana* in OT. Only the latter species by itself increased aboveground biomass with MAP ($r^2 = 0.497$, $P < 0.001$, Fig. 3b).

3.2 Effects of species and precipitation on plant element concentrations

Precipitation had a negative and significant relationship with leaf nitrogen concentration (LNC) and leaf phosphorus concentration (LPC), as well as significant overall effects on proportion of aboveground C, P and K content (Table 1). There were significant interactions for species by precipitation on most plant traits analysed (Table 1).

Although LNC decreased along a gradient of MAP for all three species, the effect was stronger in *S. grandis* (slope $-1.24\,mg\,g^{-1}$ per 100 mm yr^{-2}, $r^2 = 0.718$, $P < 0.001$) and *A. ordosica* (slope $-0.79\,mg\,g^{-1}$ per 100 mm yr^{-2}, $r^2 = 0.426$, $P = 0.011$) than in *S. bungeana* (slope $-0.22\,mg\,g^{-1}$ per 100 mm yr^{-2}, $r^2 = 0.167$, $P = 0.059$). All three species had similar leaf carbon concentration (LCC)

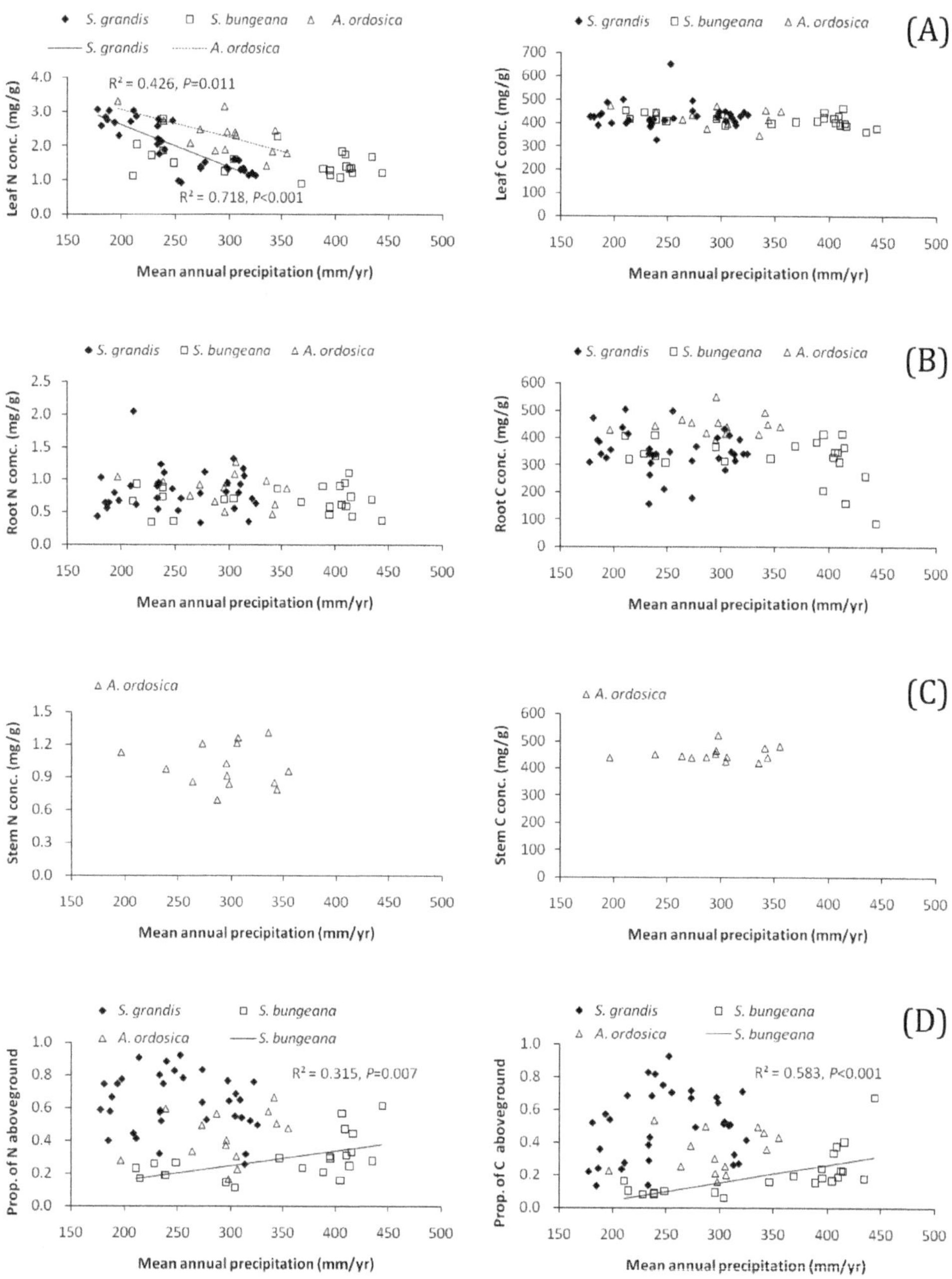

Figure 4. The relationship between mean annual precipitation and plant traits: **(a)** leaf nitrogen (LNC) and leaf carbon concentration (LCC), **(b)** root nitrogen (RNC) and root carbon concentration (RCC), **(c)** stem nitrogen (SNC) and stem carbon concentration (SCC), and **(d)** proportion of N and C aboveground.

values, which increased steadily with MAP (Fig. 4a). MAP had no significant effect on root carbon concentration (RCC) or root nitrogen concentration (RNC). The three species had similar RNC, while *A. ordosica* had a higher RCC than the two *Stipa* species (Fig. 4b). Stem nitrogen concentration (SNC) of *A. ordosica* showed no trend with increasing MAP in OT (Fig. 4c, $r^2 = 0.011$, $P = 0.719$).

A. ordosica had higher LPC and leaf potassium concentration (LKC) than the two *Stipa* species. MAP decreased LPC in all three species (but only significantly in *S. grandis*, slope $-0.175\,\%$ LPC per $100\,\mathrm{mm\,yr^{-2}}$, $r^2 = 0.343$, $P < 0.001$), and had no significant negative effect on LKC in any species (Fig. S1a in the Supplement). MAP had a significant positive effect on root phosphorus concentration (RPC) in *S. grandis* (slope $0.125\,\%$ RPC per $100\,\mathrm{mm\,yr^{-2}}$, $r^2 = 0.128$ and $P = 0.038$), a marginal negative effect in *A. ordosica* (slope $-0.224\,\%$ RPC per $100\,\mathrm{mm\,yr^{-2}}$, $r^2 = 0.205$, $P = 0.10$) and no effect in *S. bungeana* (Fig. S1b, $r^2 = 0.084$, $P = 0.189$).

Table 1. Effects of species and mean annual precipitation on each plant trait in the two transects. Bold values indicate $P<0.05$.

	Plant traits	Species			Precipitation			Interaction		
		df	F	P	df	F	P	df	F	P
	Leaf N conc.	2	14.665	**<0.001**	1	52.915	**<0.001**	2	16.711	**<0.001**
	Stem N conc.	–	–	–	1	0.136	0.719	–	–	–
N	Root N conc.	2	0.598	0.553	1	0.360	0.550	2	0.393	0.677
	Proportion of N aboveground	2	10.543	**<0.001**	1	2.540	0.116	2	3.336	**0.042**
	Leaf C conc.	2	0.581	0.562	1	1.031	0.314	2	0.703	0.499
	Stem C conc.	–	–	–	1	0.406	0.536	–	–	–
C	Root C conc.	2	0.147	0.864	1	0.705	0.404	2	0.397	0.674
	Proportion of C aboveground	2	6.774	**0.002**	1	9.185	**0.004**	2	1.141	0.326
	Leaf P conc.	2	6.122	**0.004**	1	7.860	**0.007**	2	2.927	0.061
	Stem P conc.	–	–	–	1	0.481	0.501	–	–	–
P	Root P conc.	2	7.158	**0.002**	1	1.325	0.254	2	5.448	**0.007**
	Prop. of P aboveground	2	23.905	**<0.001**	1	8.187	**0.006**	2	7.930	**0.001**
	Leaf K conc.	2	1.978	0.147	1	0.921	0.341	2	0.056	0.945
	Stem K conc.	–	–	–	1	3.96	0.07	–	–	–
K	Root K conc.	2	7.414	**0.001**	1	1.602	0.210	2	3.833	**0.027**
	Prop. of K aboveground	2	22.785	**<0.001**	1	12.185	**0.001**	2	5.272	**0.008**

With increasing MAP, root potassium concentration (RKC) appeared to decrease in *S. bungeana* (slope 0.082 % RKC per 100 mm yr^{-2}, $r^2=0.151$, $P=0.074$) but showed no clear trend in *S. grandis* or *A. ordosica* (Fig. S1a). Stem phosphorus concentration (SPC) of *A. ordosica* showed no trend ($r^2=0.039$, $P=0.50$), while stem potassium concentration (SKC) significantly decreased with increasing MAP in OT (Fig. S1, slope −2.904 % SKC per 100 mm yr^{-2}, $r^2=0.248$ and $P=0.070$).

3.3 Proportion of aboveground plant N, C, P and K pools

Precipitation had a significant effect on the proportion of N and C pool aboveground, and this effect differed among species (significant interactions, see Table 1). *S. grandis* had the highest proportion of N and C pool aboveground, followed by *A. ordosica* and *S. bungeana* respectively. There was a significant positive relationship between MAP and proportion of N (slope 0.129 % per 100 mm yr^{-2}, $r^2=0.315$, $P=0.007$) and C aboveground content (slope 0.239 % per 100 mm yr^{-2}, $r^2=0.583$ and $P<0.001$) in *S. bungeana*, while *A. ordosica* showed only a weaker trend and *S. grandis* none at all (Fig. 4d).

Both species and precipitation had significant effects on the proportion of P and K pool aboveground and also the interactions were significant (Table 1). Comparable with N pool, there were significant positive relationships between MAP and proportion of P (slope 0.212 % per 100 mm yr^{-2}, $r^2=0.631$, $P<0.001$) and K aboveground content (slope 0.239 % per 100 mm yr^{-2}, $r^2=0.620$, $P<0.001$) in *S. bungeana*, but not in the other two species (Fig. S1d).

4 Discussion

The aboveground versus belowground pool size and concentrations of carbon and nitrogen are linked, and understanding the shifts across a precipitation gradient requires a whole-plant perspective (Van der Putten et al., 2001; Wardle et al., 2004; Bardgett et al., 2005; De Deyn et al., 2008). Because of the many fungi and animals associated with the plants – mutualists, pathogens, herbivores, decomposers – the whole plant C and N strategy is crucial for carbon storage, food webs, soil water and nitrogen retention and cycling (Hooper et al., 2000; Bardgett and Wardle, 2003). Changes in the stochiometry and balance between these pools owing to differences or changes in precipitation regime are therefore likely to impact greatly on ecosystem function and biodiversity.

Two quantities affect the proportion of N above- versus belowground: the relative allocation of biomass and the concentration of N in that biomass. A priori we expected the proportion of biomass aboveground to be greater in mesic versus dry places (Orians and Solbrig, 1977; Mooney et al., 1978). We also expected Inner Mongolia to show similar tissue N concentration patterns to China at larger scale (Han et al., 2011), Australia (Wright et al., 2001, 2002), Hawaii (Cornwell et al., 2007), Panama (Santiago et al., 2005), and western North America (Maherali and DeLucia, 2001) – that is, the concentration of N in leaves would be greater in dry versus mesic environment. For the pool size of N, because these two components were expected to change in opposing directions, we expected the result to be a function of the relative effect sizes of the two trends. Because, to the best of our knowledge, the effect sizes for biomass and N concentration have not previously been measured in one system, we started from the null hypothesis that the two trends would cancel each other out, resulting in similar proportions of N aboveground versus belowground pool size across aridity gradients.

We found that like other aridity gradients across the world (Maherali and DeLucia, 2001; Austin and Sala, 2002; Santiago et al., 2005; He et al., 2008), there was a shift from more aboveground towards more belowground biomass allocation with decreasing precipitation in *S. bungeana* (Fig. 3b). Also consistent with our expectation, there was a significant increase in leaf N concentration from the drier towards the more mesic sites in *S. grandis* and *A. ordosica*, and a weaker increase in leaf N concentration in *S. bungeana* ($r^2 = 0.167$, $p = 0.059$) (Fig. 4a). Integrating both patterns, we found that for one species, *S. bungeana*, there was a significant positive trend in proportion of N aboveground with precipitation (Fig. 4d). The other two species did not show relationships that differed significantly from flat. This was neither due to differing biomass allocation, the slopes of which relationships were very similar (Fig. 3b); nor to root N concentration, which also showed similar pattern for the three plant species (Fig. 4b). Instead, *S. bungeana* did not increase leaf N concentration in response to aridity to nearly the extent that the other two species did (Fig. 4a). This meant that unlike in the two other species where N and biomass shifts came close to equal effect sizes, in *S. bungeana* the biomass effect dominated and the proportion of N aboveground increased with precipitation.

Species differ in their intraspecific variation – both for biomass allocation and the extent to which they adjust their N concentration. While intraspecific variation in leaf traits (including N concentration and leaf mass per area) has been relatively widely studied (e.g. Shipley and Almeida-Cortez, 2003; McGill et al., 2006; Ackerly and Cornwell, 2007; Cornwell and Ackerly, 2009; Albert et al., 2010; Messier et al., 2010), there is still no consensus on why species differ in the magnitude of their intraspecific variation. We suggest that intraspecific variation in a few key traits, especially N and C allocation, is crucial for understanding how much of N and biomass cycling occurs belowground versus aboveground as living biomass and then on the surface of the soil as leaf and stem litter.

In our study of three predominant species in Inner Mongolia, in two cases (clearly in *S. grandis*, borderline in *A. ordosica*) intraspecific variation in biomass allocation balanced shifts in the concentration of leaf N. In the third species, *S. bungeana*, biomass allocation changed, but the shift in N concentration did not fully compensate. Where *S. bungeana* was predominant, a greater proportion of N was held aboveground in mesic versus dry places. In contrast, where *S. grandis* was predominant, the pool sizes of N above- versus belowground were similar in mesic versus dry places, but the pool size of C was not. There was much more C aboveground in mesic versus dry places, but that C was accompanied with a lower concentration of N.

We propose one potentially important consequence of the decreasing aboveground allocation to biomass that is richer in N as aridity gets more severe. Logically, at the level of the plant individual, a given proportion of consumption of aboveground biomass by herbivores in a more arid site will remove a relatively large proportion of the plant's total N pool, which has to be compensated for to the extent possible by N uptake from the soil. Under strong grazing regimes, uptake may not keep up with N removal and this might lead to exhaustion of the aboveground N pool and eventually lead to shoot death, with consequences for the soil-stabilizing function of the vegetation in these erosion-prone areas. Thus, increasing divergence of above- versus belowground N and C pool with increasing aridity could accelerate land degradation; as many semi-arid and arid lands are grazed by livestock, often heavily, the dominant vegetation could drive a positive feedback on land degradation by livestock at low precipitation regimes. This situation will not be unique to Inner Mongolia, as extensive regions in a similar climate zone in at least West, Central and East Asia and North America (Fig. 2, insert) host similar vegetation types and similarly heavy use by livestock. This possibility may deserve in-depth study that would have to extend to vegetation regrowth patterns and N and C pathway and pool via livestock at different animal densities. It would also be important to test whether and how the possible feedback suggested above- might be influenced by possible changes in preference of livestock for different food plants as related to changing leaf nutritional value for the different species, with consequences for their abundance hierarchies at different points along the precipitation gradient.

5 Conclusions

The C and N cycle are usually tightly coupled, because the proportion of N that is needed in specific tissue types is relatively constrained – N is usually low in structural tissue and high in absorbing (roots) and photosynthetic (leaves) tissue (Sterner and Elser, 2002). Therefore, decoupling of the C and N pool across different climates is a paradox. Changing water availability potentially decouples the C and the N pool because of greater allocation to high N leaves in dry sites. Here we have fully decomposed above- and belowground C and N allocation and stoichiometry. We have found examples of consistent N pool sizes above- and belowground and a shift to a greater proportion of belowground N in drier sites depending on the species. We suggest that precipitation gradients do potentially decouple the C and N pool, but the exact nature of the decoupling depends on the dominant species' capacity for intraspecific variation. Even in cases where the above- versus belowground N and C pool themselves do not change proportionally, smaller biomass allocation to shoots with higher N concentrations in drier areas might facilitate a positive feedback to land degradation by livestock.

Author contributions. M. Dong directed, coordinated and funded this study with intellectual input from J. H. C. Cornelissen, W. K. Cornwell and X. Ye; X. Ye, X. Pan, S. Gao and M. Dong carried out the fieldwork and lab analyses; X. Ye, W. K. Cornwell and J. H. C. Cornelissen did the data analysis and wrote the first draft. All authors commented on the paper and approved the submitted version.

Acknowledgements. This research was supported by the "Strategic Priority Research Program" of the Chinese Academy of Sciences (CAS) (XDA05050406), the National Natural Science Foundation of China (30570296), and the Chinese Academy of Sciences Visiting Professorship for Senior International Scientists (2010T2S24). Many thanks to Renqiang Li for his help with our field work in NECT and Jianjiang Qiao in OT.

Edited by: F. Wittmann

References

Ackerly, D. D. and Cornwell, W. K.: A trait-based approach to community assembly: Partitioning of species trait values into within- and among-community components, Ecol. Lett., 10, 135–145, 2007.

Aerts, R. and Chapin, F. S.: The mineral nutrition of wild plants revisited: A re-evaluation of processes and patterns, Adv. Ecol. Res., 30, 1–67, 2000.

Albert, C. H., Thuiller, W., Yoccoz, N. G., Soudant, A., Boucher, F., Saccone, P., and Lavorel, S.: Intraspecific functional variability: Extent, structure and sources of variation, J. Ecol., 98, 604–613, 2010.

Austin, A. T. and Vitousek, P. M.: Precipitation, decomposition and litter decomposability of *Metrosideros polymorpha* in native forests on Hawai'i, J. Ecol., 88, 129–138, 2000.

Austin A. T. and Sala O. E.: Carbon and nitrogen dynamics across a natural precipitation gradient in Patagonia, Argentina, J. Veg. Sci., 13, 351–360, 2002.

Bardgett, R. D. and Wardle, D. A.: Herbivore-mediated linkages between aboveground and belowground communities, Ecology, 84, 2258–2268, 2003.

Bardgett, R. D., Bowman, W. D., Kaufmann, R., and Schmidt, S. K.: A temporal approach to linking aboveground and belowground ecology, Trends Ecol. Evol., 20, 634–641, 2005.

Brouwer, R.: Some aspects of the equilibrium between overground and underground plant parts, Jaarb IBS Wageningen, 213, 31–39, 1963.

Buckley, T. N., Miller, J. M., and Farquhar, G. D.: The mathematics of linked optimization for water and nitrogen use in a canopy, Silva Fenn., 36, 639–669, 2002.

Chapin, F. S., McFarland, J., McGuire, A. D., Euskirchen, E. S., Ruess, R. W., and Kielland, K.: The changing global carbon cycle: Linking plant-soil carbon dynamics to global consequences, J. Ecol., 97, 840–850, 2009.

Cornwell, W. K. and Ackerly, D. D.: Community assembly and shifts in plant trait distributions across an environmental gradient in coastal California, Ecol. Monogr., 79, 109–126, 2009.

Cornwell, W. K., Bhaskar, R., Sack, L., Cordell, S., and Lunch, C. K.: Adjustment of structure and function of Hawaiian *Metrosideros polymorpha* at high vs. Low precipitation, Funct. Ecol., 21, 1063–1071, 2007.

Craine, J. M., Tilman, D., Wedin, D., Reich, P., Tjoelker, M., and Knops, J.: Functional traits, productivity and effects on nitrogen cycling of 33 grassland species, Funct. Ecol., 16, 563–574, 2002.

De Deyn, G. B., Cornelissen, J. H. C., and Bardgett, R. D.: Plant functional traits and soil carbon sequestration in contrasting biomes, Ecol. Lett., 11, 516–531, 2008.

Dong, M., Wang, Y. F., Kong, F. Z., Jiang, G. M., and Zhang, Z. B. (Eds.): Survey, observation and analysis of terrestrial biocommunities Standard methods for observation and analysis in Chinese ecosystem research network, Standards Press of China, Beijing, China, 152–153, 1996.

Field, C. and Mooney, H. A.: The photosynthesis-nitrogen relationship in wild plants, in: On the economy of plant form and function, edited by: Givnish, T. J., Cambridge University Press, Cambridge, UK, 25–55, 1986.

Field, C., Merino, J., and Mooney, H. A.: Compromises between water-use efficiency and nitrogen-use efficiency in five species of California evergreens, Oecologia, 60, 384–389, 1983.

Gleeson, S. K. and Tilman, D.: Plant allocation and the multiple limitation hypothesis, Am. Nat., 139, 1322–1343, 1992.

Han, W. X., Fang, J. Y., Reich, P. B., Woodward, F. I., and Wang, Z. H.: Biogeography and variability of eleven mineral elements in plant leaves across gradients of climate, soil and plant functional type in china, Ecol. Lett., 14, 788–796, 2011.

He, J. S., Wang, L., Flynn, D. F. B., Wang, X. P., Ma, W. H., and Fang, J. Y.: Leaf nitrogen: phosphorus stoichiometry across Chinese grassland biomes, Oecologia, 122, 301–310, 2008.

Hijmans, R. J. and van Etten, J.: Raster: Geographic analysis and modeling with raster data, 166 pp., 2011.

Hooper, D. U., Bignell, D. E., Brown, V. K., Brussaard, L., Dangerfield, J. M., Wall, D. H., Wardle, D. A., Coleman, D. C., Giller, K. E., Lavelle, P., Van der Putten, W. H., De Ruiter, P. C., Rusek, J., Silver, W. L., Tiedje, J. M., and Wolters, V.: Interactions between aboveground and belowground biodiversity in terrestrial ecosystems: Patterns, mechanisms, and feedbacks, Bioscience, 50, 1049–1061, 2000.

IPCC: Climate change 2007: Synthesis report, in: International Panel for Climate Change, edited by: Rkara, P., Geneva, Switzerland, 2007.

Lambers, H., Chapin, F. S., and Pons, T. L.: Plant physiological ecology, Springer-Verlag, New York, USA, 1998.

Maherali, H. and DeLucia, E. H.: Influence of climate-driven shifts in biomass allocation on water transport and storage in ponderosa pine, Oecologia, 129, 481–491, 2001.

McGill, B. J., Enquist, B. J., Weiher, E., and Westoby, M.: Rebuilding community ecology from functional traits, Trends Ecol. Evol., 21, 178–185, 2006.

Messier, J., McGill, B. J., and Lechowicz, M. J.: How do traits vary across ecological scales?, A case for trait-based ecology, Ecol. Lett., 13, 838–848, 2010.

Mooney, H. A., Ferrar, P. J., and Slatyer, R. O.: Photosynthetic capacity and carbon allocation patterns in diverse growth forms of eucalyptus, Oecologia, 36, 103–111, 1978.

New, M., Lister, D., Hulme, M., and Makin, I.: A high-resolution data set of surface climate over global land areas, Clim. Res., 21, 1–25, 2002.

Ni, J. and Zhang, X. S.: Climate variability, ecological gradient and the northeast china transect (NECT), J. Arid Environ., 46, 313–325, 2000.

Niu, J. M. and Li, B.: Multivariate analysis of vegetation and ecological factors on the Ordos Plateau, Inner Mongolia, Acta Ecol. Sin., 2, 105–112, 1992.

Orians, G. H. and Solbrig, O. T.: A cost-income model of leaves and roots with special reference to arid and semiarid areas, Am. Nat., 111, 677–690, 1977.

Perez-Harguindeguy, N., Diaz, S., Cornelissen, J. H. C., Vendramini, F., Cabido, M., and Castellanos, A.: Chemistry and toughness predict leaf litter decomposition rates over a wide spectrum of functional types and taxa in central Argentina, Plant Soil, 218, 21–30, 2000.

Poorter, H. and Nagel, O.: The role of biomass allocation in the growth response of plants to different levels of light, CO_2, nutrients and water: A quantitative review, Aust. J. Plant Physiol., 27, 595–607, 2000.

Reynolds, J. F. and Chen, J. L.: Modelling whole-plant allocation in relation to carbon and nitrogen supply: Coordination versus optimization: Opinion, Plant Soil, 185, 65–74, 1996.

Santiago, L. S., Schuur, E. A. G., and Silvera, K.: Nutrient cycling and plant-soil feedbacks along a precipitation gradient in lowland panama, J. Trop. Ecol., 21, 461–470, 2005.

Shaver, G. R. and Chapin, F. S.: Production: Biomass relationships and element cycling in contrasting arctic vegetation types, Ecol. Monogr., 61, 1–31, 1991.

Shipley, B. and Almeida-Cortez, J.: Interspecific consistency and intraspecific variability of specific leaf area with respect to irradiance and nutrient availability, Ecoscience, 10, 74–79, 2003.

Sterner, R. W. and Elser, J. J.: Ecological stoichiometry: The biology of elements from molecules to the biosphere, Princeton University Press, Princeton, NJ, 2002.

Swift, M. J., Heal, O. W., and Anderson, J. M.: Decomposition in terrestrial ecosystems, University of California Press, Berkeley, CA, 1979.

Tilman, D.: Relative growth rates and plant allocation patterns, Am. Nat., 138, 1269–1275, 1991.

Van der Putten, W. H., Vet, L. E. M., Harvey, J. A., and Wackers, F. L.: Linking above- and belowground multitrophic interactions of plants, herbivores, pathogens, and their antagonists, Trends Ecol. Evol., 16, 547–554, 2001.

Wardle, D. A., Bardgett, R. D., Klironomos, J. N., Setala, H., van der Putten, W. H., and Wall, D. H.: Ecological linkages between aboveground and belowground biota, Science, 304, 1629–1633, 2004.

Wright, I. J., Reich, P. B., and Westoby, M.: Strategy shifts in leaf physiology, structure and nutrient content between species of high- and low-rainfall and high- and low-nutrient habitats, Funct. Ecol., 15, 423–434, 2001.

Wright, I. J., Westoby, M., and Reich, P. B.: Convergence towards higher leaf mass per area in dry and nutrient-poor habitats has different consequences for leaf life span, J. Ecol., 90, 534–543, 2002.

Wright, I. J., Reich, P. B., and Westoby, M.: Least-cost input mixtures of water and nitrogen for photosynthesis, Am. Nat., 161, 98–111, 2003.

Zhang, X. S.: Principles and optimal models for development of Maowusu sandy grassland, Acta Phyt. Sinica, 18, 1–16, 1994.

Zhang, X. S., Gao, Q., Yang, D. A., Zhou, G. S., Ni, J., and Wang, Q.: A gradient analysis and prediction on the northeast china transect (NECT) for global change study, Acta Bot. Sin., 39, 785–799, 1997.

Zheng, S. X., Ren, H. Y., Lan, Z. C., Li, W. H., Wang, K. B., and Bai, Y. F.: Effects of grazing on leaf traits and ecosystem functioning in Inner Mongolia grasslands: scaling from species to community, Biogeosciences, 7, 1117–1132, doi:10.5194/bg-7-1117-2010, 2010.

9

Origin and fate of particulate and dissolved organic matter in a naturally iron-fertilized region of the Southern Ocean

L. Tremblay[1], J. Caparros[2,3], K. Leblanc[4], and I. Obernosterer[2,3]

[1]Department of Chemistry and Biochemistry, Université de Moncton, Moncton, NB, E1A 3E9, Canada
[2]CNRS, UMR 7621, Laboratoire d'Océanographie Microbienne, Observatoire Océanologique, 66650 Banyuls/mer, France
[3]Sorbonne Universités, UPMC Univ Paris 06, UMR 7621, Laboratoire d'Océanographie Microbienne, Observatoire Océanologique, 66650 Banyuls/mer, France
[4]Aix Marseille Université, CNRS, Université de Toulon, IRD, Mediterranean Institute of Oceanography (MIO), UM 110, 13288, Marseille, France

Correspondence to: L. Tremblay (luc.tremblay@umoncton.ca)

Abstract. Natural iron fertilization of high-nutrient low-chlorophyll (HNLC) waters induces annually occurring spring phytoplankton blooms off the Kerguelen Islands (Southern Ocean). To examine the origin and fate of particulate and dissolved organic matter (POM and DOM), D- and L-amino acids (AA) were quantified at bloom and HNLC stations. Total hydrolyzable AA accounted for 21–25 % of surface particulate organic carbon (%POC_{AA}) at the bloom sites, but for 10 % at the HNLC site. A marked decrease in %POC_{AA} with depth was observed at the most productive stations leading to values between 3 and 5 % below 300 m depth. AA contributed to only 0.9–4.4 % of dissolved organic carbon (%DOC_{AA}) at all stations. The only consistent vertical trend was observed at the most productive station (A3-2) where %DOC_{AA} decreased from ~ 2 % in the surface waters to 0.9 % near 300 m. These AA yields revealed that POM and DOM were more rapidly altered or mineralized at the bloom sites compared to the HNLC site. Alteration state was also assessed by trends in C / N ratio, %D-AA and degradation index. Different molecular markers indicated that POM mostly originated from diatoms and bacteria. The estimated average proportion of POM from intact phytoplankton cells in surface waters was 45 % at the bloom station A3-2, but 14 % at the HNLC site. Estimates based on D-AA yields indicated that ~ 15 % of POM and ~ 30 % of DOM was of bacterial origin (cells and cell fragments) at all stations. Surprisingly, the DOM in HNLC waters appeared less altered than the DOM from the bloom, had slightly higher dissolved AA concentrations, and showed no sign of alteration within the water column. Unfavorable conditions for bacterial degradation in HNLC regions can explain these findings. In contrast, large inputs of labile organic molecules and iron likely stimulate the degradation of organic matter (priming effect) and the production of more recalcitrant DOM (microbial carbon pump) during iron-fertilized blooms.

1 Introduction

The Southern Ocean plays a key role in the global carbon cycle because it represents an important physical and biological CO_2 sink (Le Quéré et al., 2007). The magnitude of the biological contribution to the CO_2 sink is driven by primary productivity and the fate of the organic matter (OM) produced. It was univocally shown that iron controls primary production in the Southern Ocean having the largest high-nutrient low-chlorophyll (HNLC) zone (Boyd et al., 2007). The extent to which biological degradation of particulate (POM) and dissolved organic matter (DOM) affect their potential export to depth and preservation is far from understood (Ebersbach and Trull, 2008).

Natural iron fertilization induces intense phytoplankton blooms in spring, especially around the Kerguelen Islands (Blain et al., 2007). Blooms induced by natural iron fertilization likely represent major events for heterotrophic bacteria which are reported to be limited by carbon in the HNLC

Southern Ocean (Church et al., 2000). Indeed, pronounced responses of heterotrophic bacteria are reported from natural iron-fertilization studies (Zubkov et al., 2007; Christaki et al., 2008, 2014). Rapid bacterial mineralization of ~45 % of the primary production was observed during the decline of the spring phytoplankton bloom above the Kerguelen Plateau (Obernosterer et al., 2008). Furthermore, the efficiency of the carbon export, defined as the ratio of particulate organic carbon (POC) in deep waters to primary production, was about 2 times lower above the Kerguelen Plateau than in HNLC waters (Savoye et al., 2008). These findings suggest that the POM produced during the bloom is relatively labile, and rapidly turned over. Although DOM production is an essential step during bacterial mineralization, the effect of iron fertilization on the concentration and reactivity of DOM has been far less studied.

To better understand to role of the Southern Ocean in the cycles of carbon and other vital elements, such as nitrogen, it is important to study the origin and fate of its OM in regions having contrasting conditions and productivities. To do so, molecular level analyses of POM and DOM provide unique and valuable pieces of information. Amino acid enantiomers (L- and D-AA) are among the few markers that can be analyzed in bulk seawater at ambient concentrations and provide numerous indicators on OM origin and fate (e.g., Kaiser and Benner, 2008; Dittmar et al., 2009). AA contain most of the nitrogen in living organisms and are important constituents of detrital POM and DOM (Cowie and Hedges, 1992). While L-AA are ubiquitous (e.g., all proteins), D-AA are only produced in important proportions in bacteria (Asano and Lübbehüsen, 2000).

AA are generally selectively utilized by heterotrophic organisms compared to bulk OM. As a result, the proportions of bulk carbon and nitrogen in the form of AA, or AA yields, decrease with time and are useful indicators of POM and DOM alteration state (Cowie and Hedges, 1994; Davis et al., 2009). Other independent diagenetic markers are the molar percentage (mol%) of D-AA versus total hydrolyzable AA (THAA) (Tremblay and Benner, 2009) and the degradation index (DI) calculated based on the relative distribution of individual AA (Dauwe et al., 1999; Peter et al., 2012). An increase in mol% D-AA during degradation is attributed to the bacterial contribution of D-AA to the OM and to a lower degradation rate of the bacterial biomolecules rich in D-AA compared to proteins having only L-AA (Nagata et al., 2003). During degradation some AA are selectively degraded (the others are selectively preserved) and these changes reduce the DI value. AA yields are known to be more sensitive to the first alterations of the OM compared to mol% D-AA and DI which require more intense alterations before showing consistent changes (Davis et al., 2009; Bourgoin and Tremblay, 2010).

C- and N-normalized yields of individual D-AA (e.g., D-alanine) were shown to have a similar dynamic as bulk bacterial C and N, respectively (Tremblay and Benner, 2006; Kaiser and Benner, 2008). Thus, these yields can be used to estimate the contribution of bacteria (from intact cells and cell fragments or detritus) to POM and DOM. The description of this approach and its limitations are presented in previous studies (Tremblay and Benner, 2006, 2009; Kaiser and Benner, 2008; Kawasaki et al., 2011). In addition, plots of the sum of serine and threonine versus the ratio aspartic acid / glycine have been previously utilized to discriminate different carbon sources, specifically diatoms, coccolithophores and bacteria (Ittekkot et al., 1984; Gupta and Kawahata, 2000; Salter et al., 2010).

The objective of this work was to evaluate the origin and fate of the OM produced during spring phytoplankton blooms in the naturally iron-fertilized Southern Ocean. To do so, bulk and AA enantiomeric analyses were done on unfiltered and filtered waters collected at different bloom sites and in HNLC waters during the KEOPS2 (Kerguelen Ocean and Plateau compared Study 2) expedition. This work provides a detailed description of OM characteristics and new evidences on the distinct fates of the OM from naturally iron-fertilized regions.

2 Materials and methods

2.1 Site description and sampling

Among the sampling sites visited during the KEOPS2 campaign in October and November 2011, five stations were selected for the present study (Table 1) (map available in Blain et al., 2014). Station R-2 (bottom depth = 2532 m) was located in a HNLC region outside the western boundary of the Kerguelen Plateau. In contrast, station A3-2 (bottom depth = 527 m) was in the center of the phytoplankton bloom that occurs annually above the Kerguelen Plateau. Stations E-1 (bottom depth = 2056 m) and E-5 (bottom depth = 1920 m) were located in offshore waters within a stationary meander south of the polar front and these stations were sampled in a quasi-Lagrangian manner before (E-1) and during the bloom (E-5). TEW-2 is a shallow station (total depth = 85 m) above the eastern continental shelf of Kerguelen Island and it is characterized by continental inputs and relatively high productivity. Sampling depths were selected to cover important features of water masses revealed by CTD (conductance, temperature, depth) profiling. The surface mixed layer depth was 105 m at station R-2, 153 m at station A3-2, 46 m at station E-5, 72 m at station E-1, and 40 m at station TEW-2. Because of water quantity and time limitations, only the first 350 m of the water column were sampled at E-5.

Water samples for AA, dissolved organic carbon (DOC) and nitrogen (DON) analyses were collected using 10 L Teflon-lined Niskin-1010X bottles mounted on a 1018 rosette system adapted for trace metal clean work (Bowie et al., 2014). Subsamples were transferred to 500 mL Schott glass bottles that were rinsed with acid (HCl, 10 %) and then ultrapure water before use. Part of each water sample

Table 1. Sampling sites, chlorophyll *a* (Chl *a*) concentrations, particulate and dissolved organic carbon (POC, DOC) and nitrogen (PON, DON) concentrations, and amino acid (AA) concentrations in particulate and dissolved (including free AA) phases.

Stations	Coordinates		Depth (m)	[Chl *a*] ($\mu g\,L^{-1}$)[a]	[POC] (μM)[a]	[DOC] (μM)	[PON] (μM)[a]	[DON] (μM)[b]	[AA] (nM)			Free AA[e] (%)
	Latitude °S	Longitude °E							partic.[c]	dissolv.[d]	free	
R-2 (HNLC)[f]	50.383	66.683	40	0.32	7.8	48	0.8	6.0	162	335	38	11.3
27 Oct 2011[g]			70	0.27	nd	47	nd	6.1	262	216	nd	
			100	0.23	5.2	48	0.7	6.9	498	276	nd	
			200	0.02	2.9	47	0.3	nd	45	173	nd	
			300	nd	nd	45	nd	nd	23	198	nd	
			500	nd	1.5	44	0.1	nd	77	186	nd	
			900	nd	1.6	45	0.1	nd	92	192	11	5.7
			1300	nd	nd	44	nd	nd	33	151	nd	
A3-2	50.617	72.050	40	1.7	14.8	52	2.8	6.0	821	277	52	18.8
17 Nov 2011[g]			70	2.1	14.3	51	2.6	5.1	716	292	69	23.6
			110	2.4	13.9	50	2.6	3.9	656	237	nd	
			150	1.4	10.8	48	2.1	5.3	400	173	nd	
			210	0.15	3.5	47	0.4	3.9	44	144	10	6.9
			300	0.05	2.0	48	0.3	2.7	22	128	nd	
E-5	48.400	71.884	25	1.2	11.0	48	2.2	6.5	539	208	28	13.5
18 Nov 2011[g]			70	0.95	9.1	47	2.2	7.0	333	234	nd	
			110	0.36	nd	47	0.6	6.8	71	160	nd	
			200	0.12	3.5	46	0.3	6.9	45	191	6	3.1
			350	nd	3.0	45	0.2	nd	27	146	nd	
E-1	48.450	72.183	20	0.99	7.7	49	1.3	4.8	223	406	nd	
30 Oct 2011[g]			40	0.96	8.9	49	1.6	5.0	486	352	nd	
			70	0.86	4.5	47	1.0	5.2	124	495	nd	
			120	0.78	6.5	46	1.2	6.4	387	289	nd	
			200	0.07	5.1	45	1.0	nd	203	314	nd	
			500	nd	1.6	45	0.1	nd	nd	383	nd	
			1300	nd	nd	45	nd	nd	44	199	nd	
TEW-2 (shelf)	48.884	70.650	15	1.8	9.3	47	1.5	7.4	627	413	37	9.0
31 Oct 2011[g]			30	1.4	5.7	46	1.0	7.1	338	489	nd	
			62	0.6	6.4	45	0.7	7.6	205	403	11	2.7

[a] Data are from Lasbleiz et al. (2014).
[b] Data are from Blain et al. (2014).
[c] Particulate AA = total hydrolyzable amino acids (THAA).
[d] Dissolved AA = dissolved combined and free amino acids.
[e] % free amino acids versus dissolved AA.
[f] High-nutrient low-chlorophyll.
[g] Dates of sampling.
nd: not determined.

was filtered through two combusted GF/F filters (glass fiber, 0.7 μm nominal pore size, Whatman) using a Hamilton glass syringe and PTFE (polytetrafluoroethylene) tubing. For AA analyses of the dissolved fraction, 20 mL of the GF/F filtered water were stored in HDPE (high-density polyethylene) bottles at −20 °C. For DOC analyses, 15 mL of the GF/F filtered water were acidified (H_3PO_4, final pH = 2) and stored in the dark in combusted and sealed glass ampoules. DON analyses were done with an aliquot of the GF/F filtered water that was stored poisoned (100 μL of $HgCl_2$; 4 $g\,L^{-1}$, working solution) in PTFE bottles (Blain et al., 2014). Total AA (dissolved and particulate) were determined in 20 mL of unfiltered water stored in HDPE bottles at −20 °C. The concentration of particulate AA was calculated by subtracting the AA concentration in filtered water from the total AA concentration.

Water samples for POC and PON analyses were collected with regular 12 L Niskin bottles. A total of 1 L of seawater was filtered on combusted GF/F filters, and these were stored in the dark until analyses (Lasbleiz et al., 2014). The sampling depths of these water samples did not always perfectly match the depths of the samples used for the other analyses. In these cases, the depths that were the closest to those shown on Table 1 were used for POC and PON.

2.2 Organic carbon and nitrogen analysis

DOC concentrations were measured on acidified aliquots using a Shimadzu TOC-VCP analyzer with a Pt catalyst at 680 °C (Benner and Strom, 1993). All DOC analyses were run in duplicate and 3–5 injections were done for each sample. Analytical precision was 0.2–1 % (CV). Reference materials (http://www.rsmas.miami.edu/groups/biogeochem/CRM.html) were injected every 15 samples to insure stable operating conditions. DON concentrations were determined after wet oxidation and calculated as the difference between total N concentration and the sum of inorganic N (i.e., nitrate, nitrite, ammonium) (Blain et al., 2014). Inorganic N was determined by the automated colorimetric technique on a Skalar autoanalyzer (Aminot et Kérouel, 2007). POC and PON analyses of particles on filters were done on a PerkinElmer 2400 CHN analyzer (Lasbleiz et al., 2014). The filter blank values for POC and PON corresponded to 1.27 μM ± 0.26 and 0.06 μM ± 0.02, respectively.

2.3 Amino acid analysis

AA enantiomers (L and D) and achiral AA were measured after liquid-phase acid hydrolysis and reversed-phase high-performance liquid chromatography (HPLC) according to a newly developed procedure (Escoubeyrou and Tremblay, 2014). This procedure is similar to traditional ones, except that a high volume (100 μL) of undiluted samples is injected and a robust hybrid C-18 column is used. These changes increase the precision of the measurements while providing a more consistent separation.

Briefly, 366 μL of seawater (filtered or unfiltered) were transferred to an ampoule and mixed with 634 μL of HCl 30 % (Merck, Suprapur grade). Procedural blanks were also prepared with 634 μL of HCl 30 %. The ampoules were then vacuum sealed and placed at 110 °C for 20 h. After hydrolysis, 300 μL of hydrolyzate (190 μL for blanks) were evaporated on a Genevac Personal Evaporator (EZ-2 + HCl compatible). A total of 150 μL of ultrapure water were then added and evaporated to remove remaining traces of HCl. Samples consisting of 110 μL of nonhydrolyzed seawater were also evaporated for the analysis of free AA. Samples and blanks were then dissolved in 120 μL of borate buffer (0.4 M, Merck, Suprapur) with a pH adjusted to 10 with NaOH 6 M (Merck, Suprapur). Strong vortex agitation and sonication were necessary to completely dissolve the dried residue. The vials were then centrifuged (4300 rpm, 5 min) to separate the flocculate formed in borate buffer.

HPLC analyses were performed on an Ultimate 3000 Dionex system equipped with an autosampler and a fluorescence detector (excitation at 335 nm, emission at 450 nm). The separation was done with a Phenomenex Gemini C18 column (250 x 4.6 mm, 5 μm, 110 Å) and a SecurityGuard column (4 × 3 mm). Automated precolumn derivatization with o-phthaldialdehyde (OPA) and N-isobutyryl-L-cysteine (IBLC, L-run) or with OPA and N-isobutyryl-D-cysteine (IBDC, D-run) was performed. The reagents were prepared by dissolving 2 mg of OPA and 5 mg of IBLC or IBDC in 200 μL of methanol and diluting with 1.8 mL of ultrapure water. The derivatization mixture consisted of 100 μL of sample or blank, 10 μL of OPA/IBD(L)C reagent and 10 μL of S-methyl-L-cysteine (SMC, 0.74 μM) used as internal standard. After a reaction time of 2 min at 8 °C, 100 μL of this mixture were injected. OPA/IBLC (run L) and OPA/IBDC (run D) were used alternatively for every sample and blank. The elution order of L-AA and D-AA is reversed during the second run, which increases the accuracy of peak identification and quantification (Brückner et al., 1994). The largest peak of the two runs was not considered because it may be caused by a coelution. The column temperature was maintained at 44 °C. A gradient of organic phase (93 % methanol, 7 % acetonitrile) and KH_2PO_4 aqueous solution (40 mM, pH 6.15) was used with a flow rate of 0.8 mL min^{-1} (Escoubeyrou and Tremblay, 2014).

Before injecting samples, blanks or standards, several analyses of borate buffer were performed to check the level of contamination from free AA. AA identification and quantification in unknown samples and blanks were achieved by comparing the retention times and peak areas with those of standard AA solutions. Coefficients of determination of calibration curves were always between 0.99 and 1.0. Standard solutions contained 23 AA: L- and D-aspartic acid (L/D-Asp), L- and D-glutamic acid (L/D-Glu), L- and D-serine (L/D-Ser), L- and D-alanine (L/D-Ala), L- and D-valine (L/D-Val), L- and D-leucine (L/D-Leu), L-threonine (L-Thr), glycine (Gly), L-arginine (L-Arg), L-isoleucine (L-Ileu), L-phenylalanine (L-Phe), L-lysine (L-Lys), L-histidine (L-His), L-tyrosine (L-Tyr), L-methionine (L-Met), γ-aminobutyric acid (Gaba), and β-alanine (BAla).

Peak areas of AA measured in blanks were subtracted from those in samples. The borate buffer solution was used as the blank for the quantification of free AA. Individual AA concentrations in injected samples were then calculated using the slope of calibration curve of each AA. The average racemization rates of individual free and protein AA occurring during hydrolysis and measured by Kaiser and Benner (2005) were used to correct the concentrations of chiral AA. Asp and Glu peaks included the contribution of L-glutamine (Gln) and L-asparagine (Asn), respectively, which were deaminated during hydrolysis. They were thus identified as Asx and Glx.

Particulate THAA were quantified by the difference between the peak areas obtained with unfiltered and filtered hydrolyzed samples. The chromatograms of filtered samples were used to quantify dissolved AA, which are the sum of free and dissolved combined AA. Dissolved combined THAA can be determined by the difference between the peak areas obtained with filtered hydrolyzed and nonhydrolyzed samples. The limits of detection were 0.007–3.57 nM depending on the AA (Escoubeyrou and Tremblay, 2014). Most particulate AA (determined by difference) were near their limit of detection in samples having low POC concentrations. This increases the uncertainty of particulate AA, POC and PON values in these samples.

2.4 Chlorophyll analysis

Concentrations of chlorophyll *a* were determined by HPLC following the protocol of Van Heukelem and Thomas (2001) modified by Ras et al. (2008). Seawater samples (1–2.23 L) were filtered on 25 mm GF/F filters (Whatman) which were then stored in cryotubes in liquid nitrogen until further processing in the home lab, as described in detail in Lasbleiz et al. (2014).

3 Results

3.1 Organic carbon and nitrogen

Striking differences in OM concentrations and compositions were observed between stations (Table 1). As expected, station R-2 located in the HNLC region exhibited the lowest chlorophyll *a* ($< 0.32\,\mu g\,L^{-1}$) and POC concentrations (1.5–7.8 μM) compared to the other stations for similar depths. The highest concentrations of chlorophyll *a* (up to $2.4\,\mu g\,L^{-1}$) and POC (up to 14.8 μM) were observed in the mixed layer at station A3-2. Station E-1, sampled before the bloom, had relatively low POC concentrations in its surface waters (7.7 μM at 20 m). After 19 days, the POC concentrations increased to 11 μM at E-5. There was a marked decrease in POC concentrations with depth, especially at station A3-2 showing a 7.4-fold decrease. The lowest POC concentrations, 1.5–3 μM, were measured below 210 m.

DOC represented 78 (surface water of A3-2)–97 % (deep water of R-2) of total organic carbon (Table 1). Despite a slight decreasing trend with depth at all stations, DOC concentrations were much less variable than those of POC. The highest concentrations (51–52 μM) were measured in the surface waters of station A3-2, the bloom site above the plateau. The lowest concentration (44 μM) was found in the deep waters of the less-productive station R-2.

PON and DON concentrations also varied with location and depth (Table 1). They generally followed the same trends as their carbon counterparts, although not always to the same extent as indicated by the POC / PON and DOC / DON ratios in representative samples (Fig. 1). A preferential degradation of N-rich molecules in particles lead to a general increase in POC / PON ratios with depth (Fig. 1a). As expected, the particles in the surface waters of the bloom at stations A3-2 and E-5 were enriched in N. In contrast, the HNLC region (station R-2) had more N-poor particles (Fig. 1a) and 2–3.5 times lower PON concentrations (Table 1). DON concentrations varied from 7.6 μM at station TEW-2 to 2.7 μM in the deep waters of station A3-2. Station A3-2 was the only site where DON concentrations decreased (Table 1) and DOC / DON ratios increased with depth (Fig. 1b). This indicates that N-rich dissolved organic molecules at station A3-2 are more easily or rapidly degraded than those from the other stations.

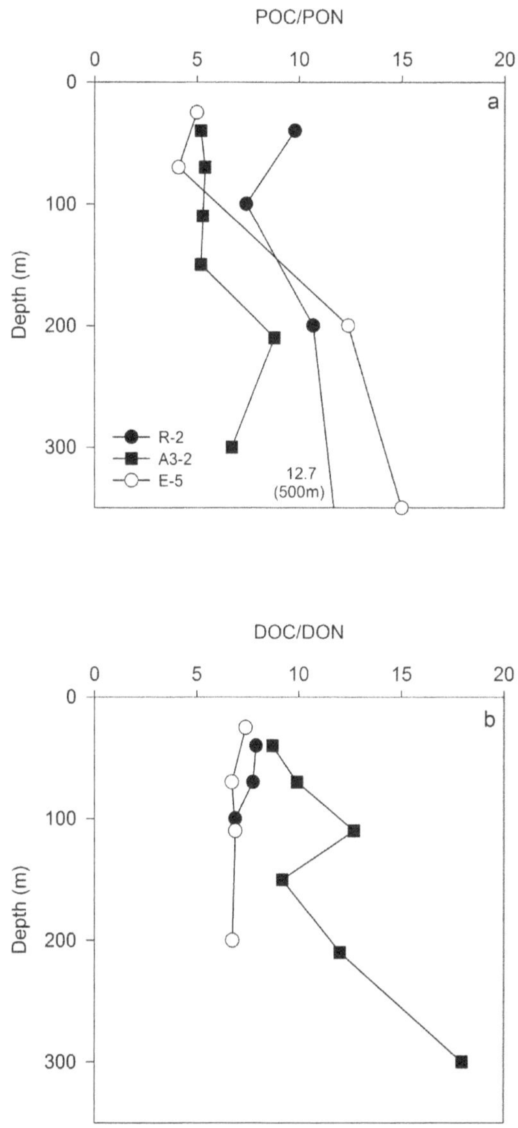

Figure 1. Carbon to nitrogen atomic ratios in the **(a)** particulate and **(b)** dissolved fractions for three representative stations (R-2, A3-2, E-5) at 0-350 m depth. The value close to the station R-2 profile is the next data point at 500 m. Ratios cannot be quantified at certain depths because elementary data were not available (Table 1).

3.2 Amino acid concentrations

Table 1 presents AA concentrations measured in the particulate and dissolved fractions. As expected, the most productive stations (A3-2, E-5, and TEW-2) had the highest concentrations of particulate THAA. The water collected at the surface of these stations contained between 539 and 821 nM of particulate THAA, while concentrations were 162–223 nM in the surface waters of stations R-2 and E-1. Though deep waters contained less particulate THAA than surface waters at all stations, the decrease with depth was much faster at the bloom sites. These concentrations dropped by a factor of 20 and 37 at stations E-5 and A3-2, respectively, but only by a factor of ~5 at stations R-2 and E-1. At stations R-2 and E-1, particulate THAA concentrations were maximal in the subsurface layer (40–100 m), and not at the surface like for bloom locations. This suggests a production of particulate THAA in the subsurface or a localized removal mechanism (e.g., photo-oxidation) in the first meters of the water column at stations R-2 and E-1.

In contrast to particulate THAA, concentrations of dissolved AA (combined and free AA) were generally higher at the less productive locations (R-2 and E-1; up to 495 nM) than at the bloom sites (A3-2 and E-5; 128–292 nM) for similar depths (Table 1). The shallow station TEW-2 also had high concentrations (403–489 nM). However, a greater proportion of dissolved AA were free AA at the surface of stations A3-2 and E-5 compared to station R-2 (13.5–23.6 % vs. 11.3 %; Table 1). Station A3-2 exhibited the highest concentrations of free AA (52–69 nM; Table 1), but lower dissolved combined AA (e.g., in proteins and peptides) concentrations than station R-2.

Concentrations of total dissolved AA decreased with depth, by a factor 1.4–2.2 (Table 1), except in the shallow water column at station TEW-2. The decrease in free AA was more important (factor 3.4–5.2). These trends were more pronounced at the bloom station A3-2, but were not as marked as for particulate THAA.

3.3 Amino acid yields and diagenetic markers

AA are the most abundant organic N-molecules in nature. We thus expect to see similar trends in the proportions of N and of THAA in POM as indicated by POC / PON (Fig. 1a) and $\%POC_{AA}$ (particulate organic carbon quantified as amino acids) (Fig. 2a), respectively. Both parameters indicated higher proportions of N-molecules in POM at stations A3-2 and E-5 and a decrease in these proportions with depth at all stations. The decrease in $\%POC_{AA}$ was particularly marked at stations A3-2 and E-5 where THAA accounted for 22–25 % of POC in the surface waters, but only for 4.2–4.9 % below 300 m (Fig. 2a). Particulate THAA accounted for 30–51 % of PON in the surface of the more productive stations and for 27 % at stations R-2, and these proportions decreased with depth at all stations, especially at station A3-2 (not shown). The degradation of particulate THAA was thus more rapid than that of other N molecules.

Dissolved AA contributed to only 0.86–2.7 % of DOC at stations R-2, A3-2 and E-5 (Fig. 2b). $\%DOC_{AA}$ (dissolved organic carbon quantified as amino acids) were slightly higher in the surface sample of stations E-1 and TEW-2 (2.4–4.4 %, E-1 and TEW-2 data not shown). The only consistent vertical trend was observed at station A3-2 where $\%DOC_{AA}$

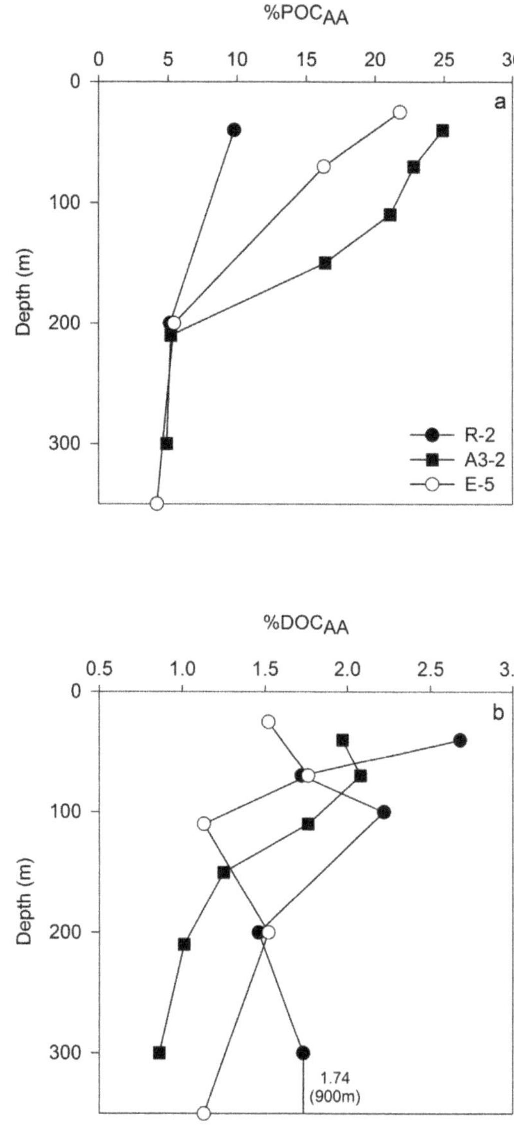

Figure 2. Proportion of **(a)** particulate and **(b)** dissolved organic carbon quantified as amino acids ($\%POC_{AA}$ and $\%DOC_{AA}$) for three representative stations (R-2, A3-2, E-5) at 0–350 m depth. The value close to the station R-2 profile is the next data point at 900 m.

decreased with depth from ~2 to 0.86 %. This suggests a preferential removal of dissolved AA with depth at this bloom station. This trend is not observed or is less marked at the other stations. Dissolved AA contributed to 2.3–8 % of DON with no trend according to station or depth (not shown).

During degradation, the mol% D-AA should increase and the DI should decrease. These markers indicated that POM from station R-2 was more altered than the POM from the other stations (Figs. 3a, 4a), in agreement with its lower N and THAA contents. However, mol% D-AA and DI did not show consistent trends with depth, except for mol% D-AA at station R-2 (Fig. 3a) and for DI in deep waters only (Fig. 4a). Clearly, these markers were not as sensitive, or effective, as

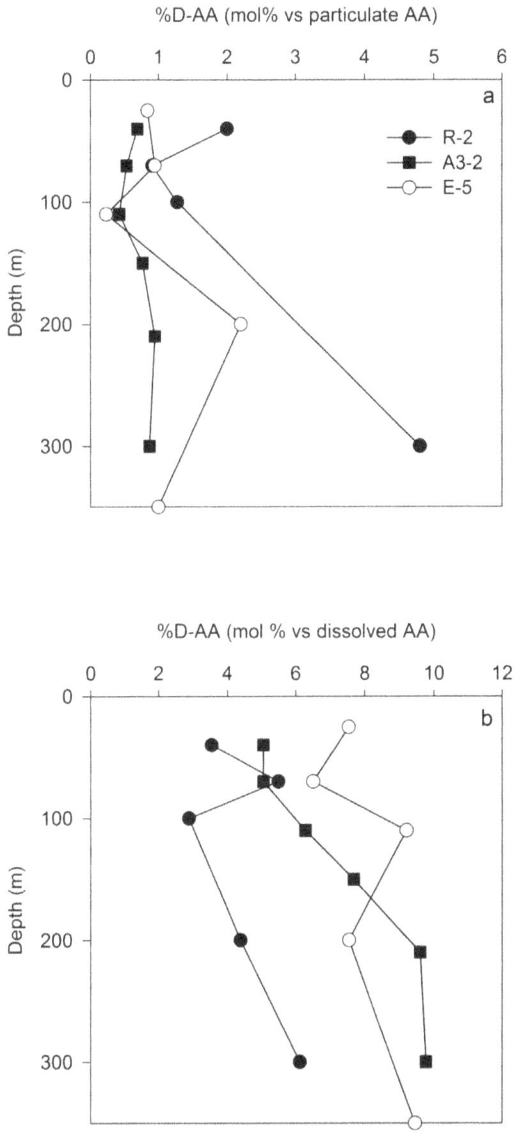

Figure 3. Proportion of D-amino acids (%D-AA) versus total amino acids (AA) in the **(a)** particulate and **(b)** dissolved fractions for three representative stations (R-2, A3-2, E-5) at 0–350 m depth.

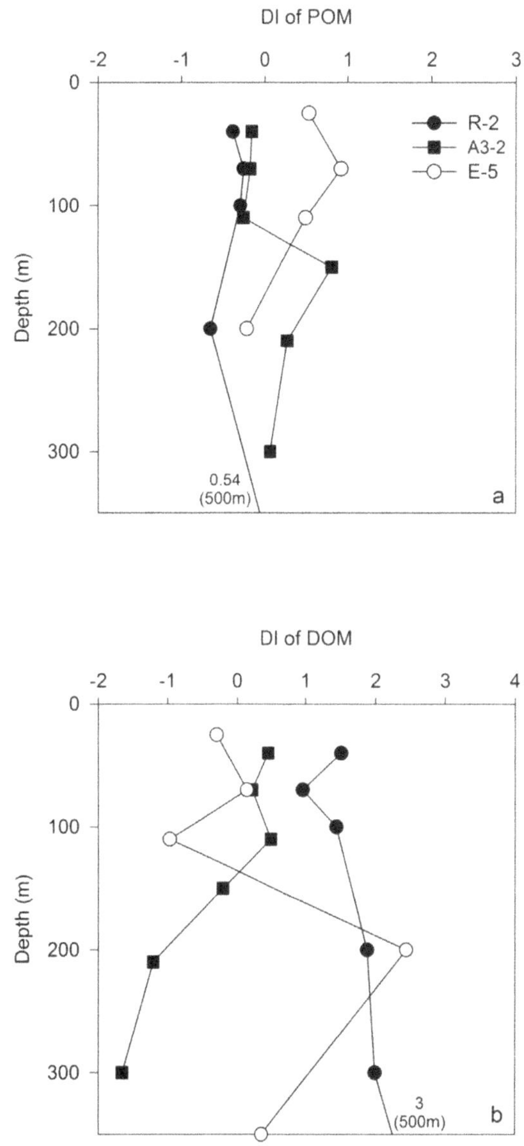

Figure 4. DI of **(a)** POM (Dauwe et al., 1999) and **(b)** DOM (Peter et al., 2012) calculated for three representative stations (R-2, A3-2, E-5) at 0–350 m depth. Values close to the station R-2 profiles are the next data point at 500 m.

$\%POC_{AA}$ (or POC / PON) for tracking POM degradation with depth in the present study. In the dissolved fraction, mol% D-AA and DI values suggest, once again, that DOM from station R-2 was less altered (Figs. 3b, 4b). However, no sign of alteration with depth was observed at this station. In fact, the only station showing a clear increase of DOM alteration state with depth was, once again, A3-2 (Figs. 3b, 4b). From 40 to 300 m, mol%D-AA went from 5 to 10 % and the DI from 0.5 to -1.7 in DOM. The increase in mol%D-AA in total OM (POM and DOM) with depth at station A3-2 was even more important (from ~ 2 to 8 %, not shown).

3.4 Markers of organic matter origin

The average values of Ser + Thr and Asp / Gly found in diatoms, coccolithophores and bacteria (Muller et al., 1986; Cowie and Hedges, 1992) are presented in Fig. 5, along with the values measured here at stations A3-2 and E-5. All the data points for POM (open circles) are near average values (open boxes) found in diatoms or between those of diatoms and bacteria, suggesting that the POM mostly originated from diatoms and bacteria. Except for one data point at each of the other stations having an Asp / Gly ratio of over 1.5, the same conclusion can be stated for the other studied

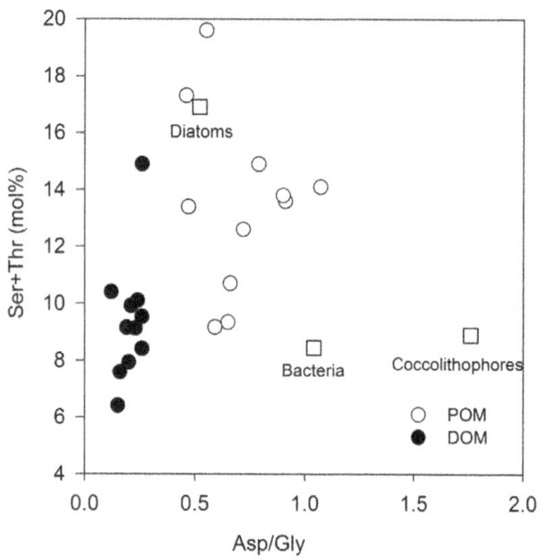

Figure 5. Molar contribution of serine + threonine (Ser + Thr) versus the molar ratio of aspartic acid and glycine (Asp / Gly) in the particulate and dissolved organic matter (POM and DOM) of stations A3-2 and E-5. Average values for diatoms, coccolithophores, and bacteria are based on Muller et al. (1986) and Cowie and Hedges (1992).

locations (not shown). No trend with depth was observed for Ser + Thr, but Asp / Gly generally decreased with depth (not shown). Gly is known to be selectively preserved during POM degradation (Dauwe et al., 1999). Changes in the composition of AA during degradation (e.g., microbial) are certainly responsible for the lower Ser + Thr and Asp + Gly values measured in DOM (Fig. 5).

Chlorophyll *a* and D-AA can be used to estimate the proportions of organic matter associated with phytoplankton (cells) and bacteria (cells and detritus), respectively. The ratios of chlorophyll *a* to POC concentrations (concentrations in Table 1) at stations A3-2 and E-5 are more than twice those measured in the HNLC station R-2. This difference can be due to a higher proportion of phytoplanktonic cells in the POM from the bloom and/or to lower intracellular chlorophyll content in the phytoplankton of the HNLC region. Assuming that diatom-dominated phytoplankton communities have 40 times more carbon than chlorophyll *a* on a per cell basis (Sathyendranath et al., 2009), calculations indicated that phytoplankton biomass could account for 14 % of POC in the surface water of station R-2, but for 35–63 % in the surface waters of the other stations (not shown), except at 30 m of TEW-2 where 81 % was calculated. Estimates were much lower in the deep waters, with 2.3 % calculated at station R-2 (200 m) and 8.3 % at station A3-2 (300 m). This approach takes mostly into account phytoplankton cells (vs. altered fragments) considering that a rapid decrease in chlorophyll *a* content is expected in phytoplankton detritus.

D-AA can be used to estimate the proportions of OM coming from bacterial cells and altered cell fragments or detritus. D-Asx yields varied from < 2 to 31.7 nmol mg^{-1} POC and D-Glx yields varied from < 2 to 18.1 nmol mg^{-1} POC (Table 2). In both cases, the highest values were measured above 200 m in the water columns. D-Ala yields were highly variable, < 2–68.8 nmol mg^{-1} POC with no clear trend with depth. The sample collected at 200 m at station R-2 exhibited relatively high values for each D-Asx, D-Glx and D-Ala (31.7, 18.1 and 53.7 nmol mg^{-1} POC, respectively), suggesting a greater contribution of bacteria to POC in this sample. Average yields reported in cultured marine bacteria (Kaiser and Benner, 2008) represent 15, 48 and 50 nmol mg^{-1} POC for D-Asx, D-Glx and D-Ala, respectively (Table 2). Comparisons between these values and those measured here in POM suggest that a significant, although variable, fraction of POC can be from bacteria. For instance, at station A3-2, where higher POC concentrations increase estimation accuracy and precision, D-Glx and D-Ala yields suggest that ~ 5–20 % of the POC was from bacteria. However, the lack of correlations, or consistent trends, between the D-AA biomarkers and the large ranges of D-AA yields in POM (and bacteria) makes it difficult to provide estimates of the percentages of POM coming from bacteria based on this approach.

D-AA yields in the dissolved fraction were much more consistent (Table 2). Figure 6 shows average values measured in DOM produced by bacteria (Kaiser and Benner, 2008) and in DOM from surface (0–120 m) and deep waters (> 120 m) at stations impacted (A3-2 + E-5) or not (R-2 + E-1) by the bloom. D-Asx and D-Ala yields suggest that 16.5–35 % of this DOC is from bacteria, with the lowest estimates found in the deep waters of stations R-2 and E-1. Similar trends, but lower estimates (10.4–17.5 %) were obtained when D-Glx yields are used. D-Glx yields in bacterial DOM were highly variable (Kaiser and Benner, 2008), and the reference average value used here might be higher than that of the bacterial DOM produced in the studied waters.

4 Discussion

4.1 Amino acids in the Southern Ocean

This study presents the first detailed quantification of particulate and dissolved AA in the Southern Ocean. The measured concentrations, yields, and relative distributions of AA in POM and DOM were in the range of values previously found in the open equatorial Pacific (Lee et al., 2000), the coastal waters of the Arctic (Dittmar et al., 2001; Shen et al., 2012) and North Atlantic Ocean (Bourgoin and Tremblay, 2010), and in the polar front zone of the Southern Ocean (Panagiotopoulos et al., 2002). THAA represented between 21 and 25 % of POC in the surface waters of bloom sites, which is close to the 27 % measured in suspended particles at the chlorophyll maximum depth in the equatorial Pacific (Lee et al., 2000). Such high values are expected when protein-rich plankton represents an important part of POM (see Sect. 4.2

Table 2. Yields of D-AA in particulate and dissolved fractions along with average yields measured in cultured bacteria and bacterial DOM.

Stations	Depth	D-Asx[a]	D-Glx[b]	D-Ala[c]	D-Asx	D-Glx	D-Ala
	(m)	nmol (mg POC)$^{-1}$			nmol (mg DOC)$^{-1}$		
R-2 (HNLC)	40	10.2	2.83	13.9	5.31	1.77	10.6
	70	nd	nd	nd	6.09	1.59	9.75
27 Oct 2011[g]	100	7.06	6.36	68.8	6.52	1.81	4.70
	200	31.7	18.1	53.7	4.27	0.81	7.17
	300	nd	nd	nd	6.58	1.63	11.3
	500	nd	nd	nd	3.80	1.06	7.69
	900	5.48	8.22	nd	4.85	1.84	10.5
	1300	nd	nd	nd	5.30	1.82	9.53
A3-2	40	13.5	9.33	5.34	6.36	2.61	11.0
17 Nov 2011[g]	70	9.80	8.99	2.17	6.88	2.56	12.1
	110	6.12	7.65	2.62	7.29	2.63	11.9
	150	7.30	6.25	6.77	6.59	2.62	11.0
	210	nd	2.40	7.51	6.88	2.61	11.6
	300	nd	6.57	nd	6.80	2.09	10.5
E-5	25	14.8	10.5	4.25	7.44	3.24	13.5
18 Nov 2011[g]	70	9.67	4.37	10.4	8.14	3.38	12.9
	110	nd	nd	nd	7.32	2.87	13.0
	200	7.79	1.91	10.8	6.95	2.50	13.0
	350	2.13	0.55	nd	6.82	2.73	14.1
E-1	20	1.70	7.13	nd	5.64	2.45	12.1
30 Oct 2011[g]	40	5.27	4.62	14.0	5.39	2.47	11.2
	70	nd	2.70	4.34	7.47	2.92	11.4
	120	nd	7.09	nd	6.51	2.44	11.3
	200	nd	2.17	nd	2.69	2.44	9.94
	500	nd	nd	nd	1.94	1.88	5.63
	1300	8.39	nd	nd	2.63	2.25	8.17
TEW-2	15	nd	2.85	nd	8.39	4.13	14.7
31 Oct 2011[g]	30	21.3	14.3	nd	8.46	4.24	15.8
	62	9.26	2.74	nd	8.12	4.10	16.0
Bacteria[d]		15 (22)	48 (26)	50 (19)	24 (3)	17 (7)	35 (7)

[a, b, c] D-aspartic acid or D-asparagine, D-glutamic acid or D-glutamine, and D-alanine, respectively.
[d] From Kaiser and Benner (2008). POC-normalized values are averages from three phototrophic and five heterotrophic marine bacteria assuming 20 % phototrophic and 80 % heterotrophic. DOC-normalized values are averages measured in three DOM samples freshly produced by bacteria. Values in parentheses are standard deviations.
nd: not determined or below detection (for POC or D-AA measurements).

below) (Cowie and Hedges, 1992; 1994). Concentrations of chlorophyll *a* and POM were highest in the surface water of the bloom station A3-2, consistent with a greater production of biomass induced by iron inputs. In contrast, THAA represented 10 % of POC at station R-2, which is expected in more-altered POM (Cowie and Hedges, 1994).

A particularity of the Southern Ocean surface waters are the lower DOC concentrations compared to other oceans (i.e., ~50 vs. 70–80 μM) (Hansell, 2013). Here, AA accounted for only 0.9–4.4 % of DOC, which is in the range of values found in other regions (Benner, 2002; Bourgoin and Tremblay, 2010; Shen et al., 2012). However, AA yields in the DOC of HNLC surface water (2.7 %) were about twice that measured in the oligotrophic North Pacific near Hawaii (Kaiser and Benner, 2008). This suggests more favorable conditions for AA accumulation/preservation in the HNLC of the Southern Ocean.

4.2 Origin and fate of POM

Molecular AA markers (i.e., Ser + Thr, Asp / Gly and D-AA yields) indicated, at least qualitatively, that the POM at the bloom sites mostly originated from diatoms and bacteria. Different approaches can be used to estimate the proportion of each of these components of the plankton community. Pigment analysis and microscopic observations confirmed that diatoms were abundant in surface waters of the bloom sites

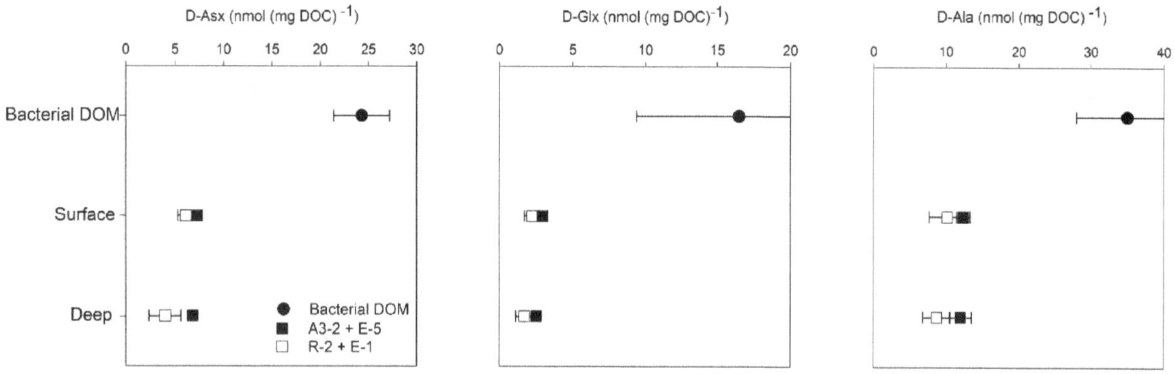

Figure 6. Average C-normalized yields of D-aspartic acid or D-asparagine (D-Asx), D-glutamic acid or D-glutamine (D-Glx), and D-alanine (D-Ala) in bacterial DOM (Kaiser and Benner 2008) and DOM from surface (0–120 m) and deep waters (> 120 m) at stations impacted (A3-2 + E-5) or not (R-2 + E-1) by the bloom. Error bars represent standard deviations.

(Lasbleiz et al., 2014). Species-specific biovolume determination in the euphotic layer (between the surface and 0.01 % PAR – photosynthetically active radiation) revealed that diatoms accounted, on average, for 44 % of POC at station A3-2, while their contribution to POC was only 5 % at the HNLC site (Lasbleiz, 2014) (Table 3). These observations are consistent with the present estimates based on chlorophyll *a* concentrations, indicating that phytoplankton cells could account on average for 45 % of POC in the surface waters of station A3-2, but 14 % of POC in the surface waters of station R-2 (Table 3). The lower proportion of diatoms among the phytoplankton community at station R-2 is in agreement with direct observations (Lasbleiz et al., 2014). These estimates were 5–7 times lower below 200 m at both locations (Table 3). In deep waters, the phytoplankton contribution to POC is thus likely dominated by detrital material (vs. intact cells). Laurenceau et al. (2014) used microscopic observations of material collected in gel traps and estimated that phytodetrital aggregates represented 41 and 34 % of POC at stations A3-2 and R-2, respectively, in sinking particles collected at 210 m (Table 3). Both types of estimates show the same trend of a higher phytoplankton contribution at station A3-2. The higher percentage contribution obtained for the gel traps is most likely due to the fact that this approach considers both living and dead phytoplankton cells, and that particles defined as phytodetrital aggregates can contain POC from other origins (e.g., bacteria).

D-AA yields were used to estimate the proportion of POC from bacteria (cells and altered debris of bacteria). However, some extreme and inconsistent values were measured in the present dataset. Extreme values or the disagreement between D-AA yields are probably caused by the fact that yields in POM were calculated by the difference between the chromatographic peaks measured in unfiltered and filtered water samples. In many samples this difference was very small because of the very low POM concentrations, increasing the uncertainty of these yields. Different dynamics between D-Asx, D-Glx et D-Ala in the studied POM could also explain inconsistent values. This appears to be the case at station A3-2, where D-Glx and D-Ala markers were not always in agreement, despite the increased estimation accuracy and precision due to the higher POC concentrations. For instance, at 70 m, at this site, estimates based on D-Glx and D-Ala were 19 and 5 %, respectively (not shown). Averages suggest that ~ 15 % of the POC originated from bacteria at all depths at station A3-2 (Table 3, Bacteria total). Using the same approach and markers, Kaiser and Benner (2008) estimated that about 25 % of POC was of bacterial origin in oligotrophic areas of the North Pacific and North Atlantic.

An independent approach, based on cell counts of autotrophic (unpublished data) and heterotrophic bacteria (Christaki et al., 2014) and conversion of cell counts to carbon units using 250 fg C cell^{-1} for autotrophic (Campbell et al., 1994) and 12.4 fg C cell^{-1} for heterotrophic bacteria (Fukuda et al., 1998), revealed that only 2–3 % of POC are accounted for by bacterial cells as determined by flow cytometry in the surface waters at station A3-2 (Table 3, Bacteria biomass). However, there was a steady increase of this proportion with depth at station A3-2, reaching 10 % at 300 m. When compared to the D-AA estimate of ~ 15 % of POC originating from bacteria, it appears that intact bacterial cells represent a small proportion (< 20 %) of the total bacterial contribution (living + detritus) in the surface water, which is in agreement with previous studies (e.g., White and Howes, 1994; Bourgoin and Tremblay, 2010). However, intact bacterial cells appear to dominate (67 %) the bacterial contribution in the deep waters at the bloom site. This increase in bacterial cell contribution with depth was not observed at the HNLC site where proportions fluctuated between 4 and 7 % of POC. Different trends have been reported in previous studies. As observed here at the bloom site, the proportion of POC as bacterial biomass increased with depth in the St. Lawrence Gulf (Bourgoin and Tremblay, 2010). This trend was caused by a depletion of bacterial POM with depth

Table 3. Summary of the characteristics of POM and DOM in HNLC waters (station R-2) and at the bloom site above the Kerguelen Plateau (station A3-2). Surface refers to 10–90 m depth layer and deep refers to 210–300 m depth. Contributions to POM and DOM are average values given as percent of bulk organic carbon.

Depth	Parameters	POM		DOM	
		HNLC	Bloom	HNLC	Bloom
Surface					
	Phytoplankton				
	diatom cells[a]	5 %	44 %	nd	nd
	total cells[b]	14 %	45 %	nd	nd
	Bacteria				
	cells[c]	5 %	3 %	nd	nd
	total[d]	> 15 %[g]	15 %	30 %	30 %
	Alteration state	high	low	low	med
	Alterations with depth[e]	med	high	low	high
Deep					
	Phytoplankton				
	total cells[b]	2 %	8 %	nd	nd
	phytodetritus [f]	34 %	41 %	nd	nd
	Bacteria				
	cells[c]	7 %	10 %	nd	nd
	total[d]	> 15 %[g]	15 %	25 %	30 %
	Alteration state	high	high	med	high

[a] From microscopic counts and biovolume determination (Lasbleiz, 2014).
[b] Total cells assuming that phytoplankton cells have 40 times more carbon than chlorophyll *a* (Sathyendranath et al., 2009).
[c] From cell counts of autotrophic and heterotrophic bacteria and conversion factors (Christaki et al., 2014).
[d] Total bacterial contribution estimated from D-AA yields (see text), for POC it includes living cells and cell fragments or detritus.
[e] Alterations promoted by different organic matter inputs or environmental conditions (see text).
[f] Estimated based on direct observation of sinking particles defined as phytodetrital aggregates (Laurenceau et al., 2014).
[g] Estimates were inconsistent but generally > 15 % (see text).
nd: not determined.
med: medium.

that was more intense than the decrease of bacterial biomass with depth. In contrast, Kawasaki et al. (2011) estimated that bacterial detritus represented a greater proportion of POC (vs. living bacterial biomass) at 100 m (∼ 7 %) than at 5 m (∼ 4.5 %) in an oligotrophic area of the North Pacific gyre. However, autotrophic bacteria accounted for ∼ 13 % of their POC which was about 50 % of the total bacterial contribution (living + detritus). In the present study, autotrophic bacteria (*Synechococcus*) had low cell numbers ($2–10 \times 10^5$ cells L^{-1}; data not shown). It thus appears that a high lability of bacterial detritus and a low proportion of autotrophic bacteria increase the relative proportion of living bacteria versus bacterial detritus in POC in deep waters as in the present study (see below).

A large fraction of the POM remains of unknown origin, especially in the HNLC site. This fraction could be mostly from detritus of various origins including zooplankton. Laurenceau et al. (2014) estimated that zooplankton fecal pellets represented on average 56 ± 19 % of POC flux below 110 m at the studied locations. These pellets contain materials from different origins including phytoplankton (Silver and Gowing, 1991).

The reactivity of POM was markedly higher at the bloom site A3-2 compared to the HNLC site, as revealed by a much more rapid decrease of POM and particulate THAA concentrations with depth. In addition, THAA yields in POM ($\%POC_{AA}$ and $\%PON_{AA}$), reliable indicators of POM alteration or diagenetic state (Cowie and Hedges, 1994; Tremblay and Benner, 2009), indicated that the POM from the bloom was relatively fresh or unaltered, but was more rapidly degraded during sinking than the POM from the HNLC site (Table 3). These findings could explain the lower efficiency of carbon export at 100 m at the bloom sites A3-2 (5 %) and E-5 (11 %) than in HNLC waters (station R-2, 34 %) (Planchon et al., 2014). Laurenceau et al. (2014) also found that export efficiencies (at 200 m) were negatively correlated to net primary productivity during the same sampling campaign.

In contrast to THAA yields, no consistent trend in mol% D-AA and DI values were observed with depth, probably because more intense degradation of bulk POM is required to be detectable with these markers (Davis et al., 2009; Bourgoin and Tremblay, 2010). Our data show that a large fraction of particulate AA is rapidly utilized in surface waters of the bloom. The AA that remain in the POM below the surface layer are probably in more recalcitrant structures. These structures may be too recalcitrant to show significant change in AA composition (i.e., mol% D-AA and DI values) with depth. In addition, an apparent mismatch between DI values and the expected trend during degradation has been previously observed in the Southern Ocean and was attributed to the unusual AA composition of diatoms, the dominant source of POM (Ingalls et al., 2003).

4.3 Origin and fate of DOM

Although an important fraction of the primary production is transferred to the DOM pool (e.g., 30–50 % as short lived labile DOM: Pomeroy et al., 2007; Obernosterer et al., 2008; Hansell, 2013), it was not possible to estimate the phytoplankton contribution to bulk DOC with the markers used in the present study. By contrast, the bacterial contribution to DOC was estimated using D-AA yields, indicating that 16.5–35,% of bulk DOC was of bacterial origin with no consistent trend with depth or location. These contributions were near 30 % in surface waters, and similar between Fe-fertilized waters and the HNLC site (Table 3). Similarly, DOC concentrations were not substantially different among sites (i.e., 3–4 μM greater in the bloom). Taken together, these results suggest that DOM production was counterbalanced by bacterial consumption (Christaki et al. 2014) of DOM from bacteria and other sources. The relatively constant bacterial contribution to DOC measured here (~ 30 %) or in the North Atlantic and North Pacific (~ 25 %, Kaiser and Benner, 2008) indicates that bacterial DOM has an average reactivity similar to bulk DOM and contributes to all DOM fractions, from labile to refractory DOM.

This study shows for the first time the contrasting fates of the DOM produced during natural iron fertilization and the DOM from an HNLC region (Table 3). The bloom site A3-2 clearly shows intense DOM alteration within the water column. At station A3-2, all the molecular diagenetic markers (%DOC_{AA}, mol% D-AA, DI values) indicated that surface DOM was less altered than deep water DOM. In contrast, DOM in HNLC surface waters revealed a lower alteration state that did not change with depth (Table 3). The DOM produced during the bloom was thus more rapidly degraded and mineralized, and this pattern reflects well that observed for POM. Mixing of deep and surface water layers at station A3-2 could explain the gradual character of the trends observed in profiles of DOM molecular markers at this location. Despite the fact that a water-column stratification was present during sampling, with a surface mixed layer of 153 m, station A3-2 is located above the Kerguelen Plateau which is surrounded by deeper areas and thus a more intense or frequent vertical mixing may occur. However, this probable mixing cannot explain why station A3-2 had more altered DOM (and lower dissolved AA) than in the HNLC area even in deep waters.

A surprising finding of the present study is the lower total dissolved AA concentrations and yields in surface water DOM in the bloom sites as compared to HNLC waters. These differences were due to ~ 30 % more dissolved combined AA in HNLC waters. By contrast, free AA had higher concentrations and represented a greater proportion of dissolved AA at the bloom sites (13.5–23.6 % vs. 11.3 % at HNLC) which likely reflects the higher phytoplankton activity and associated DOM release mechanisms (Fuhrman, 1987; Sarmento et al., 2013).

We propose four possibilities that could explain the lower alteration state of DOM and the higher concentration of dissolved combined AA in less productive sites (i.e., stations R-2 and E-1) compared to the bloom sites (i.e., stations A3-2 and E-5). First, the higher quantities of labile DOM produced at the bloom sites could induce a priming effect (Bianchi, 2011). These large quantities of labile DOM (e.g., free AA) induce a strong microbial response (Christaki et al., 2014) and the energy gained through labile DOM may allow for the degradation of more recalcitrant DOM including dissolved combined AA. This degradation was particularly marked within the water column of the bloom site A3-2. As a consequence, the DOM at the bloom sites appears more altered than the DOM from the HNLC site (Table 3). The higher concentration of more altered DOM in the bloom sites is consistent with the microbial carbon pump concept in which marine microbes transform labile DOM into refractory DOM (Jiao et al., 2011). Important inputs of labile DOM likely stimulate bacterial activity leading to more intense DOM degradation and production of more recalcitrant DOM. The more rapid degradation of the DOM from the bloom is consistent with a recent study in Arctic fjords (Osterholz et al., 2014). Their findings suggest that DOM produced during the spring/summer bloom is degraded in a few weeks by a specialized resident microbial community. Second, another reason for the slower DOM degradation and the accumulation of combined dissolved AA in HNLC waters is an iron limitation for heterotrophic bacteria (Obernosterer et al., 2014). Third, the DOM in HNLC waters can be involved in physicochemical processes (e.g., inter-molecular interactions, partial photooxidation) that protect DOM such as dissolved combined AA from microbial attack (Keil and Kirchman, 1994). Protection mechanisms may be too slow to significantly reduce the microbial degradation rate of the labile DOM produced during the bloom. Finally, a fourth explanation could be that HNLC waters produce DOM molecules, including ones that contain combined AA, which are more recalcitrant to degradation.

The first three explanations are associated with unfavorable conditions for DOM degradation in HNLC areas, namely low quantity of labile or priming DOM, iron limitation, and DOM protection. The last explanation implies that the DOM produced in the HNLC region is intrinsically more recalcitrant than the DOM produced during the bloom. Though the DOM from the HNLC site appears less degraded, the relative lability or recalcitrance of this DOM remains unknown. This notion of substrate quality "may only be valid in the context of the ambient environment" or conditions (Bianchi, 2011). Freshly produced and highly altered or old molecules can persist in unfavorable conditions, but be degraded in another environment.

5 Conclusions

Important differences in POM and DOM origins, compositions, and alterations were observed between the HNLC region and the areas where a phytoplankton bloom was induced by natural iron fertilization in the Southern Ocean. Phytoplankton cells represent an important fraction of POC ($\sim 45\,\%$) in the surface waters at the bloom sites, while the bacterial total (cells and detritus) contribution was much lower ($\sim 15\,\%$). Molecular indicators demonstrate that POM originating from the bloom was more rapidly degraded during sinking than the POM from the HNLC site. The fraction of the bloom POM that was transferred to DOM is unknown, but $\sim 30\,\%$ of the DOM appears to come from bacteria. As for POM, the DOM from the bloom sites was involved in more intense degradation processes. In contrast, the DOM in HNLC waters was less altered and showed no significant alteration throughout the water column. The bloom induced by natural iron fertilization probably leads to favorable conditions (e.g., large quantities of labile OM) supporting high microbial activity (Christaki et al., 2014) and enhanced POM and DOM degradation. The more intense degradation in bloom sites reduces the carbon export efficiency and the net impact of such blooms on the global carbon budget. However, a higher microbial activity appears to also lead to a greater production of more recalcitrant DOM or a more active microbial carbon pump (Jiao et al., 2011).

Acknowledgements. We thank Louise Oriol and Karine Escoubeyrou for their help during sampling and analyses. We also thank our many colleagues who participated in the collection of various data sets, the KEOPS coordinator S. Blain, the chief on board, B. Quéguiner, and the crew aboard the R/V *Marion Dufresne* for their help in the successful completion of the cruise. C. Panagiotopoulos and an anonymous reviewer provided valuable comments that improved the manuscript. This research was supported by the project KEOPS2 (ANR-10-Blanc-614, LEFE INSU, CNES, IPEV) and the Natural Science and Engineering Research Council of Canada (NSERC). L. Tremblay received support from the UPMC during his sabbatical stay at the Laboratoire d'Océanographie Microbienne. We are grateful to the BIO2MAR platform (http://bio2mar.obs-banyuls.fr) for providing access to instrumentation.

Edited by: B. Quéguiner

References

Aminot, A. and Kérouel, R.: Dosage Automatique des Nutriments dans les Eaux Marines: Meìthodes en Flux Continu, Ifremer, France, 188 pp., 2007.

Asano, Y. and Lübbehüsen, T. L.: Enzymes acting on peptides containing d-amino acid, J. Biosci. Bioeng., 89, 295–306, 2000.

Benner, R.: Chemical Composition and Reactivity, in: Biogeochemistry of Marine Dissolved Organic Matter, edited by: Hansell, D. and Carlson, C., Academic Press, San Diego, 59–90, 2002.

Benner, R. and Strom, M.: A critical-evaluation of the analytical blank associated with DOC measurements by high-temperature catalytic-oxidation, Mar. Chem., 41, 153–160, 1993.

Bianchi, T. S.: The role of terrestrially derived organic carbon in the coastal ocean: a changing paradigm and the priming effect, P. Natl. Acad. Sci. USA, 108, 19473–19481, 2011.

Blain, S., Carlotti, F., Christaki, U., Corbière, A., Durand, I., Ebersbach, F., Fuda, J., Garcia, N., Gerringa, L., Griffiths, B., Guigue, C., Quéguiner, B., Guillerm, C., Jacquet, S., Jeandel, C., Laan, P., Lefèvre, D., Lo Monaco, C., Malits, A., Mosseri, J., Obernosterer, I., Park, Y., Armand, L., Picheral, M., Pondaven, P., Remenyi, T., Sandroni, V., Sarthou, G., Savoye, N., Scouarnec, L., Souhaut, M., Thuiller, D., Timmermans, K., Belviso, S., Trull, T., Uitz, J., van Beek, P., Veldhuis, M., Vincent, D., Viollier, E., Vong, L., Wagener, T., Bombled, B., Bopp, L., Bowie, A., Brunet, C., and Brussaard, C.: Effect of natural iron fertilization on carbon sequestration in the Southern Ocean, Nature, 446, 1070–1074, 2007.

Blain, S., Capparos, J., Guéneuguès, A., Obernosterer, I., and Oriol, L.: Distributions and stoichiometry of dissolved nitrogen and phosphorus in the iron fertilized region near Kerguelen (Southern Ocean), Biogeosciences Discuss., 11, 9949–9977, doi:10.5194/bgd-11-9949-2014, 2014.

Bourgoin, L. and Tremblay, L.: Bacterial reworking of terrigenous and marine organic matter in estuarine water columns and sediments, Geochim. Cosmochim. Ac., 74, 5593–5609, 2010.

Bowie, A. R., van der Merwe, P., Quéroué, F., Trull, T., Fourquez, M., Planchon, F., Sarthou, G., Chever, F., Townsend, A. T., Obernosterer, I., Sallée, J.-B., and Blain, S.: Iron budgets for three distinct biogeochemical sites around the Kerguelen archipelago (Southern Ocean) during the natural fertilisation experiment KEOPS-2, Biogeosciences Discuss., 11, 17861–17923, doi:10.5194/bgd-11-17861-2014, 2014.

Boyd, P. W., Jickells, T., Law, C. S., Blain, S., Boyle, E. A., Buesseler, K. O., Coale, K. H., Cullen, J. J., de Baar, H. J. W., Follows, M., Harvey, M., Lancelot, C., Levasseur, M., Owens, N. P. J., Pollard, R., Rivkin, R. B., Sarmiento, J., Schoemann, V., Smetacek, V., Takeda, S., Tsuda, A., Turner, S., and Watson, A. J.: Mesoscale iron enrichment experiments 1993–2005: synthesis and future directions, Science, 315, 612–617, 2007.

Brückner, H., Haasmann, S., Langer, M., Westhauser, T., Wittner, R., and Godel, H.: Liquid chromatographic determination of d-

and l-amino acids by derivatization with o-phthaldialdehyde and chiral thiols, J. Chromatogr. A, 666, 259–273, 1994.

Campbell, L., Nolla, H. A., and Vaulot, D.: The importance of Prochlorococcus to community structure in the central North Pacific Ocean, Limnol. Oceanogr., 39, 954–961, 1994.

Christaki, U., Obernosterer, I., Van Wambeke, F., Veldhuis, M., Garcia, N., and Catala, P.: Microbial food web structure in a naturally iron-fertilized area in the Southern Ocean (Kerguelen Plateau), Deep-Sea Res. Pt. II, 55, 706–719, 2008.

Christaki, U., Lefèvre, D., Georges, C., Colombet, J., Catala, P., Courties, C., Sime-Ngando, T., Blain, S., and Obernosterer, I.: Microbial food web dynamics during spring phytoplankton blooms in the naturally iron-fertilized Kerguelen area (Southern Ocean), Biogeosciences, 11, 6739–6753, doi:10.5194/bg-11-6739-2014, 2014.

Church, M. J., Hutchins, D. A., and Ducklow, H. W.: Limitation of bacterial growth by dissolved organic matter and iron in the Southern ocean, Appl. Environ. Microb., 66, 455–466, 2000.

Consensus Reference Materials Project: http://www.rsmas.miami.edu/groups/biogeochem/CRM.html, last access: 16 January 2015.

Cowie, G. and Hedges, J.: Sources and reactivities of amino-acids in a coastal marine environment, Limnol. Oceanogr., 37, 703–724, 1992.

Cowie, G. and Hedges, J.: Biochemical indicators of diagenetic alteration in natural organic matter mixtures, Nature, 369, 304–307, 1994.

Dauwe, B., Middelburg, J. J., Herman, P. M. J., and Heip, C. H. R.: Linking diagenetic alteration of amino acids and bulk organic matter reactivity, Limnol. Oceanogr., 44, 1809–1814, 1999.

Davis, J., Kaiser, K., and Benner, R.: Amino acid and amino sugar yields and compositions as indicators of dissolved organic matter diagenesis, Org. Geochem., 40, 343–352, 2009.

Dittmar, T., Fitznar, H. P., and Kattner, G.: Origin and biogeochemical cycling of organic nitrogen in the eastern Arctic Ocean as evident from D- and L-amino acids, Geochim. Cosmochim. Ac., 65, 4103–4114, 2001.

Dittmar, T., Cherrier, J., and Ludwichowski, K.: The analysis of amino acids in seawater, in: Practical guidelines for the analysis of seawater, edited by: Wurl, O., CRC Press, 67–77, 2009.

Ebersbach, F. and Trull, T.: Sinking particle properties from polyacrylamide gels during the KErguelen Ocean and Plateau compared Study (KEOPS): Zooplankton control of carbon export in an area of persistent natural iron inputs in the Southern Ocean, Limnol. Oceanogr., 53, 212–224, 2008.

Escoubeyrou, K. and Tremblay, L.: Quantification of free, dissolved combined, particulate, and total amino acid enantiomers using simple sample preparation and more robust chromatographic procedures, Limnol. Oceanogr.-Meth., 12, 421–431, 2014.

Fuhrman, J.: Close coupling between release and uptake of dissolved free amino acids in seawater studied by an isotope dilution approach, Mar. Ecol.-Prog. Ser., 37, 45–52, 1987.

Fukuda, R., Ogawa, H., Nagata, T., and Koike, I.: Direct Determination of Carbon and Nitrogen Contents of Natural Bacterial Assemblages in Marine Environments, Appl. Environ. Microb., 64, 3352–3358, 1998.

Gupta, L. and Kawahata, H.: Amino acid and hexosamine composition and flux of sinking particulate matter in the equatorial Pacific at 175 degrees E longitude, Deep-Sea Res. Pt. I, 47, 1937–1960, 2000.

Hansell, D.: Recalcitrant Dissolved Organic Carbon Fractions, Annual Review of Marine Science, 5, 421–445, 2013.

Ingalls, A. E., Lee, C., Wakeham, S. G., and Hedges, J. I.: The role of biominerals in the sinking flux and preservation of amino acids in the Southern Ocean along 170° W, Deep-Sea Res. Pt. II, 50, 713–738, 2003.

Ittekkot, V., Deuser, W., and Degens, E.: Seasonality in the fluxes of sugars, amino acids, and amino sugars to the deep ocean: Sargasso Sea, Deep-Sea Res., 31, 1057–1069, 1984.

Jiao, N., Azam, F., and Sanders, S.: Microbial carbon pump in the ocean. Washington, DC, Sciences/AAAS, Washington, DC, 2011.

Kaiser, K. and Benner, R.: Hydrolysis-induced racemization of amino acids, Limnol. Oceanogr.-Meth., 3, 318–325, 2005.

Kaiser, K. and Benner, R.: Major bacterial contribution to the ocean reservoir of detrital organic carbon and nitrogen, Limnol. Oceanogr., 53, 99–112, 2008.

Kawasaki, N., Sohrin, R., Ogawa, H., Nagata, T., and Benner, R.: Bacterial carbon content and the living and detrital bacterial contributions to suspended particulate organic carbon in the North Pacific Ocean, Aquat. Microb. Ecol., 62, 165–176, 2011.

Keil, R. G. and Kirchman, D. L.: Abiotic transformation of labile protein to refractory protein in sea water, Mar. Chem., 45, 187–196, 1994.

Lasbleiz, M.: Dynamique nutritionnelle du phytoplancton, cycles biogéochimiques du carbone et des éléments biogènes associées (N,P,Si) et disponibilité du fer dans l'Océan Austral – programme KEOPS 2, PhD thesis, Aix Marseille Université, France, 2014 (in French).

Lasbleiz, M., Leblanc, K., Blain, S., Ras, J., Cornet-Barthaux, V., Hélias Nunige, S., and Quéguiner, B.: Pigments, elemental composition (C, N, P, and Si), and stoichiometry of particulate matter in the naturally iron fertilized region of Kerguelen in the Southern Ocean, Biogeosciences, 11, 5931–5955, doi:10.5194/bg-11-5931-2014, 2014.

Laurenceau, E. C., Trull, T. W., Davies, D. M., Bray, S. G., Doran, J., Planchon, F., Carlotti, F., Jouandet, M.-P., Cavagna, A.-J., Waite, A. M., and Blain, S.: The relative importance of phytoplankton aggregates and zooplankton fecal pellets to carbon export: insights from free-drifting sediment trap deployments in naturally iron-fertilised waters near the Kerguelen plateau, Biogeosciences Discuss., 11, 13623–13673, doi:10.5194/bgd-11-13623-2014, 2014.

Lee, C., Wakeham, S. G., and Hedges, J. I.: Composition and flux of particulate amino acids and chloropigments in equatorial Pacific seawater and sediments, Deep-Sea Res. Pt. I, 47, 1535–1568, 2000.

Le Quéré, C., Rodenbeck, C., Buitenhuis, E. T., Conway, T. J., Langenfelds, R., Gomez, A., Labuschagne, C., Ramonet, M., Nakazawa, T., Metzl, N., Gillett, N., and Heimann, M.: Saturation of the southern ocean CO_2 sink due to recent climate change, Science, 316, 1735–1738, 2007.

Muller, P. J., Suess, E., and Ungerer, C. A.: Amino-acids and amino-sugars of surface particulate and sediment trap material from waters of the Scotia Sea, Deep-Sea Res., 33, 819–838, 1986.

Nagata, T., Meon, B., and Kirchman, D.: Microbial degradation of peptidoglycan in seawater, Limnol. Oceanogr., 48, 745–754, 2003.

Obernosterer, I., Christaki, U., Lefèvre, D., Catala, P., Van Wambeke, F., and Lebaron, P.: Rapid bacterial mineralization of organic carbon produced during a phytoplankton bloom induced by natural iron fertilization in the Southern Ocean, Deep-Sea Res. Pt. II, 55, 777–789, 2008.

Obernosterer, I., Fourquez, M., and Blain, S.: Fe and C co-limitation of heterotrophic bacteria in the naturally fertilized region off Kerguelen Islands, Biogeosciences Discuss., 11, 15733–15752, doi:10.5194/bgd-11-15733-2014, 2014.

Osterholz, H., Dittmar, T., and Niggemann, J.: Molecular evidence for rapid dissolved organic matter turnover in Arctic fjords, Mar. Chem., 160, 1–10, 2014.

Panagiotopoulos, C., Sempéré, R., Obernosterer, I., Striby, L., Goutx, M., Van Wambeke, F., Gautier, S., and Lafont, R.: Bacterial degradation of large particles in the southern Indian Ocean using in vitro incubation experiments, Org. Geochem., 33, 985–1000, 2002.

Peter, S., Shen, Y., Kaiser, K., Benner, R., and Durisch-Kaiser, E.: Bioavailability and diagenetic state of dissolved organic matter in riparian groundwater, J. Geophys. Res., 117, G04006, doi:10.1029/2012JG002072, 2012.

Planchon, F., Ballas, D., Cavagna, A.-J., Bowie, A. R., Davies, D., Trull, T., Laurenceau, E., Van Der Merwe, P., and Dehairs, F.: Carbon export in the naturally iron-fertilized Kerguelen area of the Southern Ocean based on the ^{234}Th approach, Biogeosciences Discuss., 11, 15991–16032, doi:10.5194/bgd-11-15991-2014, 2014.

Pomeroy, L., Williams, P., Azam, F., and Hobbie, J.: The Microbial Loop, Oceanogr., 20, 28–33, 2007.

Ras, J., Claustre, H., and Uitz, J.: Spatial variability of phytoplankton pigment distributions in the Subtropical South Pacific Ocean: comparison between in situ and predicted data, Biogeosciences, 5, 353–369, doi:10.5194/bg-5-353-2008, 2008.

Salter, I., Kemp, A., Lampitt, R., and Gledhill, M.: The association between biogenic and inorganic minerals and the amino acid composition of settling particles, Limnol. Oceanogr., 55, 2207–2218, 2010.

Sarmento, H., Romera-Castillo, C., Lindh, M., Pinhassi, J., Sala, M., Gasol, J., Marrase, C., and Taylor, G.: Phytoplankton species-specific release of dissolved free amino acids and their selective consumption by bacteria, Limnol. Oceanogr., 58, 1123–1135, 2013.

Sathyendranath, S., Stuart, V., Nair, A., Oka, K., Nakane, T., Bouman, H., Forget, M. H., Maass, H., and Platt, T.: Carbon-to-chlorophyll ratio and growth rate of phytoplankton in the sea, Mar. Ecol.-Prog. Ser., 383, 73–84, 2009.

Savoye, N., Trull, T. W., Jacquet, S. H. M., Navez, J., and Dehairs, F.: ^{234}Th-based export fluxes during a natural iron fertilization experiment in the Southern Ocean (KEOPS), Deep-Sea Res. Pt. II, 55, 841–855, 2008.

Shen, Y., Fichot, C. G., and Benner, R.: Dissolved organic matter composition and bioavailability reflect ecosystem productivity in the Western Arctic Ocean, Biogeosciences, 9, 4993–5005, doi:10.5194/bg-9-4993-2012, 2012.

Silver, M. W. and Gowing, M. M.: The "Particle"" Flux: Origins and biological components, Progr. Oceanogr., 26, 75–113, 1991.

Tremblay, L. and Benner, R.: Microbial contributions to N-immobilization and organic matter preservation in decaying plant detritus, Geochim. Cosmochim. Ac., 70, 133–146, 2006.

Tremblay, L. and Benner, R.: Organic matter diagenesis and bacterial contributions to detrital carbon and nitrogen in the Amazon River system, Limnol. Oceanogr., 54, 681–691, 2009.

Van Heukelem, L. and Thomas, C. S.: Computer-assisted high-performance liquid chromatography method development with applications to the isolation and analysis of phytoplankton pigments, J. Chromatogr. A, 910, 31–49, 2001.

White, D. S. and Howes, B. L.: Nitrogen incorporation into decomposing litter of Spartina alterniflora, Limnol. Oceanogr., 39, 133–140, 1994.

Zubkov, M. V., Holland, R. J., Burkill, P. H., Croudace, I. W., and Warwick, P. E.: Microbial abundance, activity and iron uptake in vicinity of the Crozet Isles in November 2004–January 2005, Deep-Sea Res. Pt. II, 54, 2126–2137, 2007.

10

Lena River delta formation during the Holocene

D. Bolshiyanov[1,2]**, A. Makarov**[1,2]**, and L. Savelieva**[2]

[1]State Research Center, Arctic and Antarctic Research Institute, St.Petersburg, Russia
[2]St.Petersburg State University, St.Petersburg, Russia

Correspondence to: D.Bolshiyanov (bolshiyanov@aari.ru), A. Makarov (makarov@aari.ru), and L. Savelieva (savelieval@mail.ru)

Abstract. The Lena River delta, the largest delta of the Arctic Ocean, differs from other deltas because it consists mainly of organomineral sediments, commonly called peat, that contain a huge organic carbon reservoir. The analysis of delta sediment radiocarbon ages showed that they could not have formed as peat during floodplain bogging; rather, they accumulated when Laptev Sea water level was high and green mosses and sedges grew and were deposited on the surface of flooded marshes.

The Lena River delta formed as organomineral masses and layered sediments accumulated during transgressive phases when sea level rose. In regressive phases, the islands composed of these sediments and other, more ancient islands were eroded. Each new sea transgression led to further accumulation of layered sediments. As a result of alternating transgressive and regressive phases, the first alluvial-marine terrace formed, consisting of geological bodies of different ages. Determining the formation age of different areas of the first terrace and other marine terraces on the coast allowed the periods of increasing (8000–6000 BP (years before present), 4500–4000, 2500–1500, and 400–200 BP) and decreasing (5000, 3000, and 500 BP) Laptev Sea levels to be distinguished in the Lena Delta area.

1 Introduction

Until now the geomorphological structure of the Lena River delta has been most fully explained by the scheme proposed by Grigoriev (1993), who distinguished three terraces above the delta floodplain; the third terrace is composed of ice complex rocks, the second terrace is composed of Arga-Muora-Sise Island sands, and the first terrace to the floodplain is composed of silt and sand, enriched to varying degrees in plant remains which have often completely displaced the mineral part. Subsequent researchers in the delta proceeded from this scheme, presuming that the delta was a fluvial formation (Schwamborn et al., 2000; Schirrmeister et al., 2003). Geological and geomorphological studies by the authors (Bolshiyanov et al., 2013) showed that all delta terraces were not formed as deposits made by the flooding river. i.e. river terraces; rather, they were formed by Laptev sea-level fluctuations over a long period of time.

The aim of this paper is to elucidate the structure of the first Lena River delta terrace. This structure reflects sea-level fluctuations that resulted in periods of delta erosion alternating with periods of delta development during which organomineral sediments preferentially accumulated.

2 Study area

The Lena River delta is the largest Arctic delta with an area of about 29 630 km^2 (Bolshiyanov et al., 2013). It has a significant impact not only on the Laptev Sea regime, but also on the entire Arctic Ocean due to fresh river water discharge to the smallest and least saline of the Earth's oceans. The delta was formed by the accumulation of sediments from the river and from the erosion and abrasion of ancient landforms due to sea-level fluctuations and the tectonic movements of the Earth's crust. The huge fan-shaped delta extends for 190 km from north to south (72.0–73.8° N) and for 250 km from west to east (122.0–129.5° E). The length of the entire delta coastline, from the Olenek channel mouth to the Bykovskaya channel mouth, is about 1930 km.

The climate in the Lena River delta area is Arctic continental, characterised by a mean annual air temperature of

−13 °C, a mean January temperature of −32 °C, a mean July temperature of +6.5 °C, and annual rainfall of 190 mm. The delta is underlain by continuous permafrost 500–600 m thick. The active (annually thawed) layer is 30–50 cm thick. The entire delta lies in the tundra zone except for the southern part where the Tit-Ary Island is located. There is a north limit of larch distribution which extended to a latitude of 72°12′ N in 2006.

3 Methods

The primary method used in this study was geomorphological mapping, accompanied by a simultaneous study of the geological and geocryological delta structure. During 2000–2008, geomorphological maps were compiled for different delta areas (Bolshiyanov et al., 2013), and in 2009 the geomorphological map of the entire delta was compiled (Makarov, 2009). The mapping process consisted of direct field measurements of the altitude of various delta terraces, compiling descriptions of terrace microrelief, and deciphering the relief shown in aerial photographs. After the terrace sediments were sampled and radiocarbon dated, the relief elements distinguished in the compiled maps were assigned ages. The geological structure of the islands and the coast were studied by geologically describing key sections exposed by river erosion and sea abrasion. Samples for radiocarbon dating, spore–pollen analysis, and analysis of the plant macroremains from the organomineral masses that make up the islands were taken from these key sections. Holocene marine terraces in the delta area were also investigated and dated. Radiocarbon dating was carried out in the geochronology laboratory at St. Petersburg State University, and the results are presented below. The liquid scintillation method was used for ^{14}C dating as described by Arslanov et al. (1993). Radiocarbon dates are reported throughout the text as uncalibrated ages.

Pollen analyses were carried out at St. Petersburg State University. Peat samples were prepared according to the alkaline method of von Post described by Grichuk and Zaklinskaya (1948). Pollen and spores were identified at magnifications of 400× with the aid of published pollen keys and atlases (Kupriyanova and Alyoshina, 1972, 1978; Moore et al., 1991) and a modern pollen reference collection at St. Petersburg State University. About 400 terrestrial pollen grains per sample were counted and used for percentage calculations of each identified taxon. The Tilia and Tilia-Graph software (Grimm, 1991) was used for calculating pollen percentages and for drawing the pollen percentage diagram.

Peat samples for botanical analysis were carefully flushed with water through a 0.10–0.25 mm aperture sieve until the flushing water was completely clear. Plant fibers thus obtained were put on a large glass slide (9 × 6 or 9 × 12 cm), mixed with a little water, dispersed in a thin layer, and examined under a microscope at 100× magnification without a cover glass. When the plant species was difficult to identify, stronger magnification (up to 450×) was used to view individual parts of plant remains, and in this case the sample was covered with a cover glass. The species of all plant samples were determined by using special atlases (Katz et al., 1977; Dombrovskaya et al., 1959). The percentage of each species was visually estimated in each sample and is expressed as a percentage rounded to the nearest 5 %. Plant residues encountered in quantities of less than 5 % were marked as being single (sol.). Determination of plant macroresidues was carried out in the Department of Biogeography at St. Petersburg State University.

Figure 1. Typical outcrop of the layered sediments with radiocarbon dates and sample numbers. Bulkurskaya channel.

4 Results

4.1 Organomineral delta deposits

Our study of geological and geomorphological Lena Delta structure showed that the first terrace, which is up to 15 m above sea level, is composed partly of sands, but mostly of organomineral sediments. The latter are called layered sediments, since this term most precisely reflects the essence of the structure and the composition of these sediments (Fig. 1). These layered plant masses, enriched in sand and silt to varying degrees, are made up of undecomposed green mosses, detritus of sedges and other plants, and woody remains. Horizontal and wavy bedding of sediments indicates that the plant remains were deposited in a shallow basin. These sediments were annually frozen, as indicated by the extensive development of syngenetic massive ice wedges. In general these Holocene sediments are very similar to ice complex (IC) deposits, although IC deposits formed at a different time, at the close of the late Pleistocene, and under much drier and colder formation conditions. In both cases the mechanism of sediment formation was probably very similar and depended, to a major extent, on sea-level fluctuations.

The botanical composition of the peat that makes up the islands is of great importance for understanding the origin of the sediment deposits. On the western extremity of Sobo-Sise Island on the Bootulu-Sise channel side, Gusev (1953) described the second (as he presumed) 8 m high terrace. The lower peaty part of this terrace is composed of 50 % *Carex caespitosa* and *Carex limosa*, 35 % *Calliergon cordifolium*, *C.* sp., *Drepanocladus* sp., and *Poludella squarrosa*, and 15 % undefined herbal remains. The peat is 15 % decomposed. The middle part of this terrace consists of 30 % sedges; *Calliergon cordifolium*, *C.cuspidatum*, *Cirriphylum* sp., *Drepanocladus exanulatus*, *D.fluitans*, and *D.venicosus* make up the remaining 70 %. The peat is 5 % decomposed.

In the upper part of this terrace, new species *Drepanocladus sedtheri* and *Mnium* sp. are found, and the peat is 5 % decomposed. At 1 km downstream from Bulunkan Bay, along the Bykovskaya channel, the base of the 6 m high section is composed of strongly arenaceous peat, 70 % of which is made up of *Calliergon cordifolium*, *C.giganteum*, *Drepanocladus exannulatus*, *D. Sendtheri*, and *D. venicasus*, while 15 % is *Carex* sp. and 15 % is unidentifiable remains. Single tree trunks up to 8 cm in diametre] are found here. In the middle part of this terrace the peat is denser and the layering is less pronounced. Sixty percent of this section is composed of *Carex caespitosa*, *C. limosa*, and *C.* sp., and 20 % of *Calliergon cordifolium*, *C. giganteum*, and *Scorpidium scorpioides*. The remaining 20 % is made up of *Drepanocladus exomulatus*, *D. sencther*, and *D. venicosus*. The upper 2 m of the section are represented by horizontally layered fine-grained quartz sands with abundant interbedded peat and peat detritus.

As can be seen from the vegetation descriptions (Gusev, 1953), peat-forming plants are represented in the delta by green mosses and sedges. The peat deposits are as much as 15 % decomposed; the amount of decomposition is sometimes less in the lower part of the peat bogs than in the upper part.

Our observations indicate that section 0904 (Fig. 1) on the left slope of the Bulkur channel (72°09′02.8″ N, 126°15′57.8″ E) is of particular interest. The first terrace, up to 13.5 m above the low-water mark in the Bulkur channel, is composed of organomineral sediments which exist as isolated silt pinnacle outliers that are separated from each other by valleys which formed as ice veins melted. The width of the silt pinnacle outliers approximately corresponds to the size of the polygons on the terrace surface. In one of these silt pinnacle outliers, the following units are exposed in ascending order:

- towpath (0–3.5 m above water edge), up to 40 m wide, composed of sand with sunken wood;
- plant remains in a well-developed, 4.7 m thick horizontal layer (sample 0904-1 taken at a height of 3.5 m, sample 0904-2 at a height of 4.7 m, and sample 0904-3 at a height of 8.2 m);
- 0.8 m thick sand with interbedded plant remains;
- layered plant sediments with visible thickness of 1 m (sample 0904-4 taken from the height of 9 m above water edge, at the base of the layer);
- layered plant sediments in a 3–3.5 m thick sliding block.

As is seen from Fig. 1, the organic mass in the lower half of the section (below the sandy layer) formed in the time interval between 6300 and 3900 BP. There may have been a break in sedimentation between the two lower dated layers of sediments. In the time interval between 5100 and 3900 BP, 3.5 m of plant remains accumulated at an average rate of 3 mm yr^{-1} or 30 cm 100 yr^{-1}. Above this layer, sands and layered sediments were again deposited.

The results of pollen analysis from sediments of section 0904 are shown in the diagram (Fig. 2) The spore–pollen spectra of the three lower samples (0904-1, 0904-2, 0904-3) are uniform. Sediments are dominated by pollen of arboraceous and shrub vegetation (60–70 %), which does not grow in this location today. Vegetation is mainly represented by three taxa; *Pinus s/g Haploxilon*, *Betula* sp., and *Alnus fruticosa* account for 20 to 30 %. Pollen of *Picea* (8 %), *Larix* (3–5 %), and *Salix* (2–5 %) are also present. Herbaceous vegetation is represented by pollen of the family Cyperaceae (20–27 %) and motley grass-herb groups including, e.g. Poaceae, Brassicaceae, Artemisia, Asteraceae, Rosaceae. Spores are predominantly from sphagnum mosses, ferns, different *Lycopodium* species, and *Selaginella sibirica*. All samples contain sporadic pollen grains and spores of pre-Quaternary conifers. Pollen and spores in all samples are well preserved.

The spore–pollen spectrum of the upper sample (0904-4, from a height of 9 m above water edge) differs from those described above by the predominance of grass pollen representing the families Poaceae (38 %) and Cyperaceae (20 %), and miscellaneous herbs. There is also a high content of larch pollen (15 %), and sporadic forms of softwood stomata are recorded that could not be assigned to genera due to their poor preservation.

Among plant macroremains in the lower part of the section (sample 0904-1), sedges (30 %), cotton grass (5 %), and equisetum (5 %) were found. Mosses are represented by *Drepanocladus* (25 %), *Tomethypnym* (25 %), and *Calliergon* (5 %). There are single occurrences of *Comarum palustre*, *Bryum*, and *Aulacomnium*.

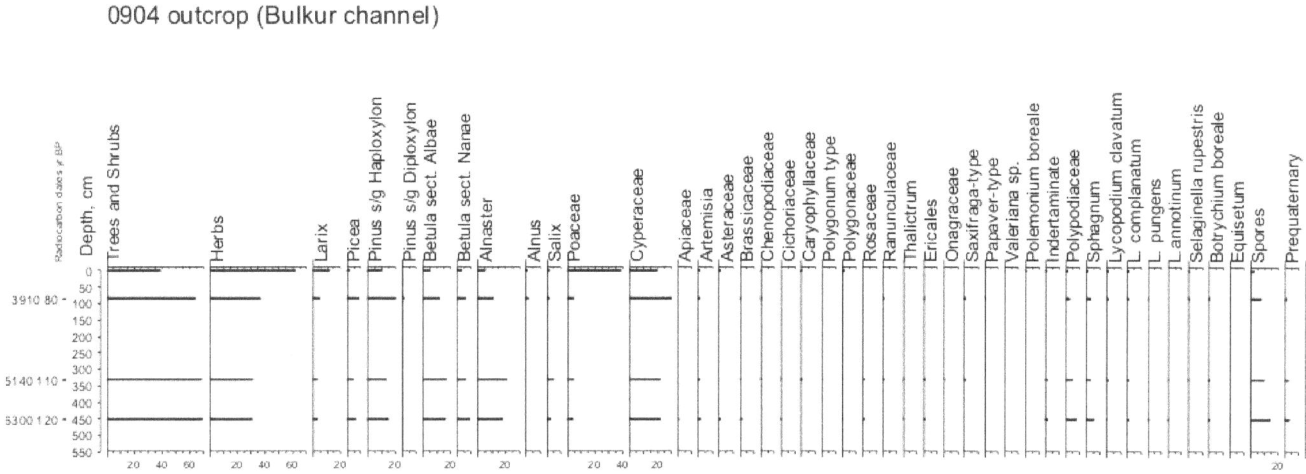

Figure 2. Spore–pollen diagram at section 0904. Bulkurskaya channel.

In sample 0904-2 the sedge species composition changes slightly. *Carex cespitosa* virtually disappears, but *Carex rariflora* appears (10 %). *Carex aguatilis* accounts for 10 %. Individual Equisetum occurs, and grasses appear (10 %). In addition, there is an increasing diversity of mosses, including *Calliergon gigantenum* (20 %), *Drepanocladus* (20 %), *Calliergon cordifolium* (15 %), and *Bryum* (5 %).

In sample 0904-3, herbaceous plants are again dominated by sedges of different species which in total account for 30 %. *Equisetum* is present in small amounts and accounts for 5 %. The remaining 65 % of remains are composed of mosses dominated by *Calliergon cordifolium* (15 %), *Drepanocladus* (15 %), and *Tomethypnym* (10 %).

Sample 0904-4 is notable for a lower share of sedges (25 %), and a rather high content of grasses (25 %) and *Drepanocladus* (30 %). Among mosses, *Calliergon cordifolium* (15 %) and Campyllum (5 %) are also recorded.

All samples are very rich in organic material that has not decomposed very much (the highest amount of decomposition, 30 %, found in the lowest sample) and is very well preserved. All the species encountered currently grow in tundra.

Results of analyses show that the vegetation composition indicated by spore–pollen spectra differs significantly from that indicated by the plant macroremains. This difference is particularly pronounced in the mosses. In layered sediments *Sphagnum* and lycopod spores are present, and layered sediments are mainly composed of green mosses. The pollen and plant remains of the herbaceous vegetation are more or less similar and sedge pollen and macroremains prevail. In the upper sample, changes in the composition of herbaceous plant pollen correspond to changes in the composition of herb remains. Both macroremains and spore–pollen analyses showed an increasing presence of grasses and a decreasing presence of sedges. Macroremains contain virtually no arboraceous species, which probably indicates that the described area remained in the tundra zone during the time interval from 6300 to 3900 BP.

Our own and previous studies suggest that the organic part of the layered sediments is virtually undecomposed. The maximum 15 % of peat decomposition degree determined by Gusev (1953) is accounted for by indefinite plant remains, i.e. by plant detritus, which is not a product of decomposition, but was transported to this location by water, as were the green mosses and herbaceous plants. As is evident from the descriptions, there are abundant remains of both sedge roots and leaves in layered sediments. Peat is generally not characterised by the presence of sedge leaves, since they decompose quickly (Denisenkov, 2000).

The mineral proportion of the layered sediments differs significantly from section to section. In general, the amount of sand is much greater in the southern part of the delta than in the northern parts. Tit-Ary is the first island of the delta, and sand makes up 30–40 % of the volume weight of these sections. In other sections (closest to the Laptev Sea) sandy material accounts for less than 10 % in layered sediment sections.

Layered sediments accumulate at a very high rate, ranging from 0.2 to 1.6 m per hundred years. In the Samoylov Island borehole, the lower metre of sediments accumulated during several dozens of years only.

In layered sediment sections of the first terrace up to 8 m thick, the upper and lower parts exhibit a nearly identical degree of moss decomposition.

Wood remains protruding from exposures are, as a rule, much younger than the accumulations of mosses and plant detritus, as was shown by radiocarbon dating. The location of the dated first terrace organomineral sediment sections is shown in Fig. 3; dates are given in Table 1.

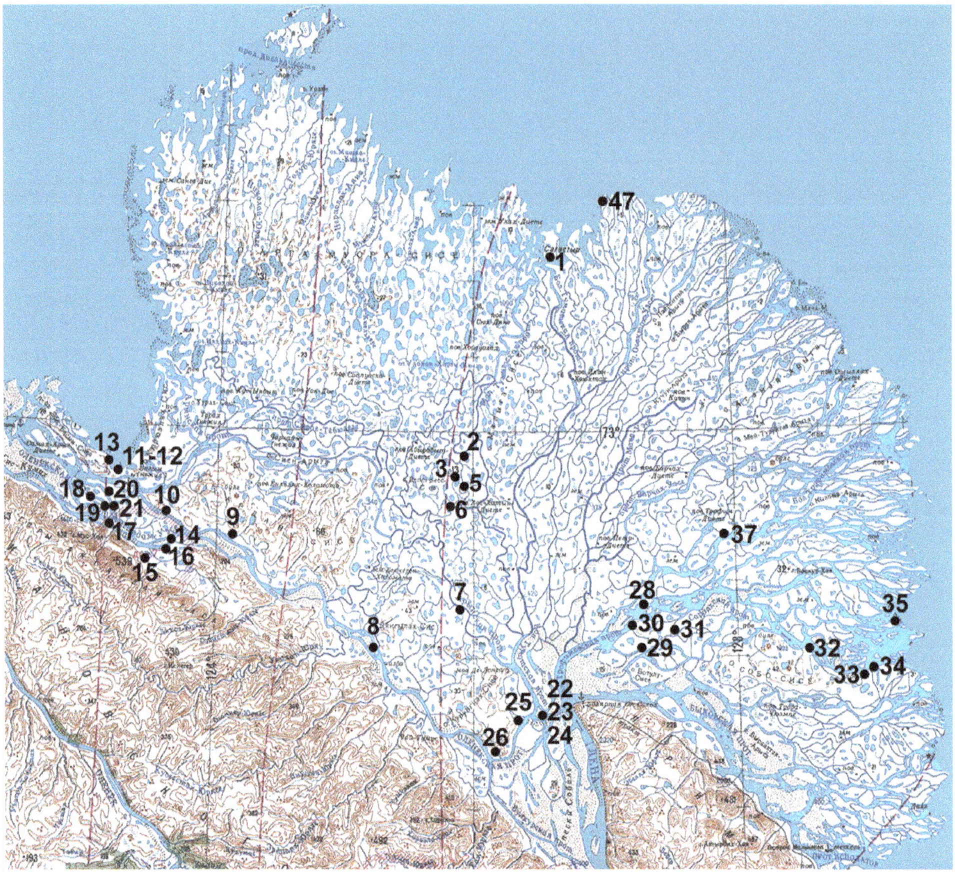

Figure 3. Map of the studied sites in the Lena River delta.

4.2 Geomorphological delta structure

The geological and geomorphological structures of delta islands obviously point to the fact that the first terrace formed during several phases of alternating accumulation and erosion of sediments. Island outlines clearly illustrate different ages of sediments on the same first terrace surface. Figure 4 presents an aerial picture of Samoylov Island. Its eastern part is polygonal tundra made up almost entirely by organic layered sediments up to several metres thick. A developed system of polygons and thermokarst lakes points to the comparatively long existence of the first terrace surface. The western part of the island is composed of sands adjoining the eastern part of the island. Surface sands on the western part of the first terrace exhibit only rudimentary polygonal forms and are covered with sparser vegetation than is the eastern part of the island. The first terrace surface is younger than a similar surface on the eastern part of the island. The western slope of the island is most affected by flood water, which leaves an annual deposit of sandy material atop it. A large amount of driftwood accumulates here, i.e. tree trunks and their fragments brought by the river and deposited as the river water recedes. Boreholes drilled in the sands of the western part of the island showed that 300–500 year old sands were deposited on 2600 year old eroded layered sediments near the modern water edge (Table 1, no. 23).

Samoylov Island was formed by the accumulation of layered sediments at 3500–2500 BP under basin conditions as indicated by horizontal and wavy sediment layers. These sediments were subsequently eroded, and at 300–500 BP the erosional outlier was joined from the west by river floodplain deposits composed of sands and plant detritus. Such major differences in the accumulation mode of organic vs. sandy deposits can only be associated with opposite phases of delta evolution: a period of still water and accumulation of light organic deposits under estuarine conditions at 3500–2500 BP, followed by a period of riverbed downcutting by erosion.

Similar ratios of differently aged island sediments composed of organomineral layered sediments and sandy sediments were recorded in the course of geomorphological surveying in many Lena River delta areas. Compilation of the geomorphological map of the delta (Makarov, 2009) revealed the unevenly aged structure of the individual alluvial fans that form the delta (Fig. 5). As a result of deciphering photos and comparing results with the relief structure description compiled in the course of geological and geomorphological investigations and the radiocarbon dates of terrace sediments, it was possible to distinguish alluvial fans of uniform proper-

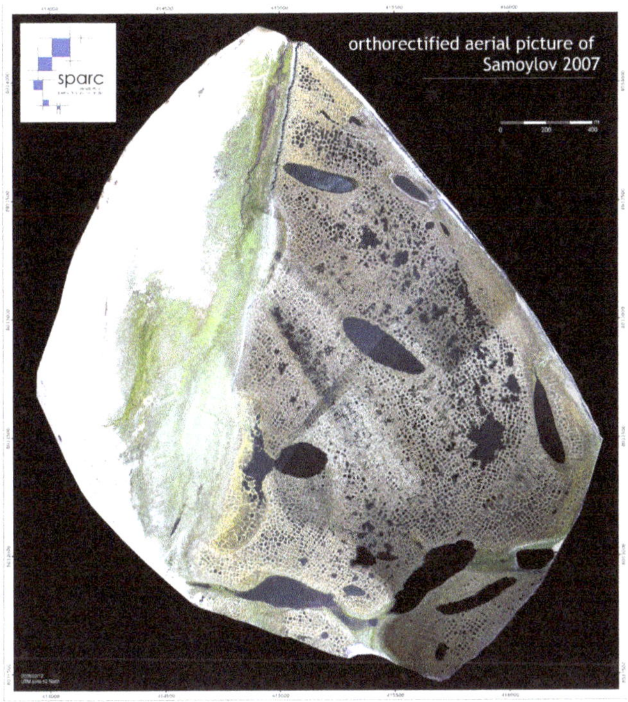

Figure 4. Aerial picture of Samoylov Island.

ties and age which together form the surface of the first Lena River delta terrace.

The most ancient parts of the first terrace are the islands lying at the boundary of the Arynskaya and Tumatskaya channels. The age of the terrace here is 8570 BP (lab. no. LU-4191, no. 2, Table 1), and the height of the islands reaches 8 m.

Another major constituent of the first terrace is a massif of islands in the Lena River delta Olenekskaya and Arynskaya channels. These islands began to form about seven thousand years ago (6890 ± 170, LU-4409; 6520 ± 130, LU-4407). The islands are 10–12 m high. The Tumatskaya alluvial fan is 2500–3000 years old (2280 ± 8, LU-5611; 3170 ± 60, LU-5612). Islands in this part of the delta are relatively low-lying at 7–8 m, and only 1–1.5 m high on the coastal area. The Sardakh or Sardakh-Trofimovsky fan formed starting around 4000 BP; islands on this fan reach 10–12 m in height.

The youngest formations in the delta are the Bykovskaya channel islands which are 10 m high, and are moving forward to the Laptev Sea through an extensive alluvial fan. These islands are not more than 2500 years old (1400 ± 100 and 2530 ± 200, MSU-773 and MSU-861, respectively; Korotaev, 1984). The surface of the entire terrace slopes from the delta head (10–12 m) to the sea edge (0.5 m). Thus, the first terrace alluvial fans accumulated first in the west, and then the formation moved to the east. The oldest delta islands lie in its western part; the young ones, in the eastern part. This peculiar delta formation can be accounted for by the tectonic movements which are clearly recorded in sections of the some islands (Bolshiyanov et al., 2013).

5 Discussion

Islands composed of deposits of different ages and types are common in the Lena River delta. Their geological structure points to an abrupt change in the sedimentation environment, both laterally and vertically, and with time. This process of island formation contradicts the accepted doctrine of sedimentation in deltas.

It is presumed that delta islands accrete upwards as a result of annually precipitated silt deposits that are left behind as high water recedes, as occurs in river valleys. The following is one of the traditional notions of the delta origins:

> Modern delta islands form due to a gradual rising of sandy-silty bars at the sea edge or during breakage of channel bends. Shoals are gradually transformed into above-water 'sands', which are covered with bush-like grass vegetation strengthening their surface against weathering and promoting their growth in height due to detaining drift during high water. With further rising, the vegetation becomes thicker, and the islands become dry sandy tundra. With age, ever thicker moss-turf and peat surface layers accrete on the islands. A moss cushion forms, bogging-up of soil starts, and a cover of hydrophilic plants appears. Thickness of the active thawing layer decreases, and typical tundra of the permafrost area appears. Frost cracks form (Samoilov, 1952).

A key issue of delta formation is the origin of organomineral masses, the layered sediments that make up most of delta islands. Virtually all researchers regard these sediments as peat.

> Over the vast plain land areas with severe continental climate favourable conditions set in for the development of bog-tundra vegetation. The lower peat bogs formed as a result of overgrowth of water bodies in near-Arctic regions, whereas the upper peat bogs (Mammoth Horizon) are characteristic of wet marshes and overgrowing water bodies of tundra type. Peat accumulation on all three terraces of the delta inevitably proceeded following the type of lowland hypnum or sedge hypnum bogs (Gusev, 1959).

The three terraces proposed by Gusev (1959) are, in fact, a single terrace, the first delta terrace which is composed of organomineral sediments.

In general, the described deposits are quite dissimilar from peat deposits in composition and texture.

Table 1. Radiocarbon and Calibration was made using OxCal programme v.4.2 (Ramsey et al., 2010) and calibration curve IntCal13 (Reimer et al., 2013).

No.	Location, channel, island	Coordinates	Geomorphic position	Height (m above water level)	Lab no., reference	C14 age	C14 age calibrated
1	Bol'shaya Tumatskaya channel, Sagastyr Island	73°22′17″ N, 125°50′22″ E	Alluvial-estuarine terrace 3–4 m height	3.1	LU-4201	1400 ± 80	1319 ± 80
2	Malaya Tumatskaya channel, Chenchike-Sise Island	72°57′09″ N, 125°55′06″ E	Alluvial-estuarine terrace 4–7 m	2.5	LU-4191	8570 ± 160	9625 ± 223
3	Malaya Tumatskaya channel, Dgipiries Island	72°54′43″ N, 125°52′30″ E	Alluvial-estuarine terrace 10–11 m on the Pleistocene basis	9.0	LU-4412	6460 ± 100	7370 ± 92
4	Malaya Tumatskaya channel, Chenchike-Sise Island	72°51′58″ N, 125°55′08″ E	Alluvial-estuarine terrace 9–10 m	7.8	LU-4197	1880 ± 120	1819 ± 146
5	Malaya Tumatskaya channel, Dgipiries-Sise Island, N–E coast of the Lake Yugus-Dgie-Kyuele	72°51′14″ N 125°50′22″ E	Alluvial-estuarine terrace 10–11 m on the Pleistocene basis	7.5	LU-4198	6430 ± 120	7337 ± 117
6	Malaya Tumatskaya channel, Dgipiries-Tubelege, Tyuba-Aryta Island	72°49′45″ N, 125°49′40″ E	Alluvial-estuarine terrace 8.5–9m	5.8 7.0	LU-4193 LU-4199	2690 ± 100 1320 ± 80	2812 ± 129 1226 ± 81
7	Arynskaya channel, Buor-Syr-Aryta Island	72°36′50″ N 125°53′30″ E	Alluvial-estuarine terrace 8.5 m	5.2	LU-4413	3930 ± 90	4368 ± 141
8	Olenekskaya channel, Tyuyarenkey-Tubelege (Gusinka) Dzhangilakh-Sise Island	72°29′33″ N 125°18′45″ E	Alluvial-estuarine terrace 9–13 m	7.6 10.7	LU-4409 LU-4407	6890 ± 170 6520 ± 130	7752 ± 140 7418 ± 116
9	Olenekskaya channel, Khardang Island	72°45′26″ N 124°05′26″ E	Alluvial-estuarine terrace 5–7 m	2.0	LU-4411	5100 ± 140	5862 ± 162
10	Olenekskaya channel, Khardang Island	72°48′06″ N 123°37′28″ E	Alluvial-estuarine terrace 4–5 m	2.0	LU-4414	2850 ± 200	3016 ± 243
11	Kuba Bay Ebe-Bsyn-Sise Island	72°55′06″ N 123°57′50″ E	Marine terrace 1.2–4 m	2.5	LU-4548	1170 ± 50	1096 ± 69
12	Kuba Bay Ebe-Bsyn-Sise Island	72°53′31″ N 123°11′20″ E	Marine terrace 3.5 m setup wood	3.2 4.0	LU-4545 LU-4547	750 ± 40 180 ± 50	691 ± 29 161 ± 92
13	Kuba Bay Ebe-Bsyn-Sise Island	72°53.796′ N 123°10.186′ E	Marine terrace 5–8.5 m Marine terrace 2.5–3 m	3.5 2.3	LU-4551 LU-4544	3620 ± 60 1220 ± 40	3940 ± 88 1151 ± 61
14	Angardam channel, Kharayalakh-Ary Island	72°44′50″ N 123°38′00″ E	Alluvial-estuarine terrace 5.5 m	2.8	LU-4564	410 ± 60	434 ± 67

No.	Location, channel, island	Coordinates	Geomorphic position	Height (m above water level)	Lab no., reference	C14 age	C14 age calibrated
15	Angardam channel, Dzhakhse- Tubelege Angardam Ridge bottom	72°42.813′ N 123°29.970′ E	Alluvial-estuarine terrace 5–5.5 m	1.7	LU-4569	2390 ± 50	2472 ± 106
16	Angardam channel, Ulakhan-Ary Island	72°43.777′ N 123°29.442′ E	Alluvial-estuarine terrace 8–9 m	0.5, 1	LU-4541 LU-4542	2600 ± 50 790 ± 40	2700 ± 86 716 ± 31
17	Angardam channel, Olenek-Kumaga sands	72°47.353′ N 123°07.511′ E	Sand island in a channel, 1.7 m	−0.35	LU-4563	3400 ± 60	3657 ± 84
18	Angardam channel, Belkey Island	72°49.531′ N 123°05.264′ E	Bulguniagkh 12 m on the surface of 5 m flood-plain	5.2	LU-4561	6020 ± 110	6887 ± 145
19	Angardam channel–Olenekskaya channel, Belkey Island	72°49.063′ N 123°09.557′ E	Driftwood, 4 m bottom of termokarst lake	4	LU-4546	220 ± 40	199 ± 104
20	Belkey-Aryn-Uesya channel, Soduom-Aryta Island	72°50.254′ N 123°11.596′ E	Alluvial-estuarine terrace 1.5–2 m	0.9	LU-4572	1770 ± 50	1690 ± 68
21	Samoim-Uesya channel, Tyalyr-Mengenyakh-Ary Island	72°45.481′ N 123°10.463′ E	Alluvial-estuarine terrace 7 m	4	LU-4568	1710 ± 60	1627 ± 75
22	Bol'shaya Tumatskaya channel, Samoylovsky channel	72°22.103′ N 126°29.581′ E	Alluvial-estuarine terrace 8.5–10 m	2.5	LU-4577	3220 ± 70	3451 ± 81
23	Bol'shaya Tumatskaya channel, Samoylovsky channel (borehole)		Alluvial-estuarine terrace 8.5–10 m	7.5 6.6 6.2 1.3 1.2 0.4	Schwamborn et al. (2000) KIA-8169 KIA-8170 KIA-8171 KIA-8172 KIA-8173 KIA-8174	435 ± 30 230 ± 25 500 ± 25 2605 ± 30 2530 ± 30 2635 ± 35	492 ± 35 219 ± 89 526 ± 14 2743 ± 25 2622 ± 81 2763 ± 25
24	Bol'shaya Tumatskaya channel, Samoylovsky channel		Alluvial-estuarine terrace 8.5–10 m	6.2 4.2 3.6	Kuptsov and Lisitsin (1996) IORAN-4167 IORAN-4164 IORAN-4101	3700 ± 260 4220 ± 240 2140 ± 110	4091 ± 349 4783 ± 336 2134 ± 135
25	Olenekskaya channel, Arga-Aryta	72°21.713′ N 126°19.588′ E	Alluvial-estuarine terrace 8–10 m	7 3.5	LU-4565 LU-4609	540 ± 60 3170 ± 50	572 ± 47 3391 ± 61
26	Olenekskaya Kurungnakh Island. Bulkurskaya channel mouth area		Alluvial-estuarine terrace 8–12 m		Korotaev (1984) MGU-862	3480 ± 500	3877 ± 651
27	Olenekskaya channel, Gagariy Island				Korotaev (1984) MGU-808	4200 ± 250	4756 ± 349

No.	Location, channel, island	Coordinates	Geomorphic position	Height (m above water level)	Lab no., reference	C14 age	C14 age calibrated
28	Malaya Trofimovskaya channel, Gogolevsky Island	72°36.980′ N 127°15.708′ E	Alluvial-estuarine terrace 7 m	0.5	LU-4911	3550 ± 60	3835 ± 87
29	Maastakh-Uesya channel, Sardakh-Aryta Island.	72°31.333′ N 127°05.167′ E	Alluvial-estuarine terrace 7 m Beach ridge	4.0 9.4	LU-4914 LU-4910	1840 ± 70 modern	1768 ± 84
30	Maastakh-Uesya channel, Sardakh-Aryta Island.		Alluvial-estuarine terrace 8–9 m	8.5 8.1 6.6 3.9 1.7 0.6 −0.2	Schwamborn et al. (2000) KIA-6759 KIA-6760 KIA-6761 KIA-6762 KIA-6763 KIA-6764 KIA-6765	2755 ± 25 1369 ± 20 2525 ± 30 3460 ± 30 3025 ± 35 3420 ± 35 2830 ± 35	2841 ± 36 1293 ± 9 2616 ± 79 3736 ± 56 3229 ± 63 3677 ± 57 2936 ± 50
31	Noname channel, in the line of NW from the Sardakh Island, Dzhielyakh-Ordono-Aryta Island	72°33.184′ N 127°32.845′ E	Alluvial-estuarine terrace 10 m	2.0	LU-5080	3830 ± 90	4234 ± 131
32	Sardakhskaya channel, Sobo-Sise Island, Boruo hole	72°27.382′ N 128°56.741′ E	Alluvial-estuarine terrace 4 m	0.5	LU-4916	5220 ± 60	6007 ± 89
33	Sardakhskaya channel, Sobo-Sise Island.	72°27.382′ N 128°56.741′ E	Alluvial-estuarine terrace 15–16 m	5.0 13.5	LU-4913 LU-4917	16 630 ± 130 2320 ± 70	20 068 ± 180 2357 ± 131
34	Sardakhskaya channel, Kuba-Ayanyn-Aryta Island, NW of the Sobo-Sise Island	72°27.441′ N 128°57.252′ E	Past delta of the Bykovskaya channel 2–2.5 m	0.6, wood 1.0 detritus	LU-5073 LU-4919	1140 ± 40 2920 ± 70	1057 ± 61 3070 ± 101
35	Toyonokh-Uesya channel, mouth of the Trofimovskaya channel area, Arangastakh Island	72°33.203′ N 129°12.701′ E	Island on the sea edge of the delta 1.5 m	0.1 0.5	LU-4915 LU-4912	2970 ± 70 2230 ± 70	3137 ± 104 2227 ± 83
36	River Lena, main channel, Tit-Ary Island	71°58.382′ N 127°06.326′ E	Flood-plain 12–15 m	5.4 10.7	LU-4909 LU-4918	4050 ± 110 830 ± 70	4560 ± 162 771 ± 69
37	Trofimovskaya channel, Trofim-Diete Island.	72°45.963′ N 127°56.252′ E	Alluvial-estuarine terrace 6–7 m	0.5	LU-5075	1510 ± 40	1410 ± 56
38	Bykovskaya channel, Lagutin Island				Korotaev (1984) MGU-773 MGU-861	1400 ± 100 2530 ± 200	1318 ± 105 2603 ± 243
39	Olenekskaya channel, Batyyalakh-Aryta Island	72°34′52.5″ N 124°56.027″ E	Terrace 8–10 m	0.7 detritus 0.7 wood 7.5 detritus	LU-5497 LU-5491 LU-5498	4800 ± 80 modern 1961-1981 2700 ± 70	5513 ± 96 2825 ± 68
40	Olenekskaya channel, Dzhangilakh Island	72° 30′20.9″ N 125°17.104″ E	Terrace 9–11 m	1.5 8.5	LU-5494 LU-5496	4380 ± 70 3610 ± 260	5007 ± 123 3976 ± 346

No.	Location, channel, island	Coordinates	Geomorphic position	Height (m above water level)	Lab no., reference	C14 age	C14 age calibrated
41	Sardakhskaya channel, Omsukor-Belkee Island	72°38′16.5″ N 127°34.280″ E	Terrace 3–4 m	2.2 2.7	LU-5495 LU-5492	3490 ± 180 880 ± 60	3797 ± 234 808 ± 65
42	Bol'shaya Tumatskaya channel, Syrdakh Island		High flood-plain 1.6–2.0 m	0.9 0.5	LU-5607 LU-5608	950 ± 50 2590 ± 70	855 ± 52 2658 ± 111
43	Bol'shaya Tumatskaya channel, Kerdiepime Island.	72°58.889′ N 126°29.837′ E	Alluvial-estuarine terrace 5–7 m	0.7 0.5	LU-5610 LU-5609	3210 ± 70 9010 ± 110	3439 ± 83 10109 ± 171
44	Osokhtokh channel, Bezymianny-Oburo Island	73°06.642′ N 127°01.159′ E	Alluvial-estuarine terrace 5–6 m	4.2 1.3	LU-5611 LU-5612	2280 ± 80 3170 ± 60	2288 ± 123 3389 ± 75
45	Osokhtokh channel, Sutura Island	73°10.142′ N 127°12.271′ E	Alluvial-estuarine terrace 4–5 m	2.2 1.2	LU-5614 LU-5613	2440 ± 50 4260 ± 80	2527 ± 107 4802 ± 131
46	Osokhtokh channel, Kegeliakh Island	73°20.979′ N 127°22.520′ E	Alluvial-estuarine terrace 3–4 m	2.0	LU-5605	2920 ± 100	3076 ± 133
47	Osokhtokh channel, Alkhan Island – marine edge of delta	73°31.546′ N 127°35.864′ E	Alluvial-estuarine terrace 1 m	0.25	LU-5606	540 ± 80	566 ± 63
48	Bykovskaya channel, Chay-Ary Island	72° 24′53.3″ N 127°27′14″ E	Alluvial terrace 5 m	3.7	KIA-38588	> 1954	
49	Bykovskaya channel, Malsyva Island	72° 03′46.1″ N 128°37′11.7″ E	Terrace, 5 m	0.7 1.7 2.3	LU-6002 LU-6003 LU-6004	2110 ± 70 1790 ± 60 900 ± 60	2104 ± 103 1714 ± 77 822 ± 63
50	Lena river, main channel, Sordokh-Ary Island	72° 22′09.7″ N 126°31′04.7″ E	Holocene sediments boderinf with ice complex	0.3	LU-5604	3830 ± 110	4233 ± 158
51	Olenekskaya channel		Alluvial-estuarine terrace 10 m	9.0 7.0 6.8 6.0 4.9	LU-5817 LU-6006 LU-6005 LU-5818 LU-5819	2230 ± 70 2750 ± 70 3710 ± 80 2860 ± 100 3470 ± 140	2227 ± 83 2867 ± 75 4065 ± 119 3007 ± 128 3758 ± 179

Peat is fossil fuel assigned to gummites and representing the first stage of plant material transformation as it is converted to coal. It accumulates in swamps from the remains of dead plants subject to partial decomposition under a high humidity and poor access of air. Peat can have a fibre-like structure at a low (25 %) degree of decomposition and an amorphous structure at a high (50–65 %) decomposition degree. Texture is mostly non-layered. Porosity of peat of a low decomposition degree is very high (70–80 %), and that of strongly decomposed one is usually negligible (Dictionary of Geology, 1973).

The sediments studied in these sections and boreholes differ strongly from peat sediments in several ways:

1. Layering is the most characteristic feature of these sediments and it is clearly pronounced everywhere. It is usually horizontal, but there are also wavy layered sediment sequences. Peat, as defined (Dictionary of Geology, 1973), exhibits little or no layering.

2. Layered sediments show no or very little decomposition of the accumulated plants; this lack of decomposition is not typical for peat. The decomposition degree indicates the content of amorphous matter in peat. Amorphous matter consists of decomposition products of the initial plant matter and very fine tissue fragments which have lost their cellular structure. Depending upon environmental conditions, the degree of decomposition varies from 1 to 70 %. Arboraceous and arboraceous–herbaceous peat types exhibit the maximum decompo-

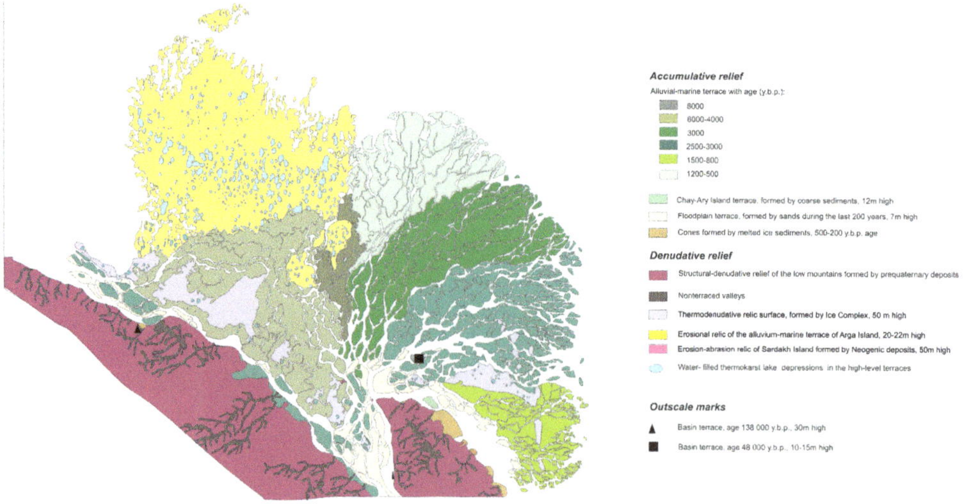

Figure 5. Geomorphic map of the Lena Delta.

Figure 6. High-floodplain sediments. Samoylov Island.

sition degree, while moss types exhibit the minimum. Under the highest decomposition degree of about 70 % not more than 17 % of initial organic matter is preserved; the remaining 83 % of organic matter is lost due to mineralization to gaseous products and water or from washout. Plant fibre remains account for less than 5 % of the initial organic mass (Lishtvan and Korol, 1975).

3. Organomineral Lena River delta sediments are extremely thick as compared to peat accumulations in northern Yakutia. The maximum recorded thickness of layered sediments composing the first delta terrace is 8 m, and the age of these accumulations ranges from hundreds to 2000 years (the minimum accumulation rate of organic material is 40 cm in 100 years). According to Skobeyeva and Simakova (1983), peat deposits no more than 1 m thick rarely form under the harsh conditions of northern Yakutia. Under similar conditions in the Anadyr lowland, Smirnov (1981) points to the impossibility of long-term thick peat deposit formation and also presumes that maximum peat thickness cannot exceed 1 m. In the Kola Peninsula forest zone, peat accumulation rates during the sub-Boreal were 3–8 cm in 100 years, and slightly later, from 1310 to 1100 BP, 18 cm in 100 years (Pavlova et al., 2005). Peat accumulation rates during the Holocene in the southern taiga of West Siberia ranged from 3.5 to 11.3 cm 100 yr^{-1} (Borren et al., 2004) under conditions favourable for peat accumulation. In the zone of severe permafrost in the Lena River delta with a seasonally thawed active soil layer depth of only 30–40 cm, the organic material accumulation rate is 1 or 2 orders of magnitude higher than the known peat accumulation rates in peat deposits of the temperate zone.

4. Under present delta conditions, peat bogs do not form from green mosses in any of the studied water bodies. In numerous drained thermokarst lakes, only a few dozen centimetres of rather well-decomposed plant remains occur, along with mineral material in the form of sand and silt in various proportions.

Thus, referring to the described features of layered sediments, it can be concluded that these organomineral masses did not accumulate in swamps, since the island soils thaw only a few dozen centimetres deep during ≤ 3 months a year. The age of these layered sediments points to a very recent origin in a period when natural conditions did not differ drastically from modern ones.

Layered sediments are not floodplain sediments. In the modern floodplain with significant river water-level fluctuation amplitudes, the alluvial floodplain facies that are deposited are not at all similar to the organomineral accumulations found on delta islands. Floodplain sediments are characterised by sandy composition, horizontal and cross bedding, and a significant content of plant detritus marking the sediment texture (Fig. 6). The high flow rates that accompany high water levels would make any significant deposition of light mosses and plant detritus unlikely.

Horizontal and wavy fibrous textures point to deposition of layered sediments under basin conditions. In more or less stagnant water bodies, where flow rates were not high, mosses, plant detritus, and alternating sands and silts were deposited, indicating flowing water in the deposition basins. Water level fluctuations would be necessary to deposit mosses and sedges that had earlier been flushed from the tundra surface or redeposited from more ancient sections of delta islands. When fluctuating water level conditions (high vs. low tides, surges, floods vs. low water) occur in temporary dewatered squares, plant detritus accumulates at present. Thus, in order to explain past deposition of such sediments, sea-level fluctuations of synoptic scale are required, which are recorded only in the zone of sea basin influence.

The structure of the first Olenek River terrace points to deposition of organomineral sediments in the headwater estuary. At a 100 km distance from the mouth, layered sediments making up the islands and part of the river banks occur in the lower part of the valley. This area of the river is presently influenced by the sea (Bolshiyanov et al., 2013).

Layered moss sediments are also exposed in a 300 m section along the eastern coast of Ebelyakh Bay (coordinates 72°47′24″ N, 140°49′11.5″ E), i.e. on the Laptev Sea coast without modern river influence. These sediments are an outlier of the former thick accumulations of such material under conditions of a basin that was protected from the direct influence of the sea, but maintained a hydraulic connection with the sea. Similar ≤ 15 m thick accumulations of undecomposed vegetation in modern tundras were described by Tarakanov and Biryukov (1974) in accumulative terraces cut by an erosion scarp in Vankina Bay (Shirokostan Peninsula in the south-western Laptev Sea). These researchers also called these sequences of organogenic deposits peat.

So-called peat bogs are also widespread in the New Siberian Islands. An investigated peat monolith on Kotelny Island (Anisimov, 2010) consisting of undecomposed *Hypnum* mosses and herbs is 3 m thick; the age at the top is 3340 and at the base, 5460 radiocarbon years. There are virtually no changes in spore–pollen composition based on analysis of 51 samples which indicates that the sediment sequence was deposited quickly.

Organomineral sediments that are deposited under summer conditions freeze in winter under drain conditions or in a water body under ≤ 2 m deep ice that reaches the bottom. That is why the first terrace syngenetic massive wedge ice deposits are so widespread and thick. The development of a polygon system with vein ice has been recorded when the water levels of Lake Taimyr are abnormally low, and the polygons are seen under water in its shallow areas. Due to significant water level fluctuations Lake Taimyr also provides an example of massive wedge ice formation under the ice cover which reaches the bottom and persists under a water layer throughout several summer months (Bolshiyanov 2006).

The source of so much moss vegetation has yet to be elucidated. The green mosses and sedges that dominate the botanical composition of organomineral deltaic deposits were not brought from a long distance away. These are typical tundra plants. These plants might have grown on the surface of vast marshes submerged periodically. The following composition of modern vegetation was described on different Samoylov Island cryogenic polygons (Sachs et al., 2008).

- Polygon 1 (low) with water in the centre: *Drepanocladus revolvens* (90 %); *Carex chordorrhiza* (10 %).
- Polygon 2 (high) without water in the centre: *Hylocomus splendens* (85 %); *Tomentypnum nitens* (10 %).
- Polygon 3, with a massive ridge and water in the centre, one side of the polygon ridge has crumbled: *Drepanocladus revolvens* (90 %); *Carex chordorrhiza* and *Carex concolor* (10 %). Ridge of polygon 3: *Hylocomus splendens* (60 %); *Rhytidium rugosum* (30 %); *Carex concolor* (4 %).
- Polygon 4 with well-developed ridges, high standing water, and complete isolation from the external water sources feeding the other polygons: *Scorpidium scorpidoides* (89 %); *Carex chordorrhiza* (8 %); *Carex concolor* (3 %).

As can be seen from the above, the composition of modern vegetation is close to that of the fossil deposits found in layered sediments; the vegetation mix is dominated by green mosses and sedges. Commonly, during high water the Lena River carries more tree trunks and small woody and vegetative detritus than mosses and sedges.

The conditions under which the tundra vegetation cover is washed away can occur not only during floods and surges, but as a result of longer-term (dozens to hundreds of years) high levels of the receiving water body. Throughout the last several thousand years, the Lena River delta has been represented by numerous streams and estuaries which were isolated from the receiving water body, because the river and the sea continued to erode significant remnants of land that once existed in the Laptev Sea (Bolshiyanov et al., 2008), particularly in the territory of the modern delta. Backwater from the sea side caused the washout of plant cover from land and its redeposition into the layered sediment sequences. In addition to modern plants, exactly the same green mosses, which are of great importance in the structure of the rock IC, were also incorporated into the newly formed layered sediments. It is the rock IC that composed the islands and land masses in the Laptev Sea which have now disappeared.

Direct evidence of sedimentation under marine conditions is recorded at the edge of the delta in the area of Sagastyr Village (Fig. 2, no. 1), where sediments are salinized and have an age of 1400 years (Schwamborn et al., 2000).

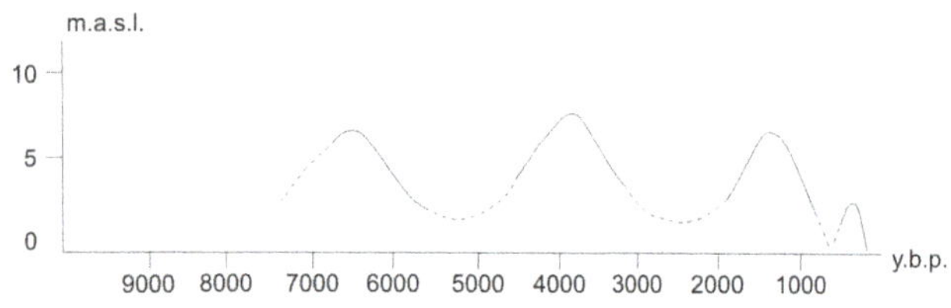

Figure 7. Laptev sea-level fluctuation during the middle and late Holocene.

Thus, the deposits of organomineral masses in the Lena River delta are a record of the Laptev sea-level fluctuations. Numerous radiocarbon dates of these deposits show that the periods of layered sediment accumulation occurred at 8600, 4500–4000, 2500–1500, and 400–200 BP during Holocene sea-level highstand stages, and sediment washout occurred at 5300 and 500 BP. These ages of marine terraces in the Lena River delta region allowed a curve of sea-level fluctuations in this region to be plotted (Fig. 7). At present the period of deposit erosion is coming to an end and Bykovskaya channel is already in an estuary development phase due to the sea level rise on the eastern side of the delta. In the west, in the Olenek channel mouth, sea level decrease continues (Ashik et al., 2010). In accordance with these periods, it was possible to construct a series of palaeo-geographic delta evolution schemes at the close of the late Pleistocene and into the Holocene (Fig. 8).

Around 30 000–40 000 BP, the IC accumulation continued, and by the end of this stage significant IC masses had formed along the Primorsky and Chekanov ridges (Bolshiyanov et al., 2008). No evidence has been unearthed to disprove that at 30 000–40 000 BP the Lena River flowed into the sea in the same place as it does today; thus, the Lena apparently cut through the IC masses as it flowed out from the narrow "Lena tube" caused by a crustal fault in the Earth. From 17 000 to 15 000 BP the sea level was much lower than it is today, and the river delta or deltas were hundreds of kilometres north of the modern alluvial fan. In place of the modern delta there existed eroded IC rocks. The first layered plant sediments formed in the delta 8000 BP which may indicate an elevated basin stand. During the sea's retreat around 6000 BP the earlier accumulated sediments were eroded, including the IC deposits and the marine alluvial sand mass of Arga-Muora-Sise Island which had been exposed on the surface. At that time, the Lena River apparently acquired the Olenekskaya channel, located along a crustal fault in the Earth, as its main channel. Delta channels again entered an estuary evolution stage at 5000 BP, and layered plant sediments accumulated actively at this time. A stage of washout and accumulation of alluvial sands occurred around 4500 BP, and around 3500 BP this stage graded into a stage during which the shallow delta estuaries filled with plant material and sandy sediments. A washout stage occurred again from 1000 to 500 BP, when IC outliers in the delta assumed their final appearance; the

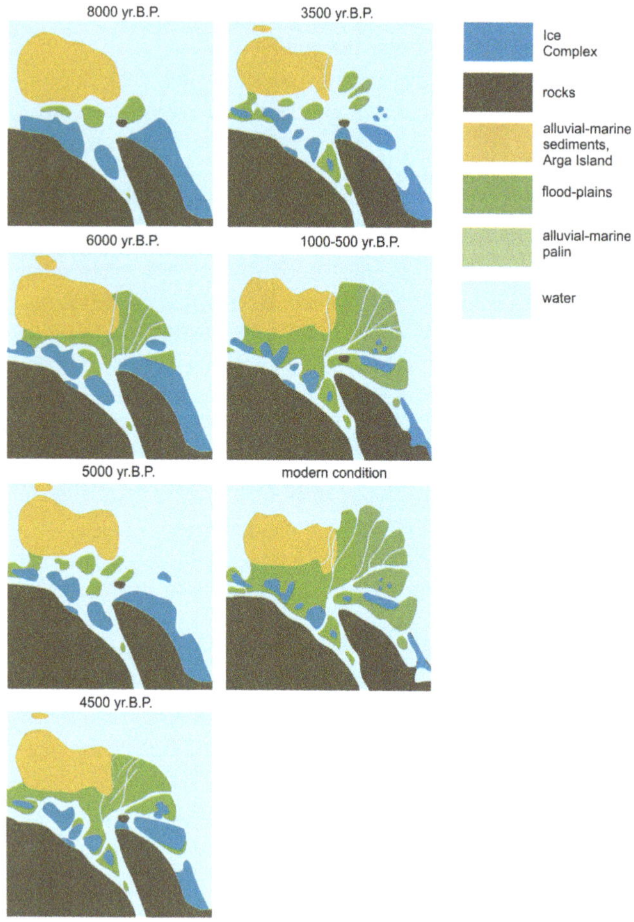

Figure 8. Palaeo-reconstructions of the Lena Delta area.

Bykovskaya channel was formed at the same time. A short-term sea level rise occurred around 400 BP, and then a modern stage of erosion began with another short-term sea level rise around 200 BP. Due to tectonic uplifting of the Earth's surface at the delta head the water surface was slanted; as a result the Olenekskaya channel and the central delta channels were cut off from the sea and runoff was redistributed in favour of the eastern and south-eastern delta channels (Bolshiyanov et al., 2013).

6 Conclusions

In the Lena River delta, organomineral sediments are extremely widespread and comprise the first alluvial-marine terrace. Results of radiocarbon dating, lithological description, and analysis of spores, pollen, and plant macroremains indicate that these deposits could not have accumulated in floodplain swamps. A layered sediment is a specific sediment type that accumulates under stagnant estuarine conditions when sea level is rising. Since layered sediments accumulated under conditions of high sea level, their dating can show when transgressive phases of delta evolution occurred.

Geomorphological analysis and compilation of the geomorphological map of the Lena River delta showed that the delta did not form through a simple protrusion into the Laptev Sea. Rather, it evolved by stages under estuarine conditions when it was essentially cut off from the sea; delta formation progressed from west to east during the Holocene. Against this background, sea transgressions occurred during which the organomineral masses accumulated, creating the first terrace. During stages of regression these sediments were partly washed out, as were more ancient islands. Transgressive stages occurred during 8000–6000, 4500–4000, 2500–1500, and 400–200 BP, and periods of sea regression took place at 5300 and 500 BP. The regressive periods were shorter than the stages of sea level rise. Currently, the south-eastern part of the delta (Bykovskaya channel) has developed as an estuary under conditions of rising sea level, and the south-western part (Olenekskaya channel) as a meandering arm under conditions of falling sea level.

Acknowledgements. The authors thank to E. A. Raschke for botanical analysis of peat samples. We are grateful to the Alfred Wegener Institute for Polar and Marine Research (Germany) for the financing the Russian–German "Laptev Sea system" project under which these investigations were conducted. Part of the funds for conducting the research was allocated by the Russian Foundation for Basic Research grant no. 05-05-64419. Great support was given to the authors by the entire Russian–German team of researchers in the course of joint expeditions from 1998 to 2012. However, not all the participants will agree with the interpretation of delta origin proposed here by the authors.

Edited by: P. Overduin

References

Anisimov, M. A.: Development of natural environment of the New Siberian Islands in Holocene (in Russian), PhD.Thesis, Herzen State Pedagogical University of Russia, 154 pp., 2010.

Arslanov, Kh. A., Tertychnaya, T. V., and Chernov, S. B.: Problems and methods of dating low-activity samples by liquid scintillation counting (in Russian), Radiocarbon, 35, 393–398, 1993.

Ashik, I. M., Makarov, A. S., and Bolshiyanov, D. Yu.: Development of the Russian Arctic coasts related with sea level fluctuations (in Russian), Meteospektr., 2, 23–27, 2010.

Bolshiyanov, D. Yu.: Passive glaciation in the Arctic and in the Antarctica (in Russian), AARI, St. Petersburg, Russia, 296 pp., 2006.

Bolshiyanov, D. Yu., Makarov, A. S., Schneider, V., and Gusev, E. A.: Problems of origin of Ice Complex sediments and existence of "Sannikov's Lands" in the past (in Russian), Problemy Arktiki i Antarktiki, 1, 151–160, 2008.

Bolshiyanov, D. Yu., Makarov, A. S., Schneider, V., and Stoof, G.: Origin and development of the Lena River Delta (in Russian), AARI, St. Petersburg, Russia, 267 pp., 2013.

Borren, W., Bleuten, W., and Lapshina, E.: Holocene peat and carbon accumulation rates in the southern taiga of Western Siberia. Quaternary Res., 61, 42–51, 2004.

Denisenkov, V. A.: Fundamentals of Bog Science: Handbook (in Russian), SPb, Publishing House of SPb Univ., 224 pp., 2000.

Dictionary of Geology (in Russian), Nedra, Moscow, Nedra, V. 2., 320 pp., 1973.

Dombrovskaya, A. V., Koreneva, M. M., and Tyuremnov, S. N.: Atlas of plant residues encountered in peat (in Russian), Moscow-Leningrad, 90 pp. and 137 pp. Illustrated, 1959.

Grichuk, V. P. and Zaklinskaya, E. D.: Analysis of pollen and spores and its application in paleogeography (in Russian), Geographgiz, Moscow, 122–123, 1948.

Grigoriev, M. N.: Cryomorphogenesis of the Lena River mouth (in Russian), Yakutsk Institute of Permafrost Studies of the SB RAS, 176 pp., 1993.

Grimm, E. C.: Tilia 1.12, Tilia-Graph 1.18. Illinois State Museum, Research and Collection Center, Springfield, Illinois, 1991.

Gusev, A. I.: Relief of terraces and stages of their development in the Lena River Delta (in Russian), Transactions of NIIGA, Leningrad, 72, 214–223, 1953.

Gusev, A. I.: Quaternary stratigraphy of the Lena-Indigirka coastal plain (in Russian), in: Transactions of Conference of North-East Stratigraphy, Magadan, Russia, 465–468, 1959.

Katz, N. Ya, Katz, S. V., and Skobeeva, E. I.: Atlas of plant residues in peats (in Russian), Moscow, 376 pp., 1977.

Kupriyanova, L. A. and Alyoshina, L. A.: Pollen and spores of plants from the flora of European part of USSR (in Russian), Nauka, Leningrad, 171 pp., 1972.

Kupriyanova, L. A. and Alyoshina, L. A.: Pollen and spores of plants from the flora of European part of USSR (in Russian), Nauka, Leningrad, 184 pp., 1978.

Lishtvan, I. I. and Korol, N. T. : Main properties of peat and methods of their determination (in Russian), Science and Technology, Minsk, 320 pp., 1975.

Makarov, A. S.: Laptev Sea level fluctuations as a factor of the Lena River Delta formation in Holocene (in Russian), Ph.d thesis, Saint-Petersburg State University, faculty of geography, St. Petersburg, Russia, 142 pp., 2009.

Moore, P. D., Webb, J. A., and Collinson, M. E.: Pollen analysis, Blackwell Scientific Publications, Oxford, 216 pp., 1991.

Pavlova, E. Yu., Dorozhkina, M. V., Davydova, N. N., and Denisenkov, V. P.: History of the Ponoi Lake basin (Kola Peninsula) during Middle and Late Holocene (in Russian), in: Ecology of Fresh Water Basins of Russian North, Nauka, St. Petersburg, 169–178, 2005.

Ramsey B., C., Dee, M., Lee, S., Nakagawa, T., and Staff, R. A.: Developments in the calibration and modeling of radiocarbon dates, Radiocarbon, 52, 953–961, 2010.

Reimer, P. J., Bard, E., Bayliss, A., Beck, J. W., Blackwell, P. G., Bronk Ramsey, C., Buck, C. E., Cheng, H., Edwards, R. L., Friedrich, M., Grootes, P. M., Guilderson, T. P., Haflidason, H., Hajdas, I., Hatté, C., Heaton, T. J., Hoffmann, D. L., Hogg, A. G., Hughen, K. A., Kaiser, K. F., Kromer, B., Manning, S. W., Niu, M., Reimer, R. W., Richards, D. A., Scott, E. M., Southon, J. R., Staff, R. A., Turney, C. S. M., and van der Plicht, J.: IntCal13 and Marine13 radiocarbon age calibration curves 0–50,000 years cal BP, Radiocarbon, 55, 1869–1887, 2013.

Sachs, T., Giebels, M., Wille, Ch., Kutzbach, L., and Boike, J.: Methane emission from Siberian wet polygonal tundra on multiple spatial scales: Vertical flux measurements by closed chambers and eddy covariance, Samoylov Island, Lena River Delta in Proceeding of the 9th International Conference of permafrost, 29 Jun 2008–3 July 2008, Fairbamks,USA, 1549–1554, 2008.

Samoilov, I. V.: River mouths (in Russian), Geografizdat, Moscow, 527 pp., 1952.

Schirrmeister, L., Grosse, G., Schwamborn, G., Andreev, A., Meyer, H., Kunitsky, V., Kuznetsova, T., Dorozhkina, M., Pavlova, Y., Bobrov, A., and Oezen, D.: Late Quaternary history of the accumulation plain north of the Chekanovsky Ridge (Lena Delta, Russia): A multidisciplinary approach, Polar Geogr., 27, 277–319, 2003.

Schwamborn, G., Rachold V., and Grigoriev, M.: Late Quaternary sedimentation history of the Lena Delta, 38 GeoLines, 2000.

Skobeyeva, E. I. and Simakova, L. A.: Peat deposits of Yakutia and their features (in Russian), in: Properties and Methods of Studying Peat and Sapropel Deposits, Kalinin State Univ., Kalinin, 3–9, 1983.

Smirnov, I. P.: Peat formation in tundra of Amguem Valley and Anadyr Lowland (in Russian), in: Properties and Methods of Studying Peat and Sapropel Deposits, Tver, 4–15, 1981.

Tarakanov, L. V. and Biryukov, V. Yu.: Geomorphological features of modern Laptev Sea ingression in the area of Shirokostan Peninsula (in Russian), Geomorphologiya, no. 4, 98–100, 1974.

11

Impacts of soil moisture on de novo monoterpene emissions from European beech, Holm oak, Scots pine, and Norway spruce

C. Wu[1], I. Pullinen[1], S. Andres[1], G. Carriero[2], S. Fares[3], H. Goldbach[4], L. Hacker[1], T. Kasal[4], A. Kiendler-Scharr[1], E. Kleist[5], E. Paoletti[2], A. Wahner[1], J. Wildt[5], and Th. F. Mentel[1]

[1]Institut für Energie und Klimaforschung, IEK-8, Forschungszentrum Jülich, 52425 Jülich, Germany
[2]Istituto Protezione Sostenibile Piante, Consiglio Nazionale Ricerche (CNR), Via Madonna del Piano10, 50019 Sesto Fiorentino, Florence, Italy
[3]Consiglio per la Ricerca e la sperimentazione in Agricoltura (CRA), Research Centre for the Soil-Plant System, Rome, Italy
[4]Department of plant nutrition (INRES), Universität Bonn, Karlrobert Kreiten Str. 13, 53115 Bonn, Germany
[5]Institut für Bio- und Geowissenschaften, IBG-2, Forschungszentrum Jülich, 52425 Jülich, Germany

Correspondence to: J. Wildt (j.wildt@fz-juelich.de)

Abstract. Impacts of soil moisture on de novo monoterpene (MT) emissions from Holm oak, European beech, Scots pine, and Norway spruce were studied in laboratory experiments. The volumetric water content of the soil, Θ, was used as the reference quantity to parameterize the dependency of MT emissions on soil moisture and to characterize the severity of the drought.

When Θ dropped from $0.4\,m^3 \times m^{-3}$ to $\sim 0.2\,m^3 \times m^{-3}$ slight increases of de novo MT emissions were observed but with further progressing drought the emissions decreased to almost zero. In most cases the increases of MT emissions observed under conditions of mild drought were explainable by increases of leaf temperature due to lowered transpirational cooling. When Θ fell below certain thresholds, MT emissions decreased simultaneously with Θ and the relationship between Θ and MT emissions was approximately linear. The thresholds of Θ (0.044–$0.19\,m^3 \times m^{-3}$) were determined, as well as other parameters required to describe the soil moisture dependence of de novo MT emissions for application in the Model of Emissions of Gases and Aerosols from Nature, MEGAN.

A factorial approach was found appropriate to describe the impacts of Θ, temperature, and light. Temperature and Θ influenced the emissions largely independently from each other, and, in a similar manner, light intensity and Θ acted independently on de novo MT emissions. The use of Θ as the reference quantity in a factorial approach was tenable in predicting constitutive de novo MT emissions when Θ changed on a time scale of days. Empirical parameterization with Θ as a reference was only unsuccessful when soil moisture changed rapidly

1 Introduction

Biogenic volatile organic compounds (BVOC) are important atmospheric trace gases. They are involved in photochemical ozone- and particle-formation, and they impact the oxidation capacity of the troposphere (e.g. Chameides et al., 1988; Jacob and Wofsy, 1988; Derwent et al., 1996; Kulmala et al., 2004). On a global scale, the source strength of BVOC is estimated to be ~ 1000 Tg per year (Guenther et al., 1995, 2012) which exceeds the source strengths of anthropogenic volatile organic compounds by about an order of magnitude. The estimates of the global source strengths originate from model calculations which are often based on the same general procedure: BVOC emissions for standard conditions and for representative plant species (plant functional types) are used as basic model input. Dependencies of the BVOC emissions on variables such as temperature, light intensity (PAR = photosynthetic active radiation) and soil moisture are considered by applying phenomenological algorithms that describe the respective dependencies.

One of the first algorithms was developed by Tingey et al. (1980, 1991), who showed that monoterpene (MT) emissions from Slash pine exponentially depend on temperature but are independent of PAR. The underlying reason is that MT emissions from conifers originate from MT diffusion out of pools, which depends on temperature and not directly on PAR. In a modified syntax, Tingey's algorithm reads as follows:

$$\Phi = \Phi^{S} \times \exp(\beta \times (T - T_{S})). \quad (1)$$

In Eq. (1), Φ is the emission rate of the MT at the actual temperature T, Φ^{S} is the standard emission rate or emission activity factor, i.e. the emission rate measured at standard temperature T_{S}. $\beta$is the parameter describing the temperature dependence. $\beta$is in the range of $0.09\,K^{-1}$.

Isoprene is emitted directly after its biosynthesis. Isoprene emissions thus are de novo emissions and directly related to the rate of isoprene biosynthesis. Biosynthetic activity generally depends on T and PAR. Hence, both variables are required to describe isoprene emissions as shown in the algorithm by Guenther et al. (1993). In a modified syntax, and neglecting decreases of isoprene emissions induced by high temperature stress (e.g. Guenther et al., 1993), the algorithm of Guenther et al. (1993) reads as follows:

$$\Phi = \Phi^{S} \times \frac{c_{L} \times \alpha \times \mathrm{PAR}}{\sqrt{1 + \alpha^{2} \times \mathrm{PAR}^{2}}} \times \exp(\beta \times (T - T_{S})). \quad (2)$$

In Eq. (2), Φ is the emission rate at temperature T and at light intensity PAR. Φ^{S} is the emission activity factor i.e. the emission rate measured at standard light intensity (often set to $1000\,\mu\mathrm{mol} \times \mathrm{m}^{-2} \times \mathrm{s}^{-1}$) and at standard temperature, T_{S} (often set to 30 °C). α is the parameter describing the PAR dependence of the emission, and c_{L} is a normalization factor used to obtain $\Phi = \Phi^{S}$ when PAR is equal to the standard light intensity.

Equation (2) consists of three factors: first the emission activity factor, second the factor describing the PAR dependence and third the factor describing the temperature dependence. A factorial approach requires the individual factors to be independent of each other; i. e. Guenther et al. (1993) postulated that the PAR dependence is independent of the actual temperature and vice versa, that the temperature dependence is independent of the actual PAR.

Schuh et al. (1997) found interdependent impacts of temperature and PAR on MT emissions from sunflower which could not be described by a factorial approach as given in Eq. (2). They ascribed the observed interdependency to the two different mechanisms of MT emissions, MT emissions from pools that are independent of the actual light intensity and de novo emissions that depend on the actual light intensity. Depending on the species, both mechanisms can act in parallel. Schuh et al. (1997) suggested two additive terms to describe T and PAR dependence of MT emissions. One of them was similar to Eq. (2) and was used to describe de novo MT emissions. The other one was similar to Eq. (1) and describes the temperature dependence of MT emissions from pools. By using additive terms the factorial approach for de novo emissions was retained. This was confirmed by Shao et al. (2001) who used this algorithm to describe the emissions from Scots pine. Scots pine exhibits pure pool emissions, mixed pool and de novo emissions as well as a pure de novo emission (Kleist et al., 2012).

Schuh et al. (1997) noted the high variability of MT standard emission rates. Measuring emissions from different individuals of a given species under identical conditions of T and PAR still resulted in highly variable emission rates. This clearly shows that MT emissions also depend on other variables than T and PAR alone. Among such other variables are leaf phenology, CO_2 concentrations and soil moisture. We here focus on soil moisture.

In the Model of Emissions of Gases and Aerosols from Nature (MEGAN, Guenther et al., 2006, 2012), impacts of soil moisture are only considered for isoprene emissions. Guenther et al. (2006) implemented the empirical algorithm by Pegoraro et al. (2004) who used the volumetric water content, Θ, as the reference quantity for characterizing the impacts of soil moisture. Three different regimes of Θ are used in MEGAN to define the factor that describes the impact of soil moisture on isoprene emissions. Above a threshold Θ_{1}, isoprene emissions are not affected by soil moisture and the factor is unity. Below Θ_{1}, isoprene emissions linearly decrease with decreasing Θ until the wilting point Θ_{W} is reached. The wilting point Θ_{W} is the soil moisture below which plants cannot extract water from the soil. At and below Θ_{W} isoprene emissions are set to zero. In MEGAN, Θ_{W} is taken from a database by Chen and Dudhia (2001). The difference: $\Theta_{1} - \Theta_{W}$, $\Delta\Theta_{1}$, is the empirical parameter used to describe the dependence of isoprene emissions on soil moisture. Its value ($\Delta\Theta_{1} = 0.06\,\mathrm{m}^{3} \times \mathrm{m}^{-3}$) is taken from Pegoraro et al. (2004).

Compared to isoprene there are less studies on impacts of soil moisture on MT emissions. Some studies show increasing emissions with decreasing soil moisture (Bertin and Staudt, 1996; Blanch et al., 2007; Ormeño et al., 2007), others show decreasing emissions with increasing severity of drought (Lavoir et al., 2009; Šimpraga et al., 2011). Besides this, different reference quantities have been used to characterize the soil moisture level. Among these are the plant water potential (Ormeño et al., 2007; Lavoir et al., 2009) and the diurnal variation of the radial stem diameter (Šimpraga et al., 2011). As these plant parameters are highly variable between individuals and influenced by a larger number of factors, they can hardly be used for up-scaling.

Aim of our study was to provide data that can be used for modelling the impacts of soil moisture on de novo MT emissions with MEGAN. For this purpose we performed laboratory measurements with different plant species exposed to drought. We tested whether the volumetric water content Θ can be used as a reference quantity and whether a factorial approach is justified. Results of our experiments and the limitations of our approach are described below.

2 Methods

2.1 Laboratory set up

Experiments were performed at the Jülich Plant Atmosphere Chamber facility (JPAC). The setup of the chambers has already been described elsewhere in detail (e.g. Schimang et al., 2006; Mentel et al., 2009, 2013). The two plant chambers used for these experiments were made of borosilicate glass and operated as continuously stirred tank reactors (CSTR). To keep the temperature constant, the CSTRs were mounted in separate walk-in climate chambers. Each CSTR was equipped with a Teflon fan providing homogeneous mixing and diminishing the boundary layer resistance at leaf surfaces. The chambers were equipped with several connections to introduce temperature sensors (Newport Omega, HTMTSS), a light-intensity sensor (LI-COR, LI-189) and to connect the tubes for gas phase analysis and air supply.

For the experiments described here two plant chambers with volumes of 1150 and 164 L were used. The chambers were equipped with 11 and 7 discharge lamps (HQI 400 W/D; Osram, Munich, Germany), respectively, resulting in PAR at full illumination and at typical mid-canopy heights of 440 and 700 μmol × m^{-2} × s^{-1}, respectively. Infrared radiation (between 750 and 1050 nm) from the lamps was reflected by filters (type IR3, Prinz Optics GmbH, Stromberg, Germany) placed between the lamps and the chambers in order to minimize radiative heating of the plants. To minimize wall losses, all tubes either consisted of Teflon (PTFE or PFA) or glass.

Ambient air was purified by an adsorptive drying device (Zander Aufbereitungstechnik GmbH and Co. KG, Essen, Germany, KEA 70) and by a palladium catalyst operating at 450 °C. Ozone, NO, NO_2, and volatile organic compounds (> C_3) were removed after the air had passed the purification system. Concentrations of CO_2 and water vapour were also reduced by the adsorption dryer. CO_2 was added to the inlet air to keep the CO_2 concentrations similar to those in the environment. CO_2 concentrations at plant chamber inlet were about 385 ppm. Uptake by the plants reduced the CO_2 concentrations in the chamber to about 350 ppm when the plants were well watered. Progressing drought caused lowered net photosynthesis and CO_2 concentrations in the chamber increased near to those at chamber inlet. The air flow through the chambers was kept constant by mass flow controllers (Brooks Instruments). Typical air flows were in the range of 20–30 L × min^{-1} when using the small chamber and 50–100 L × min^{-1} when using the large chamber.

Also the analytical equipment at JPAC is described in detail in previous publications (Schuh et al., 1997; Wildt et al., 1997; Schimang et al., 2006). O_3 concentrations were measured by UV absorption (Thermo Environmental Instruments, model 49). Concentrations of NO were measured by chemiluminescence (Tecan, CLD 770 AL PPT) and for the measurements of NO_2 the analysed air was led through a photolytic converter (Tecan, PLC 760). During all the measurements presented here, O_3 mixing ratios were below 1 ppb. NO_x mixing ratios were about 300 ppt.

Differences in mixing ratios of H_2O and CO_2 between chamber inlet and outlet were measured by IR absorption (Rosemount Binos 100 4P). Absolute H_2O concentrations were determined with dew point mirrors (Walz, MTS-MK1). Mixing ratios of BVOC were measured using gas chromatography–mass spectrometry (GC–MS, Agilent GC–MSD-system HP5890 Series II + MS HP5972A, or GC–MSD system HP6890 + HP5973, equipped with thermal desorption, TDSG, Gerstel, Mülheim, Germany). Calibration of the GC–MS systems was performed using a diffusion source containing pure chemicals in individual vials in combination with a dynamic dilution system. Concentrations of the compounds released from the calibration source were determined from the mass loss rates of the individual compounds and the dilution fluxes. The VOC mixing ratios were in the lower ppb to ppt range. For details on the GC systems and calibration procedure see, e.g. Heiden et al. (2003).

The CSTR contained shoots and leaves of the plants. The upper part of the setup was separated from the lower part containing roots and soil by PTFE sheets with a hole at the centre to accommodate the stem. The stem duct was closed by a flexible bag.

The pot containing the plants' roots was positioned in a shallow dish allowing for collecting excess water from the pot. Plant, pot, and dish were mounted on a balance (Sartorius, MC1) that was used to measure the weight of the investigated plant together with the soil and the water in the soil. The flexibility of the bag that sealed the gap between stem and stem duct allowed weighing the plant/soil system on-line. The balance had a nominal resolution of 1 g, but variations of the chamber's slight overpressure (5–10 mbar) imposed noise in the range of 20–30 g limiting the precision of weight measurements.

2.2 Determination of flux densities

Flux densities for the compound X, $\Phi(X)$, were calculated using the respective differences of the mixing ratios between chamber inlet and outlet and the leaf area, A_{leaf}, as the basis:

$$\Phi(X) = \frac{F_{air} \times ([X]_i - [X])}{A_{leaf}}. \quad (3)$$

In Eq. (3), F_{air} is the air flow through the chamber (in units of mol s^{-1}), $[X]_i$ is the mixing ratio of compound X at chamber inlet and $[X]$ is the mixing ratio of compound X in the

Table 1. List of experiments and chamber settings.

Plant	Purpose of experiment	Leaf area (m^2)	Chamber temperature (°C)	PAR ($\mu mol\,m^{-2}\,s^{-1}$)
European beech	Θ-dependence	0.91	23	440
Holm oak 1	Θ-dependence	0.25	25	600
Holm oak 2	Θ-dependence	0.27	22	440
Norway spruce	Θ-dependence	0.7	23	440
Scots pine	Θ-dependence	0.45	25	400
Holm oak 3	Θ- and T-dependence	0.18	15	500
Holm oak 3	Θ- and T-dependence		20	500
Holm oak 3	Θ- and T-dependence		25	500
Holm oak 4	Θ- and PAR-dependence	0.25	20	700
Holm oak 4	Θ- and PAR-dependence		20	400
Holm oak 4	Θ- and PAR-dependence		20	200

chamber. Flux densities are termed as MT emission rates for $X = \text{MT}$, transpiration rates for $X = H_2O$, and net assimilation rates for $X = CO_2$. According to the results of several tests (e.g. Schuh et al., 1997; Heiden et al., 2003; Schimang et al., 2006), wall losses and chemical reactions were unimportant for all compounds investigated here and were therefore neglected. As convention, flux densities are positive when compounds are emitted (e.g. water, MT) and negative when compounds are taken up (e.g. CO_2). To allow better visual comparison in diagrams, net assimilation is multiplied by -1.

Stomatal conductance was calculated using Eq. (4).

$$c\,(H_2O) = \frac{\phi(H_2O)}{\Delta[H_2O]} \tag{4}$$

Herein, $\Delta[H_2O]$ is the difference between water vapour mixing ratios in the substomatal cavity and in the chamber air, the former calculated from leaf temperature assuming saturation vapour pressure in the substomatal cavity.

After finishing an experiment, about 5 % of the leaves from broadleaf species were dissected from the plant, taking care to select representative leaves. The leaves were scanned for leaf area determination and evaluated by image analysis. Total leaf area was calculated by multiplying the averaged area measured per dissected leave by the total number of leaves.

Needle area for the conifers were determined as described in Shao et al. (2001). The number of needles per 10 cm of branches with green needles was counted for representative branches, needles were removed from these branches, scanned and the area of dark pixels was determined. Upscaling from the measured needle area to the total needle area was performed by considering the total lengths of branches with green needles for the respective conifer.

Uncertainty in leaf/needle area determination was estimated to be approximately 10 %. Note that the needle areas given in Table 1 are projected needle areas and leaf areas for broadleaf species are one-sided leaf areas.

2.3 Experiments with plants

Experiments were conducted with individual plants using species representative for European climate zones. European beech (*Fagus sylvatica* L.) and Holm oak (*Quercus ilex* L.) represented species without storage pools for monoterpenes. Scots pine (*Pinus sylvestris* L.) and Norway spruce (*Picea abies* L.) mainly exhibited MT pool emissions but also showed de novo MT emissions.

In total seven plants were used for the measurements: one individual each for beech, spruce, and pine and four individuals of Holm oak. One to two-year old beech-, pine-, and spruce-seedlings were taken from the forest, potted in buckets of 15 L volume and stored outside for about a year before they were used for the experiments. Seedlings of Holm oak were obtained from the forest nursery of the Castelporziano estate, Rome, central Italy. These have been successively potted in 15 L buckets and stored for several weeks in a growth room before using them in the CSTR. All plants were potted in the same soil that was a mix of peat (Einheitserde ED73 containing $2\,g \times L^{-1}$ nutrient salts and $1\,g \times L^{-1}$ clay) with quartz sand, volumetric ratio 5 : 1, density $0.46 \pm 0.07\,kg \times L^{-1}$.

To control for the de novo nature of the 1,8-cineol emission from the conifers we exposed the plants to $^{13}CO_2$. The chamber was flushed with synthetic air (Air Liquide, Germany) and $^{13}CO_2$ (99 % ^{13}C, ~350 ppm) was added for ~3 h in each of both experiments.

Individual plants were investigated for time periods of 3–10 weeks. In most of our studies, the impact of soil moisture was investigated (Table 1, European beech, two experiments with Holm oak – experiment Holm oak 1 and Holm oak 2 – one experiment with Norway spruce and one with Scots pine). In these experiments the plants were exposed to a diurnal rhythm of 11 h illumination and 11 h darkness, and simulation of twilight by switching on or off individual lamps within 1 h in the morning and evening, respectively.

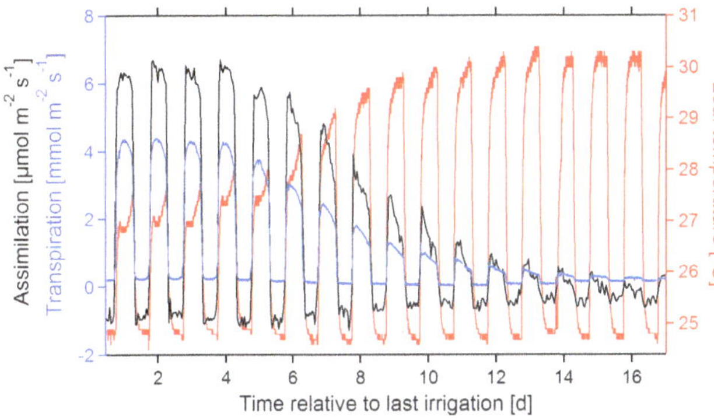

Figure 1. Temporal development of transpiration (blue trace, left hand y axis) and leaf temperature (red trace, right hand y axis, average of three leaves) at the example of Holm oak, experiment 1. The plant was irrigated at day 0 ($\Theta = 0.4\,m^3 \times m^{-3}$). At the end of the drought period Θ had fallen to $0.05\,m^3 \times m^{-3}$. Chamber temperature 25 °C, PAR $= 600\,\mu mol\,m^{-2}\,s^{-1}$.

In additional studies, we varied T or PAR, respectively, in parallel to soil moisture to investigate the potential limitations of the factorial approach. In the third experiment with Holm oak the chamber temperature was systematically changed during periods of illumination. Except for 2 days, the chamber temperature was set to 15 °C during the night and kept at 15 °C for the next 6–7 h of the following illumination period (from 3 o'clock to ~ 10 o'clock). Then the temperature was set to 20 °C for 4 hours (from ~ 10 to ~ 14 o'clock) and thereafter to 25 °C (from ~ 14 to ~ 18 o'clock, all data local time). The exact timing of temperature settings was adapted to the start of GC runs. The period of illumination was elongated to 15 h allowing measuring at the three different temperatures, but at the same PAR for each day. When Θ had fallen below $0.02\,m^3 \times m^{-3}$, the systematic temperature variations were interrupted for 2 days to follow the drought-induced decrease of MT emissions without changes of temperature and PAR. After these 2 days, the chamber temperatures were again varied in the same manner as before these days.

In the fourth experiment with Holm oak, PAR was changed systematically during periods of illumination. Every second day PAR was set to $700\,\mu mol\,m^{-2}\,s^{-1}$ for 6 h (from 4 o'clock to ~ 10 o'clock), thereafter to $400\,\mu mol\,m^{-2}\,s^{-1}$ for 5 h (from ~ 10 to ~ 15 o'clock), and then to $200\,\mu mol\,m^{-2}\,s^{-1}$ for 5 h (from ~ 15 to ~ 20 o'clock). From ~ 20 o'clock to 4 o'clock PAR was zero. During the other days, PAR was held constant at $400\,\mu mol\,m^{-2}\,s^{-1}$ from 4:00 to 20:00 LT. Table 1 lists the experiments and the respective conditions of PAR and chamber temperatures.

Due to the radiation from the lamps, the chamber temperatures and the leaf temperatures varied with illumination. During periods of darkness leaf temperatures and chamber temperatures were similar. During periods of illumination leaf temperature was higher than chamber temperature. For well watered plants leaf temperature was about 2 °C higher than the chamber temperature. Progressing drought stress decreased transpiration and leaf temperatures increased by additional 2–3 °C relative to the chamber temperature (Fig. 1).

2.4 Determination of Θ and the Θ-dependence of MT emissions

The volumetric water content, Θ, was determined from the mass loss of water during the respective experiments (Eq. 5).

$$\Theta = \frac{M_{act} - M_{dry}}{V_{soil}} \times \frac{1}{\rho} \tag{5}$$

In Eq. (5), M_{act} is the actual mass of the soil and M_{dry} is the dry mass of the soil. V_{soil} is the volume of the soil in the pots neglecting the volume of the roots. M_{act} was measured online and M_{dry} was estimated from soil samples taken from the top of the pots and oven-dried at 110 °C for 5 days. The measured mass loss was converted to volume loss by using a water density, ρ, of $1\,kg\,L^{-1}$.

The dependence of MT emission on Θ was parameterized in the following manner: maximum MT emissions were reached when Θ had fallen near to the threshold below which MT emissions started to decrease (Θ_1). The average of the emission rates measured around the maximum emissions was set as standard emission rate, Φ^S. Emission rates measured at the same light intensity and chamber temperature were normalized by dividing them by Φ^S. The normalized data were used to determine the relationship between MT emissions and Θ. Although the decrease of emissions with decreasing Θ appeared to be exponential, we applied a linear approximation to be comparable to other parameterization approaches. Linear fits to the normalized data allowed to determine Θ_1 as well as the volumetric water content Θ_0 below which the extrapolated emissions became zero. Only data with $\Theta < \Theta_1$ and measured at the same PAR and the same chamber temperature were taken for fitting. Furthermore, all data where $\Phi/\Phi^S < 0.05$ were discarded to diminish the impact of the exponential behaviour. Similar to the notation used in MEGAN, we calculated $\Delta\Theta_1$, the difference $\Theta_1 - \Theta_0$. This procedure of determining the Θ dependence of MT emissions was performed for all data sets.

2.5 Error determination and conversion from Θ to RWC

Experimental errors in the determination of Θ were due to the noise on M_{act}, due to the uncertainties of M_{dry} and due to the error in the determination of V_{soil}. The error in volume determination was negligible and, compared to the error in M_{dry}, the error by the noise was of minor importance. The statistical noise of 20 – 30 g (peak to peak noise, deviation from average at maximum ± 15 g) added an uncertainty of ± 15 mL to the volume of water, which is low compared to the total amount of added water (5 – 6 L). At soil volumes of about 13 L, the uncertainty produced by the noise on Θ was $\pm 0.0012\,m^3 \times m^{-3}$.

The error due to our procedure of taking soil samples was higher. Taking samples from the same pot caused differences in the dry mass. Extrapolated to the total mass of the soil, the maximum deviation was 420 g, leading to an uncertainty of ±210 g for M_{dry}. This is more than an order of magnitude higher than the uncertainty added by the noise on M_{act}. The uncertainty we give for Θ is therefore mainly based on the uncertainty of M_{dry} which, converted to the water volume is ±210 mL. The possible error added to Θ by the uncertainty of M_{dry} is: $\pm 0.016\,m^3\,m^{-3}$.

It has to be noted that the error caused by the different error sources have different qualities; while errors due to the noise are statistical errors, the error caused by uncertainty of M_{dry} is a systematic error for each individual experiment. Errors in M_{dry} mainly cause a systematic shift of the Θ axis. Errors in the zero point may therefore be high and in one case, the measured M_{act} was lower than M_{dry}. This led to slightly negative values for Θ which is physically impossible. However, since the deviation from zero was quite low, we left the negative values.

Erroneous determination of M_{dry} does not impose important uncertainties on $\Delta\Theta_1$. Due to the systematic nature of this error, its main effect is a systematic shift of the Θ axis. As $\Delta\Theta_1$ is the difference between Θ_1 and Θ_0, systematic shifts in Θ cancel out. Hence, total uncertainties in $\Delta\Theta_1$ which is the main parameter for modelling with MEGAN were quite low and thus acceptable.

Besides uncertainties of Θ caused by errors in the determination of M_{act} and M_{dry}, there is also an uncertainty due to our fitting procedure. Also MT emissions contain uncertainties and hence, data obtained from fits using the MT emissions as base conserve these uncertainties. Nevertheless, the statements we give on the soil moisture dependence of MT emissions and on the differences observed with this respect between MT emissions on the one hand and net photosynthesis and transpiration on the other hand are not substantially affected by the errors in Θ.

To allow using our data also for models that use the relative water content of the soil as reference we give a conversion factor from Θ to RWC. According to Rambal et al. (2003), RWC is the ratio of current water content to water content at field capacity. Using our mass based data, RWC can be calculated according to Eq. (6):

$$\mathrm{RWC} = \frac{M_{act} - M_{dry}}{M_{FC} - M_{dry}}. \qquad (6)$$

In Eq. (6), M_{FC} is the mass of the soil at field capacity, M_{act} is the actual mass and M_{dry} the dry mass as in Eq. (5). Field capacity is reached when the micropores of the soil are filled with water and the macropores filled with air after water is lost by gravity. According to our procedure of waiting some hours before measuring the weight of pot and plant, the water in the macropores should have been lost. We therefore

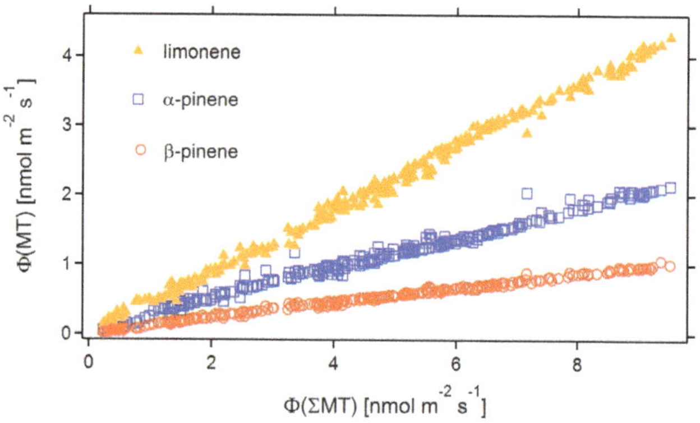

Figure 2. Correlation plot of emission rates measured for Holm oak, experiment 2. Only data measured at a PAR of 440 μmol $m^{-2}\,s^{-1}$ and a chamber temperature of 22 °C are plotted. The variation of emission rates is caused by changes in soil moisture. The high coefficients of determination ($R^2 > 0.95$) indicate that only the strengths of the emissions was changed but not the emission pattern.

approximate M_{FC} from the weight shortly after removing the excess water from the dishes below the pots and after subtracting the mass of the empty pot. Setting RWC $= x \times \Theta$ it follows:

$$x = \frac{V}{M_{FC} - M_{dry}} \times \rho. \qquad (7)$$

As factor pooled from all measurements we obtained $x = 2.6\,kg \times kg^{-1} \times m^{-3} \times m^3$ if soil water content is measured in $kg(H_2O)_{act} \times kg(H_2O)^{-1}_{FC}$. Note that the conversion factor is only valid for our type of soil and cannot be transferred to other experiments or field conditions.

3 Results

3.1 Emission patterns

All investigated plants emitted monoterpenes (MT) while emissions of isoprene and sesquiterpenes were low, if detectable at all. Neither stress-induced emissions of phenolic volatiles originating downstream of the shikimate pathway nor stress-induced emissions originating from the octadecanoid pathway were observed. Even for drought-stressed plants such non-MT emissions were absent indicating that the plants did not substantially suffer from other unintended stresses than drought.

The MT emission patterns were constant for each individual Holm oak and European beech. Relating the emission rates of a given MT to those of other MT (cross correlations) emitted from the same plant yielded significant correlations with coefficients of determination always above $R^2 > 0.95$ (Fig. 2). On the one hand the high correlation showed that all MT had the same basic emission mechanism: all of them were de novo emissions. On the other hand, the

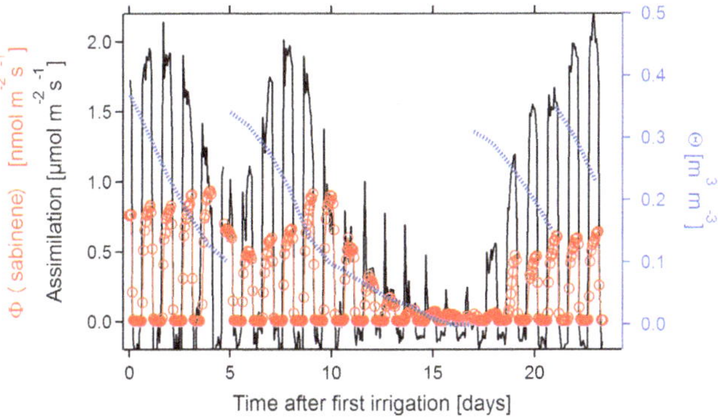

Figure 3. Long term time series of sabinene emissions from a European beech seedling (red circles, left hand *y* scale), assimilation (multiplied by -1, black line, left hand *y* scale) and volumetric water content of the soil Θ (blue dashed line, right hand *y* scale).

excellent correlation was obtained including the data during severe drought. This implies that for a given plant the impacts of drought were exactly the same for emissions of each individual MT species. Therefore the effects of drought on de novo MT emissions can be shown at the example of a single MT.

Sabinene was the dominant emission from European beech, comprising nearly 1/3 of the total emissions. Other MT emitted in substantial amounts were the following: γ-terpinene, α-terpinene, β-phellandrene, and α-terpinolene. We will show the response of European beech to drought at the example of sabinene. Main MT emissions from Holm oak were those of α-pinene, limonene, β-pinene, sabinene and myrcene. In two of the experiments with Holm oak α-pinene was the dominant emission (experiment 1: $\sim$48 % of total MT emissions; experiment 3: $\sim$42 % of total MT emissions), in the two other experiments the emissions of limonene were the strongest (experiment 2: limonene $\sim$45 %, α-pinene $\sim$23 % of total MT emissions; experiment 4: limonene $\sim$39 %, α-pinene $\sim$25 % of total MT emissions). To demonstrate the behaviour of MT emissions from Holm oak in response to drought we choose α-pinene as pivotal substance. It has to be noted that strong emissions of the acyclic ocimenes have been absent from Holm oaks. Emissions of ocimenes may show different behaviour than emissions of the cyclic MT (Staudt and Bertin, 1998). On the one hand, correlations as shown in Fig. 2 may be disturbed in the presence of strong ocimene emissions. On the other hand we cannot report on the drought impacts on ocimene emissions which may be different from that shown here for the cyclic MT. During our experiments, the ocimene emissions were too low to significant influence the sum of all MT emissions (maximum $\sim$6 % of total MT emissions).

For spruce and pine cross correlations as shown in Fig. 2 were useless. Both conifers exhibited pure pool emissions, mixed pool and de novo emissions as well as one pure de novo emission (compare Shao et al., 2001; Kleist et al., 2012). The only pure de novo MT emission was that of 1,8-cineole. We therefore show only the data we obtained for 1,8-cineole although these emissions made only a minor contribution to the total emissions. For spruce the emissions of 1,8-cineole contributed to less than 1 % to the total, for pine they contributed to about 10 % to the total MT emissions.

3.2 Impacts of soil moisture

3.2.1 European beech

Figure 3 shows the time series of sabinene emissions from European beech over a period of about 3 weeks. According to the de novo nature of the emissions a significant diurnal variation was observed with nearly no emissions during darkness and high emissions during periods of illumination. As can be seen in the Θ-trace, the plant was watered 4 times during the experiments. Between 2.5 and 5.6 L water were added after soil moisture had reached values of $\Theta \sim 0.1\ m^3 \times m^{-3}$ (dry conditions) and Θ near to $0\ m^3 \times m^{-3}$ (severe drought), respectively.

In addition to the strong light-dependent variation, impacts of soil moisture on MT emissions were also observable. During severe drought (2nd period, days 5–17 in Fig. 3) MT emissions decreased near to zero. After re-watering, MT emissions increased again. Such increases were not instantaneous but appeared on a time scale of few days.

Transpiration (not shown in Fig. 3) and net assimilation showed similar responses to Θ as MT emissions but with different response times. As can be seen, changes in net assimilation appeared earlier than changes in MT emissions, in particular during the phase of the second drought period (Fig. 3, days 5–17).

Figure 4 shows the dependence of normalized sabinene emissions and net assimilation on Θ. Only data taken at PAR $= 440\ \mu mol \times m^{-2} \times s^{-1}$ are shown. For Θ between 0.35 and $0.1\ m^3 \times m^{-3}$, MT emissions showed a considerable scatter that was mainly caused by different response times for the emissions to increase to pre-drought levels. With Θ falling below $0.1\ m^3 \times m^{-3}$ emissions decreased to minute amounts. As obvious from Fig. 4, changes of net assimilation as response to decreasing soil moisture were different from those of MT emissions. Net assimilation already decreased at higher Θ of $\sim 0.22\ m^3 \times m^{-3}$.

3.2.2 Holm oak

Two experiments with Holm oak were set up to investigate whether the reduction of MT emissions with decreasing Θ was determined by the soil moisture or by some other time constant of the plant itself (Table 1, experiment Holm oak 1 and Holm oak 2, temporal shape of α-pinene emissions, assimilation and Θ see Supplement Fig. S1). Holm oak 1 was investigated at higher PAR and at higher T and showed higher rates of net photosynthesis and higher transpiration

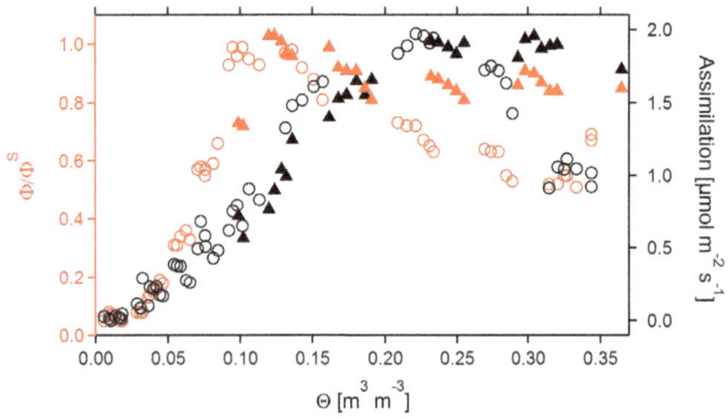

Figure 4. Normalized sabinene emissions from beech (red symbols, left scale) and rates of net photosynthesis (black symbols, right scale, multiplied by -1) in dependence on Θ. Closed triangles represent data taken during a first drought period that was stopped when Θ had fallen to $0.1\,m^3 \times m^{-3}$ (days 0–5 in Fig. 3), open circles represent data taken during the following period of recovery (days 6–10 in Fig. 3) until the end of the severe drought at day 17. Only data taken at a chamber temperature of 23 °C and a PAR of $440\,\mu mol\,m^{-2}\,s^{-1}$ are considered.

(data not shown). The soil and the plant roots were also exposed to higher temperatures than Holm oak 2. Due to the combination of higher evaporation and transpiration rates, Θ dropped from $\sim 0.12\,m^3 \times m^{-3}$ to $\sim 0.03\,m^3 \times m^{-3}$ within 8 days, faster than for Holm oak 2 where it took about 18 days.

The MT emissions decreased also on different time scales in the two experiments (see Fig. 5a). In contrast, the decrease of MT emissions with decreasing Θ was similar for both individuals although they were investigated at different dynamics of drought progression (Fig. 5b). The substantial difference observed for the temporal behaviour of the emissions (Fig. 5a) did not cause significant changes in decrease of emission rates for $\Theta < 0.1\,m^3 \times m^{-3}$ (Fig. 5b). Similar as for beech, increasing MT emissions were observed for Holm oak when Θ declined from ~ 0.4 to $0.15\,m^3 \times m^{-3}$. For Holm oak 1, at higher T and PAR, the emissions increased nearly threefold. For Holm oak 2 the increase was about 40 %.

In the third experiment with Holm oak chamber temperature was changed systematically during the progressing drought (Table 1, experiment Holm oak 3). This experiment was designed to measure the impact of Θ on MT emissions at different temperatures as well as the temperature dependence at different stages of drought.

The temperature dependence of MT emissions was determined by linear regression analysis of $\ln(\Phi)$ versus leaf temperatures (see Eq. 1 or Eq. 2 third factor). During the first 9 days with varying temperature at $\Theta > 0.1\,m^3 \times m^{-3}$ the temperature dependence remained constant with $\beta = 0.12 \pm 0.007\,K^{-1}$ (mean, 1σ standard deviation).

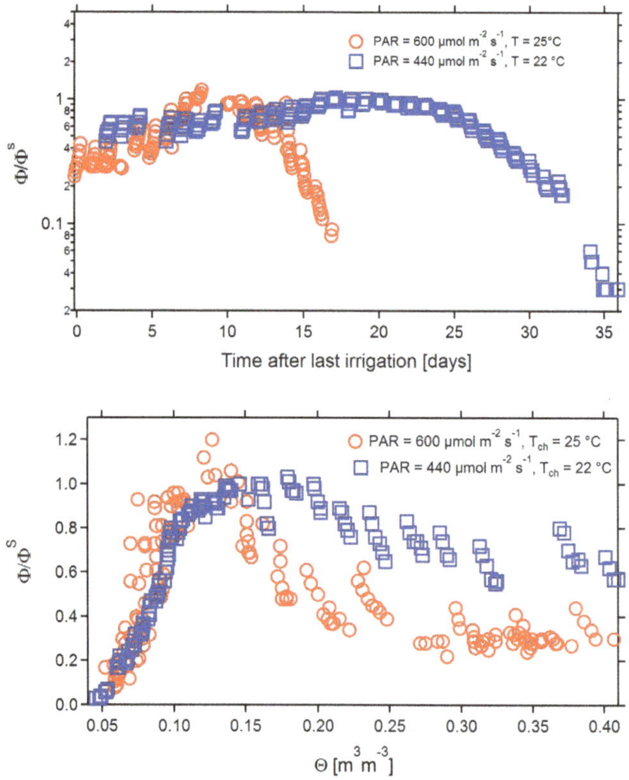

Figure 5. **(a)** Temporal shape of normalized α-pinene emissions from two individuals of Holm oak, experiments Holm oak 1 and Holm oak 2. Red circles show the data obtained at higher PAR and at higher chamber temperature. Blue squares show the data obtained for the plant investigated at lower temperature and PAR. Only data taken during periods of full illumination are shown. For better comparison the emissions were separately normalized to the emission rates measured for the respective individual at $\Theta \sim 0.12\,m^3 \times m^{-3}$. **(b)** Normalized α-pinene emissions from Holm oak in dependence of Θ. Red circles show the data measured for the plant investigated at higher PAR and higher chamber temperature, blue squares show data taken for the plant investigated at lower PAR and lower temperature. Same data as in **(a)**.

Similar to the observations in the other experiments with Holm oak, MT emissions decreased when Θ fell below $0.06\,m^3 \times m^{-3}$. During this measurement period the impacts of progressing drought and variations of temperature superimposed each other. Although temperature increased systematically over the day, emissions did not increase substantially. Consistent with the findings reported by Bertin and Staudt (1996). In the raw data typical log-linear relationships between emissions and temperature were not easily observable. We therefore had to remove the drought-induced decrease of MT emissions over the day.

Approaches to determine the Θ dependence from the data obtained at the days when temperature was held constant were not successful. The small dynamic range in which Θ changed during these 2 days (from ~ 0.025 to

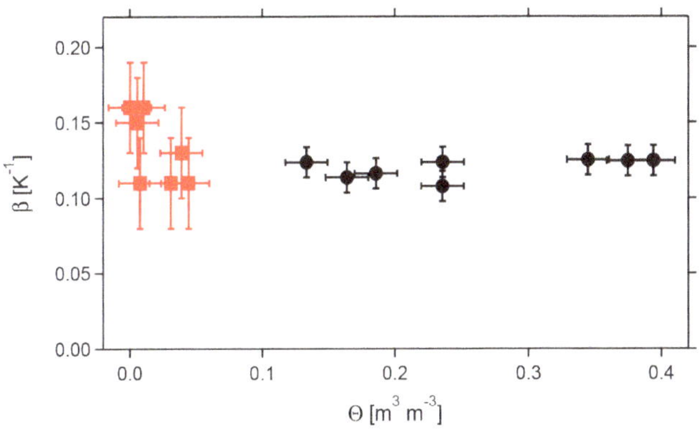

Figure 6. Temperature coefficient β for α-pinene emissions from Holm oak (experiment 3) at different soil moisture. Black circles represent the data obtained without impacts of drought on MT emissions. Red squares represent the data obtained for $\Theta < 0.06\,m^3 \times m^{-3}$ after correcting for the drought-induced decrease of emissions. Errors in β were about $\pm 0.01\,K^{-1}$ for the data obtained without drought stress. For the data obtained during drought stress the errors from the normalization procedure had to be taken into account and errors in β are estimated to be $\pm 0.03\,K^{-1}$. Error in Θ is estimated to be $\pm 0.016\,m^3 \times m^{-3}$.

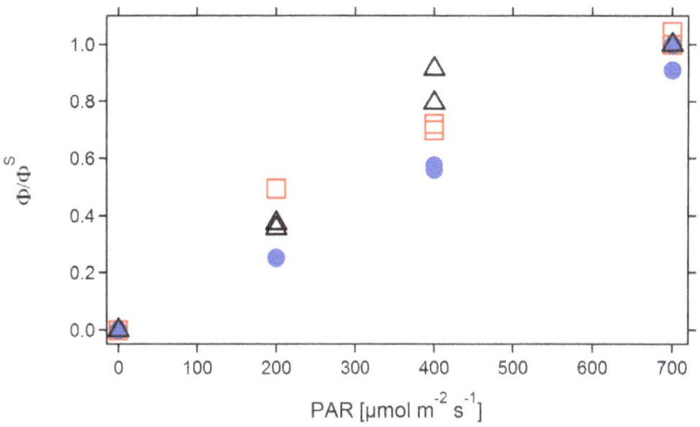

Figure 7. Emission rates corrected for the temporal decrease due to the progressing drought and normalized to the emission rates measured the respective day at PAR $= 700\,\mu mol\,m^{-2}\,s^{-1}$ as a function of PAR. Blue circles represent data taken at $\Theta \sim 0.055\,m^3 \times m^{-3}$, black triangles show data taken at $\Theta \sim 0.043\,m^3 \times m^{-3}$ and red squares show the data obtained at $\Theta \sim 0.018\,m^3 \times m^{-3}$. Experiment Holm oak 4, absolute emission rates measured at $700\,\mu mol\,m^{-2}\,s^{-1}$ dropped by a factor of about 40 with Θ falling from 0.055 to $0.018\,m^3 \times m^{-3}$.

$\sim 0.013\,m^3 \times m^{-3}$) prevented a reliable analysis. On the other hand, the assumed linear decrease of MT emissions with decreasing Θ is only an approximation because the decrease appeared to be exponential (compare Figs. 4, 5b). We therefore directly used the temporal decay of the emissions during the 2 days when temperature and PAR were held constant to develop the drought correction considering that the temporal decay of emissions reflects the decay in Θ (compare Fig. 5a and b).

An exponential function was fitted to the temporal decay measured during the respective days yielding a decay rate of $0.04 \pm 0.002\,h^{-1}$, $R^2 = 0.85$. The correction factor was set to 1 for the time when the chamber had reached steady state conditions after twilight in the morning. Correction factors were then calculated for each time when the chromatograms were taken by using the decay rate of $0.04\,h^{-1}$. The emission rates determined for the respective times were then divided by the correction factor which dropped from 1 in the morning to ~ 0.6 in the evening. Data corrected for the drought-induced decrease were used to determine the temperature coefficient β during periods with low soil moisture.

Figure 6 shows the values obtained for β in dependence of Θ. As the decay rate of 0.04 per hour was not applicable for Θ near to Θ_1, data points near to Θ_1 ($0.06 < \Theta < 0.12\,m^3 \times m^{-3}$) were discarded. Data analysis resulted in $\beta = 0.13 \pm 0.024\,K^{-1}$ for the data points determined for $\Theta < 0.06\,m^3\,m^{-3}$, which is about the same as $\beta = 0.12\,K^{-1}$, determined for stress-free conditions within the error limits. Thus, no significant differences were found between the temperature dependency of emissions under drought-stress and under stress-free conditions.

In the fourth experiment with Holm oak PAR was changed systematically during the progressing drought (Table 1, experiment Holm oak 4) to allow for determining the dependence of the emissions on Θ at different PAR and the dependence of emissions on PAR at progressing drought.

The correction was made in the same way as described above by fitting an exponential function to the temporal decay measured during the days at constant PAR and temperature. The emission rates measured the next day under conditions of variable PAR were then corrected by dividing them by the correction factors. The data already corrected for their Θ dependence were normalized using the emission rates measured at the respective day at PAR $= 700\,\mu mol\,m^{-2}\,s^{-1}$. Figure 7 shows the normalized emission as a function of light intensity at the example of 3 days. Within the precision of the data the light intensity dependence did not substantially or systematically change with progressing drought.

3.2.3 Scots pine and Norway spruce

The experiments with pine and spruce lasted for about a month each and each of the plants was exposed to 3 drought periods. At least one of these drought applications was severe with $\Theta < 0.01\,m^3 \times m^{-3}$.

Main emissions from both conifers were pool MT emissions such as α-pinene and Δ-3-carene. Besides such pool emissions the plants also emitted 1,8-cineole which is a de novo emission (Tarvainen et al., 2005; Kleist et al., 2012). This was also tested for both individuals used in our experiments by exposing them to $^{13}CO_2$. After 3 hours of $^{13}CO_2$ exposure 1,8-cineole was strongly labelled indicating that 1,8-cineole was a de novo emission with negligible contribution of pool emissions. We here focus on the Θ dependence of 1,8-cineole in order to compare with results obtained for the de novo MT emissions from the deciduous species.

Emissions of 1,8-cineole from both species behaved similarly to the de novo MT emissions from the broadleaf species (compare Fig. S2 in the Supplement). When Θ dropped from $0.35\,m^3 \times m^{-3}$ to the respective thresholds Θ_1, the emissions slightly increased ($\sim 30\,\%$ for spruce and $\sim 50\%$ for pine). Below Θ_1 the 1,8-cineole emissions decreased and approached nearly zero (compare Figs. S3 and S4). After re-watering, MT emissions recovered on a time scale of days and reached levels similar to those before the drought. All in all, the de novo emissions of pine and spruce behaved the same as the de novo MT emissions from the broadleaf species. Data of dependency of the 1,8-cineole emissions on Θ are listed in Table 2 together with the data obtained in the other experiments.

4 Discussion

4.1 Comparison to literature data

4.1.1 Mild drought stress

All de novo MT emissions from all plants investigated in our experiments were dependent on soil moisture. While mild drought ($\Theta > \Theta_1$ with $\Theta_1 = \Theta_0 + \Delta\Theta_1$, see Table 2) caused slight increases of MT emissions, severe drought ($\Theta < \Theta_1$) caused decreasing MT emissions from all investigated species.

Increasing emissions under mild drought have been reported before (Bertin and Staudt, 1996; Blanch et al., 2007; Ormeño et al., 2007). Blanch et al. (2007) studied MT emissions from Holm oak (*Quercus ilex*) and Aleppo pine (*Pinus halepensis*). They report emissions to increase twofold for Holm oak when the relative water content of the leaves decreases from ~ 90 to $80\,\%$. Ormeño et al. (2007) studied the impact of water deficit on emissions from Rosemary (*Rosmarinus officinalis*), Aleppo pine (*Pinus halepensis*), Rock rose (*Cistus albidus*), and Kermes oak (*Quercus coccifera*). Using the plant water potential as a reference, they report two- to three-fold increases of MT emissions when the water potential drops from -2 to -8 MPa.

Although these previous studies agree with our finding that MT emissions increase under mild drought, a quantitative comparison is impossible because different reference quantities are used to characterize the degree of drought stress. We assume that most of the increases in MT emissions observed during our studies were only an indirect effect of drought. In our studies leaf temperatures were reasonably well measured for the broadleaf species. Leaf temperatures increased when Θ dropped from high values ($\Theta > 0.3\,m^3 \times m^{-3}$) to $\sim 0.1\,m^3 \times m^{-3}$ even though chamber temperatures were kept constant. Obviously, the decline in transpiration lowered its cooling effect leading to higher leaf temperatures (Fig. 1). In the experiment with European beech and in three of the experiments with Holm oak (experiments 2, 3, and 4) MT emissions increased by less than $50\,\%$ when leaf temperature increased due to the progressing drought. Applying $\beta \sim 0.12\,K^{-1}$, the main fraction of increases in MT emissions from these plants was explained by the increase of leaf temperature. Only for one single Holm oak we found an increase ($\sim$ threefold, see Fig. 5b, red circles) too high to be explained by increases in leaf temperature alone.

Due to the watering–drought–watering procedure used here it is possible that parts of the increases arise from recovery from a preceding drought. In particular when Θ decreases rapidly, the effects of sequential droughts may overlap. However, the effect of strongly increasing MT emissions with Θ dropping from high levels to Θ_1 was observed in one experiment only and without further information we can only speculate on this effect.

The reason for increases of the 1,8-cineole emissions from conifers during mild drought stress (~ 30–$50\,\%$ when Θ dropped from 0.35 to $0.1\,m^3\,m^{-3}$) also remains unproven because leaf temperatures were not reliably measurable for both conifers. Nevertheless it is reasonable to assume that also in conifer species needle temperature increase due to lower transpirational cooling caused the observed slight increases in MT emissions.

For modelling of isoprene emissions in MEGAN, the impact of leaf temperature is implemented by a general temperature dependence of the emissions. Such parameterization cannot differentiate between variations of leaf temperature by variations in transpiration or by variations of air temperature. Therefore the indirect effect of drought does not appear in the factor for the dependence of isoprene emissions on soil moisture (Guenther et al., 2006). In analogy we suggest, to consider decreasing emissions caused by severe drought stress also for de novo MT emissions.

4.1.2 Severe drought stress

Similar to previous studies (e.g. Bertin and Staudt, 1996; Llusià and Peñuelas, 1998; Plaza et al., 2005; Lavoir et al., 2009; Šimpraga et al., 2011; Bourtsoukidis et al., 2014) we observed substantially reduced MT emissions during periods of severe drought as well as increasing emissions after re-watering (Peñuelas et al., 2009). Llusià and Peñuelas (1998) found strong suppression of MT emissions from non-storing species such as Holm oak and a significant relation between net assimilation and MT emissions. Lavoir et al. (2009) found exponentially decreasing emissions when the leaf-water potential fell below -2 MPa. Šimpraga et al. (2011) show strong attenuation of MT emissions from European beech with progressing drought. The temporal shape of sabinene emissions from Beech as shown here in Fig. 3 for the second severe drought was nearly identical to the

Table 2. Data from fits of emission rates vs. Θ. EB = European beech, HO = Holm oak, NS = Norway spruce, SP = Scots pine. Numbers behind species indicate experiment number according to the succession of the experiments described in Sect. 3. Data behind the slash give chamber temperatures in °C and PAR in μmol m^{-2} s^{-1}, respectively. Φ^{th} (MT) and Φ^{th} (ΣMT) are the emission rates measured for the MT listed in the second column and for the sum of all MT emissions, respectively as measured near to Θ_1. Θ_1 is the volumetric water content of the soil when emissions start to decrease ($\Theta_1 = \Delta\Theta_1 + \Theta_0$). Θ_0 = intercept of linear regression analysis when MT emissions are extrapolated to be zero. $\Delta\Theta_1$ is the range of Θ in which the emissions drop from their maximum to zero.

	MT	Φ^{th} (MT) (nmol m^{-2} s^{-1})	Φ^{th} (ΣMT) (nmol m^{-2} s^{-1})	$\Delta\Theta_1$ (m^3 m^{-3})	Θ_0 (m^3 m^{-3})
EB	Sabinene	0.9 ± 0.03	2.6 ± 0.04	0.09 ± 0.004	0.03 ± 0.002
HO1	α-pinene	4.9 ± 0.45	10.2 ± 0.47	0.044 ± 0.004	0.058 ± 0.002
HO2	α-pinene	2.1 ± 0.05	9.2 ± 0.25	0.063 ± 0.003	0.071 ± 0.002
NS	1,8-cin.	$(3.3 \pm 1.1) \times 10^{-4}$	*	0.19 ± 0.01	0.02 ± 0.007
SP	1,8-cin.	0.88 ± 0.15	*	0.068 ± 0.005	0.041 ± 0.004
HO3/15	α-pinene	3.5 ± 0.08	7.9 ± 0.74	0.055 ± 0.01	-0.006 ± 0.007
HO3/20	α-pinene	4.8 ± 0.22	11.7 ± 0.6	0.058 ± 0.008	-0.01 ± 0.004
HO3/25	α-pinene	6.5 ± 0.34	15.9 ± 0.8	0.079 ± 0.016	-0.021 ± 0.008
HO4/700	α-pinene	1.6 ± 0.18	6.7 ± 0.8	0.065 ± 0.008	0.016 ± 0.006
HO4/400	α-pinene	0.7 ± 0.1	2.6 ± 0.4	0.045 ± 0.015	0.018 ± 0.012
HO4/200	α-pinene	0.3 ± 0.04	1.2 ± 0.17	0.044 ± 0.015	0.019 ± 0.012

* For the conifers, pure de novo emissions are of minor importance for the total emissions (Φ^{th} (ΣMT) = 0.18 nmol m^{-2} s^{-1} for spruce and Φ^{th} (ΣMT) = 6.2 nmol m^{-2} s^{-1} for pine, both for the well watered individuals). Note that the formalism given here for the de novo emissions of European beech and Holm oak can be applied to individual MT as well as to Φ^{th} (ΣMT) whereas the formalism is only applicable to 1,8-cineole emissions from both conifers and not to the sum of MT emissions.

temporal shape observed by Šimpraga et al. (2011) indicating that the same general effect was found. But again, direct comparison of the results is not easy because different reference quantities are used to characterize the degree of drought.

Bourtsoukidis et al. (2014) use soil water availability to characterize the degree of drought. They measured BVOC emissions from Pedunculate oak (*Quercus robur*) and Black cherry (*Prunus serotina*). Compared to the MT emissions from regularly watered plants (soil water availability ~ 30 %) the MT emissions from plants under drought (soil water availability ~ 2.3 %) were reduced by about 65 %. This is similar to the reductions found when taking the relative water content of the soil as a reference instead of Θ.

As shown in Fig. 2, cross correlations including data obtained under severe drought led to excellent coefficients of determination implying that severe drought acts in the same manner on all MT. This can only be explained by two possibilities. Either drought suppresses all MT synthase activities in an identical manner, or affects a common precursor of all MT. While the former explanation is unlikely the latter is consistent to the findings of Grote et al. (2010) who show that MT synthase activities are unaffected by drought. The most probable explanation for the identical response of all MT emissions to the drought therefore is an impact on a common precursor.

4.2 Present state of modelling with MEGAN

While effects of soil moisture on de novo MT emissions are not yet considered in the Model of Emissions of Gases and Aerosols from Nature (MEGAN, Guenther et al., 2006, 2012) the effects on isoprene emissions are considered. As noted above, isoprene emissions are unaffected as long as Θ is above a threshold Θ_1, they decrease linearly between Θ_1 and the wilting point Θ_W and they remain zero as long as $\Theta < \Theta_W$. We found a similar behaviour for the de novo MT emissions indicating that the emission of MT on soil moisture can be modelled using the same formalism with the same basic modelling parameters Θ_W and $\Delta\Theta_1$.

The wilting point Θ_W is the soil moisture below which the plants cannot take up water. During our measurements we always observed a recovery of the plants from the drought after re-watering. Thus soil moisture during drought in our experiments must have been always slightly above a critical value causing permanent damage to the plant. Insofar Θ_0 listed in Table 1 may not directly reflect Θ_W. However, Θ_W depends on the soil and Θ_W varies by up to 2 orders of magnitude (Chen and Dudhia, 2001). For modelling, Θ_W data from Chen and Dudhia (2001) are used and such data are listed for different soil types. The data we provide for Θ_0 are therefore not needed for modelling purposes and possible differences between Θ_0 and Θ_W are not relevant for modelling.

Basic parameter for modelling is $\Delta\Theta_1$. Our average of $\Delta\Theta_1$ (average = 0.08 ± 0.05 m^3 × m^{-3}, error = 1σ standard deviation, statistical weight the same for individual plants) is similar to $\Delta\Theta_1$ used to model isoprene emissions ($\Delta\Theta_1 = 0.06$ m^3 × m^{-3}). Our data for $\Delta\Theta_1$ show substantial variability and it is unknown so far whether the high variability is caused by a different behaviour of individuals or by other reasons. Within the uncertainty of the data it was also impossible to find differences between Holm oak and the other species. No conclusions can thus be drawn whether or not more drought tolerant plants such as Holm oak show different behaviour than more drought sensitive plants such as European beech. However, all plants behaved similarly and we therefore suggest that using the average $\Delta\Theta_1$ for modelling purposes is a tenable approach.

Considering the impact of soil moisture in MEGAN, results in a 7 % reduction of global isoprene emissions only but substantial impacts of drought are expected on a regional scale (Guenther et al., 2006). In contrast Acosta Navarro et al. (2014) find significant depletion of global isoprene emissions by 24 % when considering drought impacts.

Including the impacts of soil moisture on de novo MT emissions in modelling may have different effects because isoprene emitters and de novo MT emitters are regionally differently distributed. Recent studies show that, particularly in Europe, the fraction of MT emitting species is larger than previously predicted (Oderbolz et al., 2013; Kemper Pacheco et al., 2014). The effects described here may therefore have large impacts when modelling MT emissions in Europe.

Impacts of drought on MT emissions in Europe have been modelled by Lavoir et al. (2011). The authors performed model calculations to determine the impact of soil moisture on MT emissions from Holm oak in southern France. As input for their model they used data from Lavoir et al. (2009) who determined the soil moisture dependence of MT emissions from Holm oak using the leaf-water potential as the reference quantity. For modelling they used the relative water content of the soil (RWC) as reference considering a relationship between leaf-water potential and RWC. Lavoir et al. (2011) give a threshold of $0.7\,\text{kg} \times \text{kg}^{-1}$ above which MT emissions from Holm oak are independent of soil moisture and a linear decrease to zero at $\text{RWC} = 0.4\,\text{kg} \times \text{kg}^{-1}$.

Using our data for Holm oak and applying our conversion factor (Eq. 7) we obtain $\text{RWC} = 0.24 \pm 0.1\,\text{kg} \times \text{kg}^{-1}$ as the threshold. The emissions are zero at $\text{RWC} \leq 0.09 \pm 0.1\,\text{kg} \times \text{kg}^{-1}$. Both our values are far lower than the data used by Lavoir et al. (2011) for up-scaling. Such differences are explainable by different structural properties of soils stressing the importance for regional models to consider soil properties for a more realistic emission estimate. However, the differences also show that the impacts of drought on regional MT emissions may be less pronounced than modelled by Lavoir et al. (2011).

4.3 Use of Θ as the reference quantity for modelling

We aimed at providing a data set to support modelling of soil moisture effects on de novo MT emissions with MEGAN. Therefore we used Θ as the reference quantity and we established an empirical relationship between de novo MT emissions and Θ. Nevertheless, when reflecting mechanisms of the drought impact, plant physiological processes must be considered. For isoprene the decreasing emissions as result of drought are ascribed to a general decrease of the plant's performance (e. g. Brüggemann and Schnitzler, 2002). Decreased plant performance causes reduction of isoprene biosynthesis and thus isoprene emissions. Basic parts of isoprene and MT biosynthesis pathways are identical suggesting that general mechanisms causing the reduction of isoprene and MT emissions are similar. We therefore suggest that, similar to isoprene emissions, also the decrease of de novo MT emissions is caused by a general decrease of plant performance.

As plant performance is coupled to the rate of transpiration and to net assimilation either of them may be regarded as the reference quantity. But, as already pointed out by Peñuelas et al. (2009) and by Loreto and Schnitzler (2010), drought-induced responses in transpiration or net assimilation differ from responses in MT emissions. This general behaviour was also found here in all experiments and we therefore could not use either of them as a reference quantity.

We used Θ as reference, but, from a mechanistic point of view Θ is not a direct reference quantity. In a first step, soil moisture impacts the plant performance and in a second step the plant performance affects MT synthesis and emissions. The relationship found here between de novo MT emissions and Θ therefore has to be scrutinized. In particular the temporal behaviour of MT emissions may be skewed by the time needed for the plant to respond to changes in Θ.

Indeed there were time periods with substantial time lags between changes of Θ and the plants responses in MT emissions. Re-watering caused Θ to increase to optimum conditions for the plants on time scales of hours. For the de novo MT emissions, it took multiple days until the same level was reached as before the drought. We assume that the time needed for the plants to recover from previous droughts was the reason for the decoupling of Θ and MT emissions during such time periods.

Quick changes of Θ also appear in nature in case of strong rainfall. If such rainfall appears after a severe drought period with Θ being far below the threshold affecting emissions, Θ and emissions are most probably also decoupled in nature impeding a correct description of the emission temporal behaviour during such periods.

Whereas the increase of MT emissions during recovery after severe drought cannot be described by the formalism given here, impacts of soil moisture during desiccation can be described. Desiccation appears on longer time scales than changes of soil moisture after heavy rain fall. For such longer time scales, the lag between changes of Θ and the plants' response should not carry too much weight in skewing the relationship between emissions and Θ. This assumption was confirmed by the result from the two experiments with Holm oak at different temporal progression of drought. Although the temporal decreases appeared on different time scales the Θ dependence of MT emissions remained similar (compare Figs. 5a and b). As long as desiccation appeared on a time scale of days, the plant responses to changes in Θ were obviously quick enough to allow finding the relationship between Θ and de novo MT emissions. Although Θ is only an indirect reference quantity for plant performance, it proved to be suitable for considering the impacts of soil moisture on de novo MT emissions in modelling.

We tested whether or not semi mechanistic models can be used to describe the impacts of soil moisture on MT emissions. In a first step we looked at the increase of MT emissions during recovery. Niinemets et al. (2002) couple de novo MT emissions to photosynthetic electron transport. They use the fraction of the photosynthetic electron transport necessary for MT synthesis (ε) as a surrogate for standard emissions. By keeping ε constant they closely couple isoprenoid emissions to photosynthesis. We tested this approach using the data from beech during a period of mild drought and re-watering when stomatal conductance was still reliably measurable (Fig. 3, days 0 to 11). From our data obtained during mild stress and recovery ε was calculated as described in Niinemets et al. (2002). ε was found to be constant during recovery, but increased for when $\Theta < 0.2\,m^3\,m^{-3}$, i.e. when photosynthesis already dropped but emissions were still not affected by the drought.

Constant ε during recovery indicates a close coupling between photosynthesis and MT emissions during this period. Contrary, the decoupling of MT emissions from photosynthesis observed with Θ falling from 0.2 to 0.1 $m^3\,m^{-3}$ disturbed the relationship between ε and MT emissions.

Such decoupling was also observed for isoprene emissions and photosynthetic carbon supply and has been explained by the use of alternative carbon sources for isoprene biosynthesis (Possell and Loreto, 2013 and references cited therein). This may also be the reason for decoupling of MT emissions and photosynthesis. More improved semi mechanistic models (e.g. Morfopoulos et al., 2013, 2014; Grote et al., 2014) allow varying the fraction of electron transport used for MT synthesis. Such variation is requested by our results and indeed, the dependence of MT emissions on soil moisture as shown by Grote et al. (2014) matches our findings better than a description with fixed electron transport. Nevertheless, there are still differences between our data and the model predictions. Our data show a substantial shift with sustained MT emissions at already strongly suppressed net photosynthesis. Even improved semi-mechanistic models overestimate the impacts of drought on de novo MT emissions. In particular the later reactions of MT emissions compared to the reactions in net photosynthesis (e.g. Fig. 4) should be taken into account. In units of Θ this shift is in the range of $0.09\,m^3 \times m^{-3}$ and in units of RWC about $0.23\,kg \times kg^{-1}$.

4.4 Justification of a multiplicative approach for modelling

Modelling BVOC emissions is often performed using factorial approaches. An important requirement for the validity of such a factorial approach is a negligible interdependency of the individual factors. We investigated such possible interdependencies but no substantial effects were found.

As can be seen from Fig. 6, the temperature dependence of MT emissions from Holm oak was largely independent of the actual Θ. Vice versa, the soil moisture dependence of the emissions was independent of the actual temperature (compare data in Table 2). Also the PAR dependence of MT emissions from this species was not substantially affected by the actual soil moisture (Fig. 7) and vice versa the Θ dependence was not substantially influenced by PAR (Table 2). Compared to the overall effect of drought on de novo MT emissions, possible residual interdependencies were negligible. This indicates that a factorial approach is justified at least for Holm oak investigated with this respect. From the similarity of basic processes leading to de novo MT emissions we postulate that such interdependencies are also negligible for other plant species.

5 Conclusions

All de novo MT emissions from all plants investigated in our experiments clearly depended on soil moisture. The investigated species are representative for European climate zones and they all are strong MT emitters. We therefore conclude that impacts of drought have to be re-considered for modelling.

To the best of our knowledge, the direct impacts of soil moisture on constitutive de novo MT emissions are not considered in MEGAN. Severe drought reduces the emissions and therefore models neglecting impacts of soil moisture overestimate MT emissions. Depending on the climatology in the respective regions and depending on the fraction of de novo emitted MT, modelled MT emissions may drop substantially when considering soil moisture as variable influencing de novo MT emissions.

On the other hand, to the best of our knowledge semi mechanistic models overpredict the impacts of drought on de novo MT emissions. Again, depending on the climatology of the respective regions and the fraction of de novo emitters in the respective region, modelled de novo MT emissions may increase substantially. Large differences of modelling results for MT emissions (Arneth et al., 2008) may therefore be reduced to some extent.

Although the detailed mechanisms causing the reductions in de novo MT emissions are still unknown, one statement can be made. Estimations of future alterations of MT emissions must consider such effects because climate change may induce more and longer lasting drought periods (Dai, 2013). Future long lasting drought periods will suppress constitutive de novo MT emissions. In combination with more intensive heat periods that negatively affect de novo MT emissions (Kleist et al., 2012), these abiotic stresses will have substantial impacts on regional and global BVOC emissions.

Acknowledgements. The authors would like to acknowledge financial support by the integrated EU project ECLAIRE (Contract no. 282910).

Edited by: T. Keenan

References

Acosta Navarro, J. C., Smolander, S., Struthers, H., Zorita, E., Ekman, A. M. L., Kaplan, J. O., Guenther, A., Arneth, A., and Riipinen, I., Global emissions of terpenoid VOCs from terrestrial vegetation in the last millennium, J. Geophys. Res. Atmos., 119, 6867–6885, 2014.

Arneth, A., Monson, R. K., Schurgers, G., Niinemets, Ü., and Palmer, P. I.: Why are estimates of global terrestrial isoprene emissions so similar (and why is this not so for monoterpenes)?, Atmos. Chem. Phys., 8, 4605–4620, doi:10.5194/acp-8-4605-2008, 2008.

Bertin, N. and Staudt, M.: Effect of water stress on monoterpene emissions from young potted holm oak (*Quercus ilex* L.) trees, Oecologia, 107, 456–462, 1996.

Blanch, J.-S., Peñuelas, J., and Llusià, J.: Sensitivity of terpene emissions to drought and fertilization in terpene-storing *Pinus halepensis* and non-storing *Quercus ilex*, Physiol. Plant., 131, 211–225, 2007.

Bourtsoukidis, E., Kawaletz, H., Radacki, D., Schütz, S., Hakola, H., Hellén, H., Noe, S., Mölder, I., Ammer, C., and Bonn, B.: Impact of flooding and drought conditions on the emission of volatile organic compounds of *Quercus robur* and *Prunus serotina*, Trees, 28, 93–204, 2014.

Brüggemann, N. and Schnitzler, J. P.: Comparison of isoprene emission, intercellular isoprene concentration and photosynthetic performance in water-limited oak (*Quercus pubescens* Willd. and *Quercus robur* L.) saplings, Plant Biol., 4, 456–463, 2002.

Chameides, W. L., Lindsay, R. W., Richardson, J., and Kiang, C. S.: The role of biogenic hydrocarbons in urban photochemical smog: Atlanta as a case study, Science, 241, 1473–1475, 1988.

Chen, F. and Dudhia, J.: Coupling an Advanced Land Surface – Hydrology Model with the Penn State – NCAR MM5 Modeling System. Part I: Model Implementation and Sensitivity, Monthly Weather Rev., 129, 569–585, 2001.

Dai, A.: Increasing drought under global warming in observations and models, Nat. Clim. Change, 3, 52–58, 2013.

Derwent, R. G., Jenkin, M. E., and Saunders, S. M.: Photochemical ozone creation potentials for a large number of reactive hydrocarbons under European conditions, Atmos. Environ., 30, 181–199, 1996.

Grote, R., Keenan, T., Lavoir, A.-V., and Staudt, M.: Process-based simulation of seasonality and drought stress in monoterpene emission models, Biogeosciences, 7, 257–274, doi:10.5194/bg-7-257-2010, 2010.

Grote, R., Morfopoulos, C., Niinemets, Ü., Sun, Z., Keenan, T. F., Pacifico, F., and Butler T.: A fully integrated isoprenoid emissions model coupling emissions to photosynthetic characteristics, Plant Cell Environ., 37, 1965–1980, 2014.

Guenther, A., Hewitt, C. N., Erickson, D., Geron, C., Graedel, T., Harley, P., Klinger, L., Lerdau, M., McKay, W. A., Pierce, T., Scholes, B., Steinbrecher, R., Tallamraju, R., Taylor J., and Zimmerman, P.: A global model of natural volatile organic compound emissions, J. Geophys. Res., 100, 8873–8892, 1995.

Guenther, A. B., Zimmerman, P. R., Harley, P. C., Monson, R. K., and Fall, R.: Isoprene and monoterpene emission rate variability: Model Evaluation and sensitivity analyses, J. Geophys. Res., 98, 12609–12617, 1993.

Guenther, A. B., Karl, T., Harley, P., Wiedinmyer, C., Palmer, P. I., and Geron, C.: Estimates of global terrestrial isoprene emissions using MEGAN (Model of Emissions of Gases and Aerosols from Nature), Atmos. Chem. Phys., 6, 3181–3210, doi:10.5194/acp-6-3181-2006, 2006.

Guenther, A. B., Jiang, X., Heald, C. L., Sakulyanontvittaya, T., Duhl, T., Emmons, L. K., and Wang, X.: The Model of Emissions of Gases and Aerosols from Nature version 2.1 (MEGAN2.1): an extended and updated framework for modeling biogenic emissions, Geosci. Model Dev., 5, 1471–1492, 2012.

Heiden, A. C., Kobel, K., Langebartels, C., Schuh-Thomas, G., and Wildt, J.: Emissions of oxygenated compounds from plants, Part I: Emissions from Lipoxygenase activity, J. Atmos. Chem., 45, 143–172, 2003.

Jacob, D. J. and Wofsy, S. C.: Photochemistry of biogenic emissions over the Amazon forest, J. Geophys. Res., 93, 1477–1486, 1988.

Kemper Pacheco, C., Fares, S., and Ciccioli, P.: A highly spatially resolved GIS-based model to assess the isoprenoid emissions from key Italian ecosystems, Atmos. Environ., 96, 50–60, 2014.

Kleist, E., Mentel, T. F., Andres, S., Bohne, A., Folkers, A., Kiendler-Scharr, A., Rudich, Y., Springer, M., Tillmann, R., and Wildt, J.: Irreversible impacts of heat on the emissions of monoterpenes, sesquiterpenes, phenolic BVOC and green leaf volatiles from several tree species, Biogeosciences, 9, 5111–5123, doi:10.5194/bg-9-5111-2012, 2012.

Kulmala, M., Vehkamäki, H., Petäjä, T., Dal Maso, M., Lauri, A., Kerminen, V. N., Birmili, W., and McMurry, P. H.: Formation and growth rates of ultrafine atmospheric particles: a review of observations, J. Aerosol Science, 35, 143–176, 2004.

Lavoir, A.-V., Staudt, M., Schnitzler, J. P., Landais, D., Massol, F., Rocheteau, A., Rodriguez, R., Zimmer, I., and Rambal, S.: Drought reduced monoterpene emissions from the evergreen Mediterranean oak Quercus ilex: results from a throughfall displacement experiment, Biogeosciences, 6, 1167–1180, doi:10.5194/bg-6-1167-2009, 2009

Lavoir, A. V., Duffet C., Mouillot F., Rambal S., Ratte J. P., Schnitzler J. P., and Staudt M.: Scaling-up leaf monoterpene emissions from a water limited *Quercus ilex* woodland, Atmos. Environ., 45, 2888–2897, 2011.

Llusià, J. and Peñuelas J.: Changes in terpene content and emission in potted mediterranean woody plants under severe drought, Can. J. Botany, 76, 1366–1373, 1998.

Loreto, F. and Schnitzler, J.P.: Abiotic stresses and induced BVOCs, Trends Plant Sci., 15, 154–166, 2010.

Mentel, Th. F., Wildt, J., Kiendler-Scharr, A., Kleist, E., Tillmann, R., Dal Maso, M., Fisseha, R., Hohaus, Th., Spahn, H., Uerlings, R., Wegener, R., Griffiths, P., Dinar, E., Rudich, Y., and Wahner, A.: Photochemical production of aerosols from real plant emissions, Atmos. Chem. Phys., 9, 4387-4406, 2009, http://www.atmos-chem-phys.net/9/4387/2009/.

Mentel, Th. F., Kleist, E., Andres, S., Dal Maso, M., Hohaus, T., Kiendler-Scharr, A., Rudich, Y., Springer, M., Tillmann, R., Uerlings, R., Wahner, A., and Wildt, J.: Secondary aerosol formation from stress-induced biogenic emissions and possible climate feedbacks, Atmos. Chem. Phys., 13, 8755–8770, doi:10.5194/acp-13-8755-2013, 2013.

Morfopoulos, C., Prentice, I. C., Keenan, T. F., Friedlingstein, P., Medlyn, B. E., Peñuelas, J., and Possell, M.: A unifying conceptual model for the environmental responses of isoprene emissions from plants, Ann. Bot., 112, 1223–1238, 2013.

Morfopoulos, C., Sperlich, D., Peñuelas, J., Cubells, I. F., Llusià, J., Medlyn, B. E., Niinemets, Ü., Possell M., Sun, Z., and Prentice, I. C.: A model of plant isoprene emission based on available reducing power captures responses to atmospheric CO_2, New Phytol., 203, 125–139, 2014.

Niinemets, Ü., Seufert, G., Steinbrecher, R., and Tenhunen, J. D.: A model coupling foliar monoterpene emissions to leaf photosynthetic characteristics in Mediterranean evergreen Quercus species, New Phytol., 153, 257–275, 2002.

Oderbolz, D. C., Aksoyoglu, S., Keller, J., Barmpadimos, I., Steinbrecher, R., Skjøth, C. A., Plaß-Dülmer, C., and Prévôt, A. S. H.: A comprehensive emission inventory of biogenic volatile organic compounds in Europe: improved seasonality and landcover, Atmos. Chem. Phys., 13, 1689–1712, doi:10.5194/acp-13-1689-2013, 2013,

Ormeño, E., Mévy, J. P., Vila, B., Bousquet-Mélou, A., Greff, S., Bonin, G., and Fernandez, C.: Water deficit stress induces different monoterpene and sesquiterpene emission changes in Mediterranean species. Relationship between terpene emissions and plant water potential, Chemosphere, 67, 276–284, 2007.

Pegoraro, E., Rey, A., Bobich, E. G., Barron-Gafford, G., Grieve, A., Malhi, Y., and Murthy, R.: Effect of elevated CO_2 concentration and vapor pressure deficit on isoprene emission from leaves of *Populus deltoides* during drought, Funct. Plant Biol., 31, 1137–1147, 2004.

Peñuelas, J., Filella, I., Seco, R., and Llusià J.: Increase in isoprene and monoterpene emissions after re-watering of droughted *Quercus ilex* seedlings, Biologia plantarum, 53, 351–354, 2009.

Plaza, J., Núñez, L., Pujadas, M., Pérez-Pastor, R., Bermejo, V., García-Alonso, S., and Elvira, S.: Field monoterpene emission of Mediterranean oak (*Quercus ilex*) in the central Iberian Peninsula measured by enclosure and micrometeorological techniques: Observation of drought stress effect, J. Geophys. Res., 110, D03303, doi:10.1029/2004JD005168, 2005.

Possell, M. and Loreto, F.: The Role of Volatile Organic Compounds in Plant Resistance to Abiotic Stresses: Responses and Mechanisms, in: Biology, Controls and Models of Tree Volatile Organic Compound Emissions, edited by: Niinemets, Ü. and Monson, R. K., Springer Netherlands, 209–235, 2013.

Rambal, S., Ourcival, J. M., Joffre, R., Mouillot, F., Nouvellon, Y., Reichstein, M., and Rocheteau, A.: Drought controls over conductance and assimilation of a Mediterranean evergreen ecosystem: scaling from leaf to canopy, Glob. Change Biol., 9, 1813–1824, 2003.

Schimang, R., Folkers, A., Kleffmann, J., Kleist, E., Miebach, M., and Wildt, J.: Uptake of gaseous nitrous acid (HONO) by several plant species, Atmos. Environ., 40, 1324–1335, 2006.

Schuh, G., Heiden, A. C., Hoffmann, T., Kahl, J., Rockel, P., Rudolph, J., and Wildt, J.: Emissions of volatile organic compounds from sunflower and beech: Dependence on temperature and light intensity, J. Atmos. Chem., 27, 291–318, 1997.

Shao, M., v.Czapiewski, K., Heiden, A. C., Kobel, K., Komenda, M., Koppmann, R., and Wildt, J.: Volatile organic compound emissions from Scots pine: Mechanisms and description by algorithms, J. Geophys. Res., 106, 20483–20491, 2001.

Šimpraga M., Verbeeck, H., Demarcke, M., Joó, É., Pokorska, O., Amelynck, C., Schoon, N., Dewulf, J., Van Langenhove, H., Heinesch, B., Aubinet, M., Laffineur, Q., Müller, J.-F., and Steppe, K.: Clear link between drought stress, photosynthesis and biogenic volatile organic compounds in *Fagus sylvatica* L., Atmos. Environ., 45, 5254–5259, 2011.

Staudt, M. and Bertin, N.: Light and temperature dependence of the emission of cyclic and acyclic monoterpenes from Holm Oak (*Quercus ilex* L.) leaves, Plant Cell Environ., 21, 385–395, 1998.

Tarvainen, V., Hakola, H., Hellén, H., Bäck, J., Hari, P., and Kulmala, M.: Temperature and light dependence of the VOC emissions of Scots pine, Atmos. Chem. Phys., 5, 989–998, doi:10.5194/acp-5-989-2005, 2005.

Tingey, D. T., Manning, M., Grothaus, L. C., and Burns, W. F.: Influence of light and temperature on monoterpene emission rates from Slash Pine, Plant Physiol., 65, 797–801, 1980.

Tingey, D. T., Turner, D. P., and Weber, J. A.: Factors controlling the emissions of monoterpenes and other volatile organic compounds, in: Trace Gas Emissions by Plants, edited by: Sharkey, T. D., Holland, E. A., Mooney, H. A., Academic Press, 93–119, 1991.

Wildt, J., Kley, D., Rockel, A., Rockel, P., and Segschneider, H. J.: Emission of NO from several higher plant species, J. Geophys. Res., 102, 5919–5927, 1997.

12

Effects of land management on large trees and carbon stocks

P. E. Kauppi[1], R. A. Birdsey[2], Y. Pan[2], A. Ihalainen[3], P. Nöjd[3], and A. Lehtonen[3]

[1]Department of Environmental Sciences, P.O. Box 65, 0014, University of Helsinki, Helsinki, Finland
[2]USDA Forest Service, Northern Research Station, Newtown Square, PA, USA
[3]Natural Resources Institute Finland (Luke), P.O. Box 18, 01301 Vantaa, Finland

Correspondence to: P. E. Kauppi (pekka.kauppi@helsinki.fi)

Abstract. Large trees are important and unique organisms in forests, providing ecosystem services including carbon dioxide removal from the atmosphere and long-term storage. Some reports have raised concerns about the global decline of large trees. Based on observations from two regions in Finland and three regions in the United States we report that trends of large trees during recent decades have been surprisingly variable among regions. In southern Finland, the growing stock volume of trees larger than 30 cm at breast height increased nearly five-fold during the second half of the 20th century, yet more recently ceased to expand. In the United States, large hardwood trees have become increasingly common in the Northeast since the 1950s, while large softwood trees declined until the mid 1990s as a consequence of harvests in the Pacific region, and then rebounded when harvesting there was reduced. We conclude that in the regions studied, the history of land use and forest management governs changes of the diameter-class distributions of tree populations. Large trees have significant benefits; for example, they can constitute a large proportion of the carbon stock and affect greatly the carbon density of forests. Large trees usually have deeper roots and long lifetimes. They affect forest structure and function and provide habitats for other species. An accumulating stock of large trees in existing forests may have negligible direct biophysical effects on climate through transpiration or forest albedo. Understanding changes in the demography of tree populations makes a contribution to estimating the past impact and future potential of forests in the global carbon budget and to assessing other ecosystem services of forests.

1 Introduction

Carbon, which is removed from the atmosphere by forest ecosystem processes, is stored both in vegetation and soils (e.g., McGuire et al., 2001). If carbon stocks of ecosystems build up, the carbon content of the atmosphere is reduced accordingly. Conversely, if the rate of carbon accumulation by ecosystems were to diminish, the rate of increase of carbon dioxide in the atmosphere would be much faster than currently observed (Reich, 2011). The large potential of trees for either removing CO_2 from the atmosphere or adding it was discovered in early research about forests, the carbon cycle, and climate (Dyson, 1976; Brown and Lugo, 1982; Cooper, 1983; Woodwell et al., 1983). More recently, research has highlighted other mechanisms exploring how forest canopies affect the radiative forcing of the atmosphere by modifying the albedo (Betts, 2000; Lukeš et al., 2014) and evapotranspiration (Swann et al., 2010).

Global forests are extremely diverse and provide a variety of ecosystem services such as carbon sequestration, industrial raw materials, flood and landslide protection, biodiversity preservation, and aesthetic and health benefits (Pan et al., 2013). Forests are usually defined by the presence of trees and absence of non-forest land use, even though trees are also numerous outside forests in savannas, pasture lands, and in suburban areas and green city centers (Nowak and Greenfield, 2012). Large and old trees are exceptional entities in most tree populations and they have unique and special qualities beyond their climate mitigation function; however, there are concerns about human-caused losses of old and large trees of the world (Lindenmayer et al., 2012, 2013; Blicharska and Mikusinski, 2013).

While covering only about one-quarter of the global land surface, forests dominate the net removal of CO_2 from the atmosphere into land ecosystems (Pan et al., 2011a). Live vegetation, mostly trees, accounts for three-quarters of the large and persistent sink of the global forests. The remaining one-quarter is shared by changing stocks of dead wood, litter, soils and harvested wood products (Pan et al., 2011a). Trees contain carbon in stems, branches, foliage, and roots and provide carbon sources for forest soil stocks and microbes, and for export to downstream water and sediments (Richey et al., 2002).

In mix-aged forests, large trees often contribute a significant proportion of aboveground biomass and the carbon density of the site although only a few may be present (Slik et al., 2013; Lutz et al., 2012; Martínez and Alvarez, 1998). Large trees have deep roots, which transfer carbon into the lowest layers of the forest soil. As trees die, the largest individual trees decay most slowly thus maintaining for some time the carbon stocks in snags or coarse woody debris that are standing or lying in forests (Harmon and Hua, 1991; Krankina and Harmon, 1995; Pan et al., 2011a). Harvested trees may be transformed into wood-based products, which contain significant amounts of carbon and are widely used as a sustainable raw material (Skog et al., 2004). Large trees have statistically lower mortality rates compared to small sized trees within a stand (Coomes and Allen, 2007), which affects forest carbon dynamics and successional processes.

To gain knowledge of size distribution of large trees and their long-term dynamics, we usually rely on surveying the demography of tree populations at annual or multi-annual time steps using forest inventory methodology (Lawrence et al., 2010). Forest inventory measurements are not available at the global level. However, regional data and time series are available from selected countries such as Finland and the United States, which can provide multi-decadal statistics on the demography of tree populations and changes of the size distribution of trees. The evolution of forest carbon for Finland and the United States has been described more broadly in Liski et al. (2006) and Birdsey et al. (2006), respectively. The longest time series, to our knowledge, of statistically representative measurements of timber resources is from a subregion in Finland, where the fieldwork was initiated in 1912 (Kauppi et al., 2010). The first national forest inventories from Finland were carried out in the 1920s and 1930s (Ilvessalo, 1927, 1942), and the national forest inventory in the United States was begun in the 1930s (LaBau et al., 2007).

In this study, we selected five case study regions (Fig. 1). This combination of regions was selected noting that (1) a national forest inventory has been carried out in all regions over a period of time of more than 50 years; (2) observations on the growing stock have been published by size classes of trees at intervals of 10 or less years; (3) each region is large and diverse, containing hundreds of thousands of forest stands and, therefore, disturbance anomalies affecting small regions or individual stands do not distort general regional trends, and (4) forest management history differs between the regions and can be placed in the context of forest transition theory (Mather, 1992).

We focus on large trees, broadly defined as the upper end of the size distribution of live trees (greater than 30 or 33 cm in diameter at breast height (DBH). The objective of this research is to analyze the role of large trees in the evolution of the growing stock in regions within Finland and the United States representing different land management histories using data from statistically designed sample surveys. We discuss the impact of large trees on the carbon budget, albedo and evapotranspiration of forests, and the effect of land management on the stock of large trees.

2 Materials and methods

Forest inventory is based on measurements taken from a statistically representative sample of all trees within a forest region – for details of this approach, see Tomppo et al. (2011) and LaBau et al. (2007). Historical inventory data are available at 5–10 year intervals from Finland (Kuusela, 1972, 1978; Kuusela and Salminen, 1991; Tomppo et al., 2011; Ylitalo, 2013; Korhonen et al., 2013) and decadal intervals from the United States (Smith et al., 2009). We collected data from these published inventories specifically by five regions (Fig. 1). The two regions of Finland combined equal all of Finland, whereas for the United States we selected three diverse regions. We prepared time series estimates of the growing-stock volume of large trees and the distribution of growing-stock volume by tree-size classes. Growing stock-volume (in cubic meters, m^3) refers to the volume of the tree stem as defined by common merchantability standards. The historical inventories have reported estimates of growing-stock volume by tree-size class based on consistent definitions. We chose to report the distributions as they have been published in the original literature, however, converting inches to centimeters. Future research can easily update the findings of this research as new inventory cycles become completed and published.

Data from Finland referred to nine inventory cycles as follows: 1951–1953→1960–1963→1964–1970→1971–1976→1977–1984→1986–1994→1996–2003→2004–2008→2009–2012. Measurement teams travel within and across regions. During some years measurements are taken in southern but not in northern Finland, and vice versa. DBH distributions were constructed to separate cohorts of trees representing different size classes. The total stem volume (in millions of m^3) was estimated for trees within each size cohort and each region. Forests cover a land area of about 11.5 and 11.3 million hectares; in southern Finland and northern Finland, respectively.

Even though the two regions are within one country the land management history varies greatly between the two regions. The main part of rural lands in southern Finland has

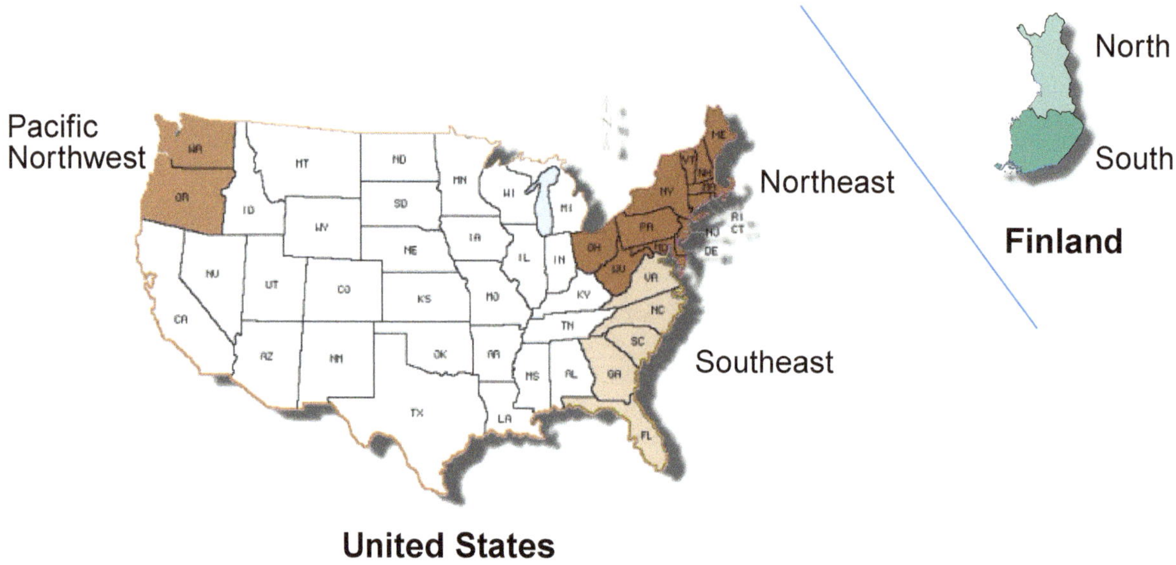

Figure 1. The location of the study regions in the United States and Finland.

been in private ownership since the middle ages. Forests in southern Finland were severely over harvested in the 19th to early 20th century. In northern Finland, the lands are largely state owned. Industrial forest harvests were common in southern Finland through the 20th century, while they rarely occurred in northern Finland before the 1950s.

The data from the United States covered the period 1953–2007, and contained five inventory cycles. Three regions within the United States were selected for more detailed analysis covering a total of 91 million hectares (Fig. 1; Table 1). For comparison, the total forest area of the United States in 2007 was 304 million hectares. The selected regions represent the diverse history of land management and impacts on large trees: the Northeast which is largely composed of forests that are re-growing on agricultural land that was abandoned over the last century; the Southeast where much of the forest land is intensively managed on short rotations for timber products; and the Pacific Northwest where old-growth forests were still being cleared and regenerated through the mid-1980s but have since been preserved.

3 Results

3.1 Results for Finland

The growing stock of Finland's forests consisted predominantly of small and medium sized trees (< 30 cm DBH). The rate of biomass accumulation was almost the same in southern Finland as in northern Finland, but there were interesting differences between the two regions in the evolution of the tree size distributions over time. The stock of large trees hardly changed in northern Finland, where the accumulation of biomass and carbon was concentrated in small trees of less than 30 cm in DBH. In contrast in southern Finland, the growing stock of large trees increased nearly five-fold from about 70 to 340 million m^3 between 1951–1953 and 2009–2012 (Fig. 2).

3.2 Results for the United States

Trees in the United States are larger on average than those in Finland. In the Pacific Northwest region which is dominated by softwood species and has the largest population of larger trees in the United States, there is a distinct pattern of change over time in the growing stock of trees greater than 33 cm compared with trees less than 33 cm. The stock declined from 3.7 billion m^3 in 1953 to 3.0 billion m^3 1987 and then nearly recovered to their prior stocking by 2007, reaching 3.4 billion m^3 (Fig. 3). In contrast, the growing stock of trees less than 33 cm increased from 1953 to 1977 and then stabilized at about 1.1 billion m^3. In the Southeast United States, the growing stock of trees greater than 33 cm doubled between 1953 and 2007, while that of trees less than 33 cm increased only until 1977 then was relatively stable (Fig. 3). These changes reflect the increasing influence of industrial plantation forestry over the period. In the Northeast United States where hardwoods predominate, the pattern is similar to the Southeast except that the growing stock of trees greater than 33 cm more than tripled between 1953 and 2007, indicating forests that are increasing in age coupled with the absence of significant harvesting or stand-replacing natural disturbances. Compared with the Southeast where growing-stock increases have leveled off, the total volume of growing stock in the Northeast has continued to increase in recent years. The Northeast has not had such a widespread conversion to intensive forest management as has happened in the Southeast.

Table 1. Statistics and forest management history of the sub regions addressed in this research. Forest area estimates for Finland include poorly productive land called scrubland (Smith et al., 2009; Ylitalo, 2013).

Region	Forest area (1000 ha)	Stem volume (million m^3)	Biomass (Tg C aboveground)	Forest management history
US Northeast	34 316	3896	2041	Most forests cleared for agriculture by 1850; current forests regrowing after agricultural abandonment
US Southeast	35 567	3589	1873	Most forests cleared for agriculture by 1850; current forests regrowing or replanted after agricultural abandonment
US Northwest	21 225	4499	1561	Most forests harvested for wood products during the 20th century; harvesting of old-growth suspended in the 1980s
Finland north	11 258	786	227	Forest land largely state owned. Forest and land management intensified after 1950
Finland south	11 501	1546	418	Most forests privately owned and recovering from severe degradation of the 19th and early 20th century

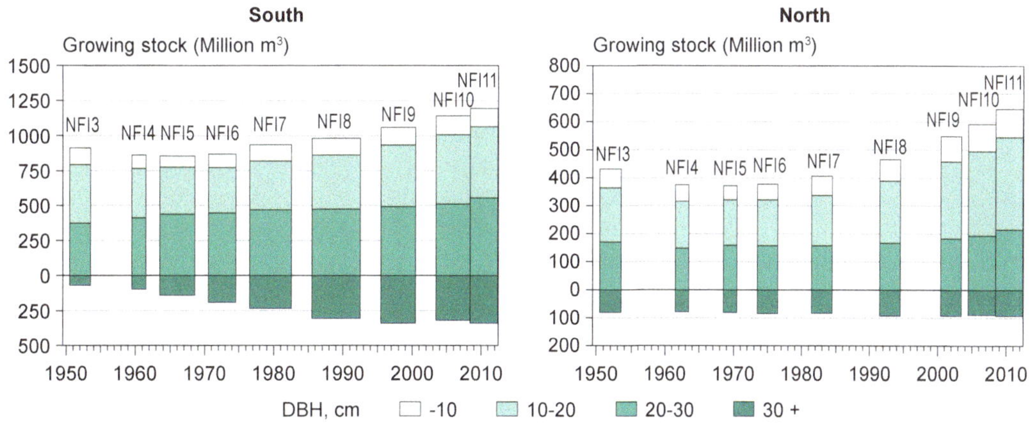

Figure 2. Dynamics of growing stock by tree size classes in regions "south" and "north" of Finland. NFI3 to NFI11 refer to the national forest inventory cycles from the 3rd to 11th cycle over the time span 1951 to 2012. The width of the bars refers to the observation years of each inventory cycle within the region. The stock of small trees is shown above the horizontal axis (DBH < 30 cm) – with the main change in north Finland beginning in the 1980s. The stock of large trees is shown in positive numbers below the horizontal axis (DBH > 30 cm) – with the main change in south Finland from 1951 to 1990.

4 Discussion

Forest biomass and the carbon stock have expanded even though the global forest area keeps decreasing (Pan et al., 2011a). Globally, forest vegetation has become increasingly dense (Rautiainen et al., 2011). This research indicates that, in Finland and the United States, carbon is sequestered in most size classes of trees, and in four of the five regions studied, the increasing volume of larger trees has made an important contribution to biomass expansion and the sequestration of carbon.

We selected our five study regions for four reasons. First, they are rare exceptions in global forests where reliable monitoring of the growing stock has occurred consistently over 6 decades based on forest inventory methodology. Second, they have been intensively managed with land use policies and practices changing over time. Third, the regions are sufficiently large and diverse for seeing the "forest from the trees", with each region containing hundreds of thousands of individual forest stands thus offering a firm basis for analyzing long-term forest trends. Fourth, they represent different biomes and climate conditions and are suitable for

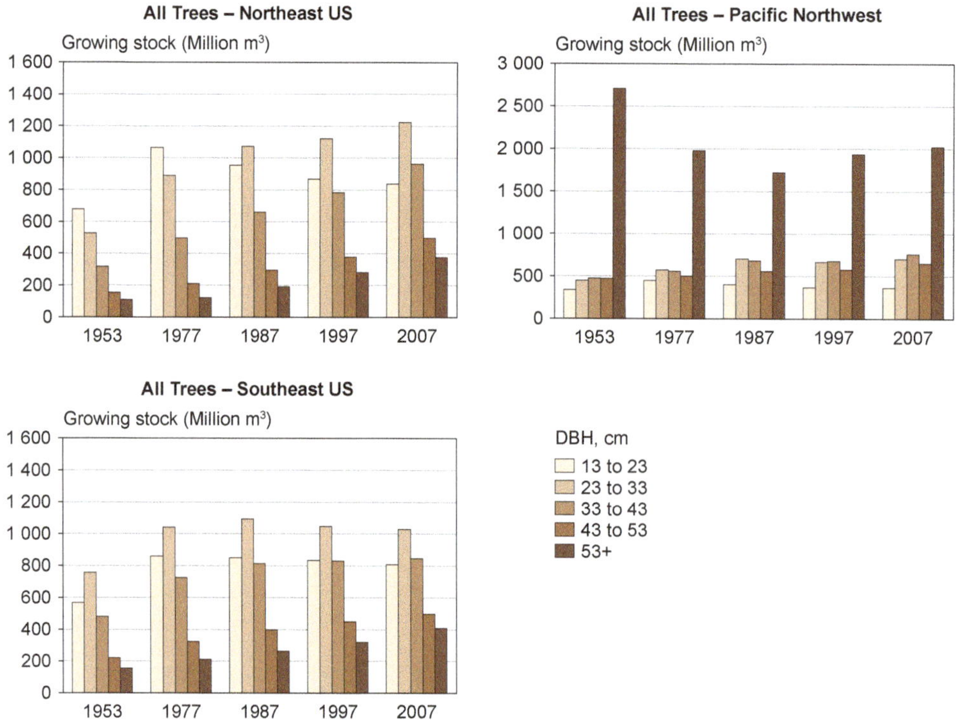

Figure 3. The distribution of growing stock by the size classes from 1953 to 2007 in regions within the United States.

discussing impacts of land management versus those of climate change or CO_2 fertilization on forest ecosystems. We used size cohort thresholds as they have been reported in the original statistical publications. This is transparent and paves the way for future studies, as the trends can be easily followed when new data become published.

In temperate zones and also in the boreal forests of Finland, the vegetation cover has been greatly altered by human activities (Pan et al., 2013; Fritzbøger and Søndergaard, 1995). The current forest demography and tree-size distributions generally reflect a disturbance legacy of land-use history (Clawson, 1979; Kauppi et al., 2006; Pan et al., 2011b). In general, the historical switch in industrial countries from subsistence agriculture to modern farming many decades ago affects contemporary forests. Forest transition has occurred from shrinking to expanding forest area (Mather, 1992). Land abandonment has no immediate impact on the populations of large trees within forests. Therefore, agricultural policy is less important than forest policy in affecting the evolution of large tree cohorts. In the United States, beginning 100 years or so ago, Northeastern forests began regrowing on abandoned agricultural land and now these forests have reached an age where large trees are becoming common once again (Pan et al., 2011b). In the Pacific Northwest, large trees were still being harvested in large quantities until about 25 years ago so their numbers were declining, but now that trend has been reversed with most remaining old-growth forest areas set aside for other purposes besides timber production.

Even though our study regions are large and diverse, they do not represent a global "average forest". For example, fire management, which are very important in many other parts of the World have been a significant issue only in parts of our study regions in the Pacific Northwest and Southeast United States. From these results we cannot extrapolate the role of large trees in the global carbon budget, because our study regions cover only 3 % of the area of global forested land as estimated by the FAO. We might try to make such an extrapolation effort, if the trends of large trees had been uniform in all regions. But we observed deviating trends between the regions and changes of biomass accumulation over time responding to land management history. All areas addressed in this study have been used intensively for industrial round wood production. This economic interest has supported the maintenance of the relatively expensive national forest inventories. Without the exceptionally persistent, accurate, and precise forest data, we could not have carried out this research.

Information on the growing stock of tree stems is directly relevant in carbon research, because the volume of growing stock is correlated with the carbon stock in forest biomass. The ratio of carbon stock/growing stock decreases with tree size; in other words, the contribution of stem biomass becomes increasingly large as trees grow in size (e.g., Lehtonen et al., 2004; Jalkanen et al., 2005; Kauppi et al., 2006). A key concept in this conversion is biomass expansion factor (BEF), which has been empirically determined for many tree species and for many regions of the world. An analysis for Finland is available in Eerikäinen (2009) and in Härkönen et al. (2011), and for the US in Smith et al. (2007). Taking Finland as an example, the biomass for each tree component is modeled by the main tree species (Repola, 2008, 2009). Thereafter, BEFs were calculated by dividing total tree biomass by stem volume within tree diameter classes (Lehtonen et al., 2004). Typically, for conifers in Finland, 1 m^3 of stem wood volume including bark corresponds to a whole tree biomass of 0.6 to 0.8 Mg dry matter. The carbon concentration of dry woody biomass is 45 to 50 % (Lehtonen et al., 2004).

Although large trees have been recognized for their special role in sequestering and storing carbon in forests (Stephenson et al., 2014), the interaction of forests with the climate system is not restricted to the exchange of carbon dioxide between the ecosystem and the atmosphere. In high latitudes as forest biomass expands in areas with low carbon density, the beneficial climate impact of carbon sequestration may be offset by the detrimental effects of decreasing albedo and increasing transpiration on radiative forcing (Betts, 2000; Bonan, 2008; Swann et al., 2010). The good news about the expansion of large trees on existing forest areas is that it favors carbon sequestration inside the surface layer of tree stems rather than adding incremental leaf biomass. The latter more directly affects canopy transpiration and albedo.

5 Conclusions

Land management has been a key driver in the change in the stocks of large trees across the regions addressed in this study, covering a total of 122 million hectares of forested land. The contribution of large tree cohorts to biomass expansion and thus carbon sequestration varied greatly in time and space. Only in north Finland was the quantitative contribution of large trees to carbon sequestration insignificant. It was important in the other four regions. For instance, in the eastern US the stock of large hardwood trees expanded and sequestered carbon effectively.

The evolution of the stock of large tree cohorts was affected by management decisions such as the harvest moratorium in the Pacific Northwest US in the early 1990s. As the patterns of large trees varied between regions and fluctuated over time they cannot be explained by changes in climate or CO_2 fertilization, which have evolved gradually over all regions. The results of this study should be interpreted with caution regarding less managed forests of the world, but similar regions exist in North America, Europe, Asia, and parts of the Southern Hemisphere where land management decisions have greatly affected the presence of large trees. Nevertheless, managing the stocks of large trees is an option for climate policy and can also be harmonized with opportunities for providing other important ecosystem services. It will be beneficial to many studies on topics such as forest carbon, forest restoration and conservation if monitoring trends of large trees can be achieved globally in the future for forested regions noting also the demography of large trees outside forests on pasture lands and in suburban areas.

Acknowledgements. The authors thank the Maj and Tor Nessling Foundation and the Finnish Society for Sciences and Letters for financial support; the national forest inventories of Finland and the US for collecting, compiling, and publishing the original data; and Sari Elomaa for finalizing the illustrations.

Edited by: N. Zeng

References

Betts, R. A.: Offset of the potential carbon sink from boreal forestation by decreases in surface albedo, Nature, 408, 187–190, 2000.

Birdsey, R., Pregitzer, K., and Lucier, A.: Forest carbon management in the United States 1600–2100, J. Environ. Qual., 35, 1461–1469, 2006.

Blicharska, M. and Mikusinski, G.: Old trees: cultural value, Science, 339, 904, 2013.

Bonan, G. B.: Forests and climate change: forcings, feedbacks, and the climate benefits of forests, Science, 320, 1444–1449, 2008.

Brown, S. and Lugo, A. E.: The storage and production of organic matter in tropical forests and their role in the global carbon cycle, Biotropica, 14, 161–187, 1982.

Clawson, M.: Forests in the long sweep of American history, Science, 204, 1168–1174, 1979.

Coomes, D. A. and Allen, R. B.: Mortality and tree-size distributions in natural mixed-age forests, J. Ecology, 95, 27–40, 2007.

Cooper, C. F.: Carbon storage in managed forests, Can. J. Forest Res., 13, 155–166, 1983.

Dyson, F. J.: Can we control carbon dioxide in the atmosphere?, Energy, 2, 287–291, 1976.

Eerikäinen, K.: A multivariate linear mixed-effects model for the generalization of sample tree heights and crown ratios in the Finnish National Forest Inventory, Forest Sci., 55, 480–493, 2009.

Fritzbøger, B. and Søndergaard, P.: A short history of forest uses, In: Multiple-Use Forestry in the Nordic Countries, edited by: Hytönen, M., METLA, Finnish Forest Research Institute, Helsinki Research Centre, 11–41, 1995.

Härkönen, S., Lehtonen, A., Eerikäinen, K., Peltoniemi, M., and Mäkelä, A.: Estimating forest carbon fluxes for large regions based on process-based modelling, NFI data and Landsat satellite images, Forest Ecol. Manag., 262, 2364–2377, 2011.

Harmon, M. E. and Hua, C.: Coarse woody debris dynamics in two old-growth ecosystems, BioScience, 41, 604–610, 1991.

Ilvessalo, Y.: Results of national forest inventory 1921–24, Communicationes Instituti Forestalis Fenniae, 11, 1–613, 1927 (in Finnish with English summary).

Ilvessalo, Y.: The forest resources and the condition of the forests of Finland: the Second National Forest Survey, Communicationes Instituti Forestalis Fenniae, 30, 1–446, 1942 (in Finnish with English summary).

Jalkanen, A., Petersson, H., Mäkipää, R., Ståhl, G., and Lehtonen, A.: Estimation of the biomass stock of trees in Sweden: comparison of biomass equations and age-dependent biomass expansion factors, Ann. Forest Sci., 62, 845–851, 2005.

Kauppi, P. E., Ausubel, J. H., Fang, J., Mather, A. S., Sedjo, R. A., and Waggoner, P. E.: Returning forests analyzed with the forest identity, P. Natl. Acad. Sci. USA, 103, 17574–17579, 2006.

Kauppi, P. E., Rautiainen, A., Korhonen, K. T., Lehtonen, A., Liski, J., Nöjd, P., and Virtanen, T.: Changing stock of biomass carbon in a boreal forest over 93 years, Forest Ecol. Manag., 259, 1239–1244, 2010.

Korhonen, K. T., Ihalainen, A., Viiri, H., Heikkinen, J., Henttonen, H. M., Hotanen, J.-P., Mäkelä, H., Nevalainen, S., and Pitkänen, J.: Suomen metsät 2004–2008 ja niiden kehitys 1921–2008, Metsätieteen aikakauskirja, 3, 269–608, 2013 (in Finnish).

Krankina, O. N. and Harmon, M. E.: Dynamics of the dead wood carbon pool in northwestern Russian boreal forests, in: Boreal Forests and Global Change, Springer Netherlands, 227–238, 1995.

Kuusela, K.: Forest resources and ownership in Finland 1964–70 and their development 1920–70, Communicationes Instituti Forestalis Fenniae, 76, 1–126, 1972 (in Finnish with English summary).

Kuusela, K.: Forest resources and ownership in Finland 1971–1976, Commun. Inst. For. Fenn., 93, 1–107, 1978 (in Finnish with English summary).

Kuusela, K. and Salminen, S.: Forest resources of Finland in 1977–1984 and their development 1952–1980, Acta Forestalia Fennica, 220, 1–84, 1991 (in Finnish with English summary).

LaBau, V. J., Bones, J. T., Kingsley, N. P., Lund, H. G., and Smith, W. B.: A history of the forest 5 survey in the United States: 1830–2004, FS-877, US Department of Agriculture, Forest Service, Washington DC, 82 pp., 2007.

Lawrence, M., McRoberts, R. E., Tomppo, E., Gschwantner, T., and Gabler, K.: Comparisons of national forest inventories, in: National Forest Inventories, Springer Netherlands, 19–32, 2010.

Lehtonen, A., Mäkipää, R., Heikkinen, J., Sievänen, R., and Liski, J.: New approach to formulate biomass expansion factors (BEF) by stand age for Scots pine, Norway spruce and birch, Forest Ecol. Manage, 188, 211–224, 2004.

Lindenmayer, D. B., Laurance, W. F., and Franklin, J. F.: Global decline in large old trees, Science, 338, 1305–1306, 2012.

Lindenmayer, D. B., Laurance, W. F., Franklin, J. F., Likens, G. E., Banks, S. C., Blanchard, W., Gibbons, P., Ikin, K., Blair, D., McBurney, L., Manning, A. D. and Stein, J. A. R.: New policies for old trees: averting a global crisis in a keystone ecological structure, Conservation Letters, 7, 61–69, 2013.

Liski, J., Lehtonen, A., Palosuo, T., Peltoniemi, M., Eggers, T., Muukkonen, P., and Mäkipää, R.: Carbon accumulation in Finland's forests 1922–2004 – an estimate obtained by combination of forest inventory data with modelling of biomass, litter and soil, Ann. For. Sci., 63, 687–697, 2006.

Lukeš, P., Rautiainen, M., Manninen, T., Stenberg, P., and Mõttus, M. Geographical gradients in boreal forest albedo and structure in Finland, Remote Sens. Environ., 152, 526–535, 2014.

Lutz, J. A., Larson, A. J., Swanson, M. E., and Freund, J. A.: Ecological importance of large-diameter trees in a temperate mixed-conifer forest, PLoS ONE, 7, e36131, doi:10.1371/journal.pone.0036131, 2012.

Mather, A. S.: The forest transition, Area, 24, 367–379, 1992.

Martínez, M. and Alvarez, E. R.: How old are tropical rain forest trees?, Trends Plant Sci., 3, 400–405, 1998.

McGuire, A. D., Sitch, S., Clein, J. S., Dargaville, R., Esser, G., Foley, J., and Wittenberg, U.: Carbon balance of the terrestrial biosphere in the twentieth century: analyses of CO_2, climate and land use effects with fourprocess-based ecosystem models, Global Biogeochem. Cy., 30, 183–206, 2001.

Nowak, D. J. and Greenfield, E. J.: Tree and impervious cover change in US cities, Urban For. Urban Gree., 11, 21–30, 2012.

Pan, Y., Birdsey, R. A., Fang, J., Houghton, R., Kauppi, P. E., Kurz, W. A., Phillips, O. L., Shvidenko, A., Lewis, S. L., Canadell, J. G., Ciais, P., Jackson, R. B., Pacala, S. W., McGuire, A. D., Piao, S. L., Rautiainen, A., Sitch, S., and Hayes, D.: A large and persistent carbon sink in the world's forests, Science, 333, 988–993, 2011a.

Pan, Y., Chen, J. M., Birdsey, R., McCullough, K., He, L., and Deng, F.: Age structure and disturbance legacy of North American forests, Biogeosciences, 8, 715–732, doi:10.5194/bg-8-715-2011, 2011b.

Pan, Y., Birdsey, A. R., Philipps, O. L., and Jackson, B. R.: The structure, distribution and biomass of the world's forests, Annu. Rev. Ecol. Evol. S., 44, 593–662, 2013.

Rautiainen, A., Wernick, I., Waggoner, P. E., Ausubel, J. H., and Kauppi, P. E.: A national and international analysis of changing forest density, PLoS ONE, 6, 19577, doi:10.1371/journal.pone.0019577, 2011.

Reich, P.: Taking stock of forest carbon, Nature Climate Change, 1, 346–347, 2011.

Repola, J.: Biomass equations for birch in Finland, Silva Fenn., 42, 605–624, 2008.

Repola, J.: Biomass equations for Scots pine and Norway spruce in Finland, Silva Fenn., 43, 625–647, 2009.

Richey, J., Melack, J., Aufdenkampe, A., Ballester, V., and Hess, L.: Outgassing from Amazonian rivers and wetlands as a large source of tropical CO2, Nature, 416, 617–620, 2002.

Skog, K. E., Pingoud, K., and Smith, J. E.: A method countries can use to estimate changes in carbon stored in harvested wood products and the uncertainty of such estimates, Environ. Manage., 33, 65–73, 2004.

Slik, J. W. F., Paoli, G., McGuire, K., Amaral, I., and Barroso, J.: Large trees drive forest aboveground biomass variation in moist lowland forests across the tropics, Global. Ecol. Biogeogr., 22, 1261–1271, 2013.

Smith, J. E., Heath, L. S., and Nichols, M. C.: US forest carbon calculation tool: forest-land carbon stocks and net annual stock change. Revised. Gen. Tech. Rep. NRS-13, U.S. Dep. Agric. For. Serv., 34 pp., Newtown Square, PA, 2007.

Smith, W. B., Miles, P. D., Perry, C. H., and Pugh, S. A.: Forest resources of the United States, RPA assessment, Gen. Tech. Rep. WO-78, US Dep. of Agric. For. Serv., Washington DC, 2009.

Stephenson, N. L., Das, A. J., Condit, R., Russo, S. E., Baker, P. J., Beckman, N. G., Coomes, D. A., Lines, E. R., Morris, W. K., Ruger, N., Alvarez, E., Blundo, C., Bunyavejchewin, S., Chuyong, G., Davies, S. J., Duque, A., Ewango, C. N., Flores, O., Franklin, J. F., Grau, H. R., Hao, Z., Harmon, M. E., Hubbell, S. P., Kenfack, D., Lin, Y., Makana, J.-R., Malizia, A., Malizia, L. R., Pabst, R. J., Pongpattananurak, N., Su, S.-H., Sun, I-F., Tan, S., Thomas, D., van Mantgem, P. J., Wang, X., Wiser, S. K.

and Zavala, M. A.: Rate of tree carbon accumulation increases continuously with tree size, Nature, 507, 90–93, 2014.

Swann, A. L., Fung, I. Y., Levis, S., Bonan, G. B., and Doney, S. C.: Changes in Arctic vegetation amplify high-latitude warming through the greenhouse effect, P. Natl. Acad. Sci. USA, 107, 1295–1300, 2010.

Tomppo, E., Heikkinen, J., Henttonen, H. M., Ihalainen, A., Katila, M., Mäkelä, H., Tuomainen, T., and Vainikainen, N.: Designing and Conducting a Forest Inventory – Case: 9th National Forest Inventory of Finland, Springer, Heidelberg, 270 pp., 2011.

Woodwell, G. M., Hobbie, J. E., Houghton, R. A., Melillo, J. M., Moore, B., Peterson, B. J., and Shaver, G. R.: Global deforestation: contribution to atmospheric carbon dioxide, Science, 222, 1081–1086, 1983.

Ylitalo, E.: Finnish Statistical Yearbook of Forestry, Finland 2013, METLA, 2013.

13

Strong stoichiometric resilience after litter manipulation experiments; a case study in a Chinese grassland

C. Xiao[1]**, I. A. Janssens**[2]**, Y. Zhou**[1]**, J. Su**[1,3]**, Y. Liang**[1,3]**, and B. Guenet**[2,4]

[1]State Key Laboratory of Vegetation and Environmental Change, Institute of Botany, the Chinese Academy of Sciences, Beijing 100093, China
[2]Department of Biology, Research Group of Plant and Vegetation Ecology, University of Antwerp, Universiteitsplein 1, 2610 Wilrijk, Belgium
[3]University of Chinese Academy of Sciences, Beijing, 100049, China
[4]CNRS-CEA-UVSQ, Laboratoire des Sciences du Climat et de l'Environnement, UMR8212, 91191 Gif-sur-Yvette, France

Correspondence to: B. Guenet (bertrand.guenet@lsce.ipsl.fr) and C. Xiao (cwxiao@ibcas.ac.cn)

Abstract. Global climate change has generally modified net primary production (NPP) which leads to increasing litter inputs in some ecosystems. Therefore, assessing the impacts of increasing litter inputs on soil nutrients, plant growth and ecological carbon (C) : nitrogen (N) : phosphorus (P) stoichiometry is critical for an understanding of C, N and P cycling and their feedback processes to climate change. In this study, we added plant above-ground litter, harvested near the experimental plots, to the 10–20 cm subsoil layer of a steppe community at rates equivalent to annual litter input of 0, 15, 30, 60 and 120 %, respectively, covering the entire range of the expected NPP increases in this region due to climate change (10–60 %). We measured the resulting C, N and P content of different pools (above- and below-ground plant biomass, litter, microbial biomass). Small litter additions, which are more plausible compared to the expected increase predicted by Earth system models, had no effect on the variables examined. Nevertheless, high litter addition (120 % of the annual litter inputs) significantly increased soil inorganic N and available P, above-ground biomass, below-ground biomass and litter. Our results suggest that while very high litter addition can strongly affect C : N : P stoichiometry, the grassland studied here is resilient to more plausible inputs in terms of stoichiometric functioning.

1 Introduction

Ecological stoichiometry is the study of the balance of multiple chemical elements in ecological interactions (Elser et al., 2000, 2010; Elser and Hamilton, 2007). Carbon (C), nitrogen (N) and phosphorus (P) are key elements in terrestrial ecosystems (Daufresne and Loreau, 2001; Elser et al., 2010; Hessen et al., 2013), and the C : N : P stoichiometry reflects complex interactions between evolutionary processes coupled to phenotypic plasticity (Sardans et al., 2012). These complex interactions are, at least partially, controlled by patterns of element supply from the environment (Hessen et al., 2004). Ecological stoichiometry provides a valuable approach in assessing possible changes in C, N and P cycling (Hessen et al., 2013). Over the past 3 decades ecological stoichiometry has expanded greatly (Hessen et al., 2013), and many studies have concentrated on understanding the variation in plant C, N and P concentrations among and within species (Schmidt et al., 1997; Aerts and Chapin, 2000; Güsewell and Koerselman, 2002; Güsewell, 2004; Frank, 2008). However, relationships among soil nutrient availability, plant growth and ecological C : N : P stoichiometry of plants, litter, soil and soil microbes under global climate change in terrestrial ecosystems remain poorly understood.

Litter is an important nutrient pool and litter decomposition rates strongly affect nutrient availability in terrestrial ecosystems with significant feedback on plant growth and on climate (Subke et al., 2004; Ryan and Law, 2005; Sayer,

2006; Cornelissen et al., 2007; Villalobos-Vega et al., 2011). The increase of atmospheric carbon dioxide (CO_2) concentration within the next 100 years (Meinshausen et al., 2011) due to continued anthropogenic carbon emissions is generally predicted to increase net primary production (NPP) (Todd-Brown et al., 2014). Nevertheless such effects do not scale linearly with increases in atmospheric CO_2 because productivity is also strongly controlled by climate. Regions with drying climates are likely to present reduced net primary production in the next decades (IPCC, 2013). Although uncertainties exists in the magnitude of the changes (e.g. Campbell et al., 1991; Arnone and Körner, 1995; Gill et al., 2002), increases in NPP are probably predominant and will simultaneously increase litter inputs to soils. Modification of atmospheric CO_2, of climate and of nutrient cycling may also modify the chemical composition of the litter (Cotrufo et al., 1999). Elevated CO_2 generally increases the lignin content and reduces the N concentration of plant tissues, although Norby et al. (2001) showed that this response depends on the experimental system used (open top chamber, free-air CO_2 enrichment, etc.).

In past decades, the effects of litter addition on plant growth, soil C content and cycling, microbial biomass and priming of decomposition rates have been reported and confirmed (e.g. Sulzman et al., 2005; Guenet et al., 2010; Jin et al., 2010; Chemidlin Prévost-Bouré et al., 2010; Sayer et al., 2011; Villalobos-Vega et al., 2011; Ma et al., 2012, Xiao et al., 2014). However, the impacts of litter addition on soil nutrients, plant growth and ecological C : N : P stoichiometry of plants, litter, soil and soil microbes remain highly uncertain. Moreover, climate change and increased atmospheric CO_2 would not only affect NPP but also environmental conditions for decomposers in soils (soil moisture, temperature), inducing modification of microbial community structure and activity (Singh et al., 2010). Associated with litter additions, these effects would disturb the C cycle in soil, affecting the net ecosystem exchange particularly in grasslands where the vast majority of the C stock is stored below-ground, due to their high root : shoot biomass and productivity ratios (Mokany et al., 2006).

Grassland is one of the most important global terrestrial ecosystems, covering about 25 % of the global terrestrial area and 40 % of the land area in China (Kang et al., 2007). The semi-arid and temperate grasslands of northern China account for about 78 % of the national grassland area, where the native vegetation is predominantly characterized by the abundance of grass species such as *Stipa* spp. and *Leymus chinensis* (Trin) Tzvel., with *Stipa krylovii* Roshev. being well-represented as one of the major steppe community types (Zhao et al., 2003). Plant recruitment, growth and nutrient cycling of the region often are limited by soil water, nitrogen and phosphorus, and the regional soil fertility and productivity are maintained by recycling of nutrients through plant litter decomposition as an essential mechanism (Liu et al., 2006) with little natural nitrogen deposition. A better understanding of the effects of litter addition on plant growth and ecological C : N : P stoichiometry of plants, litter, soil and soil microbes could help to reduce uncertainties in our predictions of C, N and P balance as well as cycling, and structure and function in grassland ecosystems under global climate change.

We conducted a field experiment in which we artificially added litter inputs to subsoils (i.e. 10–20 cm) under a *S. krylovii* steppe community in a temperate grassland of northern China to assess the effects on soil inorganic N and available P, plant growth, litter, C, N and P pools and the C : N : P stoichiometry of plant, litter, soil and soil microbes. The litter added was obtained under current atmospheric CO_2 concentration, and therefore the impact of elevated CO_2 on the litter stoichiometry (Cotrufo et al., 1999) was not represented in our case. The primary objectives of our study were to determine whether the studied grassland was resilient to litter addition or whether litter additions would increase soil inorganic N and available P and thereby affect plant growth, litter, and the C, N, P pools and the C : N : P stoichiometry of plants, litter, soil and soil microbes. Here, we define the term resilience as the absence of a significant effect of litter additions on the measured variables (plant biomass, C : N : P ratio, etc.).

In a previous study (Xiao et al., 2014), we observed that litter additions in the temperate grassland of northern China accelerate decomposition of soil organic matter through a priming effect. Priming is defined here as a modification of the soil organic matter decomposition rate induced by an input of litter and mediated by the altered activity of the microbial community (Blagodatskaya and Kuzyakov, 2008). The main objective of this study was to analyse how additional litter inputs and associated priming of soil organic matter decomposition alter nutrient availability and thereby plant stoichiometry and productivity.

2 Materials and methods

2.1 Study site

This study was conducted at a field site of the Duolun Restoration Ecology Experimentation and Demonstration Station of the Institute of Botany, the Chinese Academy of Sciences, located in southeastern Inner Mongolia, northern China (Latitude 42°02′ N, longitude 116°16′ E, elevation 1350 m a.s.l.). The climate is temperate and semiarid with a dry spring and a wet summer. The long-term mean annual temperature at the site is 2.1 °C, with monthly mean temperatures ranging from −17.5 °C in January to 18.9 °C in July. Mean annual precipitation is about 380 mm, with 90 % of the precipitation falling in the growing season between May and October.

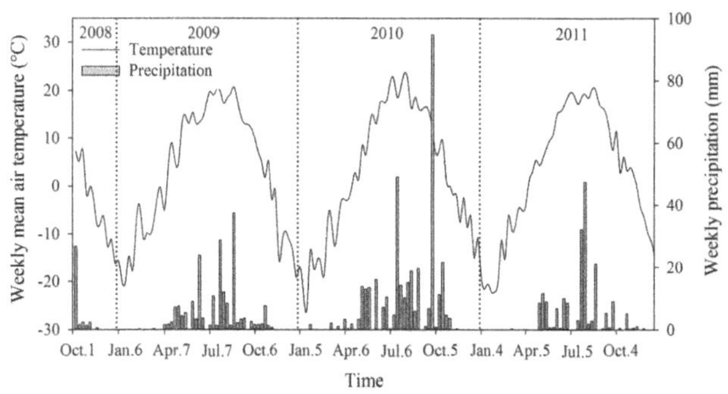

Figure 1. Seasonal variations of weekly mean air temperature and weekly precipitation during the experimental period from 1 October 2008 to 31 September 2011.

Total precipitation was 196, 369 and 187 mm in 2009, 2010 and 2011, respectively, and the precipitation between 1 May and 30 September was 164, 314 and 159 mm in 2009, 2010 and 2011, respectively. Mean annual air temperature was 2.96, 2.43 and 2.11 °C in 2009, 2010 and 2011, respectively (Fig. 1).

Soil type was classified as chestnut soil (Chinese classification) or Calcic Luvisols according to the FAO Classification (FAO, 1974). The soil characteristics were measured in the 0–30 cm layer. Soils are composed of 63 % sand, 20 % silt and 17 % clay (Niu et al., 2010), with concentrations of soil organic C, N and P of 1.55, 0.17 and 0.03 %, respectively. The C : N : P ratio of soil is about 51.7 : 5.7 : 1. Compared to the global database of Xu et al. (2013), the soil of the study site is a nutrient-rich soil. Mean bulk density is 1.31 g cm^{-3}, and the soil pH is 7.7. The native vegetation is represented by typical steppe communities, where *Stipa krylovii* Roshev., a perennial bunchgrass, dominates. Other common species include *Leymus chinensis* (Trin) Tzvel, *Cleistogenes squarrosa* (Trin.) Keng, *Agropyron cristatum* (L.) Gaertner, *Artemisia frigid* Willd., *Potentilla acaulis* L., and *Carex duriuscula* CA Mey. Total vegetation cover was relatively sparse, ranging from 85 to 90 %. Annual plant biomass production at the site was ca. 1000 g (dry mass) m^{-2} year^{-1} (Li et al., 2004).

2.2 Experimental design

On 1 October 2008, at the end of the growing season, we established twenty-five 2 m × 1 m treatment plots. Treatments involved the addition of fresh organic matter to the soil in the 10–20 cm soil layer. The applied fresh organic matter consisted of senescent above-ground tissues from an abundance-weighted mix of plant species occurring at the site. For this purpose, senescent plant biomass was harvested from an adjacent field, air-dried, fragmented, and passed through a sieve with a 2 mm mesh size. We used fragmented and sieved litter mainly to standardize the litter area available to microbes. The additions were equivalent to 0 (control treatment), 150, 300, 600 and 1200 g (dry mass) m^{-2}, with all arranged as a complete randomized-block design including five replications corresponding to litter input increases of 0, 15, 30, 60 and 120 % respectively. It must be noted that NPP is assumed to increase between 10 and 60 % at the end of the 21st century (Arora and Boer, 2014; Todd-Brown et al., 2014), which means that the first three addition amounts are in the range of the expected increase. The plant litter used had a C concentration of 400.8 mg g^{-1} (standard error (SE) = 1.3 mg g^{-1}, $n = 5$), an N concentration of 9.72 mg g^{-1} (SE = 0.04 mg g^{-1}, $n = 5$) and a P concentration of 0.768 mg g^{-1} (SE = 0.01 mg g^{-1}, $n = 5$), corresponding to a C : N : P ratio of 521.9 : 12.6 : 1. Lignin concentration in litter was 190.9 mg g^{-1} (SE = 0.9 mg g^{-1}, $n = 5$).

Adding litter to the uppermost soil layers is impossible without drastically disturbing the soils. To minimize disturbance, we carefully removed the top 10 cm soil blocks, containing 60 % of the root system (Zhou et al., 2007), with a sharp spade, keeping the soil blocks and vegetation as intact as possible. The soil underneath was loosened to a depth of 20 cm, and a predetermined quantity of plant litter was mixed homogeneously with the soil in the 10–20 cm layer. The surface soil blocks were then placed back into their original positions. Remaining fissures between the soil blocks were carefully filled with soil from the 0–10 cm soil layer and gently compacted by hand. To create consistent soil disturbance across treatments, the plots with zero litter addition were processed in the same manner as the plots that received plant litter. We did not add the litter at the soil surface, but inserted it in the soil to reduce export due to wind or rain and thereby better control the amount of litter added.

2.3 Field sampling and measurements

Field sampling of above-ground biomass, root biomass, litter and soils was conducted from 1 to 3 August in 2009, 2010 and 2011. We choose this period because August is the peak of the growing season.

Above-ground biomass and litter were simultaneously sampled in each plot using a 1 m × 0.3 m quadrat. Live and dead above-ground biomass were measured by clip-harvesting and dead parts were combined with the litter. Below-ground biomass in the 0–20 cm soil layer was determined by soil coring to a depth of 20 cm using a cylindrical root sampler (8 cm inner diameter). Roots were manually removed from the soil samples. All samples of above-ground biomass, below-ground biomass and litter were oven-dried at 65 °C to constant mass.

Soils were sampled in three different points of each plot within the 0–20 cm soil layer with a soil sampler of 3 cm inner diameter. The samples were pooled and mixed to produce one composite sample. The fresh samples were sieved using a 2 mm sieve and visible plant tissues were removed. Then the samples were divided into two groups. The first group was immediately transported to the laboratory with a portable ice box and stored at 4 °C before microbial biomass analysis and the second group was dried by air in the shade. Additionally, soil bulk densities of the 0–20 cm soil layers of each plot were determined concurrently with soil sampling by a special coring device (volume = 100.0 mL).

In the lab, chemical analysis was performed on samples of above-ground biomass, below-ground biomass, litter and soil in the 0–20 cm soil layer for organic C and total N using an automatic elemental analyser Vario EL III (Elementar Analysen Systeme Comp., Hanau, Germany). Total P was determined by the H_2SO_4–$HClO_4$ fusion method (Sparks et al., 1996). Soil microbial C and N biomass was measured by the fumigation–extraction method (Vance et al., 1987). Briefly, the fresh soil samples were adjusted to approximate 60 % of water holding capacity and then incubated for 1 week in the dark at 25 °C to reactivate soil microbes following the original procedure published by Vance et al. (1987). Next 20 g (dry weight equivalent) of fumigated with CH_3Cl for 24 h and non-fumigated soil samples were both extracted with 0.5 M K_2SO_4. The extracts were filtered through 0.45 μm filters and the extractable C and N was analysed by dichromate and Kjeldahl digestion as described by Lovell et al. (1995). Soil microbial C and N biomass was calculated as the difference in extractable C and N contents between the fumigated and the unfumigated samples using conversion factors (k_{ec} and k_{en}) of 0.38 and 0.45 (Lovell et al., 1995), respectively. Mass ratios of C : N, C : P and N : P in plant, litter and soil samples and of C : N in soil microbial biomass were calculated and used to facilitate comparisons with previous studies (He et al., 2008). Additionally, 10 g dry soil samples in the 0–20 cm layer were extracted with 50 mL of 2 M KCl. Inorganic N (NH_4^+-N and NO_3^--N) of the filtered extracts of soils in the 0–20 cm soil layer were determined using a flow injection autoanalyser (FIAstar 5000 Analyser, Foss Tecator, Denmark). Available soil P in soils in the 0–20 cm soil layers was measured using the Olsen method (Olsen et al., 1954).

2.4 Statistical analysis

Data management and statistical analyses were performed using the SPSS software package (SPSS, Chicago, IL, USA). Two-way analysis of covariance (ANCOVA) was used to detect the effects of litter addition and year (sampling time) on soil inorganic N and available P, above-ground and below-ground biomass, total biomass, litter, the C, N, P pools and C : N : P stoichiometry of plant, litter and soil, and C and N pools and C : N of soil microbial biomass. Multiple comparisons were also performed to permit separation of effect means using the Duncan test at a significance level of $P < 0.05$.

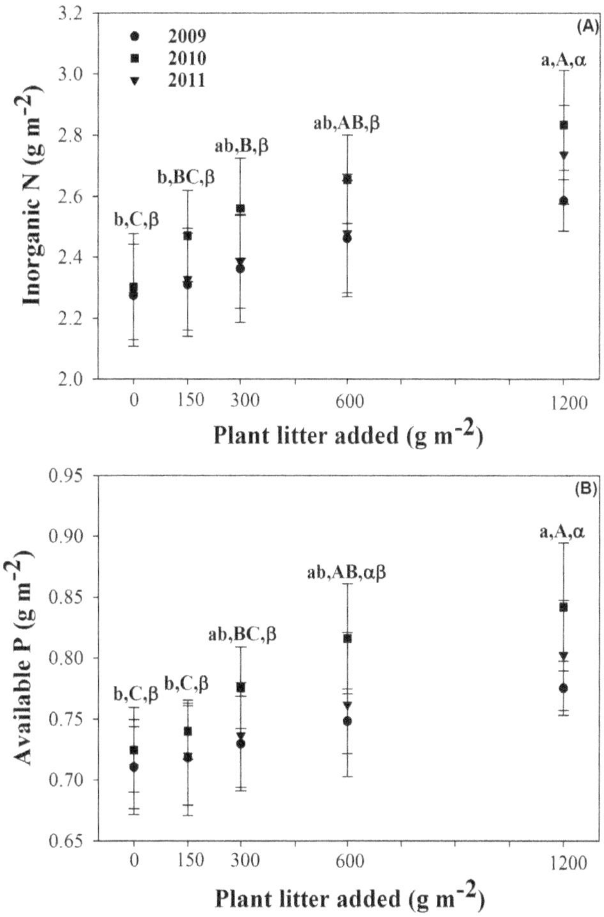

Figure 2. Soil inorganic N (**a**) and available P (**b**) in 2009, 2010 and 2011 under different amounts of litter addition. Vertical bars indicate one standard error about the mean ($n = 5$). The soil inorganic N and the available P contents were different each year ($P < 0.01$) but no interaction occurred between litter addition and year. Treatments with different letters are significantly different ($P < 0.05$) according to the Duncan test, small letters, capital letters and Greek letters correspond to the year 2009, 2010 and 2011, respectively; "ns" implies that no significant differences were detected.

3 Results

3.1 Soil inorganic N and available P

Litter addition significantly enhanced soil inorganic N and available P in 2009, 2010 and 2011 ($P < 0.05$; Fig. 2), and there were significant differences in soil inorganic N and available P among different years ($P < 0.01$; Fig. 2). Nevertheless, these effects were mainly due to the highest input treatments. Indeed, for the years 2009 and 2011, only the highest litter inputs, corresponding to 1200 g DM m^{-2}, induced significant highest inorganic N and available P contents in soils. For the year 2010, the two highest litter inputs treatments induced significant increases of the inorganic N and available P contents in soils. There were no significant interactions between litter addition and year for soil inorganic N and available P ($P > 0.05$; Fig. 2), indicating that the effect of treatment was conservative over the year.

3.2 Plant biomass, carbon allocation and litter production

Significant effects of litter addition were observed for above-ground biomass and litter in 2009, 2010 and 2011 as well as for below-ground biomass and total biomass in 2010 and 2011 ($P < 0.05$; Fig. 3). In our study site, the below-ground biomass was 6 times higher than the above-ground biomass. The ratio of below-ground biomass to above-ground biomass in 2010 was significantly affected by litter inputs ($P < 0.05$; Fig. S1 in the Supplement). However, it must be noted that the Duncan post hoc test showed that the effects described above are mainly due to the highest-input treatments, which is generally the sole treatment significantly different from the control when significant effects were detected by the two-way ANCOVA. The highest litter addition increased biomass in all the compartments except for the below-ground biomass and the total biomass in 2009. Furthermore, the highest litter addition treatment decreased the ratio of below-ground biomass to above-ground biomass in 2010, while other treatments produced no effect. There were no significant interactions between litter addition and year on above-ground biomass, below-ground biomass, total biomass, litter and ratio of below-ground biomass to above-ground biomass ($P > 0.05$; Figs. 3 and 4). Therefore, for all these variables the treatment was conservative over the years.

3.3 C, N and P pools in plants, litter, soil and soil microbial biomass

Litter addition did not affect significantly the C, N and P pools of above-ground biomass and litter as well as soil C pools in 2009, 2010 and 2011 for all but the highest treatment. The C, N and P pools of below-ground biomass in 2010 and 2011, and the C and N pools of soil microbial biomass were also not affected by litter addition except for the highest treatment ($P < 0.05$; Fig. 4). There were no significant differences in the C, N and P pools of above-ground biomass, the N and P pools of litter, the C pool of soil and the C and N pools of soil microbial biomass among different years for all treatment but the highest treatment ($P < 0.05$; Fig. 4). There were no significant interactions between litter addition and year, indicating a conservative treatment effect over the year on the C, N and P pools of above-ground biomass, below-ground biomass, litter, soil and the C and N pools of soil microbial biomass ($P > 0.05$; Fig. 4). Even though C, N and P contents were significantly different for the highest treatment, for almost all the pools, they tended to increase when litter was added.

3.4 C, N and P stoichiometry in plant, litter, soil and soil microbe

Litter addition did not significantly modify C : N and C : P ratios of above-ground biomass and litter in 2010 and 2011 and the C : N ratio of soil microbial biomass in 2009, 2010 and 2011 for all but the highest treatment where a decrease was observed. The highest treatment significantly increased soil C : N and C : P ratios in 2009, 2010 and 2011 ($P < 0.05$; Fig. 5). But no effect was detected for the other treatments. Litter addition did not affect N : P of above-ground biomass, below-ground biomass, litter and soil ($P > 0.05$; Fig. 5). There was significant difference in the C : N of above-ground biomass and soil microbial biomass among different years ($P < 0.05$; Fig. 5). There were no significant interactions between litter addition and year in effects on the C : N : P stoichiometry of above-ground biomass, below-ground biomass, litter, soil and soil microbial biomass ($P > 0.05$; Fig. 5) indicating that the effect of treatment was conservative over the year.

4 Discussion

4.1 Effect on litter additions on plant growth

Plant growth is limited by the rate of resource supply, for example nutrients and water (Enquist et al., 2003). Furthermore, soil N and P are the main nutrient sources for plant growth (Elser et al., 2007; Vitousek et al., 2010; Alvarez-Clare et al., 2013; Fageria et al., 2013). The NPP in our study sites is expected to increase between 10 and 60 % due to climate change and atmospheric CO_2 increase (Arora and Boer, 2014; Todd-Brown et al., 2014). Moreover, temperature is also expected to increase in China during the next decades (Piao et al., 2010), likely accelerating litter and soil organic matter decomposition and nutrient release. Future predictions about the evolution of precipitation are still highly uncertain (Piao et al., 2010). The litter amendments we applied are substantial supply of nutrients, suggested to release nutrients during decomposition. Results show that availability of N and P were significantly modified only for the two highest input treatments. Additionally, high litter addition also greatly increased soil microbial biomass C and N, indicating that soil microbial biomass plays an active role in nutrient transformation, conservation, and availability to plants (Wardle, 1992; Zaman et al., 1999; Tu et al., 2003). Notably, for more moderate litter additions, the observed effect on plant biomass was limited, suggesting that the plant did not benefit from these inputs. Litter addition significantly increased above-ground biomass in 2009, 2010 and 2011, and below-ground biomass and total biomass in 2010 and 2011 only for the highest input level.

One plausible explanation for the lack of effect on plant biomass in spite of the important amount of organic nutrients

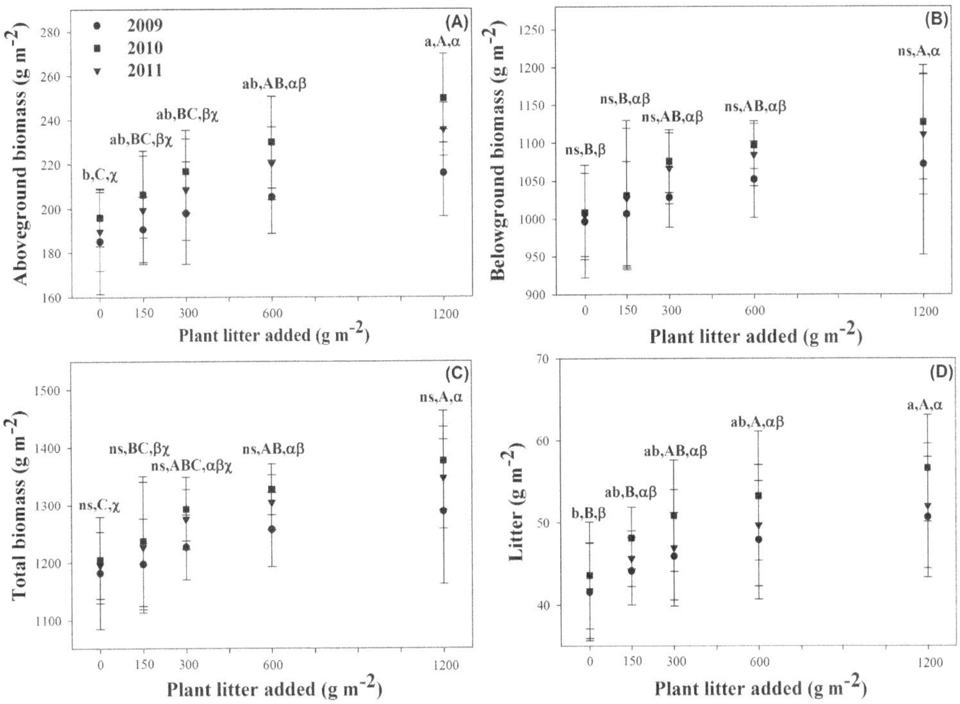

Figure 3. Above-ground biomass (**a**), below-ground biomass (**b**), total biomass (**c**) and litter (**d**) in 2009, 2010 and 2011 under different amounts of litter addition in a steppe community of northern China. Vertical bars indicate one standard error about the mean ($n = 5$). A significant year effect ($P < 0.05$) was detected by ANCOVA except for the below-ground biomass. Interactions between litter addition and year were not significant. See Fig. 2 legend for letter signification.

added might be the imbalance between the C : N : P ratio of the litter added and the C : N : P ratio of the soil, the plants and the microbial community. For instance, the C : N and C : P ratios of the added litter were 41.3 and 521.9, respectively. Compared to the C : N ratio of microbial biomass around 12 and the C : N and C : P ratio of soil around 9 and 55, respectively, added litter decomposition might immobilize available mineral nutrients. As a consequence, the competition between plants and microorganisms for mineral N and available P would increase. Furthermore, a priming effect was observed during this experiment (Xiao et al., 2014) but the nutrient release associated might be not sufficient to increase the plant growth. Since plant biomass only responded to the highest litter treatments, it suggests that, in our case, microorganisms might be more efficient than plants in using nutrient resources and that plant growth might be still nutrient limited. Nevertheless, soil microorganisms benefited only slightly from the litter inputs, as suggested by the small observed increase in microbial biomass (Fig. 4m, n). Therefore, carbon use efficiency probably decreased when litter was added, increasing heterotrophic respiration. Such modification of the carbon use efficiency might be due to physiological modifications at the individual level, to a modification of the microbial community structure or to both mechanisms combined. With our data, we cannot estimate the carbon use efficiency at the individual level but, based on the same experiment, Xiao et al. (2014) showed that the microbial community structure was different between treatments. It must be noted that the litter used here was harvested under natural conditions. Thus, the modifications of the litter chemical composition expected under climate change are not taken into account here. Litter C : N ratio may increase in the future (Norby et al., 2001) inducing a higher microbial N demand. Microorganisms may still decrease their carbon use efficiency as assumed here or may increase the N uptake. Since the microbial N uptake in our experiment only increased for very high and unrealistic litter inputs, we assume that, at our site, microorganisms will likely modify their carbon use efficiency in response to the modification of litter C : N ratio instead of competing with plants for N. Furthermore, when litter additions significantly affected plant biomass, above-ground biomass and below-ground biomass were higher in 2010 than in 2009 and 2011. The years 2009 and 2011 were dry compared to the year 2010. Thus, a water stress may have limited the plant growth in 2009 and 2011. Moreover, soil moisture was likely higher in 2010. Such more favourable soil moisture conditions may have caused the higher soil nutrient availability via accelerated litter decomposition.

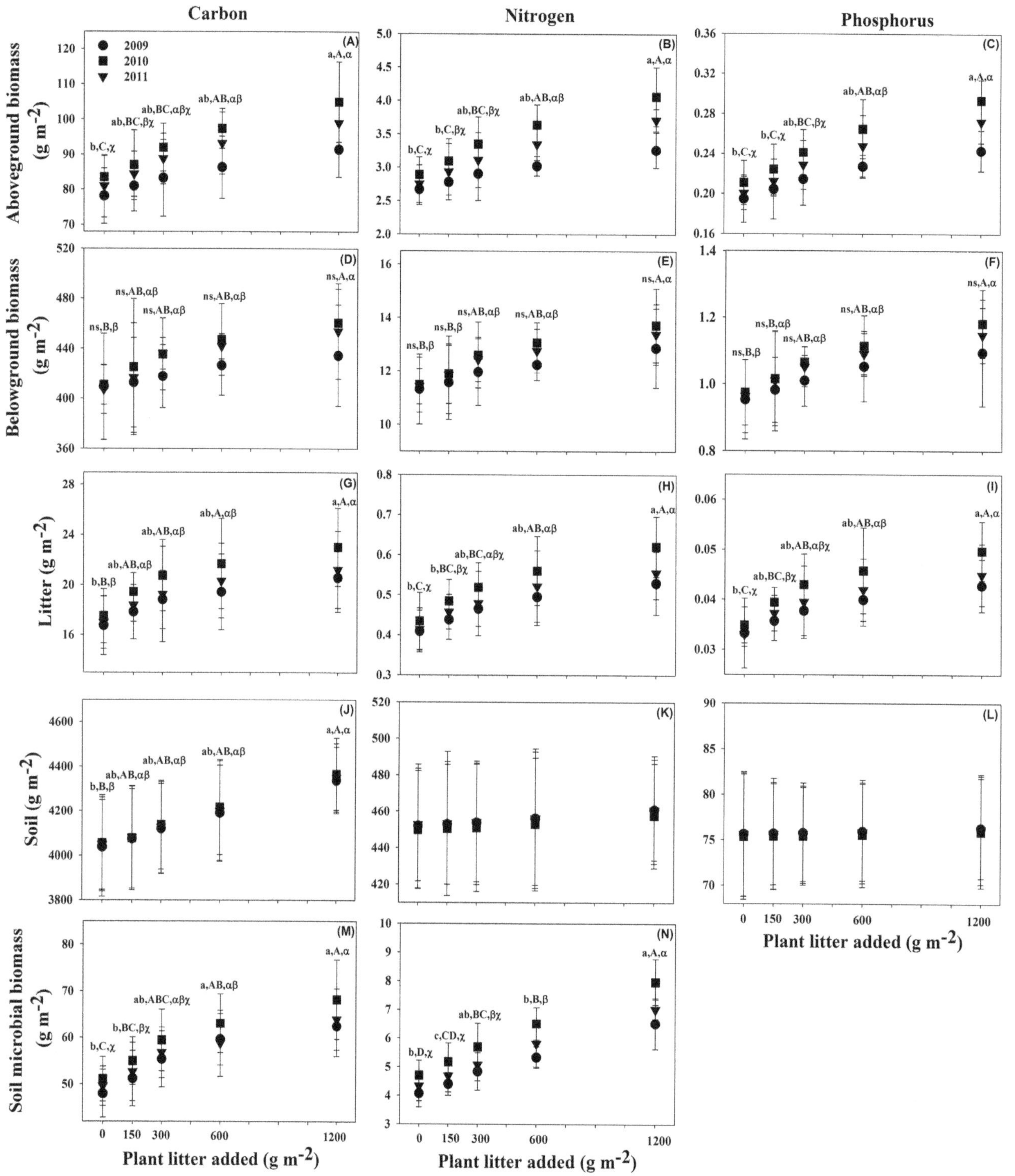

Figure 4. The C, N, and P pools of plant, litter and soil and the C and N pools of soil microbial biomass in 2009, 2010 and 2011 under different amounts of litter addition in a steppe community of northern China. Vertical bars indicate one standard error about the mean ($n = 5$). A significant year effect ($P < 0.05$) on the C, N and P content was detected by ANCOVA except for the below-ground biomass, the soil. No year effect for litter and soil microbial biomass C content was detected. Interactions between litter addition and year were not significant. See Fig. 2 legend for letter signification. Letters are only given where statistically significant differences are observed.

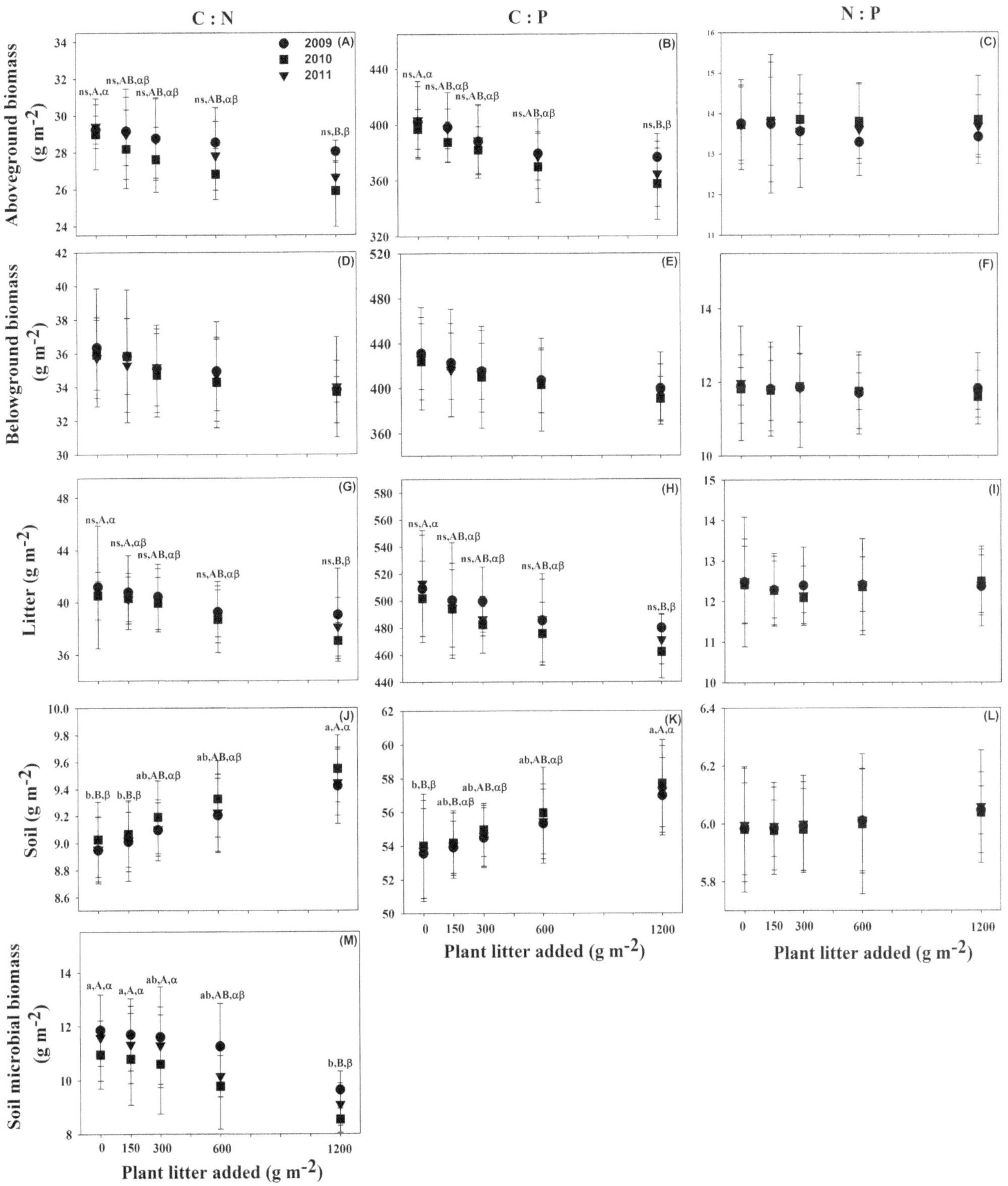

Figure 5. The C : N, C : P and N : P ratio of plant, litter and soil and C : N ratio of soil microbial biomass in 2009, 2010 and 2011 under different amounts of litter addition in a steppe community of northern China. Vertical bars indicate one standard error about the mean ($n = 5$). A significant year effect ($P < 0.05$) on the C : N ratio of above-ground biomass and soil microbial biomass was detected by ANCOVA. Interactions between litter addition and year were not significant. See Fig. 2 legend for letter signification. Letters are only given where statistically significant differences are observed.

4.2 Biomass allocation

Biomass allocation is often affected by factors such as soil nutrient conditions and plant habitat (Vogt et al., 1983; Schmid, 2002; Mokany et al., 2006). In our study, high litter addition decreased the ratio of below-ground biomass to above-ground biomass, and the decrease reached a significant level in 2010. The reason for this decrease may be that litter addition greatly increased soil organic N and available P and soil microbial biomass C and N, allowing the plant to invest more photosynthates in above-ground biomass in 2010 as already observed (Vogt et al., 1983; Schmid, 2002). Sims et al. (2012) found that adding nitrogen increased plant growth and allocated more biomass toward shoots than roots. The plastic response of increased allocation to shoots corresponds to theoretical predictions (Tilman, 1988). Such an increase in photosynthate concentration is also explained by the increase in C : N and C : P ratio in above-ground biomass and litter upon high litter addition, but not for below-ground biomass.

4.3 Effect of litter addition on plant N : P ratio

Plant N and P are essential nutrients for primary producers and decomposers in terrestrial ecosystems, and N : P ratios of plant biomass or litter have been widely used as indicators of nutrient limitation for primary production (Koerselman and Meuleman, 1996; Tessier and Raynal, 2003; Güsewell, 2004; Güsewell and Verhoeven, 2006). In our study, litter addition did not affect the N : P of above-ground biomass, below-ground biomass and litter. Our results showed that the N : P of above-ground biomass (ranging from 13.3 to 13.9 under different litter additions) is higher than that of below-ground biomass (ranging from 11.6 to 12 under different litter additions). Similar results were observed in the study of Xu et al. (2010). The N : P of above-ground biomass and below-ground biomass is lower than 14 on a community level, suggesting that our *S. krylovii* steppe community was N limited (Koerselman and Meuleman, 1996). Additionally, the N : P of litter ranged from 12.1 to 12.5 after different litter treatments and were much lower than 25 (the threshold between N and P limitation for graminoid leaf litter; Güsewell and Verhoeven, 2006), also indicating that our *S. krylovii* steppe community is subject to N limitation. This result is consistent with the conclusion of Bai et al. (2012) who found that meadow steppe, typical steppe and desert steppe communities of temperate grasslands in northern China are N-limited systems.

5 Conclusions

In our study, litter addition significantly affected the stoichiometry of the systems only when it was quite high (twice the natural inputs). Previous modelling exercises have not predicted an increase of primary production sufficient to double litter inputs (Arora and Boer, 2014; Todd-Brown et al., 2014). This suggests that the grassland studied here is quite resilient in terms of stoichiometry. This resilience is the result of complex interactions between C and nutrients cycles as well as between plants and microbial biomass.

In conclusion, our results showed that very high litter addition increased soil inorganic N and plant-available P, the C and N pools of soil microbial biomass, above-ground and below-ground plant biomass. On the other hand, plausible additions according to the models predictions had no effect. This suggests that the expected increase of NPP may not have important consequences on the stoichiometric functioning for some particular ecosystems, such as grasslands in northern China. Nevertheless, it must be noted that climate change will also affect temperature and soil moisture, which will largely affect the response of ecosystems to modifications of NPP due to the atmospheric CO_2 increase.

Acknowledgements. This study was financially supported by "Strategic Priority Research Program – Climate Change: Carbon Budget and Related Issues" of the Chinese Academy of Sciences (XDA05050407), the National Natural Science Foundation of China (No. 31370462), and State Key Laboratory of Vegetation and Environmental Change, Institute of Botany, the Chinese Academy of Sciences. We thank L. H. Li for providing weather data, and Z. Q. Pei, J. Y. Xia and J. L. Zhang for help with field sampling. We gratefully acknowledge the Duolun Restoration Ecology Experimentation and Demonstration Station for help with logistics and access permission to the study site. I. A. Janssens and B. Guenet acknowledge support from the Flemish National Science Foundation, from the European Community's Seventh Framework Programme (FP7/2007-2013) under grant agreement no. 226701 (CARBO-Extreme), from the Cost Action Terrabites (ES-0805) and the European Research Council Synergy grant ERC-2013-SyG-610028 IMBALANCE-P. The authors acknowledge M. Macgrath for his valuable comments on the paper.

Edited by: S. Zaehle

References

Aerts, R. and Chapin III, F. S.: The mineral nutrition of wild plants revisited: a re-evaluation of processes and patterns, Adv. Ecol. Res., 30, 1–67, 2000.

Alvarez-Clare, S., Mack, M. C., and Brooks, M. A.: Direct test of nitrogen and phosphorus limitation to net primary productivity in a lowland tropical wet forest, Ecology, 94, 1540–1551, 2013.

Arnone III, J. A. and Körner, C.: Soil and biomass carbon pools in model communities of tropical plants uner elevated CO_2, Oecologia, 104, 61–71, 1995.

Arora, V. K. and Boer, G. J.: Terrestrial ecosystems response to future changes in climate and atmospheric CO_2 concentration, Biogeosciences, 11, 4157–4171, doi:10.5194/bg-11-4157-2014, 2014.

Bai, Y. F., Wu, J. G., Clark, C. M., Pan, Q. M., Zhang, L. X., Chen, S. P., Wan, Q., Han, X.: Grazing alters ecosystem functioning and C : N : P stoichiometry of grasslands along a regional precipitation gradient, J. Appl. Ecol., 49, 1204–1215, 2012.

Blagodatskaya, E. and Kuzyakov, Y.,: Mechanisms of real and apparent priming effects and their dependence on soil microbial biomass and community structure: critical review, Biol. Fert. Soils, 45, 115–131, 2008.

Campbell, C. A., Lafond, G. P., Zentner, R. P., and Biederbeck, V. O.: Influence of fertilizer and straw baling on soil organic matter in a thick black chernozem in western Canada, Soil Biol. Biochem., 23, 443–446, 1991.

Chemidlin Prévost-Bouré, N., Soudani, K., Damesin, C., Berveiller, D., Lata, J.-C., and Dufrêne, E.: Increase in aboveground fresh litter quantity over-stimulates soil respiration in a temperate deciduous forest, App. Soil Ecol., 46, 26–34, 2010.

Cornelissen, J. H. C., Van Bodegom, P. M., Aerts, R., Callaghan, T. V., Van Logtestijn, R. S. P., Alatalo, J., Stuart Chapin, F., Gerdol, R., Gudmundsson, J., Gwynn-Jones, D., Hartley, A. E., Hik, D. S., Hofgaard, A., Jónsdóttir, I. S., Karlsson, S., Klein, J., Laundre, J., Magnusson, B., Michelsen, A., Molau, U., Onipchenko, V. G., Quested, H. M., Sandvik, S. M., Schmidt, I. K., Shaver, G. R., Solheim, B., Soudzilovskaia, N., Stenström, A., Tolvanen, A., Totland, Ø., Wada, N., Welker, J. M., Zhao, X., Brancaleoni, L., Brancaleoni, L., De Beus, M., Cooper, E. J., Dalen, L., Harte, J., Hobbie, S. E., Hoefsloot, G., Jägerbrand, A., Jonasson, S., Lee, J., Lindblad, K., Melillo, J. M., Neill, C., Press, M. C., Rozema, J., and Zielke, M.: Global negative vegetation feedback to climate warming responses of leaf litter decomposition rates in cold biomes, Ecol. Lett., 10, 619–627, 2007.

Cotrufo, M., Raschi, A., Lanini, M., and Ineson, P.: Decomposition and nutrient dynamics of Quercus pubescens leaf litter in a naturally enriched CO_2 Mediterranean ecosystem, Func. Ecol., 13, 343–351, 1999.

Daufresne, T. and Loreau, M.: Ecological stoichiometry, primary producer-decomposer interactions, and ecosystem persistence, Ecology, 82, 3069–3082, 2001.

Elser, J. J. and Hamilton, A. L.: Stoichiometry and the new biology: the future is now, PLoS. Biol., 5, e181, doi:10.1371/journal.pbio.0050181, 2007.

Elser, J. J., Sterner, R. W., Gorokhova, E., Fagan, W. F., Markow, T. A., Cotner, J. B., Harrison, F., Hobbie, S. E., Odell, G. M., Lawrence, J.: Biological stoichiometry from genes to ecosystems, Ecol. Lett., 3, 540–550, 2000.

Elser, J. J., Bracken, M. E. S, Cleland, E. E., Gruner, D. S., Harpole, W. S., Hillebrand, H., Ngai, J. T., Seabloom, E. W., Shurin, J. B., Smith, J. E.: Global analysis of Nitrogen and Phosphorus limitation of primary producers in freshwater, marine and terrestrial ecosystems, Ecol. Lett., 10, 1135–1142, 2007.

Elser, J. J., Fagan, W. F., Kerkhoff, A. J., Swenson, N. G., and Enquist, B. J.: Biological stoichiometry of plant production: metabolism, scaling and ecological response to global change, New Phytol., 186, 593–608, 2010.

Enquist, B. J., Economo, E. P., Huxman, T. E., Allen, A. P., Ignace, D. D., and Gillooly, J. F.: Scaling metabolism from organisms to ecosystems, Nature, 423, 639–642, 2003.

Fageria, N. K., Baligar, V. C., Moreira, A., and Moraes, L. A. C.: Soil phosphorous influence on growth and nutrition of tropical legume cover crops in acidic soil, Commun. Soil Sci. Plan, 44, 3340–3364, 2013.

FAO-UNESCO.: Soil map of the world 1 : 5 000 000. Paris, France: UNESCO, 1974.

Frank, D. A.: Ungulate and topographic control of nitrogen: phosphorus stoichiometry in a temperate grassland; soils, plants and mineralization rates, Oikos, 117, 591–601, 2008.

Gill, R. A., Polley, H. W., Johnson, H. B., Anderson, L. J., Maherall, H., and Jackson, R. B.: Nonlinear grassland responses to past and future atmospheric CO_2, Nature, 417, 279–282, 2002.

Guenet, B., Neill, C., Bardoux, G., and Abbadie, L.: Is there a linear relationship between priming effect intensity and the amount of organic matter input?, Appl. Soil Ecol., 46, 436–442, 2010.

Güsewell, S.: N : P ratios in terrestrial plants: variation and functional significance, New Phytol., 164, 243–266, 2004.

Güsewell, S. and Koerselman, W.: Variation in nitrogen and phosphorus concentrations of wetland plants, Perspect. Plant Ecol., 5, 37–61, 2002.

Güsewell, S. and Verhoeven, J. T. A.: Litter N:P ratios indicate whether N or P limits the decomposability of graminoid leaf litter, Plant Soil, 287, 131–143, 2006.

He, J. S., Wang, L., Flynn, D. F. B., Wang, X. P., Ma, W. H., and Fang, J. Y.: Leaf nitrogen:phosphorus stoichiometry across Chinese grassland biomes, Oecologia, 155, 301–310, 2008.

Hessen, D. O., Ågren, G. I., Anderson, T. R., Elser, J. J., and de Ruiter, P.: Carbon sequestration in ecosystems: the role of stoichiometry, Ecology, 85, 1179–1192, 2004.

Hessen, D. O., Elser, J. J., Sterner, R. W., and Urabe, J.: Ecological stoichiometry: An elementary approach using basic principles, Limnol. Oceanogr., 58, 2219–2236, 2013.

IPCC.: Fifth Assessment Report: Climate Change. The physical science basis, 2013.

Jin, H., Sun, O. J., and Liu, J.: Changes in soil microbial biomass and community structure with addition of contrasting types of plant litter in a semiarid grassland ecosystem, J. Plant Ecol., 3, 209–217, 2010.

Kang, L., Han, X., Zhang, Z., and Sun, O. J.: Grassland ecosystems in China: review of current knowledge and research advancement, Philos. T. Roy. Soc. B., 1482, 997–1008, 2007.

Koerselman, W. and Meuleman, A. F. M.: The vegetation N:P ratio: a new tool to detect the nature of nutrient limitation, J. Appl. Ecol., 33, 1441–1450, 1996.

Li, L., Li, X., Bai, W. M., Wang, Q. B., Yan, Z. D., Yuan, Z. Y., and Dong, Y. S.: Soil carbon budget of a grazed Leymus chinensis steppe community in the Xilin River Basin of Inner Mongolia, Acta Phytoecologica Sinica, 28, 312–317, 2004 (in Chinese).

Liu, P., Huang, J., Han, X., Sun, O. J., and Zhou, Z.: Differential responses of litter decomposition to increased soil nutrients and water between two contrasting grassland plant species of Inner Mongolia, China, Appl. Soil Ecol., 34, 266–275, 2006.

Lovell, R. D., Jarvis, S. C., and Bardgett, R. D.: Soil microbial biomass and activity in long-term grassland: effects of management changes, Soil Biol. Biochem., 27, 969–975, 1995.

Ma, L. N., Huang, W. W., Guo, C. Y., Wang, R. Z., and Xiao, C. W.: Soil microbial properties and plant growth responses to carbon and water addition in a temperate steppe: the importance of nutrient availability, PLoS ONE, 7, e35165, doi:10.1371/journal.pone.0035165, 2012.

Meinshausen, M., Smith, S. J., Calvin, K., Daniel, J. S., Kainuma, M. L. T., Lamarque, J.-F., Matsumoto, K., Montzka, S. A., Raper, S. C. B., Riahi, K., Thomson, A., Velders, G. J. M., and Vuuren, D. P. P.: The RCP greenhouse gas concentrations and their extensions from 1765 to 2300, Climatic Change, 109, 213–241, 2011.

Mokany, K., Raison, R. J., and Prokushkin, A. S.: Critical analysis of root:shoot ratios in terrestrial biomass, Global Change Biol., 12, 84–96, 2006.

Niu, S., Wu, M., Han, Y., Xia, J., Zhang, Z., Yan, H., and Wan, S.: Nitrogen effects on net ecosystem carbon exchange in a temperate steppe, Global Change Biol., 16, 144–155, 2010.

Norby, R. J., Cotrufo, M. F., Ineson, P., O'Neill, E. G., and Canadell, J. G.: Elevated CO_2, litter chemistry, and decomposition: a synthesis, Oecologia, 127, 153–165, 2001.

Olsen, S. R., Cole, C. V., Watanabe, F. S., and Dean, L. A.: Estimation of available phosphorus in soils by extraction with sodium bicarbonate, Circular No. 939, USDA, 1954.

Piao, S., Ciais, P., Huang, Y., Shen, Z., Peng, S., Li, J., Zhou, L., Liu, H., Ma, Y., Ding, Y., Friedlingstein, P., Liu, C., Tan, K., Yu, Y., Zhang, T., and Fang, J.: The impacts of climate change on water resources and agriculture in China, Nature, 467, 43–51, 2010.

Ryan, M. G. and Law, B. E.: Interpreting, measuring, and modeling soil respiration, Biogeochemistry, 73, 3–27, 2005.

Sardans, J., Rivas-Ubach, A., and Peñuelas, J.: The C : N : P stoichiometry of organisms and ecosystems in a changing world: A review and perspectives, Perspectives in Plant Ecology, Evolution and Systematics, 14, 33–47, 2012.

Sayer, E. J.: Using experimental manipulation to assess the roles of leaf litter in the functioning of forest ecosystems, Biol. Rev., 81, 1–31, 2006.

Sayer, E. J., Heard, M. S., Grant, H. K., Marthews, T. R., and Tanner, E. V.: Soil carbon release enhanced by increased tropical forest litterfall, Nat. Clim. Change, 1, 304–307, 2011.

Schmid, I.: The influence of soil type and interspecific competition on the fine root system of Norway spruce and European beech, Basic Appl. Ecol., 3, 339–346, 2002.

Schmidt, I. K., Michelsen, A., and Jonasson, S.: Effects of labile soil carbon on nutrient partitioning between arctic graminoid and microbes, Oecologia, 112, 557–565, 1997.

Sims, L., Pastor, J., Lee, T., and Dewey, B.: Nitrogen, phosphorus and light effects on growth and allocation of biomass and nutrients in wild rice, Oecologia, 170, 65–76, 2012.

Singh, B. K., Bardgett, R. D., Smith, P., and Reay, D. S.: Microorganisms and climate change: terrestrial feedbacks and mitigation options, Nature reviews, Microbiology, 8, 779–90, 2010.

Sparks, D. L., Page, A. L., Loeppert, P. A., Soltanpour, P. N., Tabatabai, M. A., Johnston, C. T., and Sumner, M. E.: Methods of Soil Analysis Part 3: Chemical methods, Soil Science Society of America and American Society of Agronomy, Madison, WI, 1996.

Subke, J. A., Hahn, V., Battipaglia, G., Linder, S., Buchmann, N., and Cotrufo, M. F.: Feedback interactions between needle litter decomposition and rhizosphere activity, Oecologia, 139, 551–559, 2004.

Sulzman, E. W., Brant, J. B., Bowden, R. D., and Lajtha, K.: Contribution of aboveground litter, belowground litter, and rhizosphere respiration to total soil CO_2 efflux in an old growth coniferous forest, Biogeochemistry, 73, 231–256, 2005.

Tessier, J. T. and Raynal, D. Y.: Use of nitrogen to phosphorus ratios in plant tissue as an indicator of nutrient limitation and nitrogen saturation, J. Appl. Ecol., 40, 523–534, 2003.

Tilman, D.: Plant strategies and the dynamics and structure of plant communities, Princeton University Press, Princeton, 1988.

Todd-Brown, K. E. O., Randerson, J. T., Hopkins, F., Arora, V., Hajima, T., Jones, C., Shevliakova, E., Tjiputra, J., Volodin, E., Wu, T., Zhang, Q., and Allison, S. D.: Changes in soil organic carbon storage predicted by Earth system models during the 21st century, Biogeosciences, 11, 2341–2356, doi:10.5194/bg-11-2341-2014, 2014.

Tu, C., Koenning, S. R., and Hu, S.: Root-parasitic nematodes enhance soil microbial activities and nitrogen mineralization, Microb. Ecol., 46, 134–144, 2003.

Vance, E. D., Brookes, P., Jenkinson, D. S.: An extraction method for measuring soil microbial biomass C, Soil Biol. Biochem., 19, 703–707, 1987.

Villalobos-Vega, R., Goldstein, G., Haridasan, M., Franco, A. C., Miralles-Wilhelm, F., Scholz, F. G., and Bucci, S. J.: Leaf litter manipulations alter soil physicochemical properties and tree growth in a Neotropical savanna, Plant Soil, 346, 385–397, 2011.

Vitousek, P. M., Porder, S., Houlton, B. Z., and Chadwick, O. A.: Terrestrial phosphorus limitation: mechanisms, implications, and nitrogen-phosphorus interactions, Ecol. Appl., 20, 5–15, 2010.

Vogt, K. A., Moore, E. E., Vogt, D. J., Redlin, M. J., and Edmonds, R. L.: Conifer fine root and mycorrhizal root biomass within the forest floors of Douglas-fir stands of different ages and productivities, Can. J. For. Res., 13, 429–437, 1983.

Wardle, D. A.: A comparative assessment of factors which influence microbial biomass carbon and nitrogen levels in soil, Biol. Rev. Camb. Philos. Soc., 67, 321–358, 1992.

Xiao, C., Guenet, B., Zhou, Y., Su, J., and Janssens, I. A.: Priming of soil organic matter decomposition scales linearly with microbial biomass response to litter input in steppe vegetation, Oikos, doi:10.1111/oik.01728, 2014.

Xu, B., Cheng, Y. X., Gan, H. J., Zhou, W. J., and He, J. S.: Correlations between leaf and fine root traits among and within species of typical temperate grassland in Xilin River Basin, Inner Mongolia, China, Chinese J. Plant Ecol., 34, 29–38, 2010.

Xu, X., Thornton, P. E., and Post, W. M.,:. A global analysis of soil microbial biomass carbon, nitrogen and phosphorus in terrestrial ecosystems, Glob. Ecol. Biog., 22, 737–749, 2013.

Zaman, M., Di, H. J., and Cameron, K. C.: A field study of gross rates of N mineralization and nitrification and their relationships to microbial biomass and enzyme activities in soils treated with dairy effluent and ammonium fertilizer, Soil Use Manage., 15, 188–194, 1999.

Zhao, M. L., Han, B., Wu, J. F., Zhang, H. L., Zhen, H., and Suo, P. F.: A preliminary study on the population variation of *Stipa krylovii*, Acta Agrestia Sininca, 11, 163–169, 2003.

Zhou, Z., Sun, O. J., Huang, J., Li, L., Liu, P., and Han, X.: Soil carbon and nitrogen stores and storage potential as affected by land-use in an agro-pastoral ecotone of northern China, Biogeochemistry, 82, 127–138, 2007.

Distributions and stoichiometry of dissolved nitrogen and phosphorus in the iron-fertilized region near Kerguelen (Southern Ocean)

S. Blain[1,2], J. Capparos[1,2], A. Guéneuguès[1,2], I. Obernosterer[1,2], and L. Oriol[1,2]

[1]Sorbonne Universités, UPMC Univ Paris 06, UMR7621, Laboratoire d'Océanographie Microbienne, Observatoire Océanologique, 66650 Banyuls/mer, France
[2]CNRS, UMR7621, Laboratoire d'Océanographie Microbienne, Observatoire Océanologique, 66650 Banyuls/mer, France

Correspondence to: S. Blain (stephane.blain@obs-banyuls.fr)

Abstract. During KEOPS2 (Kerguelen Ocean and Plateau Compared Study 2), we determined dissolved inorganic and organic nitrogen and phosphorus species in the naturally fertilized region of Kerguelen Island (Southern Ocean). Above 150 m, stations were clearly separated by the polar front (PF), with concentrations of NO_3^-, NO_2^- and PO_4^{3-} overall lower north of the PF than south. Though less pronounced, a similar trend was detectable for dissolved organic nitrogen (DON) and dissolved organic phosphorus (DOP). At all stations offshore and above the plateau, a subsurface maximum of NH_4^+ was observed between 50 and 150 m. We examined nutrient stoichiometry by calculating the linear combination $N^* = [NO_3^-]\text{-}16\ [PO_4^{3-}]$. The majority of stations and depths revealed N^* close to $-3\,\mu$M; however, for surface waters north of the PF, N^* increased up to $6\,\mu$M. This suggests a preferential uptake of PO_4^{3-} versus NO_3^- by fast-growing diatoms. Using the tracer $TN_{xs} = [TDN]\text{-}16[TDP]$ (TDN, total dissolved nitrogen; TDP, total dissolved phosphorus) revealed that the dissolved organic fraction significantly contributed to changes in TN_{xs}. TN_{xs} values were negative for most stations and depths, and relatively constant in the 0–500 m layer. As for N^*, the stations north of the PF had higher TN_{xs} in the 0–100 m layer. We discuss this stoichiometric anomaly with respect to possible external sources and sinks of N and P. Additional data collected in February 2013 at two sites revealed the occurrence of a subsurface minimum of N^* located just below the pycnocline, which denotes a layer where remineralization of particulate organic matter with low N : P ratio P, possibly associated with preferential remineralization of P versus N, persists throughout the season.

1 Introduction

The first scientific expeditions in the Southern Ocean discovered high concentrations of major nutrients such as nitrate (NO_3^-) and phosphate (PO_4^{3-}) in surface waters south of 50° S (Hart, 1942). The general meridional overturning circulation that brings deep water to the surface at the southern limits of the Antarctic Circumpolar Current (Marshall and Speer, 2012) is the major mechanism supplying surface waters with NO_3^- and PO_4^{3-}. Most of the nutrient-rich upwelled waters are transported northward and they leave the surface north of the polar front through their transformation into intermediate and mode waters. Despite the several-month-long northward transport during which the NO_3^-- and PO_4^{3-}-rich waters are exposed to sunlight, not much phytoplankton biomass develops. This system was characterized as high-nitrate low-chlorophyll (HNLC). The major consequence of the HNLC status of the Southern Ocean is that large amounts of unused nutrients are transported back into the ocean interior where they feed the main thermocline and finally supply low- and mid-latitude surface waters with essential nutrients (Sarmiento et al., 2004). Another consequence is that similarly to NO_3^- and PO_4^{3-}, large amounts of upwelled dissolved inorganic carbon (DIC) are not converted to particulate organic carbon (POC) and remain in contact with the atmosphere for time periods long enough to degas carbon dioxide (CO_2) with important consequences for climate (Sigman and Boyle, 2000).

The iron hypothesis (Martin and Fitzwater, 1988) was a major advancement for our understanding of the HNLC paradox. More than 2 decades of intense research have confirmed

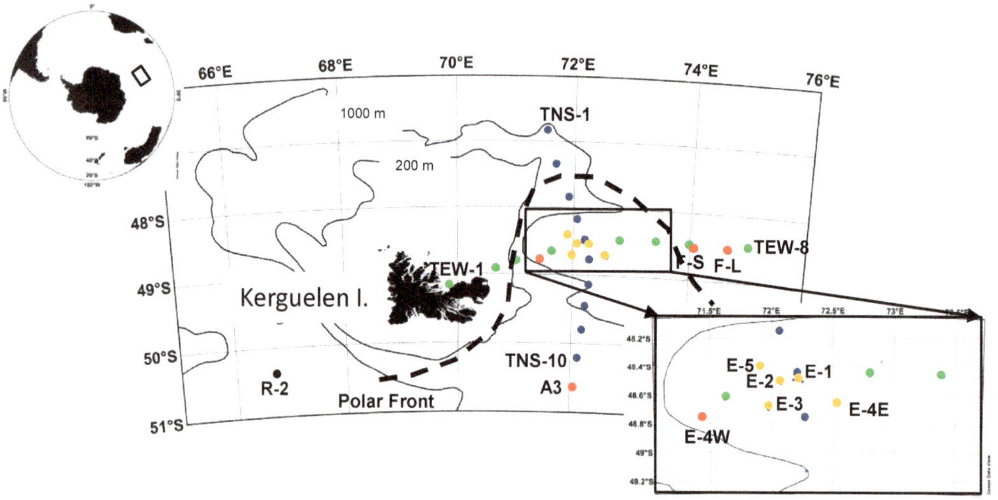

Figure 1. Map of the KEOPS2 study area. The locations of the stations are marked by colored dots. Blue indicates the stations of the north–south transect (TNS), green indicates the stations of the east–west transect (TEW), orange indicates the stations E located in the meander of the polar front (zoom panel). Red shows other stations located in the fertilized region and black shows the station located in the high-nitrate low-chlorophyll (HNLC) region. Detailed positions of the stations are given in Table S1.

that increasing iron supply stimulates primary production, major nutrient utilization and the air-to-sea flux of CO_2 in surface waters. Nutrient utilization in surface waters is therefore a diagnostic of the efficiency of the biological pump of CO_2. Nitrate utilization has also received much attention in paleoceanographic studies, because it can be inferred from the isotopic composition of N in bulk material or specific compounds of fossil organisms preserved in the sediment. Recent results provide support to the enhanced NO_3^- utilization related to higher dust deposition during the ice ages in the sub-Antarctic region (Martinez-Garcia et al., 2014).

Early modeling studies on the iron hypothesis were conducted using models that did not explicitly represent the iron cycle. The effect of iron fertilization was mimicked using the extreme assumption that iron fertilization results in the complete depletion of N or P in surface waters (Gnanadesikan et al., 2003; Sarmiento and Orr, 1991). However, this was never observed during artificial iron fertilization (Boyd et al., 2007), iron addition during deck incubations (Moore et al., 2007) or in naturally iron-fertilized regions (Blain et al., 2007). For most previous research in this context, it was assumed that NO_3^- and PO_4^{3-} behave in a similar way. This is only true at first glance as interesting differences have been noticed (Jenkins et al., 1984; Minas and Minas, 1992; Lourey and Trull, 2001). Weber and Deutsch (2010) used zonal mean distributions of NO_3^- and PO_4^{3-} in the Southern Ocean to reveal that the differential utilization of both nutrients is likely related to the composition of the phytoplankton community. Detailed investigations of blooms in varying regions of the Southern Ocean confirm different utilization of NO_3^- and PO_4^{3-} depending on the dominant species in the phytoplankton community (Arrigo, 1999; De Baar et al., 1997; Moore et al., 2007). In addition, the possible role of dissolved organic nitrogen (DON) and dissolved organic phosphorus (DOP) for N and P decoupling has not been investigated, although modeling studies suggest that these organic forms may significantly contribute to the cycling of N and P in the Southern Ocean (Wang et al., 2003).

Our work presents new data on dissolved inorganic and organic nitrogen and phosphorus concentrations from the iron-fertilized regions near the Kerguelen archipelago. We present their spatial and temporal distributions, and discuss their stoichiometry.

2 Material and methods

2.1 Sampling

During KEOPS2 (Kerguelen Ocean and Plateau Compared Study 2), the samples were collected at the stations presented on the map in Fig. 1. The coordinates and date of sampling are summarized in Table S1. Additional samples were collected during the cruise KEOPSMOOR (Kerguelen Ocean and Plateau compared Study, Moorings) in February 2013 at stations A3 and at station TNS-6 (Table S1). The samples for dissolved nitrogen and phosphorus analyses were collected with twenty-two 12 L Niskin bottles mounted on a rosette equipped with a Seabird SBE911-plus CTD (conductivity, temperature and depth) unit. In this work, we used potential temperature (θ) and density anomaly (σ) to characterize the hydrology of the stations. A more complete description of the hydrology and the circulation is presented in Park et al. (2014).

For NO_3^-, PO_4^{3-} and nitrite (NO_2^-), syringes (50 mL) were directly connected to the spigot of the Niskin bottles. The samples were drawn through a 0.45 μm Uptidisc (Whatman) adapted for the syringe. Duplicate samples were collected. The second sample (25 mL) was poisoned with mercuric chloride ($HgCl_2$, 20 mg L^{-1}, final concentration) and stored in the dark at room temperature for later analysis.

For ammonium (NH_4^+), samples were collected from Niskin bottles in two 50 mL Schott glass bottles. Following rinsing, the bottles were filled with 40 mL of seawater and closed immediately to avoid contamination by air. Back at the onboard laboratory, the oxidative reagent (Holmes et al., 1999) was added. Samples for NH_4^+ determination were incubated for at least 3 h in the dark, at ambient temperature, before fluorescence measurements ($\lambda_{exc} = 370$, $\lambda_{emi} = 460$ nm) with a fluorimeter (Jabsco).

For dissolved organic nitrogen (DON) and phosphorus (DOP) analysis the samples were collected from Niskin bottles in 100 mL Schott glass bottles. The Schott glass bottles were rinsed with HCl (10 %) once and several times with ultrapure water (prepared by deionization and UV sterilization) between casts. The samples were then filtered through two combusted GF/F filters. 20 mL of the filtered samples were transferred to 20 mL PTFE vials and poisoned with 100 μl of $HgCl_2$ (4 g L^{-1}, working solution) before storage at 4 °C. All analyses were performed aboard as described below.

2.2 Analytical methods

For NO_3^-, NO_2^-, PO_4^{3-}, one sample was immediately analyzed aboard with a segmented flow analyzer (Skalar) equipped with colorimetric detection using methods described in (Aminot and Kérouel, 2007). The accuracy of the methods was assessed using reference material (Certipur, Merck). The precisions were in the range of 1–4 %, and the limit of detection was 0.02 μM for NO_3^- and NO_2^-, and 0.03 μM for PO_4^{3-}.

Samples for DON and DOP determination were spiked with 2.5 mL of the oxidative reagent (boric acid + sodium hydroxide + potassium peroxodisulfate), and then heated at 120 °C for 30 min. After cooling, the concentrations of NO_3^- and PO_4^{3-} were determined as mentioned above. This provides the concentrations of total dissolved nitrogen (TDN) and total dissolved phosphorus (TDP). The concentrations of DON and DOP were calculated as follows; DON = TDN-[NO_3^-]-[NO_2^-] and DOP = TDP-[PO_4^{3-}].

3 Results

Most of the stations are located south of the polar front (PF), with the exception of the coastal stations TEW-1–2 and the offshore stations TNS-1–2, TEW-7–8 and F-L, which were located north of the PF (Fig. 1). Station R-2, located west of the plateau, had low chlorophyll concentrations in surface water throughout the season ($\sim$0.3 mg m^{-3}) (Lasbleiz et al., 2014), an observation that is explained by the low iron supply (Quéroué et al., 2015). In contrast, all other stations were characterized by the development of large spring blooms consistent with higher iron supply (Lasbleiz et al., 2014). However, the development of the blooms within the iron-fertilized region was not homogenous in time and space. A3-1 and stations TNS-1 to TNS-10 of the north–south transect, sampled at the beginning of the spring bloom, were characterized by low chlorophyll concentrations only slightly higher than that at station R-2. Stations TEW-1 to TEW-8 of the east–west transect, stations E-2 to E-5, and station A3-2 (the second visit to station A3), were sampled a few days later, when the bloom rapidly developed with large spatial heterogeneity. The largest stocks of chlorophyll *a* within the 0–200 m layer were observed at stations F-L, north of the PF, and at station A3-2, above the plateau. Based on the trajectories of two surface drifters (Zhou et al., 2014), stations E-1, E-2, E-3, E4-E and E-5 are assumed to evolve in a quasi-Lagrangian framework, and their succession in time can be considered approximately as a time series.

3.1 Two-dimensional distributions of dissolved nitrogen and phosphorus.

In the upper 200 m of the water column, concentrations of NO_3^- and PO_4^{3-} were ≥ 19 and ≥ 1 μM, respectively (Figs. 2 and 3). Concentrations were higher west of the PF (transect E–W, Fig. 2) and south of the PF (transect N–S, Fig. 3) and lower in surface sub-Antarctic waters, north and east of the PF. Concentrations of NO_2^- were the highest above 150 m, and below this depth NO_2^- decreased rapidly to reach values close to the detection limit at 200 m. Above 150 m, NO_2^- concentrations were clearly higher at the stations in the polar front zone (PFZ) (NO_2^- in the range 0.3–0.4 μM) than at those in the Antarctic zone (AZ) (NO_2^- of 0.25 μM).

Along the transect E–W, the highest NO_2^- concentrations were measured at TEW-1 (0.31–0.34 μM). Contrasting with the NO_2^- distribution observed along the transect N–S, the stations of the AZ (i.e., west of the isocline $\sigma = 27$) had higher concentrations than those of the PFZ. NH_4^+ concentrations were highest at the coastal stations. At stations TEW-1, concentrations of NH_4^+ increased from 0.19 μM (at 10 m) to 1.45 μM (close to the bottom). The same trend was observed at TEW-2 (0.17 μM at 10 m and 0.39 μM close to the bottom). At all stations offshore and above the plateau, a subsurface maximum of NH_4^+ peaking at 0.5–0.6 μM was observed between 50 and 150 m. The DON distribution was characterized by a north–south gradient in the 0–150 m layer. DON concentrations above the Kerguelen plateau at stations A3-1 and TNS-10 (6.0 ± 1.0 μM) were similar to those in the meander of the PF 6.4 ± 1.7 μM (stations TNS-3 to TNS-7). But higher values were detected north of the PF (8.6 ± 1.2 μM for stations TNS-1 and TNS-2).

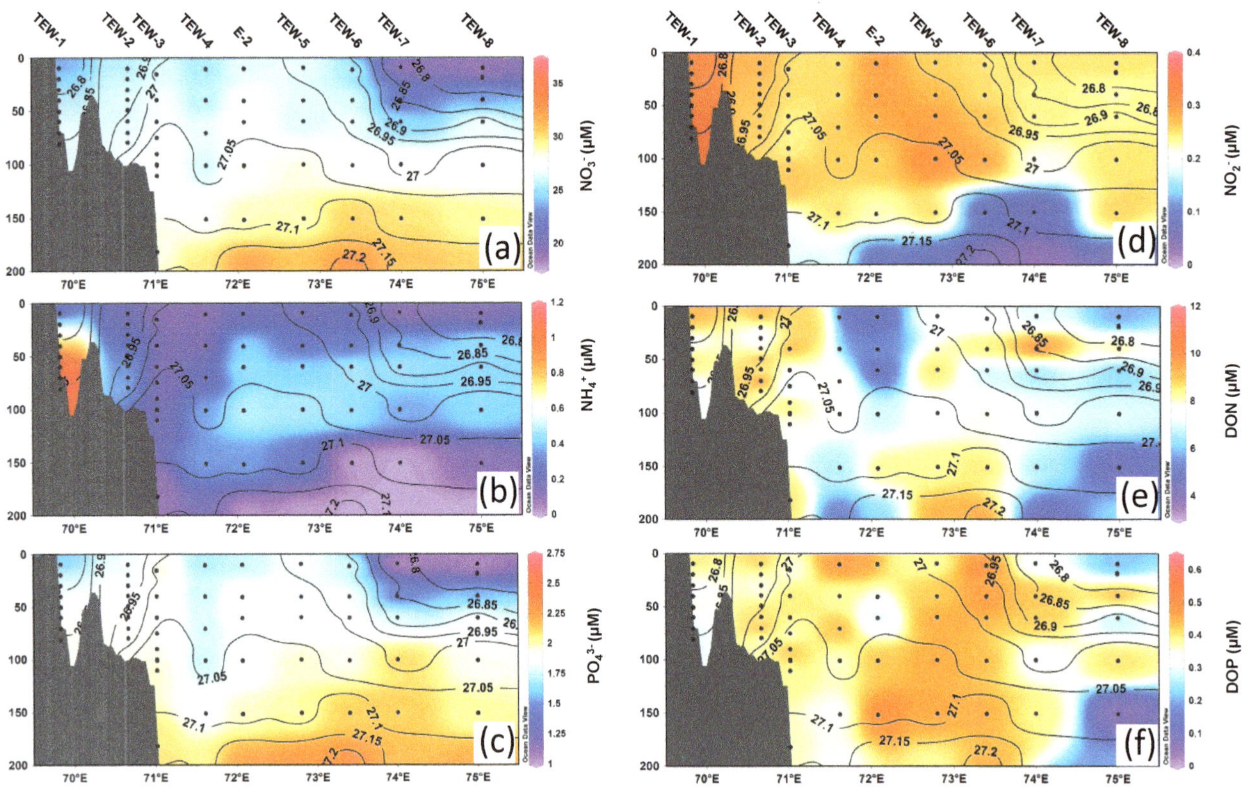

Figure 2. Vertical sections of dissolved nitrogen and phosphorus species along the East-West transect (TEW). **(a)** nitrate, **(b)** ammonium, **(c)** phosphate, **(d)** nitrite, **(e)** dissolved organic nitrogen, **(f)** dissolved organic phosphorus. The isolines for density anomaly (σ) are plotted on each panel.

For DOP, the latitudinal gradient was less pronounced, but DOP concentrations were lower above the Kerguelen plateau than at any other sites.

3.2 Speciation of dissolved nitrogen at selected sites

3.2.1 The Kerguelen plateau station A3

The vertical distribution of different chemical nitrogen species during the two visits at station A3 are detailed in Fig. 4. NO_3^- distributions are discussed in more detail in Sect. 3.3. Concentrations of NO_2^- were, during both visits, homogeneous in the mixed layer and revealed a small maximum below the mixed layer depth (MLD). NO_2^- increased from 0.27 μM at A3-1 to 0.33 μM at A3-2 (Fig. 4b). NH_4^+ concentrations roughly doubled between the two visits (0.1 μM at A3-1 to 0.2 μM at A3-2) and clear maxima were detectable at the base of the mixed layer. Concentrations of DON did not change between visits; however, DON accounted for 20 % of TDN in the mixed layer at A3-1, and this contribution increased to 25 % in the upper 40 m water layer at A3-2 (data not shown). Both NO_3^- consumption and DON release during the 4 weeks that separated the two visits explained the increase in the percent DON of TDN. Below 200 m, TDN was higher at A3-1 than at A3-2. This was mainly driven by the differences in DON concentrations that were higher at A3-1 (4.7–6.7 μM) than at A3-2 (1.8–4 μM) in the 250–300 m layer (Fig. 4).

3.2.2 Stations F-S and F-L north of the polar front

Distinct vertical profiles of NO_2^- and NH_4^+ were observed at station F-S. Concentrations of NO_2^- decreased from 0.39 μM at 10 m to 0.22 μM at 93 m. However, we note a remarkably low value of 0.15 μM at 79 m (Fig. 5a). The NH_4^+ profile presented the same anomaly, resulting in two subsurface maxima. This feature contrasts with most other stations where a single subsurface maximum was observed, as for example at station F-L (Fig. 5b) located a few nautical miles away from F-S. We suggest that this anomaly is due to the position of F-S within the polar front where a complex mixing event at a small scale could have occurred. The contribution of DON to TDN at F-S decreased continuously from 34 % at 20 m to 9 % at 120 m. However, close to the surface, the contribution of DON was only 17 % (Fig. 5d).

3.2.3 The HNLC station R-2

The vertical distribution of NO_3^- and DON revealed small variations between the surface and 200 m (Fig. 6a). DON

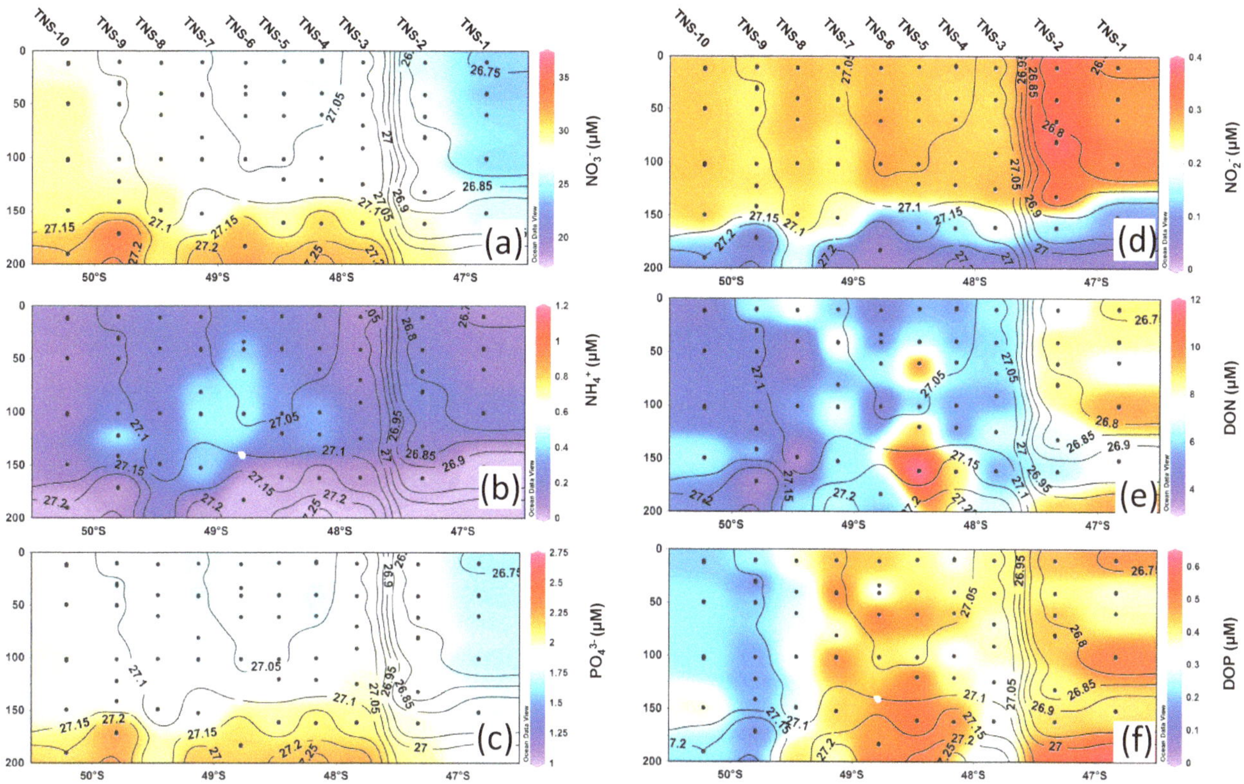

Figure 3. Vertical sections of dissolved nitrogen and phosphorus species along the North-South transect (TNS). **(a)** nitrate, **(b)** ammonium, **(c)** phosphate, **(d)** nitrite, **(e)** dissolved organic nitrogen, **(f)** dissolved organic phosphorus. The isolines for density anomaly (σ) are plotted on each panel.

accounted for 19 to 24 % of TDN, representing intermediate values as compared to the range observed in the fertilized region. Concentrations of NO_2^- and NH_4^+ presented similar vertical distributions, decreasing rapidly below the mixed layer (Fig. 6b). Concentrations of NH_4^+ in the mixed layer (0.07 μM) were at least two-fold lower than at any other stations, and NO_2^- concentrations in the mixed layer (0.3 μM) were similar to those of the mixed layers in the fertilized regions.

3.2.4 The Lagrangian sites E

All stations were characterized by similar vertical distributions of NO_2^- and NH_4^+. Concentrations in the mixed layer were in the range 0.25–0.3 μM decreasing to 0.02-0.03 μM below 200 m. The vertical distributions of NH_4^+ are characterized by a subsurface maximum with concentrations (0.4–0.65 μM) two-fold higher than at the surface (0.2–0.3 μM). NO_3^- distributions are described in more detail in the next section. The contribution of DON to TDN in the mixed layer was in the range of 15–25 %. No clear temporal evolution was detectable.

3.3 Temporal evolution of the vertical distributions of nitrate and phosphate

3.3.1 The Lagrangian sites E

The vertical profiles of NO_3^- and PO_4^{3-} concentrations in the upper 200 m of five stations located in the center of a meander of the PF are presented in Fig. 7. In addition, we show data from two other cruises. Samples collected in early October 1995 during the cruise ANTARES3 (Blain et al. 2001) provided data typical of winter conditions. Samples of the KEOPSMOOR profile were collected in February 2013, representing post-bloom conditions.

Concentrations of NO_3^- were almost identical among visits at 150 m (mean 27.5 $\pm$ 0.8 μmol L^{-1}, Fig. 7a). Above 150 m, NO_3^- concentrations change along the season. In winter, concentrations were homogenous from the surface to 150 m, resulting in a mean integrated stock of 4.22 $\pm$ 0.08 mol m^{-2}. In spring, the KEOPS2 profiles qualitatively clustered into two groups. The first cluster is composed of stations TNS-5, TNS-6, E-1, E-2 and E-3 with higher NO_3^- concentrations (mean integrated stock 0–150 m of 4.10 $\pm$ 0.05 mol m^{-2}) than in the group formed by stations E4-E and E-5 (mean integrated stock 0–150 m of 3.90 $\pm$ 0.04 mol m^{-2}). Finally, the lowest concentrations

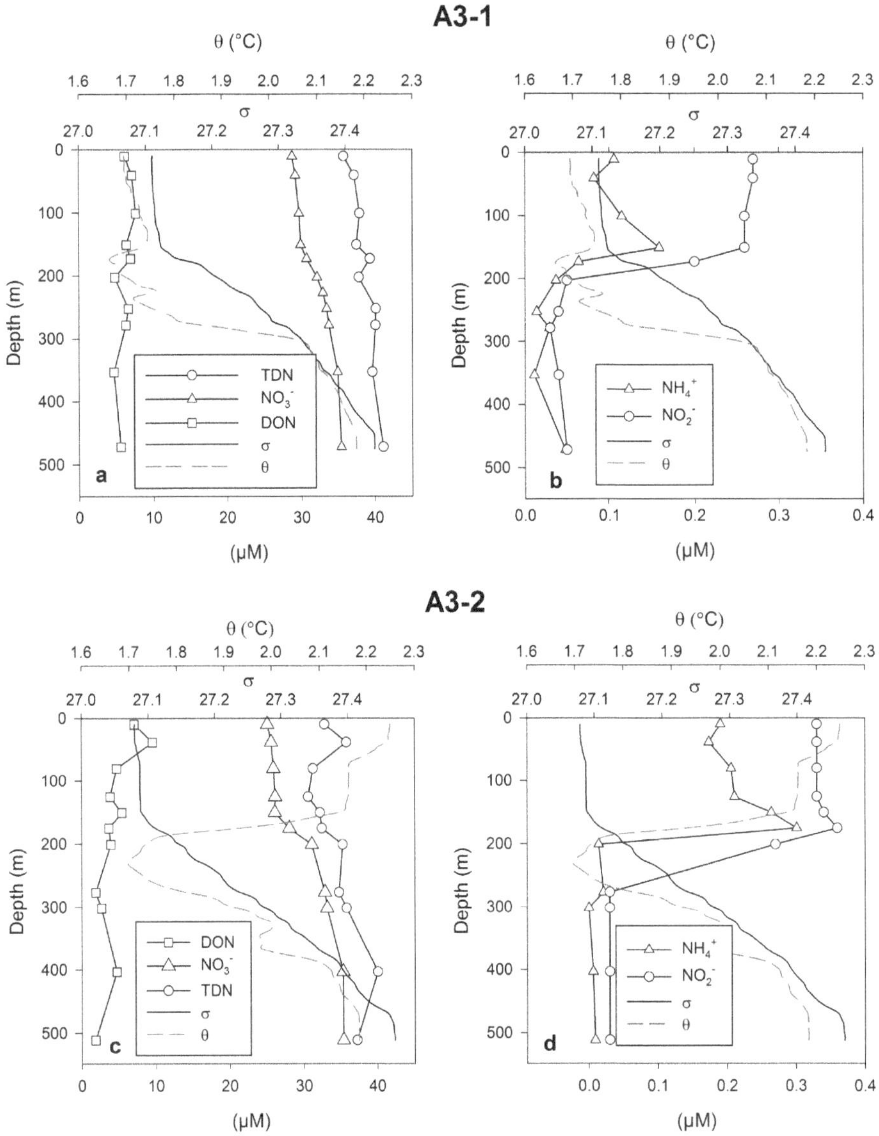

Figure 4. Dissolved nitrogen speciation at station A3-1 **(a, b)** and at station A3-2 **(c, d)** during KEOPS2. Depth profiles of temperature and σ–θ are plotted on each panel.

were measured in summer (mean integrated stock 0–150 m of 3.48 mol m^{-2}).

Vertical profiles of PO_4^{3-} presented similar characteristics as NO_3^-, with the exception of the winter profile (Fig. 7b). The winter profile indicates that PO_4^{3-} concentrations are homogenously mixed in the upper 150 m. The concentrations seem overestimated at 150 m and above. We do not think that the differences result from interannual variability because this would have also impacted NO_3^- concentrations. The high concentrations of PO_4^{3-} measured in winter 1995 lead to a NO_3^- : PO_4^{3-} ratio of 12.5 which is low. The overestimation of PO_4^{3-} could result from methodological issues. The ANTARES3 samples were not analyzed aboard, but a few months later in a laboratory by a different analytical protocol. The lack of certified international standards necessary for strong quality control of the accuracy precludes rigorous comparison of samples collected in 1995 with more recent samples.

Similarly to NO_3^-, we consider the mean concentration of PO_4^{3-} at 150 m (excluding the ANTARES3 PO_4^{3-} value) to estimate a mean winter PO_4^{3-}concentration in the surface layer of $1.93 \pm 0.09\,\mu$mol L^{-1} that yields an integrated winter stock of 0.30 ± 0.02 mol m^{-2}. The integrated stock for the group of stations E-1–E-3 (0.280 ± 0.004 mol m^{-2}) was higher than for the group E-4–E-5 (0.274 ± 0.005 mol m^{-2}). At the end of the season the integrated PO_4^{3-} stock was 0.250 mol m^{-2}.

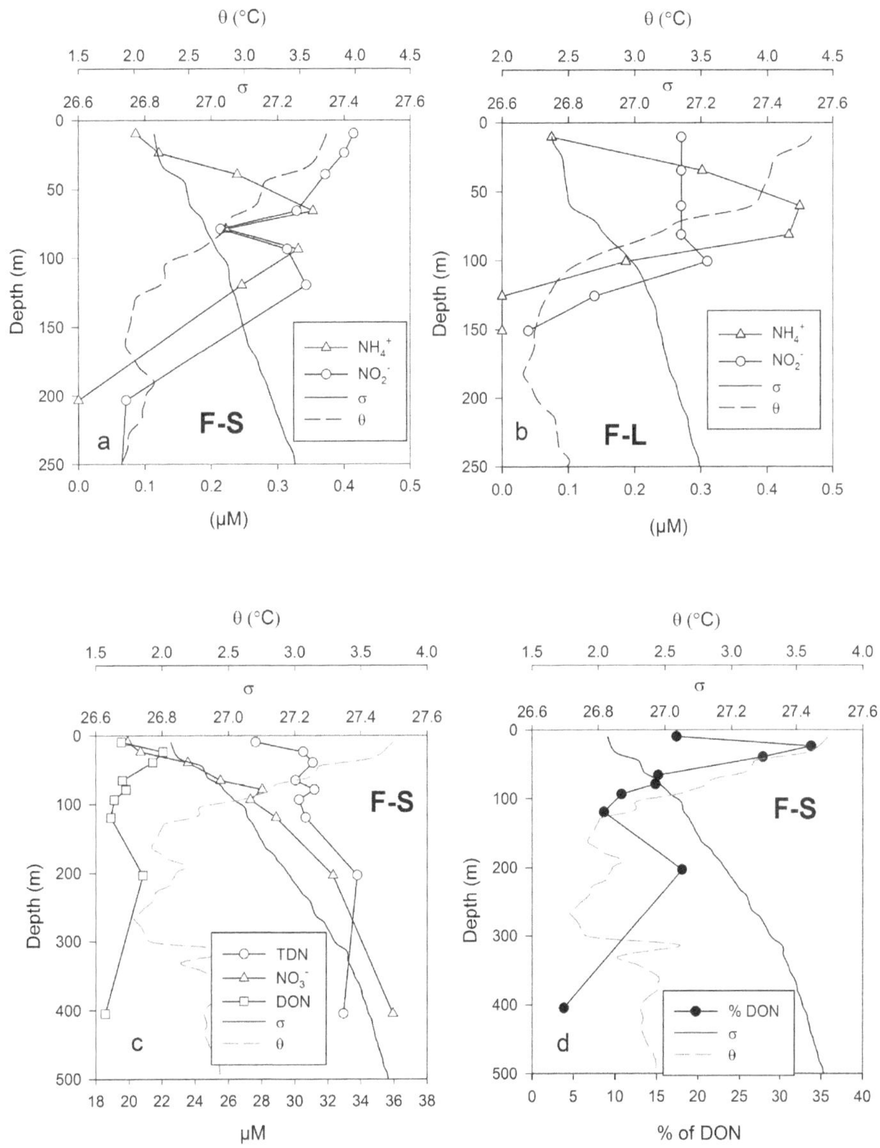

Figure 5. Dissolved nitrogen speciation at stations F-L and F-S during KEOPS2.

3.3.2 The Kerguelen plateau station A3

At station A3, vertical profiles of changes of NO_3^- and PO_4^{3-} concentrations were observed between spring and summer (Fig. 8). Although the stations were sampled in November 2011 and February 2013, we consider these variations as seasonal changes. The profiles of both nutrients merge at 200 m in early spring and summer (A3-1 and A3-2). However, during the second visit at A3 (A3-2), we observed that the surface layer was mixed down to 170 m. We propose that the concentrations at 200 m are a good estimate of the winter concentrations of NO_3^- and PO_4^{3-} at this station. Thus, winter stocks (0–200 m) were 6.27 and 0.43 mol m^{-2} for NO_3^- and PO_4^{3-}, respectively. At the first visit at station A3 the stocks had decreased to 5.96 and 0.41 mol m^{-2}. Four weeks later (A3-2) they reached 5.29 and 0.36 mol m^{-2}. Finally, in February the stocks were 4.77 and 0.35 mol m^{-2}.

4 Discussion

The distributions of NO_3^- and PO_4^{3-} in the world's oceans have been extensively studied over the past decades. A major rationale for this research is the critical role of these major nutrients for phytoplankton growth and therefore marine primary production. Further, concentrations of NO_3^- and PO_4^{3-} are used as tracers for biogeochemical processes in the ocean (Deutsch and Weber, 2012). In the Southern Ocean, south of the sub-Antarctic front, NO_3^- and PO_4^{3-} concentrations are high. They are therefore considered as non-limiting and much less attention has been paid to their distributions as

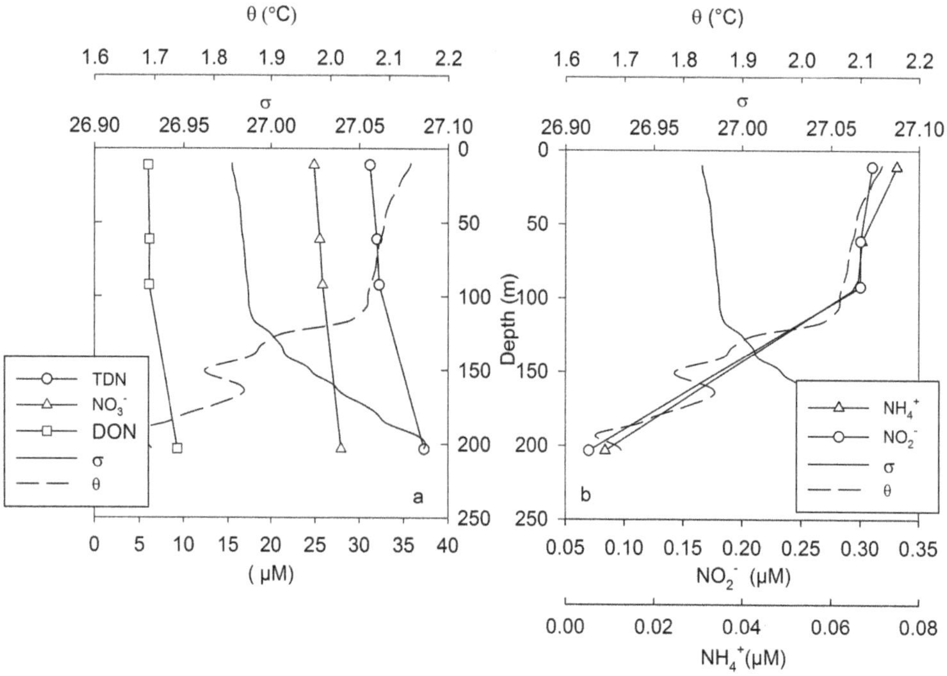

Figure 6. Dissolved nitrogen speciation at station R-2.

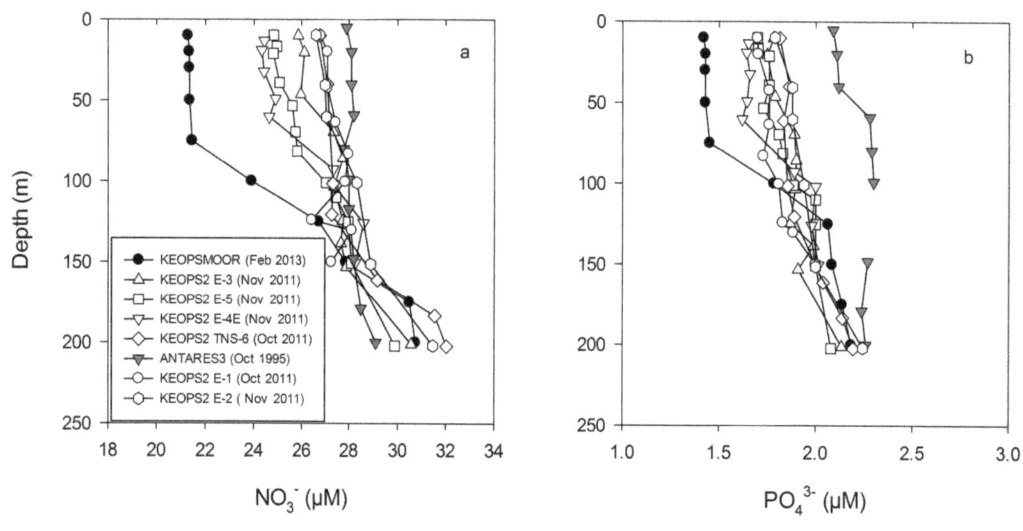

Figure 7. Temporal variability of the vertical profiles of concentrations of NO_3^- **(a)** and PO_4^{3-} **(b)** for stations located in the meander of the polar front. Details for profiles of KEOPSMOOR (February 2013) and ANTARES3 (October 1995) are provided in the text.

compared to other nutrients such as silicic acid or dissolved iron. However, the relaxation of iron limitation by natural or artificial fertilizations offers a different perspective because NO_3^- and PO_4^{3-} should be consumed as the bloom develops. This has motivated the present detailed study of dissolved N and P in the naturally fertilized region of Kerguelen.

To explore the dynamics of NO_3^- and PO_4^{3-} we examined their stoichiometry in the study region. This is commonly done by establishing the ratio $r_{N:P} = [NO_3^-] : [PO_4^{3-}]$ for comparison with the Redfield ratio of 16 (Redfield et al., 1963). However, the significance of $r_{N:P}$ is limited because this ratio is not conserved by mixing or biological processes such as uptake or remineralization (Deutsch and Weber, 2012). We therefore calculated the linear combination $N^* = [NO_3^-]$-16 $[PO_4^{3-}]$, similar to the parameter first introduced by Michaels et al., 1996, but omitting the constant term required to obtain a global average of N^* equal to 0. N^* traces the impact of processes that add or remove N and P with a stoichiometry different from the Redfield ratio of 16. At almost all stations and depths, N^* was close

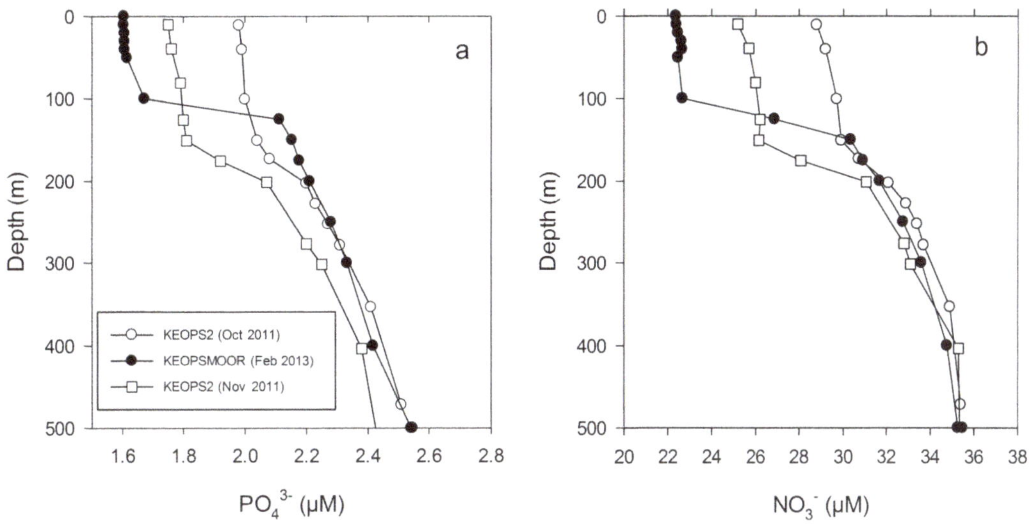

Figure 8. Temporal variability of the vertical profiles of concentrations of NO_3^- **(a)** and PO_4^{3-} **(b)** at the station A3.

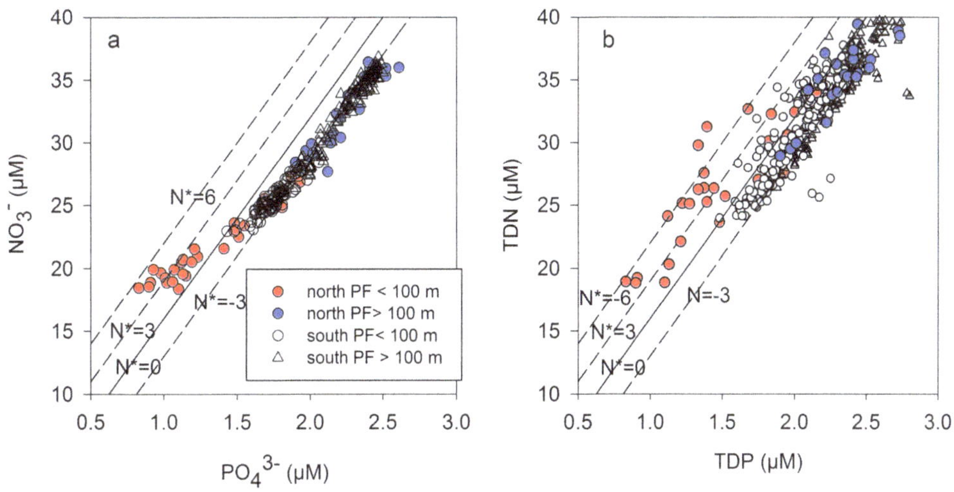

Figure 9. **(a)** Comparison of concentrations of NO_3^- versus PO_4^{3-}. Dots denote the samples, and lines show different values of $N^* = NO_3^- - r_{N:P}\,PO_4^{3-}$. **(b)** Comparison of concentrations of TDN versus TDP, dots denote the samples and lines show different values of $TN_{xs} = TDN - r_{N:P}\,TDP$.

to $-3\,\mu M$ (Fig. 9a). This value agrees well with the mean N^* computed for regions of the Southern Ocean close to the PF (Weber and Deutsch, 2010). A noticeable deviation from this value was observed for a set of data where N^* increased from $N^* = -3\,\mu M$ to $N^* = 6\,\mu M$. All data with $N^* > 0$ are for samples collected in the mixed layer north of the PF, and located in a bloom where diatoms contributed 70 % of the carbon biomass in the euphotic layer (Lasbleiz et al., 2014).

Nutrient drawdown lower than the Redfield ratio has been observed previously in the Southern Ocean. During the artificial iron fertilization experiment EIFEX, an apparent differential consumption of $\Delta(NO_3^-) : \Delta(PO_4^{3-})$ of 6.4 was reported (Smetacek et al., 2012). Arrigo (1999) and De Baar et al. (1997) determined a nutrient drawdown ratio in diatom blooms of 9.7 and 4.4–6.1, respectively. Near the Crozet Islands, the removal of NO_3^- versus PO_4^{3-} measured in situ and during iron-addition experiments revealed that the ratio was inversely related to the proportion of diatoms in the phytoplankton community (Moore et al., 2007). All these studies confirm the impact of diatom blooms on nutrient stoichiometry in the surface layer. However, the interpretations of these observations are diverse. De Baar et al. (1997) suggested that the preferential drawdown of PO_4^{3-} during the bloom of *Fragilariopsis kerguelensis* in the PF could be due to the dominance of *Fragilariopsis kerguelensis* because of the physiological trait of low N : P ratio. These hypotheses could not explain our observations because the stations with a nutrient drawdown anomaly were located in an iron-fertilized

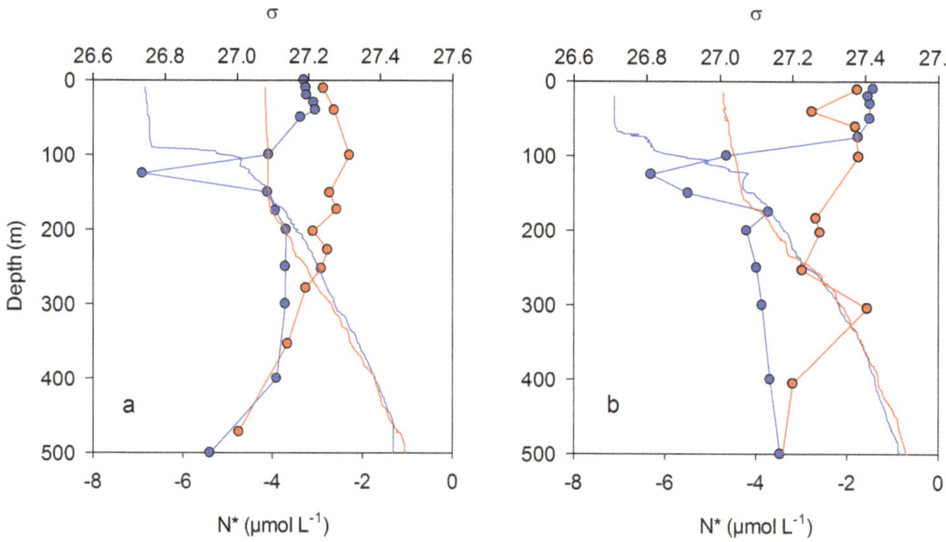

Figure 10. Depth profiles of N^* at stations A3 (**a**) and TNS-6 (**b**) for the month of November (in red) and February (in blue). Vertical profiles of density anomaly are shown with the same color.

region and the diatom community was not dominated by *Fragilariopsis kerguelensis* but rather by *Chaetoceros (Hylochaete)* spp., *Pseudo-nitzschia* spp. and *Centric* sp. (Lasbleiz et al., 2014).

Thus, we interpret the positive values of N^* as a result of the preferential uptake of PO_4^{3-} versus NO_3^- by fast-growing diatoms. Diatoms have a mean elemental N : P stoichiometry of 10 ± 4 (Sarthou et al., 2005), which differs from the Redfield value. Indeed, the elemental particulate matter composition determined at the stations with positive N^* during KEOPS2 (Lasbleiz et al., 2014) exhibits a mean ratio of particulate organic nitrogen to particulate organic phosphorus (PON : POP) of 10.5 ± 3.3 which is consistent with the observed nutrient drawdown $\Delta(NO_3^-) : \Delta(PO_4^{3-})$ of 8. We suggest that the preferential allocation of resources to the P-rich assembly of the cell machinery by exponentially growing cells is the most likely explanation for our observations (Klausmeier et al., 2004). The anomaly observed for the present data set is not linked to a particular species but to general traits of the diatom community responding to iron fertilization.

As a variant of N^* , the tracer DIN_{xs}, takes into account NO_2^- and NH_4^+ (Hansell et al., 2007), but none of those tracers consider the organic pools of N and P. Landolfi et al. (2008) have defined the tracer TN_{xs} = [TDN]-16[TDP] and have shown that the dissolved organic fraction significantly contributes to changes in TN_{xs}. For example, relying on N^* only, can lead to an underestimation of N_2 fixation at the global scale (Landolfi et al., 2008). In the case of KEOPS2, the contribution of DON and DOP to TDN and TDP reached 30 %. We have therefore considered TDN and TDP at all KEOPS2 stations where these measurements were available (Fig. 9b). Plotting TDN as a function of TDP (TDN = f(TDP)) reveals more dispersion of the data than $NO_3^- = f(PO_4^{3-})$, mainly due to the lower analytical precision for DOP and DON determinations. Still, clear trends are detectable. TN_{xs} values were negative for most stations and depths, and relatively constant in the 0–500 m layer. As for N^* the stations north of the PF had higher TN_{xs} in the 0–100 m layer.

When a water parcel is considered, N^* is affected by the redistribution of N and P between the inorganic and the organic pools, whereas TN_{xs} is only affected by net non-Redfield sources or sinks of N and P. Consequently, the positive anomaly observed for TN_{xs} in surface waters north of the PF can be explained by three possible mechanisms: deposition of N-rich material from the atmosphere, N_2 fixation and export of P-rich material. The region of Kerguelen receives low quantities of atmospheric material (Heimburger et al., 2012; Wagener et al., 2008) which is mainly of natural origin, such as desert dust, which contains little nitrogen compared to phosphorus (Zamora et al., 2013). This is confirmed by the low N deposition rate estimated around the Crozet Islands (2 nmol m^{-2} d^{-1}; Planquette et al., 2007). We can therefore refute the deposition of N-rich material as the cause of the TN_{xs} anomaly. The second hypothesis involves N_2 fixation. To date, N_2 fixation has not been reported to occur in the cold waters of the Southern Ocean. However, during KEOPS2 detectable N_2 fixation rates were measured at different stations with a few exceptionally high values ($\sim 250\,\mu$mol m^{-2} d^{-1}) in the mixed layer of station F-L (Gonzàlez et al., 2015). Such high fixation rates could contribute to an enrichment of about 1 % of TDN that is not enough to create the observed anomaly. If N_2 fixation was a dominant process driving the N : P stoichiometry at this station, particulate organic matter (POM) elemental composition should also be affected.

Generally, N_2 fixing microorganisms have a high N : P ratio (Laroche and Breitbarth, 2005). Such high ratios are at odds with the low N : P measured in the POM at station F-L (Lasbleiz et al., 2014). The third hypothesis for explaining the anomaly relies on the export of P-rich material from the mixed layer. We do not have direct measurements of N : P in the exported material, but we have already mentioned above that the elemental composition of particulate matter at station F-L yielded the lowest N : P ratio in POM (Lasbleiz et al., 2014). This provides support for the export of P-rich material resulting in high TN_{xs} values north of the PF. We propose that the anomaly of TN_{xs} results from the imprint on stoichiometry of the diatom bloom which consumed and exported phosphorus with a N : P ratio below the Redfield value.

During KEOPS2, rapidly growing diatom blooms were also sampled at other stations located south of the PF, but anomalies similar to those at F-L were not observed. We discuss here the case of stations A3 and E-4W, which had similar chlorophyll concentrations as F-L. Station A3 had a contribution of diatoms to carbon biomass and dominant diatom species similar to F-L (these observations are not available for E4-W) (Lasbleiz et al., 2014). There is no reason that the physiological features of exponentially growing diatoms as revealed for station F-L do not apply to the diatoms growing at stations A3 and E4-W. It is, however, possible that the resulting effect is not large enough to translate into N^* or TN_{xs} anomalies. A possible explanation could be the differences in the age of the blooms. The stoichiometry would be less affected in a younger bloom as compared to a bloom of longer duration. This hypothesis cannot be fully verified due to the poor temporal resolution of the satellite ocean color images available (see supplementary animations provided in Trull et al., 2014). Another or complementary explanation is the difference in the mixed layer depths that were 50 and 150 m at stations F-L and A3-2, respectively. Such a deep mixed layer as observed at station A3-2 likely resulted from a deep episodic mixing event generated by strong winds prevailing during the day preceding our visit. The deepening could have dampened the anomaly by diluting and mixing the affected water parcel with underlying water having a typical stoichiometry (e.g., N^* or TN_{xs} around −3).

In February 2013, 2 years after the KEOPS2 cruise, we were able to return to two sites visited during KEOPS2 (stations A3 and TNS-6) and obtain measurements for the concentrations of NO_3^- and PO_4^{3-}. These data, in combination with KEOPS2 data, allowed us to compare N^* during two different seasons (Fig. 10). In the mixed layer, little change in N^* was observed between spring and summer. However, in summer, N^* exhibited a clear subsurface minimum between 100–200 m at both stations. Denitrification is a process that could produce this subsurface feature. But denitrification would require low oxygen concentrations that are not observed at these stations. In a general manner, preferential remineralization of P versus N in the water column is supported by an increase of N : P in high molecular dissolved organic matter (Clark, et al., 1998) in particulate matter (Copin-Montegut and Copin-Montegut, 1978) and in supernatant of sediment trap material (Lourey et al., 2003). The observation of the N^* subsurface minimum at the end of the season, but not at the beginning implies a temporal cumulative effect. The minimum is located just below the mixed layer in the region of the pycnocline that presents the highest density gradient. This could represent a zone with a higher residence time for sinking particles resulting in an accumulation of biomass. Additional evidence for intensive remineralization at shallow depths at this location is provided by the strong attenuation of the particles fluxes as observed with moored sediment trap (Rembauville et al., 2015). Consequently, the remineralization would also be increased in this layer compared to the rest of the water column resulting in a higher accumulation of PO_4^{3-} relative to NO_3^-. This effect might be amplified by the occurrence of particulate organic matter with a low N : P ratio resulting from diatom accumulation at the pycnocline. As the subsurface minimum is located above 200 m depth, it is erased when winter mixing occurs.

To our knowledge, such a subsurface minimum has not be reported in the Southern Ocean. This could be due to the limited studies that investigate concurrently dissolved N and P biogeochemistry, and to the lack of samples collected at the appropriate vertical and temporal time scale. Our finding raises several further questions. Is the subsurface minimum of N^* a particular feature of iron-fertilized regions? What is the link between its occurrence and the strength of stratification of the water column? And what is the role of this layer in the remineralization of carbon? These questions argue for future detailed investigation of the cycling of both elements in the upper layer of the Southern Ocean.

Acknowledgements. We thank the chief scientist Bernard Quéguiner, the captain Bernard Lassiette and the crew of the R/V *Marion Dufresne* for their support aboard. We thank C. Lo Monaco for providing us with the CTD profiles of KEOPSMOOR. This work was supported by the French research program of INSU-CNRS LEFE–CYBER (Les enveloppes fluides et l'environnement – Cycles biogéochimiques, environnement et ressources), the French ANR (Agence Nationale de la Recherche, SIMI-6 program, ANR-10-BLAN-0614), the French CNES (Centre National d'Etudes Spatiales) and the French Polar Institute IPEV (Institut Polaire Paul-Emile Victor).

Edited by: T. Trull

References

Aminot, A. and Kérouel, R.: Dosage automatique des nutriments dans les eaux marines, méthodes en flux continu, Ifremer, 2007.

Arrigo, K. R.: Phytoplankton Community Structure and the Drawdown of Nutrients and CO_2 in the Southern Ocean, Science, 283, 365–367, 1999.

Blain, S., Quéguiner, B., Armand, L., Belviso, S., Bombled, B., Bopp, L., Bowie, A., Brunet, C., Brussaard, K., Carlotti, F., Christaki, U., Corbière, A., Durand, I., Ebersbach, F., Fuda, J. L., Garcia, N., Gerringa, L. J. A., Griffiths, F. B., Guigue, C., Guillerm, C., Jacquet, S., Jeandel, C., Laan, P., Lefèvre, D., Lomonaco, C., Malits, A., Mosseri, J., Obernosterer, I., Park, Y. H., Picheral, M., Pondaven, P., Remenyi, T., Sandroni, V., Sarthou, G., Savoye, N., Scouarnec, L., Souhault, M., Thuillers, D., Timmermans, K. R., Trull, T., Uitz, J., Van-Beek, P., Veldhuis, M. J. W., Vincent, D., Viollier, E., Vong, L., and Wagener, T.: Effect of natural iron fertilisation on carbon sequestration in the Southern Ocean, Nature, 446, 1070–1075, 2007.

Boyd, P. W., Jickells, T., Law, C., Blain, S., Boyle, E. A., Buesseler, K. O., Coale, K. H., Cullen, J. J., De Baar, H. J. W., Follows, M., Harvey, M., Lancelot, C., Levasseur, M., Owens, N. J. P., Pollard, D. A., Rivkin, R. B., Sarmiento, J. L., Schoemann, V., Smetacek, V., Takeda, S., Tsuda, A., Turner, D. R., and Watson, A.: Mesoscale iron enrichment experiments 1993-2005: Synthesis and future directions, Science, 315, 612–617, 2007.

Clark, L. L., Ingall, E., and Benner, R.: Marine phosphorus is selectively remineralized, Nature, 393, p. 426, 1998.

Copin-Montegut, C. and Copin-Montegut, G.: the chemistry of particulate matter from the south indien and antarctic ocean, Deep-Sea Res., 25, 911–931, 1978.

De Baar, H. J. W., Van Leeuwe, M. A., Scharek, R., Goeyens, L., Bakker, K. M. J., and Fritsche, P.: Nutrient anomalies in Fragilariopsis kerguelensis blooms, iron deficiency and the nitrate/phosphate ratio (A. C. Redfield) of the Antarctic Ocean, Deep-Sea Res. Pt. II, 44, 229–260, 1997.

Deutsch, C. and Weber, T.: Nutrient Ratios as a Tracer and Driver of Ocean Biogeochemistry, Annu. Rev. Mar. Sci., 4, 113–141, 2012.

Gnanadesikan, A., Sarmiento, J. L., and Slater, R. D.: Effects of patchy ocean fertilization on atmospheric carbon dioxide and biological production, Global Biogeochem. Cy., 17, 1050, doi:10.1029/2002GB001940, 2003.

González, M. L., Molina, V., Florez-Leiva, L., Oriol, L., Cavagna, A. J., Dehairs, F., Farias, L., and Fernandez, C.: Nitrogen fixation in the Southern Ocean: a case of study of the Fe-fertilized Kerguelen region (KEOPS II cruise), Biogeosciences Discuss., 11, 17151–17185, doi:10.5194/bgd-11-17151-2014, 2014.

Hansell, D. A., Olson, D. B., Dentener, F., and Zamora, L. M.: Assessment of excess nitrate development in the subtropical North Atlantic, Maine Chem., 206, 562–579, 2007.

Hart, T. J.: Phytoplankton periodicity in Antarctic surface water, Discov. Rep., VIII, 1–268, 1942.

Heimburger, A., Losno, R., Triquet, S., Dulac, F., and Mahowald, N.: Direct measurements of atmospheric iron, cobalt, and aluminum-derived dust deposition at Kerguelen Islands, Global Biogeochem. Cy., 26, GB4016, doi:10.1029/2012GB004301, 2012.

Holmes, R. M., Aminot, A., Keìrouel, R., Hooker, B., and peterson, B.: a simple and precise method for measuring ammonium in marine and freshwater ecosystem, 56, 1801–1808, 1999.

Jenkins, B. D., Gordon, L. I., and Nelson, D. M.: Nutrient depletion indicates high primary productivity in the Weddel Sea, Nature, 309, 51–54, 1984.

Klausmeier, C. A., Litchman, E., Daufresne, T., and Levin, S. A.: Optimal nitrogen-to-phosphorus stoichiometry of phytoplankton, Nature, 429, 171–174, 2004.

Landolfi, A., Oschlies, A., and Sanders, R.: Organic nutrients and excess nitrogen in the North Atlantic subtropical gyre, Biogeosciences, 5, 1199–1213, doi:10.5194/bg-5-1199-2008, 2008.

Laroche, J. and Breitbarth, E.: Importance of the diazotrophs as a source of new nitrogen in the ocean, J. Sea Res., 53, 67–91, 2005.

Lasbleiz, M., Leblanc, K., Blain, S., Ras, J., Cornet-Barthaux, V., Hélias Nunige, S., and Quéguiner, B.: Pigments, elemental composition (C, N, P, and Si), and stoichiometry of particulate matter in the naturally iron fertilized region of Kerguelen in the Southern Ocean, Biogeosciences, 11, 5931–5955, doi:10.5194/bg-11-5931-2014, 2014.

Lourey, M. and Trull, T. W.: seasonal nutreint depletion and carbon export in the subantarctic and polar frontal zones of the Southern Ocean, J. Geophys. Res. Oceans, 106, 31463–31487, 2001.

Lourey, M. J., Trull, T. W., and Sigman, D. M.: Sensitivity of δ^{15} N of nitrate, surface suspended and deep sinking particulate nitrogen to seasonal nitrate depletion in the Southern Ocean, Global Biogeochem. Cy., 17, 1081, doi:10.1029/2002GB001973, 2003.

Marshall, J. and Speer, K.: Closure of the meridional overturning circulation through Southern Ocean upwelling, Nat. Geosci., 5, 171–180, doi:10.1038/ngeo1391, 2012.

Martin, J. M. and Fitzwater, S. E.: Iron deficiency limits phytoplankton growth in the north-east Pacific subarctic, Nature, 331, 341–343, 1988.

Martinez-Garcia, A., Sigman, D. M., Ren, H., Anderson, R. F., Straub, M., Hodell, D. A., Jaccard, S. L., Eglinton, T. I., and Haug, G. H.: Iron Fertilization of the Subantarctic Ocean During the Last Ice Age, Science, 343, 1347–1350, 2014.

Michaels, A. F., Olson, D., Sarmiento, J., Ammerman, J. W., Fanning, K. A., Jahnke, R. A., Knap, A. H., Lipschultz, F., and Prospero, J. M.: Inputs, losses and transformations of nitrogen and phosporus in the pelagic North Atlantic Ocean, Biogeochemistry, 35, 181–226, 1996.

Minas, H. J. and Minas, M.: Net community production in "High Nutrient-Low Chlorophyll" waters of the tropical and antarctic oceans: grazing versus iron hyopthesis, Oceanol. Acta, 15, 145–162, 1992.

Moore, C. M., Hickman, A. E., Poulton, A. J., Seeyave, S., and Lucas, M. I.: Iron–light interactions during the CROZet natural iron bloom and EXport experiment (CROZEX): II – Taxonomic responses and elemental stoichiometry, Deep-Sea Res. Part II, 54, 2066–2084, 2007.

Park, Y.-H., Durand, I., Kestenare, E., Rougier, G., Zhou, M., d' Ovidio, F., Cotté, C., and Lee, J.-H.: Polar Front around the Kerguelen Islands: An up-to-date determination and associated circulation of surface/subsurface waters, J. Geophys. Res. Oceans, 119, 6575–6592, doi:10.1002/2014JC010061, 2014.

Planquette, H., Statham, P. J., Fones, G., Charette, M. A., Moore, C. M., Salter, I., Nédélec, F. H., Taylor, S. L., French, M., Baker, A. R., Mahowald, N., and Jickells, T. D.: Dissolved iron in the vicinity of the Crozet islands, Southern Ocean, Deep-Sea Res. Pt. II, 54, 1999–2019, 2007.

Quéroué, F., Sarthou, G., Planquette, H. F., Bucciarelli, E., Chever, F., van der Merwe, P., Lannuzel, D., Townsend, A. T., Cheize, M., Blain, S., d'Ovidio, F., and Bowie, A. R.: High variability of dissolved iron concentrations in the vicinity of Kerguelen Island (Southern Ocean), Biogeosciences Discuss., 12, 231–270, doi:10.5194/bgd-12-231-2015, 2015.

Redfield, A. C., Ketchum, B. H., and Richards, F. A.: The influence of organism on the composition of seawater, in: The Sea, edited by: Hill N. M., New York., 1963.

Rembauville, M., Salter, I., Leblond, N., Gueneugues, A., and Blain, S.: Export fluxes in a naturally fertilized area of the Southern Ocean, the Kerguelen Plateau: seasonal dynamic reveals long lags and strong attenuation of particulate organic carbon flux (Part 1), Biogeosciences Discuss., 11, 17043–17087, doi:10.5194/bgd-11-17043-2014, 2014.

Sarmiento, J. L. and Orr, J. C.: Three-dimensional simulations of the impact of Southern Ocean nutrient depletion on atmospheric CO_2 and ocean chemistry, Limnol. Oceanogr., 36, 1928–1950, 1991.

Sarmiento, J. L., Gruber, N., Brzezinsky, M. A., and Dunne, J. P.: High-latitude controls of thermocline nutrients and low latitude biological productivity, Nature, 427, 56–60, 2004.

Sarthou, G., Timmermans, K. R., Blain, S., and Treguer, P.: Growth physiology and fate of diatoms in the ocean: a review, J. Sea Res., 53, 25–42, 2005.

Sigman, D. M. and Boyle, E. A.: Glacial/interglacial variations in atmospheric carbon dioxide, Science, 407, 859–869, 2000.

Smetacek, V., Klaas, C., Strass, V. H., Assmy, P., Montresor, M., Cisewski, B., Savoye, N., Webb, A., d' Ovidio, F., Arrieta, J. M., Bathmann, U., Bellerby, R., Berg, G. M., Croot, P., Gonzalez, S., Henjes, J., Herndl, G. J., Hoffmann, L. J., Leach, H., Losch, M., Mills, M. M., Neill, C., Peeken, I., Röttgers, R., Sachs, O., Sauter, E., Schmidt, M. M., Schwarz, J., Terbrüggen, A., and Wolf-Gladrow, D.: Deep carbon export from a Southern Ocean iron-fertilized diatom bloom, Nature, 487, 313–319, 2012.

Trull, T. W., Davies, D. M., Dehairs, F., Cavagna, A.-J., Lasbleiz, M., Laurenceau, E. C., d' Ovidio, F., Planchon, F., Leblanc, K., Quéguiner, B., and Blain, S.: Chemometric perspectives on plankton community responses to natural iron fertilization over and downstream of the Kerguelen Plateau in the Southern Ocean, Biogeosciences Discuss., 11, 13841–13903, doi:10.5194/bgd-11-13841-2014, 2014.

Wagener, T., Guieu, C., Losno, R., Bonnet, S., and Mahowald, N.: Revisiting atmospheric dust export to the southern hemisphere ocean: biogeochemical implication, Global Biogeochem. Cy., 22, doi:10.1029/2007GB002984., 2008.

Wang, X., Matear, R. J., and Trull, T. W.: Nutrient utilization ratios in the Polar Frontal Zone in the Australian sector of the Southern Ocean: A model., Global Biogeochem. Cy., 17, 1009, doi:10.1029/2002GB001938, 2003.

Weber, T. S. and Deutsch, C.: Ocean nutrient ratios governed by plankton biogeography, Nature, 467, 550–554, 2010.

Zamora, L. M., Prospero, J. M., Hansell, D. A., and Trapp, J. M.: Atmospheric P deposition to the subtropical North Atlantic: sources, properties, and relationship to N deposition, J. Geophys. Res.-Atmos., 118, 1546–1562, 2013.

Zhou, M., Zhu, Y., d'Ovidio, F., Park, Y.-H., Durand, I., Kestenare, E., Sanial, V., Van-Beek, P., Queguiner, B., Carlotti, F., and Blain, S.: Surface currents and upwelling in Kerguelen Plateau regions, Biogeosciences Discuss., 11, 6845–6876, doi:10.5194/bgd-11-6845-2014, 2014.

15

Recent changes in the global and regional carbon cycle: analysis of first-order diagnostics

P. J. Rayner[1], A. Stavert[2], M. Scholze[3], A. Ahlström[3], C. E. Allison[2], and R. M. Law[2]

[1]School of Earth Sciences, University of Melbourne, Melbourne, Australia
[2]CSIRO Oceans and Atmosphere Flagship, Melbourne, Australia
[3]Department of Physical Geography and Ecosystem Science, Lund University, Sweden

Correspondence to: P. J. Rayner (prayner@unimelb.edu.au)

Abstract. We analyse global and regional changes in CO_2 fluxes using two simple models, an airborne fraction of anthropogenic emissions and a linear relationship with CO_2 concentrations. We show that both models are able to fit the non-anthropogenic (hereafter natural) flux over the length of the atmospheric concentration record. Analysis of the linear model (including its uncertainties) suggests no significant decrease in the response of the natural carbon cycle. Recent data points rather to an increase. We apply the same linear diagnostic to fluxes from atmospheric inversions. Flux responses show clear regional and seasonal patterns driven by terrestrial uptake in the northern summer. Ocean fluxes show little or no linear response. Terrestrial models show clear responses, agreeing globally with the inversion responses, however the spatial structure is quite different, with dominant responses in the tropics rather than the northern extratropics.

1 Introduction

The interplay of various timescales in anthropogenically forced climate change is both problematic and fascinating. It is problematic since temperature responses integrate radiative forcing and radiative forcing by greenhouse gases integrates sources. Thus changes in source processes can, if sustained, drive surprisingly large changes in the trajectory of temperature.

For the most important greenhouse gas, CO_2, this double integration gives a respectable utility to an inherently fascinating question: are there changes in the underlying processes of the carbon cycle? The utility comes from the natural carbon cycle's role in mitigating the anthropogenic perturbation by absorbing about half the anthropogenic input of carbon to the atmosphere (Le Quéré et al., 2013). Optimal policy response relies on projections of this uptake so changes in the natural carbon cycle have direct policy implications.

We have had conceptual models for these changes for many years. For the ocean these are predominantly changes in chemical buffering (Revelle and Suess, 1957) and changes in physical circulation (Sarmiento et al., 1998). For terrestrial uptake there are many countervailing factors at work such as extension of the high latitude growing season (Zhou et al., 2001; Piao et al., 2008) and the varied responses of terrestrial ecosystems to changes in temperature and rainfall. Cox et al. (2000) combined many of these responses into a reasonably complete model of the earth system and projected a strong reduction in carbon uptake with the land becoming a net source around 2050. Friedlingstein et al. (2006) showed that this was one of many possible responses. Such studies naturally prompted observational tests of the important processes such as the reaction of the Amazon forest to drying (Saleska et al., 2003). Several studies have suggested sink saturation or reduction in various regions such as Schuster and Watson (2007) for the North Atlantic, Le Quéré et al. (2007) for the Southern Ocean and Nabuurs et al. (2013) for European forests.

Meanwhile the 5-decade record of atmospheric CO_2 raises the possibility of detecting changes in the results of these processes directly. This was first taken up by Canadell et al. (2007) who suggested that sinks were saturating, at least relative to emissions. This was made more explicit by Raupach et al. (2008) who attempted to isolate the anthropogenic and natural contributions to long-term changes in CO_2 growth-rate. The statistical significance of the trends noted by (Canadell et al., 2007) and Raupach et al. (2008) was challenged by Knorr (2009). Gloor et al. (2010) also pointed out difficulties in interpreting changes in the relationship between emissions and growth-rate in terms of the response of the system. They used a linear perturbation model and developed diagnostics of the airborne fraction from it. We will use the same model but, rather than diagnosing the behaviour of a yet simpler model (airborne fraction) we will use it to diagnose the behaviour of inferred or modelled fluxes from more complex systems.

Along with this controversy over long-term changes in the sink efficiency, different questions have emerged on more recent changes. Sarmiento et al. (2010) used a combination of the atmospheric growth-rate, anthropogenic inputs and an ocean model to posit an abrupt change in the terrestrial uptake around 1988. Francey et al. (2010) and Francey et al. (2013) pointed out that, since the early 2000s, the growth-rate of atmospheric CO_2 had failed to keep pace with the acceleration in reported fossil fuel use. Their conclusion was to question the timing of this acceleration.[1]

To provide context for subsequent discussions, Fig. 1 plots the history of anthropogenic carbon fluxes and the growth-rate in atmospheric CO_2. It also shows the predicted growth-rate from two simple models to be discussed later. Data is taken from Le Quéré et al. (2013). We see a clear increase in anthropogenic fluxes and a much noisier increase in the atmospheric growth-rate. We also see an increasing divergence between these curves, connoting an increasing uptake. This uptake is a response to a range of perturbations, atmospheric CO_2 itself, nutrient input, land management and land-use change and doubtless many others. Here we analyse this uptake as a simple linear response to CO_2 concentration. We use CO_2 concentration as a surrogate for forcings with a similar time course, that is we do not attempt to separate CO_2 forcing of the response from other drivers. Rather we ask whether there has been significant departure from this linear response evident in recent years. Further we analyse regional contributions to this linear response. This provides a simple diagnostic of model responses which can be compared with inverse estimates of regional fluxes. Our focus is on the change of uptake rather than its mean value. Such analysis of trends requires reasonably long records and is hence less certain at regional than global scales.

The outline of the paper is as follows: Sect. 2 describes the simple diagnostics we use and the data. Section 3 analyses the global record in terms of this diagnostic. Section 4 applies the same diagnostic to regional fluxes from inverse estimates while Sect. 5 applies it to terrestrial models. Section 6 points out some of the caveats and implications in the preceding analysis and Sect. 7 summarizes the main points.

[1]Some confusion has arisen between the two discussions of changes in airborne fraction. In general they address different time-scales.

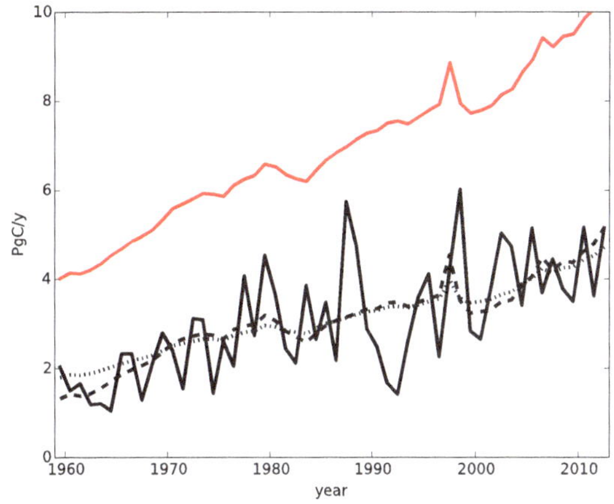

Figure 1. Anthropogenic inputs (red) and atmospheric growth rate (black) from Le Quéré et al. (2013). Anthropogenic inputs include both fossil and land-use. The dotted line shows the predicted atmospheric growth-rate from the airborne fraction model while the dashed line shows the growth-rate from the β-model.

2 Methods and tools

2.1 Defining the carbon budget

Our aim is to analyse the response of parts of the carbon budget to changes in forcing. We must therefore define which terms of the carbon budget we consider. We start with the decomposition used by the Global Carbon Project (GCP) (Le Quéré et al., 2013)

$$\frac{\partial M}{\partial t} = F_{\text{fossil}} + F_{\text{LUC}} - F_{\text{land}} - F_{\text{ocean}}, \qquad (1)$$

where M is the mass of carbon in the atmosphere, F_{LUC} is the flux due to land-use change (LUC) and all other fluxes have their usual meanings. Throughout the paper we will talk of uptakes by land and ocean so we have not followed the usual convention of writing fluxes with a single direction (towards the atmosphere or surface).

We will also frequently combine the two anthropogenic fluxes as

$$F_{\text{anthro}} = F_{\text{fossil}} + F_{\text{LUC}}. \qquad (2)$$

Some terms in Eq. (1) are ambiguous, especially the partition between F_{LUC} and F_{land}. This point is discussed by Enting et al. (2012). When discussing global budgets we will follow the GCP definitions. The atmospheric inversion studies we draw on do not separate these two fluxes. Globally we will correct F_{land} by F_{LUC} from the GCP. When we consider regional budgets we will ascribe changes in the combined flux to F_{land} and discuss the implications of this approximation.

2.2 Two models

We follow Gloor et al. (2010) in using two models for the change in atmospheric CO_2 concentration in response to anthropogenic inputs. The provenance of these two models and the relationship between them is thoroughly described by Gloor et al. (2010) so we will only summarize them here.

2.2.1 Airborne fraction model

This expresses the change in the atmospheric mass of carbon as

$$\frac{\partial M}{\partial t} = \alpha F_{\text{anthro}}, \tag{3}$$

where α is known as the airborne fraction.[2] Combining this with Eq. (1) we see

$$F_{\text{land}} + F_{\text{ocean}} = (1 - \alpha) F_{\text{anthro}}. \tag{4}$$

It is important to remember that Eq. (4) represents a relationship following from mass conservation rather than a causal relationship between anthropogenic inputs and contemporaneous uptakes. It is hard to conceive a mechanism that would link the three fluxes in Eq. (4).

2.2.2 The β-model

An alternative to the airborne fraction model is to parameterize CO_2 uptakes as a linear function of CO_2 concentration or, equivalently, CO_2 mass (Gloor et al., 2010). Thus we write

$$F_{\text{land}} + F_{\text{ocean}} = \beta(M - M_0), \tag{5}$$

where M is the mass of CO_2 in the atmosphere and M_0 is the background or equilibrium mass of CO_2 in the atmosphere. Given the near-equilibrium of the preindustrial carbon cycle evident from the data of Etheridge et al. (1996) and Francey et al. (1999) we often use the preindustrial value of M for M_0. With our focus in this paper on changes rather than mean values we are not interested in M_0 so we simplify Eq. (5) to

$$F_{\text{land}} + F_{\text{ocean}} = \beta M + F_0. \tag{6}$$

β has units of yr^{-1} and plays the role of an inverse residence time for excess carbon against the processes of land and ocean uptake.

Substituting Eq. (6) into Eq. (1) yields

$$\frac{\partial M}{\partial t} = F_{\text{anthro}} - \beta M - F_0. \tag{7}$$

With independent data available on F_{anthro} and M it is possible to estimate β and F_0 using standard statistical techniques such as linear regression. Below we apply this technique to flux estimates at several scales and from several sources. Although we consider some aspects of uncertainty in the calculation we have not applied our diagnostics to the ensembles of results available in intercomparisons of forward or inverse models.

3 Global responses

In this section we compare the behaviour of the two models introduced in Sect. 2.2. We use the data from the Global Carbon Project (Le Quéré et al., 2013) to estimate α from Eq. (3) and β from Eq. (6). Data in Le Quéré et al. (2013) comes from many sources. M (and consequently $\frac{\partial M}{\partial t}$) come from concentration measurements of CO_2 at South Pole and Mauna Loa before 1980 and a range of marine boundary layer sites thereafter. F_{fossil} is derived from inventories from the Carbon Dioxide Information Analysis Center. F_{LUC} and F_{ocean} come from a combination of inventories and models. F_{land} is derived as a residual.

We use the standard maximum likelihood least squares formulation so that

$$\boldsymbol{x} = \mathbf{K}\mathbf{J}^{\text{T}}\mathbf{R}^{-1}\boldsymbol{y}, \tag{8}$$

where $\boldsymbol{x}$ is the vector of unknowns we seek, $\boldsymbol{y}$ the data, $\mathbf{J}$ the Jacobian matrix mapping $\boldsymbol{x}$ to $\boldsymbol{y}$ and $\mathbf{R}$ the uncertainty covariance for $\boldsymbol{y}$. $\mathbf{K}$ is given by

$$\mathbf{K} = [\mathbf{J}^{\text{T}}\mathbf{R}^{-1}\mathbf{J}]^{-1}. \tag{9}$$

After some simplification the uncertainty covariance for $\boldsymbol{x}$ is given by

$$\mathbf{C}(\boldsymbol{x}) = \mathbf{K}^{-1}. \tag{10}$$

For the α-model $\mathbf{J} = F_{\text{anthro}}$ and $\boldsymbol{y} = \frac{\partial M}{\partial t}$ while for the β-model $\mathbf{J} = \frac{1}{M}$ and $\boldsymbol{y} = F_{\text{anthro}} - \frac{\partial M}{\partial t}$. For the β-model we include a constant term in the inversion (see Eq. 5). This is mathematically the uptake when $M = 0$. Physically it represents uptakes which do not vary with M, e.g. those caused by reforestation. It also contributes to the mean uptake over a period. We stress that we are not here concerned with the mean uptake over the whole or part of the study period.

For $\mathbf{R}$ there are two contributions, data uncertainties and modelling errors. The uncertainties in $\boldsymbol{y}$ are quoted in Le Quéré et al. (2013) as 5 % for F_{fossil}, 0.5 $PgC\,yr^{-1}$ for F_{LUC} and 0.7 or 0.2 $PgC\,yr^{-1}$ for $\frac{\partial M}{\partial t}$ before or after 1970. We add these quadratically. Growth-rate uncertainty dominates before 1970 while F_{LUC} is the largest contributor later. The root mean square value of the uncertainty is 0.69 $PgC\,yr^{-1}$. The errors due to the simplicity of the models can only be calculated once we have performed the fit.

Le Quéré et al. (2013) did not give clear guidance on temporal correlation for their uncertainties. The most likely form

[2] Airborne fraction can also be quoted relative to F_{fossil}.

for these is positive temporal correlation arising from systematic errors in reporting or biogeochemical models. Our analysis is concerned with trends, that is of year-to-year differences. Positive temporal correlations will increase the significance of these trends. Thus we make the conservative assumption of temporal independence. The one case where this is not true we will treat explicitly.

Figure 1 also shows the observed and predicted atmospheric growth-rate from the two models. The regression solutions give $\alpha = 0.45$ and $\beta = 0.016\,\mathrm{yr}^{-1}$. The two models produce mean-square residuals of $0.95\,\mathrm{Pg\,C\,yr}^{-1}$. Thus we use this value as the uncertainty **R** for the dependent variable in the regression for the β-model. It yields a 1σ uncertainty of $0.002\,\mathrm{yr}^{-1}$. Calculating the uncertainty of α is more difficult since the most uncertain term is the Jacobian. We can approximate it by noting that the relationship between F_{anthro} and $\frac{\partial M}{\partial t}$ can be integrated to give $M_{\mathrm{final}} - M_{\mathrm{initial}} = \alpha \sum F_{\mathrm{anthro}}$. The total change of CO_2 mass in the atmosphere is constrained by the initial and final concentration uncertainties and these concentrations are very well known. Thus the percentage error in α is the percentage error in $\sum F_{\mathrm{anthro}}$. The uncertainty in $\sum F_{\mathrm{anthro}}$ can be calculated for the limiting cases of complete independence and perfect temporal correlation. For the independent case we sum uncertainties quadratically to give 4.3 Pg C of a total of 372 Pg C, or about 1 % uncertainty. For the case of perfect correlation the 5 % uncertainty in annual values translates to a 5 % uncertainty in the total. Thus the uncertainty in α lies between 1 % and 5 %.

Figure 2 shows the residuals from the α and β models from Fig. 1. The two models produce similar residuals. Both residuals are driven by short-term changes in the atmospheric growth-rate and arise from the failure of these simple integrated models to reproduce such changes. One striking similarity is the increase in the amplitude of the residuals with time. The amplitudes grow by 60 % from the first to the second half of the period. The interannual variability has been used by Cox et al. (2013) and Wang et al. (2014) to assess the sensitivity of the carbon cycle to forcing.

By construction, β provides an optimal fit to the time course of $\frac{\partial M}{\partial t}$ but this does not mean it is optimal throughout. We can ask whether different periods suggest different magnitudes for β. Here we focus on the 11-year period 2002–2012. We repeat the calculation, obtaining $\beta = 0.057 \pm 0.018\,\mathrm{yr}^{-1}$. This is much larger but much more uncertain than the overall value of $\beta = 0.016 \pm 0.002\,\mathrm{yr}^{-1}$. The large β value is a direct result of the negative trend in the residuals evident from 2002. The difference can be considered statistically significant with a 5 % probability of a larger value occurring by chance over this period. We can also ask whether it is robust, that is how sensitive is the result to our choice of period. We repeat the analysis for every 11-year period in the record (i.e starting with 1959, 1960 etc). This yields 6 values greater than $0.057\,\mathrm{yr}^{-1}$.

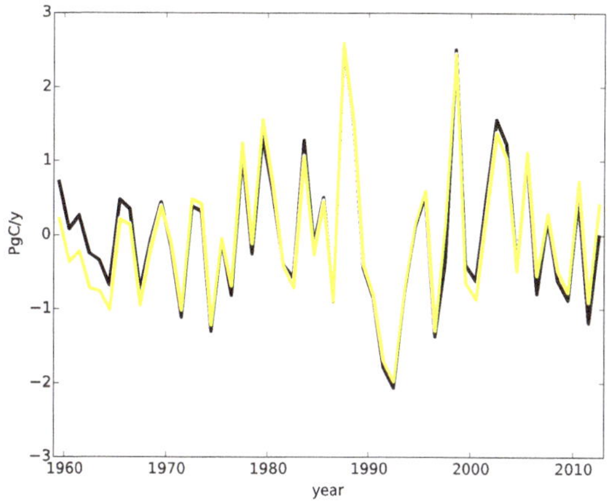

Figure 2. Residuals in the growth rate (observed – predicted) for the α model (yellow) and β model (black).

We can also ask whether the mean residual of the fit of the β-model is significantly different from 0. The mean residual for 2002–2012 is $-0.07 \pm 0.27\,\mathrm{Pg\,C\,yr}^{-1}$. The large error bar is a result of the large interannual variability. In summary, the mean CO_2 uptake over the last decade is not significantly different from that predicted by a linear response to concentration. The change in trend over that time is approaching significance but is not robust. Let us now analyse some spatially resolved estimates of fluxes to try to attribute the behaviour over the full period and more recently.

3.1 Land and ocean contributions

One useful property of Eq. (6) is that, if we can decompose fluxes as $F = F_1 + F_2 + \ldots$ we can decompose the corresponding β values as $\beta = \beta_1 + \beta_2 + \ldots$. We will show decompositions into land or ocean, by latitude band and by season. In each case we replace $\boldsymbol{y}$ in Eq. 8 with the corresponding flux.

First we calculate β for land and ocean separately using the values and uncertainties from Le Quéré et al. (2013). The uncertainties are $0.5\,\mathrm{Pg\,C\,yr}^{-1}$ for the ocean and $0.8\,\mathrm{Pg\,C\,yr}^{-1}$ for the land. We obtain $\beta = 0.010 \pm 0.001\,\mathrm{yr}^{-1}$ for ocean and $\beta = 0.006 \pm 0.002\,\mathrm{yr}^{-1}$ for land. The root mean square residuals are $0.18\,\mathrm{Pg\,C\,yr}^{-1}$ for ocean and $0.96\,\mathrm{Pg\,C\,yr}^{-1}$ for land. The calculated residuals for land have a larger magnitude than assumed which suggests we should increase the land β uncertainty to $0.003\,\mathrm{yr}^{-1}$.

Figure 3 shows the GCP estimates and the linear fits. When analysing these we must remember that the GCP land estimate is calculated as a residual from Eq. (1). The relatively small residuals from the ocean fit and the additive form of the β decomposition imply that the residuals in the land uptake resemble those in the total uptake.

Again considering the period 2002–2012 we obtain $0.047\,\mathrm{yr}^{-1}$ for land and $0.01\,\mathrm{yr}^{-1}$ for ocean. As with the total uptake, there are 6 periods of 11 years with larger β for

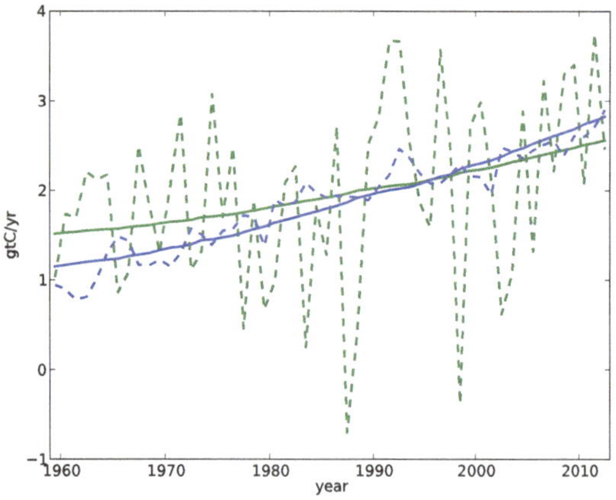

Figure 3. Ocean uptake (blue) and land uptake (green) in Pg C yr^{-1}. Dashed lines are estimates from Le Quéré et al. (2013) while the solid lines are predictions from the β model.

the land while the ocean value is at the mean and median for the set of 11-year periods. Thus even when accounting for the different interannual variability of each environment the relative changes in the land flux are much larger than for the ocean. Changes in land uptake explain all the increase in total uptake over 2002–2012 but this change cannot be regarded as robust.

4 Diagnostics for inversions

We can further decompose land and ocean fluxes into their regional contribution and calculate the related β to attribute regional contributions to trends. Regional flux estimates can come either from atmospheric inversions or models. In this section we use an update of the Cubic Conformal Atmospheric Model (CCAM) inversion used in Peylin et al. (2013) which extended the study of Rayner et al. (2008). The update extends the study period from 1992–2012. Rayner et al. (2008) required that stations must report measurements during 70% of months in their study period in order to be included. We apply the same criterion but the different study period means the network will be different from that of Rayner et al. (2008). The $^{13}CO_2$ records are also extended to 2012. Calculations for the data uncertainties are as in Rayner et al. (2008). We use only the CCAM model from the earlier study.

Before we can trust the inversion to identify regional changes we verify its ability to match the atmospheric growth-rate. This is best done by comparing the net, non-fossil flux. For the GCP this is the sum of the LUC, land and ocean fluxes while for the inversion it is the annual mean, non-fossil flux. We are interested in variability so we adjust the GCP and inverse mean fluxes to be equal for plotting purposes. The inclusion of the constant term in Eq. 5 means this will have no effect on the calculated β. Figure 4 shows the

Table 1. Land and ocean β values from the GCP budget and inversion for the periods 1992–2012 and 2002–2012.

Flux	1992–2012		2002–2012	
	β yr^{-1}	uncertainty yr^{-1}	β yr^{-1}	uncertainty yr^{-1}
Inversion land	0.025	0.007	0.050	0.017
GCP land	0.017	0.005	0.056	0.015
Inversion ocean	0.003	0.005	0.014	0.012
GCP ocean	0.008	0.005	0.009	0.011

results for the GCP and inversion. As we would hope we see good agreement for both short and long term variability. We stress that this is a necessary but not sufficient condition for successful regionalization of trends.

Next we can ask whether the GCP and inversion agree on the land–ocean division of recent sink changes. Table 1 presents the results for the inversion and GCP budget for the periods 1992–2012 and 2002–2012. For comparison we calculate the net land flux for the GCP as the difference between LUC and land uptake. Similarly we calculate the uncertainty here from the residual budget between the growth rate, fossil fuel flux and ocean uptake. Most β values agree to within their uncertainties. We see general agreement on the predominance of land over ocean responses and the much stronger response over 2002–2012. Thus, as far as we can tell from independent evidence, the inversion is partitioning reasonably the linear responses of land and ocean. We stress that this was not preordained since the ocean models which control the trend in land–ocean partition for the GCP estimates do not inform the inversion. We can proceed to discuss the regional form of these responses.

Figure 5 shows the estimated flux and fit from the β model for land and ocean and northern extratropics, tropics and southern extratropics separately. The groupings are taken from Gurney et al. (2002) rather than a latitudinal separation. This allows us to calculate the uncertainty of the regional fluxes correctly. The uncertainties used in the β-model fit are the generated annual uncertainties from the posterior covariance of the inversion. Results of the fit are shown in Table 2 for 1992–2012 and 2002–2012. Both Fig. 5 and Table 2 show strong spatial patterns in the linear response of uptake.

As one might expect from the small global β for the ocean, most ocean regions show weak response and, given their uncertainties, none could be reliably distinguished from 0. One exception is the southern extratropical ocean for 2002–2012. The large uncertainties counsel caution but the apparent increase in the response does not support findings of long-term reductions in uptake (e.g. Le Quere et al., 2009). This is in line with the results of Law et al. (2008).

For land, there is a strong positive response of uptake in the northern extratropics and near cancellation between the tropics and southern extratropics. The tropics shows a large negative β over the whole period. The tropical β depends strongly on the changes in F_{LUC}. This is particularly evident

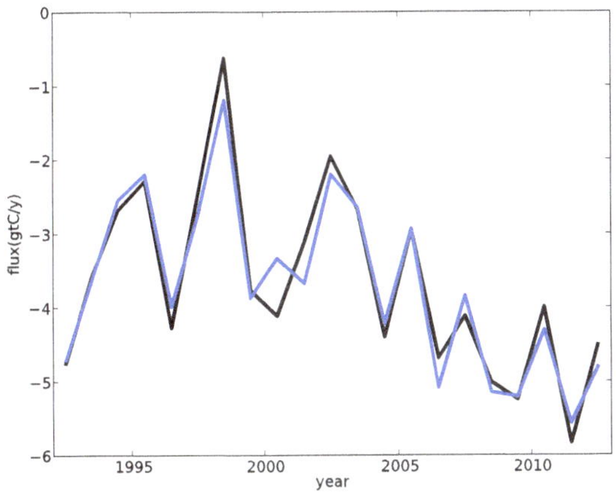

Figure 4. Net uptake from Le Quéré et al. (2013) (black) and from the inversion (blue). Means over the period have been adjusted to be equal.

Table 2. Land and ocean β values from the inversion for northern extratropics, tropics and southern extratropics for the periods 1992–2012 and 2002–2012.

Flux	1992–2012		2002–2012	
	β yr^{-1}	uncertainty yr^{-1}	β yr^{-1}	uncertainty yr^{-1}
Northern land	0.015	0.005	0.037	0.012
Northern ocean	0.002	0.002	−0.004	0.005
Tropical land	−0.014	0.008	0.002	0.021
Tropical ocean	0.001	0.003	0.002	0.006
Southern land	0.009	0.007	0.017	0.016
Southern ocean	0.001	0.003	0.007	0.007
Northern GSNF	0.014	0.027	0.034	0.069
Northern QSNF	−0.001	0.009	−0.003	0.024
Southern GSNF	0.010	0.023	0.014	0.060
Southern QSNF	−0.002	0.015	−0.003	0.039
Northern max	0.013	0.027	0.037	0.067
Southern max	0.007	0.027	0.018	0.067

for 2002–2012 where the increase in β is coincident with a sharp downward trend in F_{LUC}. Any error in trends of F_{LUC} will be aliased into the calculated β. The dipole in response between the tropics and southern extratropics raises the possibility of highly uncertain responses with strong error correlations. This was certainly the case for the mean flux noted by Jacobson et al. (2007) who reported large uncertainty correlations between these regions in atmospheric inversions.

The response with the largest signal-noise occurs in the northern extratropical land. The response is large over the whole period and much larger for 2002–2012 where it dominates the global signal. The increase in relative uncertainty as we move to smaller regions precludes a more detailed spatial examination of the signal. The results suggest that, notwithstanding the cautionary finding of Piao et al. (2008), the strong trend in greenness (e.g. Zhou et al., 2001; Xu et al., 2013) has made a strong imprint on the pattern of CO_2 uptake.

We can further decompose the β for the northern extratropical land into the positive and negative components of the flux. The growing season net flux (GSNF) is defined as the sum of all the negative (uptake) components over a year. We further define the quiescent season net flux (QSNF) as the sum of all the positive (source) fluxes. The annual uptake can be decomposed as annual flux = GSNF−QSNF and thus β can be decomposed as

$$\beta_{annual} = \beta_{GSNF} - \beta_{QSNF}. \qquad (11)$$

The term $\beta_{GSNF} + \beta_{QSNF}$ reflects a change in the integrated amplitude of the seasonal flux and hence to a likely change in the seasonal amplitude of concentration. These changes in amplitude have been noted by Keeling et al. (1995) and Graven et al. (2013) in surface and airborne measurements in the Northern Hemisphere. Roughly paraphrased, the argument of Piao et al. (2008) is that near cancellation between β_{GSNF} and β_{QSNF} means that changes in amplitude need not (and probably do not) correspond to changes in net flux.

Temporally decomposed β values for northern and southern extratropical land are also listed in Table 2. As might be expected, the uncertainties on seasonal fluxes are considerably larger than their annual means so again some caution is suggested in interpreting these values. We see a large response in the GSNF but not in the QSNF. We can hence say that the response in net flux is due to the productive part of the year but it is still a step to say the response is related to production since atmospheric inversions sense only the net flux which is always the difference between production and respiration. One further clue to the likely driver is given by a similar analysis for the maximum uptake in each year. The uncertainties are even larger here but we do see similar increases for the maximum as for the GSNF. This suggests that it is the productivity which mediates the linear response in the net flux and its change over time and that this change in productivity is the likeliest cause of the increasing annual uptake in the northern extratropics.

5 An example of model responses

If the linear diagnostic is a reasonable way to summarize the behaviour of the large-scale carbon cycle we can also apply it to models. This has the further advantage that we can create process diagnostics the same way. As examples, we analyse the linear response of the LPJ-GUESS model (Smith et al., 2001) and the LPJ model (Sitch et al., 2003). LPJ combines mechanistic treatment of terrestrial ecosystem structure (vegetation composition, biomass) and function (energy absorption, carbon cycling). Vegetation dynamics are updated annually based on the productivity, disturbance, mortality, and establishment of nine plant functional types (PFTs). Modelled potential vegetation cover (including C3-/C4-plant distribution) depends on competition and climate history. LPJ-GUESS' process formulation of plant physiology and

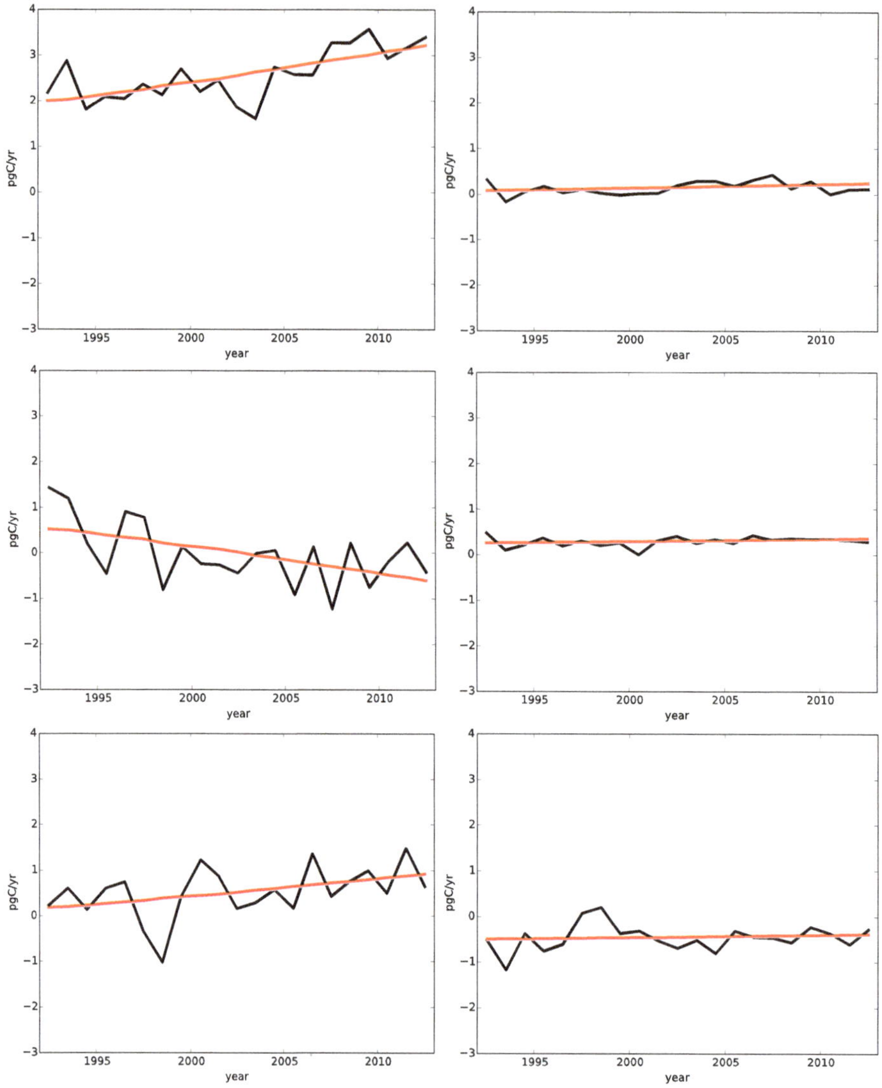

Figure 5. Estimated uptake from inversion (black) and β-model fit (red) for the north (top row), tropics (middle) and south (bottom) with land on the left and ocean on the right.

ecosystem biogeochemistry is similar to LPJ. However, in contrast to the area-based representation of vegetation structure and dynamics for mean individual plant types of LPJ, LPJ-GUESS employs a more detailed scheme that distinguishes woody plant type individuals (cohorts) and represents patch-scale heterogeneity. LPJ-GUESS explicitly models resource competition (light and water) and subsequent growth between woody plant type individuals on a number of replicate patches. Similar to LPJ, herbaceous under-storey (simulated using the grass PFT) is modelled, but individuals are not distinguished. While the LPJ simulation used here represents potential natural vegetation, LPJ-GUESS takes into account present day land use by accounting for croplands and pastures as grass PFT using the 2005 cropland and pasture map from the Hyde 3.1 database (Klein Goldewijk et al., 2011). Both LPJ and LPJ-GUESS use fixed land cover so we fit the β-model to the output directly rather than correcting with F_{LUC}.

First we compare the global fluxes for the two models with F_{land} from the GCP. Figure 6 shows the three fluxes with means adjusted to agree with F_{land}. We see that LPJ agreement is poor for the first half of the period but improves considerably after 1980. The two models do comparably well in this period.

We have fluxes computed until 2011 for LPJ and 2010 for LPJ-GUESS so we analyse the longest possible period for each model for the closest comparison with the inversion. The β values for land in the northern semi-hemisphere, tropics and southern semi-hemisphere are listed in Table 3. We note that we cut the regions at 30° here rather than the more complex boundaries from Gurney et al. (2002) used in the inversion. We see reasonable agreement for the global β for

Table 3. Global and regional β values for the LPJ and LPJ-GUESS models along with the GCP land estimates.

Flux	1992–2010/11		2002–2010/11	
	$\beta\ yr^{-1}$	uncertainty yr^{-1}	$\beta\ yr^{-1}$	uncertainty yr^{-1}
LPJ global	0.020	0.020	0.139	0.055
LPJ-GUESS global	0.020	0.020	0.051	0.063
GCP land	0.015	0.020	0.13	0.055
LPJ northern	−0.010	0.020	−0.014	0.055
LPJ-GUESS northern	−0.001	0.022	0.010	0.064
LPJ tropics	0.038	0.020	0.140	0.055
LPJ-GUESS tropics	0.032	0.022	0.066	0.064
LPJ south	−0.004	0.020	0.027	0.055
LPJ-GUESS south	−0.005	0.022	−0.000	0.064

the whole period but only LPJ shows the dramatic increase in the second half of the period.

The regional structure of the linear response in both models is quite different from that suggested by the inversions. Model responses are dominated by the tropics as is the intensification in response in the last decade. This strong positive response is offset by smaller negative responses in the extratropics. The inversion suggests positive responses in the extratropics (especially the north) with ambiguous response in the tropics.

6 Discussion

It is tempting to compare our β values with those of Gloor et al. (2010). The important difference is that Gloor et al. (2010) do not include a constant term in their linear model (their Eq. 2) while we do. This means their value of β (τ_S in their formulation) will attempt to fit the mean value of uptake while ours will not. Given the likely role of other processes in uptake we would expect that our β value would underestimate mean uptake if used without the mean term. This, indeed, is the case with a mean uptake for 1959–2010 of $2.3\,yr^{-1}$ compared to the GCP value of $3.8\,yr^{-1}$. Gloor et al. (2010) predict an uptake of $3.5\,yr^{-1}$. We stress that the formulation of Gloor et al. (2010) is valid for their purposes but that our focus made it important to separate the mean and trends.

We have analysed the CO_2 uptake throughout as a linear response to concentration. We have not, however, proposed a causal link with CO_2 concentration itself since there are many other variables (e.g. time, temperature and LUC) which are highly colinear with CO_2 concentration. The record, especially of regionally resolved fluxes, is not long enough compared to the various exponential doubling times of emission and concentration to allow a clear separation between linear and exponential changes. The evidence from the inversion of a deepening of the growing season flux minimum does suggest a role for productivity. Given the mechanistic link between productivity and concentration this does suggest increasing concentration changes have contributed to increased land uptake. The two ecosystem models we studied

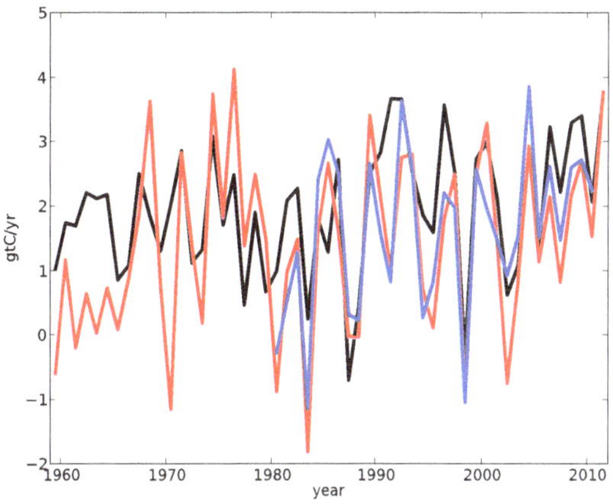

Figure 6. Terrestrial uptakes from Le Quéré et al. (2013) (black), LPJ (red) and LPJ-GUESS (blue). Means have been adjusted to give equal uptake over the whole period.

are too dissimilar in their responses to the inversion to use them as a diagnostic of the inferred flux behaviour.

The results for the most recent decade suggest a strong, but not yet robust increase in the linear response. It suggests that if there is a change in carbon-cycle behaviour, it is in a direction to mitigate rather than exacerbate climate change. We note the much weaker response of tropical uptake and the sensitivity of our result to F_{LUC} estimates.

Finally, the linear diagnostic suggests an interesting interpretation for the recent result of Wang et al. (2014). They noted a large increase in the interannual variability of the terrestrial carbon cycle over the second half of the GCP period. We noted the same thing when considering the residuals from our linear fit. While it is tempting to interpret this increase as an increase in the climate sensitivity of the carbon cycle it seems equally possible that it is a constant modulation of a more strongly forced process. As an analogy we may consider a container with a tap at the bottom which is being randomly opened and closed. The variation in flow will increase as the height of water in the container increases even if the variation in the tap is unchanged. A weakness in this argument is the difference between the location of peak variability (usually located in the tropical land) and the dominant response (located by the inversion in the extratropics).

The calculations in this paper are mainly exemplary. We have made little attempt yet to see how robust the findings are across different terrestrial models and inverse systems. The first of these is relatively easy, aided by several intercomparisons which collect model output. The specification of uncertainty is difficult for models however. For inversions the difficulty is to isolate the components of the flux which are legitimate targets for these diagnostics. F_{LUC} is often included in inverse models as part of the prior flux and it must be separated. Similarly, with an in-house inversion system it is possible to calculate uncertainty on the same scale as the flux estimates while this information is often not available for data from intercomparisons such as that of Peylin et al. (2013). That said, these diagnostics do seem a simple way of summarizing longer-term behaviour of any flux estimate. It will be interesting to see if the finding of Baker et al. (2006) that interannual variability in flux is more robust across the model ensemble than the mean flux also holds for these long-term changes.

7 Conclusions

We have characterized the global and regional response of the carbon cycle as a linear response to CO_2 concentration (or any colinear variable). We have seen that this fit works as well as the airborne fraction model with the advantage that it can be decomposed by time and space. We see an increase in the linear global response in recent years dominated by land. Inverse flux estimates show a similar response and locate it in the northern extratropics and the growing season. Terrestrial ecosystem models show a similar global response but, by contrast, locate it in the tropics.

Acknowledgements. P. J. Rayner is in receipt of an Australian Professorial Fellowship (DP1096309). This research was undertaken on the NCI National Facility in Canberra, Australia, which is supported by the Australian Commonwealth Government.

Edited by: A. Michalak

References

Baker, D. F., Law, R. M., Gurney, K. R., Rayner, P., Peylin, P., Denning, A. S., Bousquet, P., Bruhwiler, L., Chen, Y.-H., Ciais, P., Fung, I. Y., Heimann, M., John, J., Maki, T., Maksyutov, S., Masarie, K., Prather, M., Pak, B., Taguchi, S., and Zhu, Z.: TransCom 3 inversion intercomparison: impact of transport model errors on the interannual variability of regional CO_2 fluxes, 1988–2003, Global Biogeochem. Cy., 20, GB1002, doi:10.1029/2004GB002439, 2006.

Canadell, J. G., Le Quéré, C., Raupach, M. R., Field, C. B., Buitenhuis, E. T., Ciais, P., Conway, T. J., Gillett, N. P., Houghton, R. A., and Marland, G.: Contributions to accelerating atmospheric CO_2 growth from economic activity, carbon intensity, and efficiency of natural sinks, P. Natl. Acad. Sci. USA, 104, 18866–18870, doi:10.1073/pnas.0702737104, 2007.

Cox, P. D., Betts, R. A., Jones, C. D., Spall, S. A., and Totterdell, I. J.: Acceleration of global warming due to carbon-cycle feedbacks in a coupled climate model, Nature, 408, 184–187, 2000.

Cox, P. M., Pearson, D., Booth, B. B., Friedlingstein, P., Huntingford, C., Jones, C. D., and Luke, C. M.: Sensitivity of tropical carbon to climate change constrained by carbon dioxide variability, Nature, 494, 341–344, 2013.

Enting, I. G., Rayner, P. J., and Ciais, P.: Carbon Cycle Uncertainty in REgional Carbon Cycle Assessment and Processes (RECCAP), Biogeosciences, 9, 2889–2904, doi:10.5194/bg-9-2889-2012, 2012.

Etheridge, D. M., Steele, L. P., Langenfelds, R. L., Francey, R. J., Barnola, J. M., and Morgan, V. I.: Natural and anthropogenic changes in atmospheric CO_2 over the last 1000 years from air in Antarctic ice and firn, J. Geophys. Res., 101, 4115–4128, 1996.

Francey, R. J., Allison, C. E., Etheridge, D. M., Trudinger, C. M., Enting, I. G., Leuenberger, M., Langenfelds, R. L., Michel, E., and Steele, L. P.: A 1000-year high precision record of $\delta^{13}C$ in atmospheric CO_2, Tellus B, 51, 170–193, 1999.

Francey, R. J., Trudinger, C. M., Van Der Schoot, M., Krummel, P. B., Steele, L. P., and Langenfelds, R. L.: Differences between trends in atmospheric CO_2 and the reported trends in anthropogenic CO_2 emissions, Tellus B, 62, 316–328, doi:10.1111/j.1600-0889.2010.00472.x, 2010.

Francey, R. J., Trudinger, C. M., van der Schoot, M., Law, R. M., Krummel, P. B., Langenfelds, R. L., Paul Steele, L., Allison, C. E., Stavert, A. R., Andres, R. J., and Rodenbeck, C.: Atmospheric verification of anthropogenic CO_2 emission trends, Nature Clim. Change, 3, 520–524, doi:10.1038/nclimate1817, 2013.

Friedlingstein, P., Cox, P., Betts, R., Bopp, L., von Bloh, W., Brovkin, V., Doney, S., Eby, M., Fung, I., Govindasamy, B., John, J., Jones, C., Joos, F., Kato, T., Kawamiya, M., Knorr, W., Lindsay, K., Matthews, H. D., Raddatz, T., Rayner, P., Reick, C., Roeckner, E., Schnitzler, K.-G., Schnur, R., Strassmann, K., Thompson, S., Weaver, A. J., Yoshikawa, C., and Zeng, N.: Climate -carbon cycle feedback analysis, results from the C4MIP model intercomparison, J. Climate, 19, 3737–3753, 2006.

Gloor, M., Sarmiento, J. L., and Gruber, N.: What can be learned about carbon cycle climate feedbacks from the CO_2 airborne fraction?, Atmos. Chem. Phys., 10, 7739–7751, doi:10.5194/acp-10-7739-2010, 2010.

Graven, H. D., Keeling, R. F., Piper, S. C., Patra, P. K., Stephens, B. B., Wofsy, S. C., Welp, L. R., Sweeney, C., Tans, P. P., Kelley, J. J., Daube, B. C., Kort, E. A., Santoni, G. W., and Bent, J. D.: Enhanced seasonal exchange of CO_2 by northern ecosystems since 1960, Science, 341, 1085–1089, doi:10.1126/science.1239207, 2013.

Gurney, K. R., Law, R. M., Denning, A. S., Rayner, P. J., Baker, D., Bousquet, P., Bruhwiler, L., Chen, Y.-H., Ciais, P., Fan, S., Fung, I. Y., Gloor, M., Heimann, M., Higuchi, K., John, J., Maki, T., Maksyutov, S., Masarie, K., Peylin, P., Prather, M., Pak, B. C., Randerson, J., Sarmiento, J., Taguchi, S., Takahashi, T., and Yuen, C.-W.: Towards robust regional estimates of CO_2 sources and sinks using atmospheric transport models, Nature, 415, 626–630, 2002.

Jacobson, A. R., Mikaloff Fletcher, S. E., Gruber, N., Sarmiento, J. L., and Gloor, M.: A joint atmosphere–ocean inversion for surface fluxes of carbon dioxide: 2. Regional results, Global Biogeochem. Cy., 21, GB1020, doi:10.1029/2006GB002703, 2007.

Keeling, C. D., Whorf, T. P., Wahlen, M., and van der Plicht, J.: Interannual extremes in the rate of rise of atmospheric carbon dioxide since 1980, Nature, 375, 666–670, 1995.

Klein Goldewijk, K., Beusen, A., van Drecht, G., and de Vos, M.: The HYDE 3.1 spatially explicit database of human-induced global land-use change over the past 12000 years, Global Ecol.

Biogeogr., 20, 73–86, doi:10.1111/j.1466-8238.2010.00587.x, 2011.

Knorr, W.: Is the airborne fraction of anthropogenic CO_2 emissions increasing?, Geophys. Res. Lett., 36, L21710, doi:10.1029/2009GL040613, 2009.

Law, R. M., Matear, R. J., and Francey, R. J.: Comment on Saturation of the Southern Ocean CO_2 Sink Due to Recent Climate Change, Science, 319, 570, doi:10.1126/science.1149077, 2008.

Le Quéré, C., Rödenbeck, C., Buitenhuis, E. T., Conway, T. J., Langenfelds, R., Gomez, A., Labuschagne, C., Ramonet, M., Nakazawa, T., Metzl, N., Gillett, N., and Heimann, M.: Saturation of the Southern Ocean CO_2 sink due to recent climate change, Science, 316, 1735–1738, doi:10.1126/science.1136188, 2007.

Le Quere, C., Raupach, M. R., Canadell, J. G., Marland, G., Bopp, L., Ciais, P., Conway, T. J., Doney, S. C., Feely, R. A., Foster, P., Friedlingstein, P., Gurney, K., Houghton, R. A., House, J. I., Huntingford, C., Levy, P. E., Lomas, M. R., Majkut, J., Metzl, N., Ometto, J. P., Peters, G. P., Prentice, I. C., Randerson, J. T., Running, S. W., Sarmiento, J. L., Schuster, U., Sitch, S., Takahashi, T., Viovy, N., van der Werf, G. R., and Woodward, F. I.: Trends in the sources and sinks of carbon dioxide, Nat. Geosci., 2, 831–836, doi:10.1038/ngeo689, 2009.

Le Quéré, C., Peters, G. P., Andres, R. J., Andrew, R. M., Boden, T. A., Ciais, P., Friedlingstein, P., Houghton, R. A., Marland, G., Moriarty, R., Sitch, S., Tans, P., Arneth, A., Arvanitis, A., Bakker, D. C. E., Bopp, L., Canadell, J. G., Chini, L. P., Doney, S. C., Harper, A., Harris, I., House, J. I., Jain, A. K., Jones, S. D., Kato, E., Keeling, R. F., Klein Goldewijk, K., Körtzinger, A., Koven, C., Lefèvre, N., Maignan, F., Omar, A., Ono, T., Park, G.-H., Pfeil, B., Poulter, B., Raupach, M. R., Regnier, P., Rödenbeck, C., Saito, S., Schwinger, J., Segschneider, J., Stocker, B. D., Takahashi, T., Tilbrook, B., van Heuven, S., Viovy, N., Wanninkhof, R., Wiltshire, A., and Zaehle, S.: Global carbon budget 2013, Earth Syst. Sci. Data, 6, 235–263, doi:10.5194/essd-6-235-2014, 2014.

Nabuurs, G.-J., Lindner, M., Verkerk, P. J., Gunia, K., Deda, P., Michalak, R., and Grassi, G.: First signs of carbon sink saturation in European forest biomass, Nature Clim. Change, 3, 792–796, doi:10.1038/nclimate1853,2013.

Peylin, P., Law, R. M., Gurney, K. R., Chevallier, F., Jacobson, A. R., Maki, T., Niwa, Y., Patra, P. K., Peters, W., Rayner, P. J., Rödenbeck, C., van der Laan-Luijkx, I. T., and Zhang, X.: Global atmospheric carbon budget: results from an ensemble of atmospheric CO_2 inversions, Biogeosciences, 10, 6699–6720, doi:10.5194/bg-10-6699-2013, 2013.

Piao, S., Ciais, P., Friedlingstein, P., Peylin, P., Reichstein, M., Luyssaert, S., Margolis, H., Fang, J., Barr, A., Chen, A., Grelle, A., Hollinger, D. Y., Laurila, T., Lindroth, A., Richardson, A. D., and Vesala, T.: Net carbon dioxide losses of northern ecosystems in response to autumn warming, Nature, 451, 49–52, doi:10.1038/nature06444, 2008.

Raupach, M. R., Canadell, J. G., and Le Quéré, C.: Anthropogenic and biophysical contributions to increasing atmospheric CO_2 growth rate and airborne fraction, Biogeosciences, 5, 1601–1613, doi:10.5194/bg-5-1601-2008, 2008.

Rayner, P. J., Law, R. M., Allison, C. E., Francey, R. J., and Pickett-Heaps, C.: The interannual variability of the global carbon cycle (1992–2005) inferred by inversion of atmospheric CO_2 and $\delta^{13}CO_2$ measurements, Global Biogeochem. Cy., 22, GB3008, doi:10.1029/2007GB003068, 2008.

Revelle, R. and Suess, H. E.: Carbon dioxide exchange between atmosphere and ocean and the question of an increase of atmospheric CO_2 during the past decades, Tellus, 9, 18–27, 1957.

Saleska, S. R., Miller, S. D., Matross, D. M., Goulden, M. L., Wofsy, S. C., da Rocha, H. R., de Camargo, P. B., Crill, P., Daube, B. C., de Freitas, H. C., Hutyra, L., Keller, M., Kirchhoff, V., Menton, M., Munger, J. W., Pyle, E. H., Rice, A. H., and Silva, H.: Carbon in Amazon forests: unexpected seasonal fluxes and disturbance-induced losses, Science, 302, 1554–1557, 2003.

Sarmiento, J. L., Hughes, T. M. C., Stouffer, R. J., and Manabe, S.: Simulated response of the ocean carbon cycle to anthropogenic climate warming, Nature, 393, 245–249, 1998.

Sarmiento, J. L., Gloor, M., Gruber, N., Beaulieu, C., Jacobson, A. R., Mikaloff Fletcher, S. E., Pacala, S., and Rodgers, K.: Trends and regional distributions of land and ocean carbon sinks, Biogeosciences, 7, 2351–2367, doi:10.5194/bg-7-2351-2010, 2010.

Schuster, U. and Watson, A. J.: A variable and decreasing sink for atmospheric CO_2 in the North Atlantic, J. Geophys. Res., 112, C11006, doi:10.1029/2006JC003941, 2007.

Sitch, S., Smith, B., Prentice, I. C., Arneth, A., Bondeau, A., Cramer, W., Kaplan, J. O., Levis, S., Lucht, W., Sykes, M. T., Thonicke, K., and Venevsky, S.: Evaluation of ecosystem dynamics, plant geography and terrestrial carbon cycling in the LPJ dynamic global vegetation model, Glob. Change Biol., 9, 161–185, doi:10.1046/j.1365-2486.2003.00569.x, 2003.

Smith, B., Prentice, I. C., and Sykes, M. T.: Representation of vegetation dynamics in the modelling of terrestrial ecosystems: comparing two contrasting approaches within European climate space, Global Ecol. Biogeogr., 10, 621–637, doi:10.1046/j.1466-822X.2001.t01-1-00256.x, 2001.

Tarantola, A.: Inverse Problem Theory and Methods for Model Parameter Estimation, SIAM, (ISBN 0-89871-572-5), 342 pp., 2004.

Wang, X., Piao, S., Ciais, P., Friedlingstein, P., Myneni, R. B., Cox, P., Heimann, M., Miller, J., Peng, S., Wang, T., Yang, H., and Chen, A.: A two-fold increase of carbon cycle sensitivity to tropical temperature variations, Nature, 506, 212–215, 2014.

Xu, L., Myneni, R., Chapin, F., Callaghan, T., Pinzon, J., Tucker, C., Zhu, Z., Bi, J., Ciais, P., Tommervik, H., Euskirchen, E., Forbes, B., Piao, S., Anderson, B., Ganguly, S., Nemani, R., Goetz, S., Beck, P., Bunn, A., Cao, C., and Stroeve, J.: Temperature and vegetation seasonality diminishment over northern lands, Nat. Clim. Change, 3, 581–586, doi:10.1038/nclimate1836, 2013.

Zhou, L. M., Tucker, C. J., Kaufmann, R. K., Stayback, D., Shabanov, N. V., and Myneni, R. B.: Variations in northern vegetation activity inferred from satellite data of vegetation index during 1981 to 1999, J. Geophys. Res.-Atmos., 106, 20069–20083, 2001.

16

Distribution and biophysical processes of beaded streams in Arctic permafrost landscapes

C. D. Arp[1], M. S. Whitman[2], B. M. Jones[3], G. Grosse[4], B. V. Gaglioti[1,3], and K. C. Heim[5]

[1]Water and Environmental Research Center, University of Alaska Fairbanks, Fairbanks, AK 99775, USA
[2]Arctic Field Office, Bureau of Land Management, Fairbanks, AK 99709, USA
[3]Alaska Science Center, U.S. Geological Survey, Anchorage, AK 99508, USA
[4]Alfred Wegener Institute, Helmholtz Centre for Polar and Marine Research, Potsdam, Germany
[5]School of Fisheries and Ocean Sciences, University of Alaska Fairbanks, Fairbanks, AK 99775, USA

Correspondence to: C. D. Arp (cdarp@alaska.edu)

Abstract. Beaded streams are widespread in permafrost regions and are considered a common thermokarst landform. However, little is known about their distribution, how and under what conditions they form, and how their intriguing morphology translates to ecosystem functions and habitat. Here we report on a circum-Arctic survey of beaded streams and a watershed-scale analysis in northern Alaska using remote sensing and field studies. We mapped over 400 channel networks with beaded morphology throughout the continuous permafrost zone of northern Alaska, Canada, and Russia and found the highest abundance associated with medium to high ground-ice content permafrost in moderately sloping terrain. In one Arctic coastal plain watershed, beaded streams accounted for half of the drainage density, occurring primarily as low-order channels initiating from lakes and drained lake basins. Beaded streams predictably transition to alluvial channels with increasing drainage area and decreasing channel slope, although this transition is modified by local controls on water and sediment delivery. The comparisons of one beaded channel using repeat photography between 1948 and 2013 indicate a relatively stable landform, and ^{14}C dating of basal sediments suggest channel formation may be as early as the Pleistocene–Holocene transition. Contemporary processes, such as deep snow accumulation in riparian zones, effectively insulate channel ice and allows for perennial liquid water below most beaded stream pools. Because of this, mean annual temperatures in pool beds are greater than 2 °C, leading to the development of perennial thaw bulbs or taliks underlying these thermokarst features that range from 0.7 to 1.6 m. In the summer, some pools thermally stratify, which reduces permafrost thaw and maintains cold-water habitats. Snowmelt-generated peak flows decrease rapidly by two or more orders of magnitude to summer low flows with slow reach-scale velocity distributions ranging from 0.01 to 0.1 m s^{-1}, yet channel runs still move water rapidly between pools. The repeating spatial pattern associated with beaded stream morphology and hydrological dynamics may provide abundant and optimal foraging habitat for fish. Beaded streams may create important ecosystem functions and habitat in many permafrost landscapes and their distribution and dynamics are only beginning to be recognized in Arctic research.

1 Introduction

Channels with regularly spaced deep and elliptical pools connected by narrow runs are a common form of many streams that drain Arctic permafrost foothills and lowlands. These channels are often referred to as "beaded" streams because during summer low flows, pools appear as beads-on-a-string of runs (Oswood et al., 1989). Beaded streams are generally treated in scientific textbooks on permafrost (e.g., Davis, 2001), hydrology (e.g., Woo, 2012), and aquatic ecology (e.g., McKnight et al., 2008), yet to our knowledge field investigations of these systems has been limited to Imnaviat Creek in northern Alaska (e.g., Oswood et al., 1989) and the Yamal Peninsula in Siberia (Tarbeeva and Surkov, 2013).

Table 1. Summary of a circum-Arctic inventory of beaded stream networks in the zone of continuous permafrost based on a survey of high-resolution (<5 m, summer) imagery available in Google Earth™ during 2012–2013. The relative proportion of high-resolution imagery available in each region was used to estimate the total number of stream networks and drainage density assuming an average network length of 10 km.

Region	Area (km^2)	Percentage of area with high-resolution imagery (snow-free)	Identified stream networks	Estimated stream networks	Estimated drainage density ($km\,km^{-2}$)
Northern Canada	2 347 072	9	22	244	0.001
Northern Alaska (USA)	185 907	80	275	344	0.019
Northern Russia	2 123 067	11	148	1346	0.006

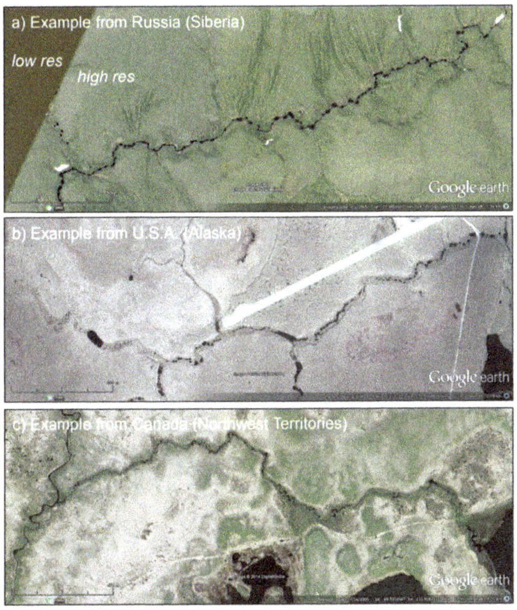

Figure 1. Examples of beaded stream networks located by scanning high-resolution (<5 m) imagery available in Google Earth in **(a)** Russia (Anabar River watershed), **(b)** USA (near Nuiqsut, Alaska), and **(c)** Canada (Tuktoyaktuk Peninsula).

Our understanding of the physical and chemical character of beaded streams mainly comes from Imnavait Creek in the Arctic foothills of Alaska (Oswood et al., 1989). Subsequent studies of this and adjacent systems suggest how beaded morphology functions in permafrost thaw (Brosten et al., 2006), hydrologic storage and hyporheic exchange (Merck et al., 2012; Zarnetske et al., 2007), and thermal regimes (Merck and Neilson, 2012). Thermal stratification in pools up to 2 m deep often occurs in beaded channels during summer low flows (Oswood et al., 1989) and this may play a role in permafrost thaw, hydrologic transport, and nutrient processing as the Arctic climate changes (Zarnetske et al., 2008; Merck and Neilson, 2012). In the winter, foothill streams freeze solid (Best et al., 2005) such that bed sediments thaw slowly and to a limited depth compared to adjacent alluvial channels (Brosten et al., 2006; Zarnetske et al., 2007). Winter analysis of multiple aquatic habitats on the Arctic coastal plain (ACP), however, shows that beaded streams can maintain liquid water under ice and potentially develop perennially thawed sediments (Jones et al., 2013). These physical regimes of water and energy flow in Arctic streams, coupled with channel morphology and drainage network organization likely also dictate how these ecosystems function as aquatic habitat (Craig and McCart, 1975). Hydrographic analysis of the Fish Creek watershed on the ACP show that beaded streams form the dominant connections between larger river systems and abundant thermokarst lakes, thus influencing both hydrology and the movement of aquatic organisms between habitats (Arp et al., 2012b).

Beaded streams are thought to be a common Arctic thermokarst landform and occur mainly in association with ice-wedge networks of polygonized tundra (Pewé, 1966). The formation of channel drainage in these streams occurs along ice-wedge troughs with mature drainage channels resulting in complete degradation of ice wedges by thermal erosion (Lachenbruch, 1966). Classification of Arctic streams place beaded channels within the tundra class as compared to springs and mountain classes (Craig and McCart, 1975). In foothill watersheds, beaded streams are typically fed by linear hillslope water tracks (McNamara et al., 1999), while on the ACP these channels initiate mainly from thermokarst lakes and drained thermokarst lake basins (DTLBs) (Arp et al., 2012b; Whitman et al., 2011). Based on existing research, it is uncertain whether high densities of beaded streams exist beyond this long-standing focal site (Imnavait Creek/Toolik Lake) and this more recent studied watershed (Fish Creek). Newly published work from Russian permafrost zones is also expanding our knowledge of beaded stream distribution (Tarbeeva and Surkov, 2013). Still, an understanding of their formative processes and the broader watershed functions they provide are currently lacking.

Knowing where beaded streams occur in permafrost landscapes and how these fluvial forms are organized within drainage networks will help advance our understanding of their broader role in watershed, ecosystem, and biological functions across the Arctic. Such analyses will also help in predicting changes in these thermokarst fluvial systems

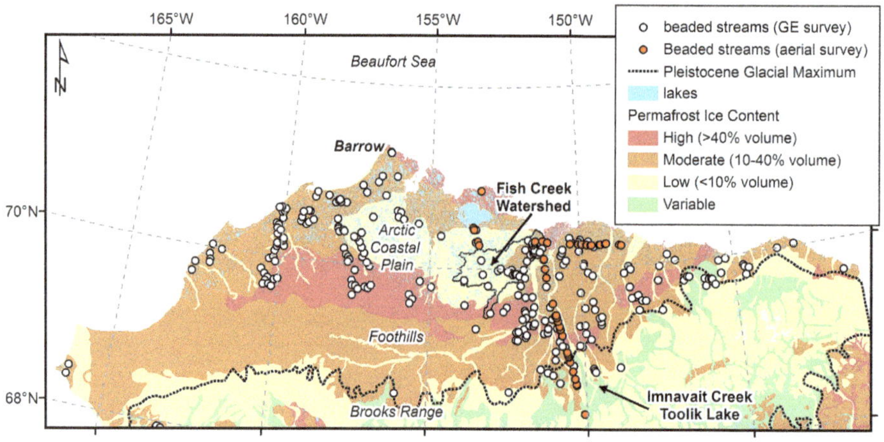

Figure 2. The distribution of beaded streams located using Google Earth and from aerial surveys across the North Slope of Alaska in relation to permafrost ice content (Jorgenson et al., 2008) and the Pleistocene glacial maximum (Manley and Kaufman, 2002). The locations of the Fish Creek watershed (focus area of this study) and Imnavait Creek (focus area of the majority of pervious work on beaded streams) are indicated.

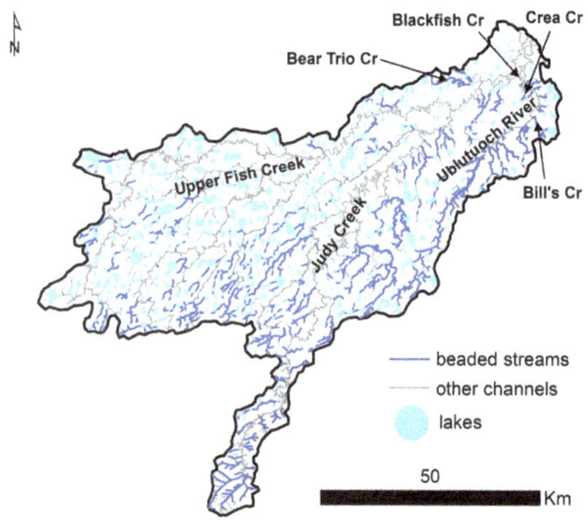

Figure 3. The drainage network of Fish Creek watershed (location shown in Fig. 2) showing all beaded stream networks that were delineated from 2.5 m CIR photography. River systems and individual beaded stream catchments where more detailed field and geospatial studies were conducted for this study are indicated.

with respect to climate and land-use changes and corresponding permafrost responses and hydrologic feedbacks. In this study, we (1) describe the distribution of beaded streams from circum-Arctic to regional scales, (2) explore whether the distribution and variation in beaded morphology helps explain physical functioning, the evolution of beaded streams, and their responsiveness to external drivers, and (3) highlight the important role that these ecosystems serve in aquatic habitat. This work expands our understanding of beaded streams beyond the foothill regions of Arctic Alaska where most of all previous work has been completed, both in terms of fundamental aspects of permafrost and fluvial processes as well as aspects relevant to fish and other aquatic biota.

2 Methods

2.1 Study areas, distribution surveys, and classification

The distribution and abundance of beaded streams were determined by using a nested survey design and a range of survey methods. These nested domains ranged from a (1) circum-Arctic assessment confined to the zone of continuous permafrost using imagery in Google Earth (GE) (Table 1, Fig. 1), (2) aerial transects across landscape gradients on the North Slope of Alaska (Fig. 2), and (3) a census of the Fish Creek watershed (4700 km^2) using high-resolution photography (Fig. 3). We also conducted field studies throughout this watershed and used data from an ongoing monitoring network at several streams in the lower portion of the watershed to characterize biophysical processes and habitat.

The circum-Arctic survey utilized imagery available in GE to identify channels with beaded morphology. This analysis focused on the continuous permafrost zone north of 66° latitude. We utilized the historical image browser function in GE to access the highest resolution imagery (< 5 m) possible for a given region. This analysis focused on portions of Alaska (USA), Siberia (Russia), and northern Canada totaling approximately 4.5 million km^2. We found that most channels with beaded morphology could be identified when scanning images at a 1 : 6000 scale when the imagery had a resolution of 5 m or finer and was mostly snow-free. The availability of high-resolution, snow-free imagery in Alaska was quite good, covering 80 % of the continuous permafrost zone surveyed. In Russia and Canada, the availability of such imagery was much lower, 11 and 9 %, respectively, as of 2013 (Table 1). Prospective beaded channels recognized while scanning were inspected more closely (finer scale) to verify their form and the course was marked as the furthest downstream network point of the continuous beaded channel.

Figure 4. Oblique photographs showing typical pool-run morphology **(a)** and examples of beaded channel forms **(b–e)** compared to alluvial channel **(f)** morphology.

Surface elevation, latitude, and classes of permafrost ground ice were attributed to each point using thematic data sets for pan-Arctic (Brown et al., 1998) and Alaska-focused permafrost and ground ice distribution (Jorgenson et al., 2008) and surface elevation. In order to compare among regions with differing extents of sufficient imagery, we extrapolated the number of surveyed streams based on the proportion of high-resolution imagery available to estimate the total number of beaded stream networks in the circum-Arctic continuous permafrost zone (Table 1). We additionally estimated drainage density of beaded channels based on assuming an average network length of 10 km, which results in only a broad regional average and definitely varies considerably on finer scales.

Regionally (Alaska North Slope) focused aerial surveys in a Cessna 185 were flown on 10 July 2011 on a clear day along three transects. One 270 km transect was from the Brooks Range divide north to the Colville River delta, which moves from glaciated terrain in the upper foothills to vast areas north of the Pleistocene glacial maximum (Fig. 3). Another transect was 130 km from Prudhoe Bay to the lower Fish Creek watershed on the ACP, and a third transect spanned 36 km of land area from Fish Creek to the lands north of Teshekpuk Lake representing an inner to outer ACP gradient. During the transect flights at approximately 150 m elevation, one observer had a sufficient view of approximately 500 m land surface to one side of the plane, thus covering approximately 220 km^2 of land surface in these surveys. During the flight each stream observed was marked with a GPS, photographed, and later these photographs were inspected to determine which streams could be classified as having beaded morphology.

The watershed census of beaded streams was conducted in the Fish Creek watershed as part of a broader effort to map, classify, and understand watershed hydrography and its role in watershed runoff processes (Arp et al., 2012b). The Fish Creek watershed is located in the northeastern portion of the National Petroleum Reserve – Alaska (NPR-A) on the ACP (Fig. 3). Surface deposits grade from marine-alluvial silt with some pebbly substrates in the east to inactive eolian sand dune fields in the west (Carter, 1981; Carter and Galloway, 2005). The sand-bedded alluvial rivers, Fish Creek (*Uulutuuq*, Iñupiat name) and its tributary Judy Creek (*Iqalliqpiq*), drain this area and form a delta in the Beaufort Sea just west of the Colville River delta. Both rivers begin as beaded streams, Judy in a narrow arm extending into the foothills and Fish in the sand sea. The Ublutuoch River (*Tingmiaqsiuqvik*) also starts as a beaded stream, but maintains this morphology for a longer distance before becoming a gravel-bedded alluvial channel near its confluence with Fish Creek (Fig. 3). All perennial channels in the Fish Creek watershed were delineated from 2002 mid-July color infrared (CIR) photography (2.5 m resolution) in a GIS environment. Streams with beaded morphology were quantified according pool density and size (measured as width perpendicular to the direction of flow) and valley gradient from a 5 m interferometric synthetic aperture radar (IfSAR) digital elevation model (DEM) at a segment scale, typically a 1–3 km length that was representative of individual drainage networks. These segments were also placed into four classes according to predominant pool (channel bead) shape and connectivity to runs: (1) elliptical (round) pools separated by distinct connecting runs (Fig. 4b), (2) coalesced pools (elliptical pools merged together) without distinct connecting runs (Fig. 4c), (3) large irregularly shaped pools often connected by long runs (Fig. 4d), and (4) connected thaw pits in degrading polygonized tundra connected by perennial or ephemeral streams (Fig. 4e). We used this classification to help evaluate if pool form of beaded morphology was correlated with landscape position within the watershed and permafrost ice-content or other thermokarst landforms (e.g., thermokarst lakes and DTLBs). We visited approximately 20 % of these stream channels in the Fish Creek watershed during late July 2011 to verify beaded morphology and classification and to collect additional field measurements, as described below in the next section.

2.2 Geospatial and field measurements

A subset of stream channels mapped and classified in the Fish Creek watershed (Arp et al., 2012b) were used for detailed geomorphic and hydrologic analysis in this study.

Specifically, we targeted a set of each channel class representing beaded streams and alluvial channels (Fig. 4), as well as points of channel initiation. During field visits, we measured stream discharge using the velocity–area method. Along stream reaches equaling 20 or more channel widths (typically 100–300 m), we surveyed the water surface elevation at 5–7 points with an engineer's level, stadia rod, and tape to measure the channel slope. At the same time, channel cross sections that bisected pools were surveyed at 2–3 locations to measure pool geometry as well as the incised zone surrounding the channel (gulch) indicated by riparian vegetation and form.

In order to better understand controls on beaded stream morphology, we conducted similar surveys in the field, and from geospatial data (CIR photography and DEMs) along a longitudinal gradient of Fish Creek and the Ublutuoch River from their headwaters downstream (Fig. 3). For each fluvial system, at least three reaches were studied in the field where the channel had distinctly beaded form and three reaches were studied downstream where the channel had transitioned to an alluvial form. Additional locations were later selected to better refine this transition including identification of sediment sinks (flow-through lakes) or clear-water inputs (lake-fed tributaries) relative to potential sediment sources including contact points with hillslopes and sand dunes, and tributaries originating from DTLBs or upland tundra. Such local controls on delivery of new water and sediment to channels were expected to help explain changes in form downstream, similar in concept to mountain drainage networks flowing through lakes (Arp et al., 2007) and as hypothesized for Arctic drainage networks (Tarbeeva and Surkov, 2013). The total length of channels analyzed for the Fish Creek watershed was about 135 km and the total length of channels analyzed for the Ublutuoch River watershed was about 70 km.

2.3 Analysis of channel change and history

To better understand the evolution of beaded channels we compared the position and morphology of one channel over a 64 a period using high-resolution (1 : 24 000 scale) photography from 1948 (black and white, Naval Arctic Research Laboratory, BW NARL) and 2013 (color-infrared at 25 cm pixel size, Aerometric Inc) located in the Fish Creek watershed. This was done to examine the hypothesis that beaded streams evolve in a manner similar to observed degradation of ice-wedge intersections, but lacking channel connectivity. The 1948 BW NARL photographs were acquired from the University of Alaska Fairbanks GeoData Center and scanned at 1200 dpi. The scanned images were georeferenced with 20 ground control points (primarily, stable ice-wedge intersections) to a light detection and ranging (lidar) data set (detailed below) using a spline transformation and converted to a pixel size of 0.5 m. The 2013 color photography was acquired, by Aerometric Inc., on 4 September to complement airborne lidar data. Manual analysis of both data sets was conducted in black and white to avoid any bias that may have arisen due to differences in film types and their separation by so many years of time. Particular attention was given to any changes in channel form (location and plan-view dimensions) relative to ambient polygonized tundra within a 100 m buffer of the channel and the presence and dynamics of thaw pits. All stream channels in both images were independently delineated manually and individual pools and ice-wedge intersections with pits marked with a central point. We tracked individual pools (beads) and thaw-pits from 1948 to 2013 and also recorded those features that were observed in one time period but not the other. The channel gulch/riparian corridor was also delineated for both periods, based primarily on the darker (greener) signature of taller sedges, willows, and dwarf birch and moister understory bryophyte communities.

In order to estimate the timing of pool initiation, long-term sedimentation rates, and the depositional environment of pools, we collected sediment cores to analyze sediment stratigraphy and estimate age–depth relationships using ^{14}C dating. In April 2012, two overlapping cores were collected from a large, deep pool in Crea Creek (Fig. 3) to a depth of 75 cm (base of unfrozen talik) using a Russian peat corer. Each core was photographed and subsampled at 5 cm increments with subsamples placed in Whirl-Pak™ bags. Here we identified what appeared to be basal sediments where the channel initiated, as indicated by an organic sediment layer with fibrous terrestrial organic remains sitting above a homogenous and thick sand layer extending down into the base of the talik. We sampled an individual twig from this basal section for ^{14}C dating. Several moss and sedge samples were also collected from above the basal layer in organic-rich, sandy sediments, similar to organic-rich *gyttja* deposited in lakes of the region, for dating as well. Another core was collected from a pool in 2013 at nearby Blackfish Creek (Fig. 3) and macrofossils were collected from above several distinct sand horizons within the core. The plant macrofossils were prepared for analysis with an acid-base treatment and analyzed for ^{14}C content using standard acceleratory mass spectrometry techniques at the NOSAMS (National Ocean Sciences Accelerator Mass Spectrometry) facility at Woods Hole Oceanographic Institute. All radiocarbon dates were calibrated to calendar ages using the IntCal 13 curve (Reimer et al., 2013) and are reported as the mean and 2σ ranges of the calibrated ages.

2.4 Hydrologic monitoring and habitat analysis

As part of an ongoing monitoring program (Fish Creek Watershed Observatory; Whitman et al., 2011), streamflow, water temperature, and other water quality parameters have been recorded at hourly intervals at five stream-lake systems since 2008. These small catchments (Fig. 3) are being monitored by the Bureau of Land Management (BLM) Arctic Field Office to collect baseline data prior to expected changes in land use, primarily new oil development, and associated

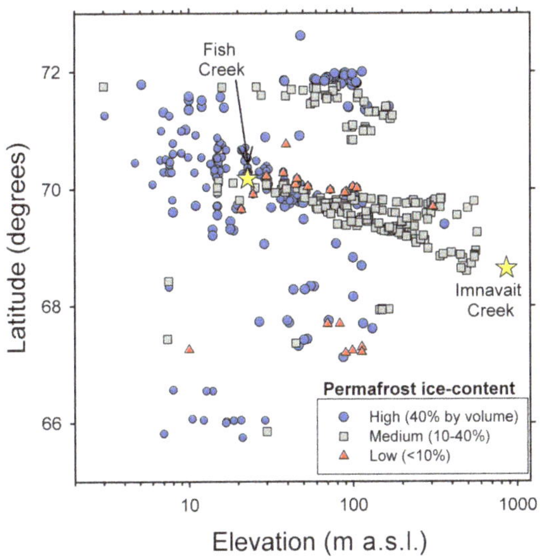

Figure 5. The distribution of beaded stream channels throughout the circum-Arctic in relation to latitude, elevation, and permafrost ground-ice content. Stream networks were identified using imagery in Google Earth in the zone of continuous permafrost where high-resolution imagery was available. The location of streams in the Fish Creek (focus area for this study) and Imnavait Creek (focus area for majority of previous beaded stream research) watersheds are indicated with yellow stars.

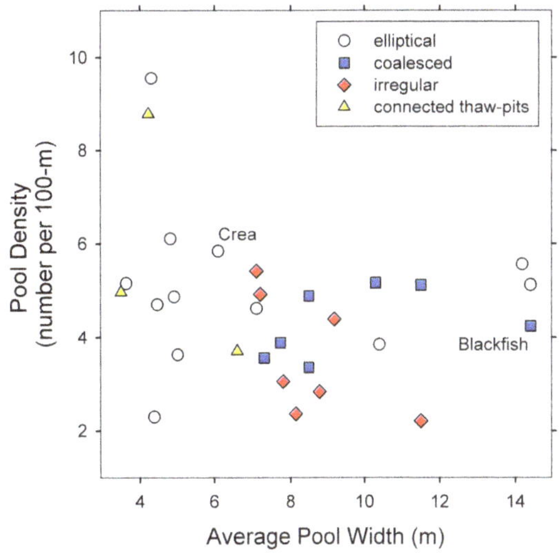

Figure 6. Morphological characteristics of beaded streams compared according to pool (bead) density, size, and shape classes (examples shown in Fig. 4 and locations shown in Fig. 3) at the segment scale (1–3 km channel length) in the Fish Creek watershed.

lake-water extraction for ice road construction and facility operations in the NE NPR-A. Stream gauging was conducted using autonomous pressure transducers (Onset U20-001-01) anchored to pool beds, which were corrected to local atmospheric pressure to measure water height. Stream discharge was measured using the velocity–area method with either an ADCP (acoustic doppler current profiler; Flowtracker™) or electromagnetic (Hach™) velocity meter mounted to a top-setting wading rod. Approximately 20 velocity measurements were made per cross section at increments spaced to not exceed 10 % of total discharge. Typically we made 3–4 measurements near the snowmelt peak flow in early to mid-June, 2–3 measurements during peak-flow recession in late June or early July and 2–3 measurements again in late July and late August. Rating curves were fit with a log or power law equation to estimate continuous discharge during the ice-free season; separate high-flow and low-flow rating curves were often required. Based on temperature sensors placed in channel runs and comparison with time-lapse cameras set during several years, we assumed that streamflow ceased during October in most years.

We tested how contrasting beaded stream morphometry and watershed features affected hydrologic residence times and velocity distributions using tracer tests on two stream reaches with contrasting morphology and flow regimes (Fig. 6). At Crea and Blackfish creeks (Fig. 3), we identified 325 and 232 m reaches, respectively, starting and ending at channel runs to ensure initial mixing and sampling of the advective flow. Rhodamine WT (RWT), a pink fluorescent dye, was used as a water tracer because it can be detected at low concentrations and only small quantities are required to reach target concentrations, which is an important practical consideration for remote field sites. RWT has low biological reactivity, yet does sorb to organic matter and begins photodegrading after several days of sunlight exposure at low concentrations (Vasudevan et al., 2001). Thus, RWT is not truly conservative,however, is widely used to characterize channel hydraulics and transient storage processes, including previous work in Arctic beaded streams (Zarnetske et al., 2007). Based on targeted downstream peak concentrations of 30 ppb (parts per billion), we made pulse additions of RWT at reach heads and monitored concentration at the reach bottom using a YSI 6600-V2 water quality sonde with a RWT probe. This experiment typically lasted a day or longer to account for all tracer moving through the system. RWT tracer data were then fit with the model One-dimensional Transport model with In-channel Storage and Parameterization (OTIS-P) to estimate advective channel area (A), storage zone area (A_S), dispersion (D), and the storage exchange coefficient (α) (Runkel, 2000). The percentage of RWT recovery averaged 81 % with an average sorption coefficient (λ) of 1×10^{-5} used to account for this loss downstream. A tracer breakthrough curve data was plotted as cumulative solute recovered downstream and converted to velocity distribution by dividing reach length by travel time. RWT injections were conducted at both Crea and Blackfish creeks in mid-June near peak flows, in early July (late peak-flow recession), and late August (low summer baseflow).

Stream thermal regimes were quantified using the same pressure transducers anchored to pool beds that also record temperature, along with thermistors (Onset U12-015) near the surface of pools (30 cm below) and in channel runs of each beaded stream; all recording at hourly intervals. These paired temperature measurements were used to assess thermal regimes and timing and extent of stratification in pools assuming that a ratio of surface temperature to bed temperature > 1.1 indicated stratification. Using this system, one pool and corresponding channel run have been monitored among five streams year-round from 2009–2013 (Fig. 3). To assess variability in thermal regimes and particularly stratification within stream systems, we selected an additional three pools of varying depth and area in both Crea and Blackfish creeks (Fig. 3) in 2012 and instrumented these with additional bed and surface thermistors. These were retrieved and downloaded in late August 2013.

During the late winters (March and April) of 2010–2013, we visited several of these same beaded stream reaches concurrent with lake-ice, snow, and water chemistry surveys. When opportunities existed, we measured snow depth either with a 3 m avalanche probe or by digging a pit, or both, above frozen pools located with a GPS. Holes were augered through the ice and ice thickness and below-ice water depth was measured using an ice-thickness gauge (Kovacs Enterprises LCC™). We also measured the depth of thawed sediment (talik) using multiple 1.2 m threaded stainless steel rods fitted with a blunt tip and driven with a slide-hammer to the depth of refusal (typically 10–20 pounds with no downward movement). When possible, these late winter surveys were done repeatedly at the same pools including measurements of dissolved oxygen, specific conductance, and pH to assess the quality of overwintering fish habitat.

3 Results and discussion

3.1 Beaded stream distribution

Using available high-resolution imagery in GE across the circum-Arctic, we found 445 individual channel networks located in northern Alaska, Russia, and Canada with beaded morphology (Table 1). This survey was restricted to land areas north of 66° latitude, which was mainly in the zone of continuous permafrost, though two streams identified were within areas classified as discontinuous and three within areas classified as sporadic permafrost. The availability of high-resolution snow-free imagery strongly reduced the number identifiable channels in Siberia and Canada. Extrapolations to the entire region of continuous permafrost based on the area we could accurately survey, suggests greater than 1900 individual beaded stream networks with 13 % in northern Canada, 18 % in Alaska, and 69 % in northern Russia (Table 1). The density of beaded streams in Alaska was estimated to be about 3 times higher than in Russia and 19 times higher than Canada, likely related to its small but wide unglaciated, ice-rich permafrost coastal plain of the Alaska North Slope relative to abundant mountain and foothills terrain of much of northern Russia and the expansive Laurentian Shield covering much of northern Canada.

In Russia, 148 beaded streams were identified and clustered mainly in several different locations. From east to west these included coastal plains along the Chukchi Sea, lake-rich valley bottoms west of the Kolyma Delta, mountainous headwaters of the Yakutia region, higher elevations of the Yamal Peninsula, and very high densities in the foothills of the Anabar River watershed (Fig. 1a). Recent field studies were completed on beaded streams on the Yamal Peninsula and these researchers also remotely identified channels with beaded morphology in other Russian taiga and steppe terrains using Google Earth (Tarbeeva and Surkov, 2013). Comparatively fewer beaded streams were identified across the Canadian Arctic (22 total) (Table 1). This is likely related to regional geology associated with the dominance of exposed bedrock and thin sediment cover and lack of ice wedges on the Canadian Laurentian Shield. From west to east, small clusters of beaded streams were found on the coastal plain east of Herschel Island and south of the Mackenzie River delta, the lake-rich Tuktoyaktuk Peninsula (Fig. 1c), the coastal plain around the Coronation Gulf and village of Kugluktuk, and the Banks Peninsula within Bathurst Inlet, where high-resolution imagery was available during this GE survey.

Because of greater availability of high-resolution imagery, over 60 % of the beaded streams we located were in Alaska even though this was a much smaller area surveyed (Table 1). The southernmost beaded streams in Alaska were found on the coastal plain of the Seward Peninsula and between Kivalina and Point Hope with an additional cluster higher in the Noatak River valley (Fig. 2). On the North Slope of Alaska, beaded streams were dense and more evenly distributed in the western foothills and along the Chukchi coastal plain. Lower densities of beaded streams were found in the central sand sea region and only a few beaded channels were found on the outer coastal plain of the Barrow Peninsula and north of Teshekpuk Lake. This lack of channels with beaded morphology on the outer coastal plain is perhaps unexpected, given the ubiquitous presence of ice-wedge polygons in which beaded drainage forms. We have observed however that most channels in this region tend to take a plane bed form without alluvial features, which may relate to very high pore-ice content that in addition to wedge-ice makes soils in this region extremely ice-rich, often exceeding 80 % by volume (Brown, 1968, Kanevskiy et al., 2013). The outer coastal plain is also extremely flat with very low drainage densities and very high coverage of thermokarst lakes and DTLBs (Hinkel et al., 2005), such that all fluvial systems are in low abundance and the ones present are strongly lake-affected. On the inner coastal plain and foothills, channels likely develop along moderately sloping terrain with varying densities of ice wedges, but otherwise low pore-ice content. Thus, bead morphology likely develops as ice-wedge networks thermally erode, yet the expansion of pools and runs is confined to the original ice-wedge casts likely because ice-poor permafrost is more resistant to thermokarst erosion. High densities of beaded streams were also found throughout the Kuparuk River watershed from the foothills to the coastal plain and on the narrower coastal plain east of the Sagavanirtok River to Barter Island (Fig. 2).

Looking at the full set of beaded streams in relation to the ground-ice content of permafrost shows that 50 % were found on high ground-ice content permafrost, 32 % on moderately high ground-ice content permafrost, and 18 % on low ground-ice content permafrost (Fig. 5). Regions with high ground-ice content were typically associated with either epigenetic permafrost along the coastal region and syngenetic yedoma permafrost in the foothills region. Approximately 50 % of all beaded streams were found below 60 m a.s.l. (above sea level) elevation and 90 % were found below 210 m a.s.l. elevation (Fig. 5). Seven beaded streams were discovered above 500 m a.s.l. These were found in both Alaska and Russia. Our survey did not identify the even higher-elevation Imnavait Creek, 861 m elevation (Figs. 2, 5), since the only high-resolution GE imagery for this area was acquired during winter snow cover when beaded morphology could not be observed, demonstrating the limitations associated with this identification approach. However, such snow-covered scenes were relatively rare in most imagery we used. Imnavait Creek, along with 12 beaded streams that were identified in our inventory, occurs above the Pleistocene glacial maximum (Fig. 2) indicating that streams with beaded morphology can readily form in glaciated terrain.

In our aerial surveys across the Alaskan North Slope, we located 43 beaded streams from three transects covering 436 km of flight lines or approximately 220 km^2, suggesting a density of 0.20 streams per square kilometer or a drainage density of roughly 0.10 $km\,km^{-2}$. Comparing transect lines to landscape classification of permafrost ground-ice content shows that these surveys covered (29 %) low, (59 %) moderate, and (12 %) high categories (Fig. 2). However, of the recognized beaded streams along these courses, a much higher proportion was associated with moderate ice-rich permafrost (76 %). Only three streams occurred on high ground-ice content permafrost, two on very flat outer coastal plain areas with glaciomarine sediments, and one in yedoma deposits of the foothills (Fig. 2). The majority of stream channels on the outer coastal plain, with very low drainage densities, would be generally classified as plane bed (Montgomery and Buffington, 1997) or F5-6 from Rosgen's classification (Rosgen, 1994), and would have also been termed lacustrine channels (Arp et al., 2012b) because they are nearly all mostly fed by lakes. Still, polygonized tundra tends to be more pronounced and uniform in this region, and so a general lack of channels with beaded morphology was unexpected.

Beaded streams in the Fish Creek watershed range in elevation from 6 to 125 m and the full range of permafrost, ground-ice contents (Jorgenson et al., 2008). We inventoried 126 beaded streams as individual catchments or drainage networks within this 4700 km^2 watershed located on the inner Arctic coastal plain of northern Alaska (Fig. 3). Based on previous analysis of lakes, streams, and river channels here (Arp et al., 2012b), beaded streams represent 1168 km of channel length or 47 % of the entire fluvial system. The equivalent drainage density of beaded stream channels is 0.25 $km\,km^{-2}$. Estimated drainage densities for the broader regions surveyed with GE were far lower compared to this watershed (Table 1).

Since the majority of beaded streams on the ACP initiate as first-order channels below thermokarst lakes or DTLBs (Arp et al., 2012b), their distribution throughout the Fish Creek watershed is linked to lake distribution (Fig. 3). The exception to this pattern is in the headwaters of Judy Creek, which form a narrow arm extending into eolian silt deposits with bedrock outcrops. In this area, lake densities are low and many streams initiate as colluvial channels (Arp et al., 2012b), which then transition to beaded morphology downstream, similar to patterns reported for the higher-elevation foothills of the Kuparuk watershed (McNamara et al., 1999). An example of this drainage pattern is also evident in Fig. 1a. Thirteen percent of all beaded streams in the Fish Creek watershed are located within this region of ice-rich eolian loess. Relatively lower densities of beaded streams occur in the eolian sand sea regions (western half of Fish Creek watershed) where permafrost is classified as having low ground-ice content (Fig. 2) and where most lakes formed between relict dunes (Jorgenson and Shur, 2007) and are up to 20 m deep (Arp et al., 2012b). The highest densities of beaded streams occur in the lower Fish Creek watershed where surface geology is dominated by alluvial and marine silts and sands with some pebbly deposits and permafrost is moderately ice-rich (Carter and Galloway, 2005). Our results suggest some variation in beaded stream distribution within the inner coastal plain, particularly with lower densities associated with eolian and alluvial sand deposits and higher densities on marine and loess silt deposits. However, we still find that beaded streams are often the dominant form of low-order channels throughout a wide range of permafrost terrain on the Alaska North Slope and this is likely the case in much of northern Russia as well (Tarbeeva and Surkov, 2013).

3.2 Morphology in relation to landscape and watershed positions

Since abundant large, deep pools are the defining characteristic of streams with beaded morphology, we initially classified and quantified these channels according to pool (bead) morphology and density (Fig. 6). On a reach scale (hundreds of meters) or segment scale (up to several kilometers between tributary junctions), pool density, form, and size was often distinct. However, on a more extensive drainage network scale, which is the scale we used for classification, pool density varied to a greater extent. Counts of pools from high-resolution CIR photography showed densities ranging from 2 pools per 100 m of channel up to 10 per 100 m (Fig. 6). Lachenbruch (1966) suggested that polygon spacings range from 5 to 50 m based on variation in ground strength and the width of stress relief zones, which approximately matches the range of beaded densities reported here. This indicates that local controls, such as size, pattern, and form (i.e., low- and high-centered polygons) of tundra or broader-scale thermokarst landforms such as DTLBs (Frohn et al., 2005; Hinkel et al., 2005), may be the main cause of such variability in channel morphology.

Of the 126 individual beaded channel networks in the Fish Creek watershed, 40 % were classified as elliptical with distinct connecting runs, 17 % had mostly coalesced pools and short or non-existent runs, 34 % had predominantly irregularly shaped pools, and the remaining 8 % were classified as connected thaw pits (Figs. 4, 6). The majority of beaded channels are shown to initiate from either lakes and DTLBs (Arp et al., 2012b) and these took a wide range of pool forms downstream. In the Fish Creek watershed, most channels with small elliptical pools were located in the higher elevation areas associated with eolian sand and loess deposits compared to lower elevation marine sand and silt deposits. Whether this pattern relates to size and form of ice-wedge networks that develop in sandy soils or how eroding sandy soils moderate expansion by infilling pools or interactions with vegetation deserves further consideration. The other channel classes were more evenly distributed throughout the watershed and by surficial geology.

Comparing channels of the entire watershed by individual slope and drainage area helps us to understand how the larger drainage network is organized from channel initiation points (channel heads) to larger alluvial sand-bedded channels (Fig. 7). This slope–area relationship is consistent with patterns more universally observed across a wide range of drainage networks (Montgomery and Buffington, 1997; Montgomery and Dietrich, 1989; Whiting and Bradley, 1993). In the Fish Creek watershed, channels initiating from hillslopes are steepest with slopes averaging 2 % and with drainage areas < 1 km^2. Channels initiating from lakes, which all form beaded streams, had average slopes of 0.4 % and drainage areas > 1 km^2 (Fig. 7). Channel initiation thresholds reported for the foothill's beaded stream Imnavait Creek are 0.02 km^2 (McNamara et al. 1999) – roughly 1 and 2 orders of magnitude smaller than hillslope- and lake-initiated channels, respectively, in this ACP watershed. Because beaded channels compose approximately half of the drainage network in the Fish Creek watershed (Arp et al., 2012b), they correspondingly have a wide range of drainage areas (2–54 km^2) and slopes ($<$0.1–0.8 %) (Fig. 7). Analysis of beaded channels in Yakutia, Russia, show a narrower range of drainage areas (3–10 km^2) with slopes smaller than 0.2 % (Tarbeeva and Surkov, 2013). Alluvial channels form the higher-order portions of most drainage networks and in the Fish Creek watershed they typically begin at drainage areas $>$40 km^2 and channel slopes smaller than 0.03 % (Fig. 7).

Table 2. Comparison of beaded stream morphology and ambient thermokarst features between black and white photography acquired in 1948 and color infrared photography acquired in 2013 for a 2.7 km segment of Crea Creek in the lower Fish Creek watershed.

Attribute compared	1948	2013
Pools		
number	132	134
total area (m^2)	7861	8334
mean area (m^2)	59.6	62.2
number unique	23	25
Gulch/riparian zone		
total area (m^2)	221 802	241 247
mean width (m)	82.1	89.4
Thaw pits		
total number	74	66
number unique	55	47

To better understand how beaded streams fit within fluvial systems of the ACP and evaluate what controls their morphology, we selected two drainage networks for more detailed analysis of longitudinal channel dynamics from headwaters downstream (Fig. 8). Fish Creek has its headwaters near the western divide of the watershed at 78 m a.s.l. It is located entirely within the eolian sand sheet and initiates from a deep depression lake (Fig. 3). This channel network first flows through several more depression lakes and in between maintains a classic beaded morphology (Fig. 4a). Over the next several kilometers, the channel cuts through both vegetated and unvegetated sand dunes, which likely supply coarse sediment. The channel also contacts steeper hillslopes that could contribute sediment as well. This portion of the channel appears transitional since reaches of beaded morphology are interspersed with more sinuous channels having point bars and meander cut banks (Fig. 8a). At kilometer 20 downstream, the channel steepens considerably below a tributary fed by a DTLB and then cuts through two more sand dunes, before taking a more even slope for the remaining 110 km with sand-bedded alluvial characteristics. Thus, Fish Creek quickly transitions from beaded to alluvial morphology likely because of ample sediment supply associated with the eolian sand landscape (Fig. 6a).

The other system we analyzed, the Ublutuoch River (Fig. 3), begins at a lower elevation than Fish Creek, 58 m a.s.l., in the southern portion of the watershed at the eastern margin of the eolian sand sheet. The channel initiates from a large set of coalesced depression lakes, totaling about 5 km^2, seen as the flat profile in Fig. 8b. The first 12 km of this stream are relatively steep with a regular density of pools typical of beaded morphology. Several oxbow lakes occur lower in this segment, indicative of channelmigration, but the

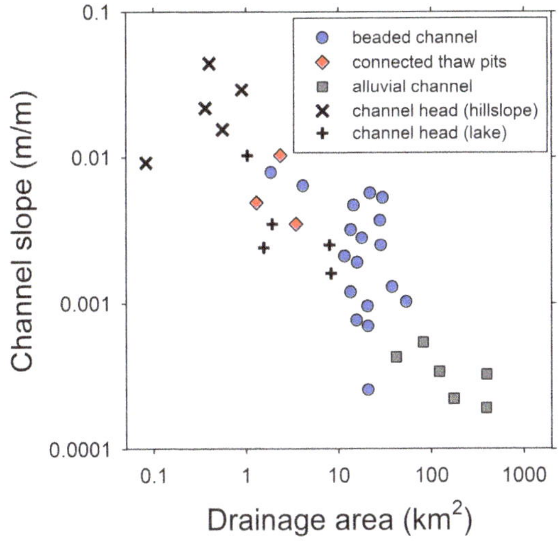

Figure 7. The organization of the major channel forms and channel initiation points (heads) in the Fish Creek watershed are shown in relation to drainage area and channel slope (measured from a 5 m DEM).

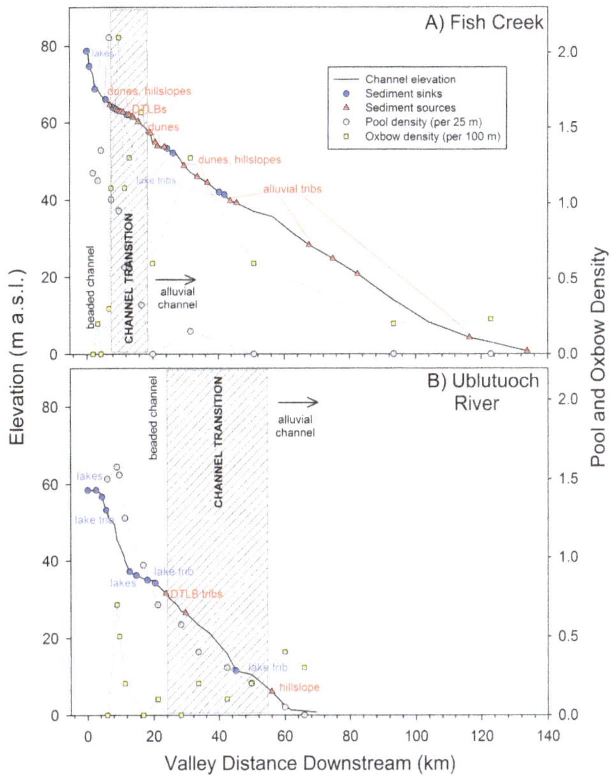

Figure 8. Headwater to downstream patterns of a beaded stream originating in the eolian sand deposits, Fish Creek **(a)**, compared with a beaded stream originating in alluvial–marine deposits, Ublutuoch River **(b)**, showing changes in channel elevation and the density of pools and oxbow (meander-cutoff) lakes relative to sediment sources and sinks.

Ublutuoch then flows through several more lakes, likely trapping all sediment and resetting the system to a beaded form with a flatter slope. At kilometer 24 downstream, a tributary from a large DTLB enters from the north and, at this point, the channel starts taking a more sinuous form with oxbow lakes and other floodplain features (Fig. 8b). We suggest that this segment of stream from kilometer 24 to 56 is transitional between beaded and alluvial morphology – a much longer transition than was observed along the upper Fish Creek. Surrounding uplands here are entirely within the zone of marine silt and sand without distinct sediment contributions from adjacent sand dunes. Near the end of the segment, the channel becomes much more sinuous with oxbows and meander scars becoming evident, yet regular pools (beads) persist. At kilometer 56, the stream contacts a distinctly higher hillslope that we think supplies sediment to the channel and after which it takes on a distinct alluvial form lacking any evenly spaced beaded morphology (Fig. 8b). During the entire transitional channel course, the slope is nearly constant at about 0.02–0.04 %. It then flattens greatly to <0.01 % over the last 5 km and becomes quite deep (exceeding 5 m in some pools) and very sinuous (2.3) with high, regular banks before its confluence with Fish Creek.

3.3 Channel change and formation

To evaluate the hypothesis that beaded streams form in ice-wedge networks and that pools progressively expand over time, more detailed studies were conducted in one system, Crea Creek, in the lower Fish Creek watershed (Fig. 3), to look at decadal-scale changes and estimate its time of formation. Using remote sensing change detection over 64 a, we found no changes in the channel position along this 2.7 km segment (Fig. 9). The total number of pools in this segment remained relatively stable, though tracking individual beads showed that 18 % disappeared or could not be observed from 1948 to 2013 and a similar number of new pools (19 %) were identified in 2013 that could not be observed in the 1948 imagery (Table 2). The majority of these were very small (diameters <4 m) and we think it is likely that changes in vegetation or variation in water levels between images may have obscured their detection. The mean pool size in 1948 was 60 m^2 compared to 62 m^2 in 2013, resulting in little net change in total pool area over this period. Tracking the size of individual pools found in both images showed that about one-third shrank by more than 10 % surface area, about one-third expanded by more than 20 % surface area, and the remaining pools were essentially unchanged. Thus, our analysis suggests progressive expansion of these thermokarst landforms, yet the channel course appeared entirely unchanged over this period. In comparison to other thermokarst landforms, thermokarst lakes in this region also progressively expand their lake basins, 0.10 m a^{-1} on average (Jorgenson and Shur, 2007), but can drain catastrophically if a shoreline expands beyond a lower gradient or is breached by another lake

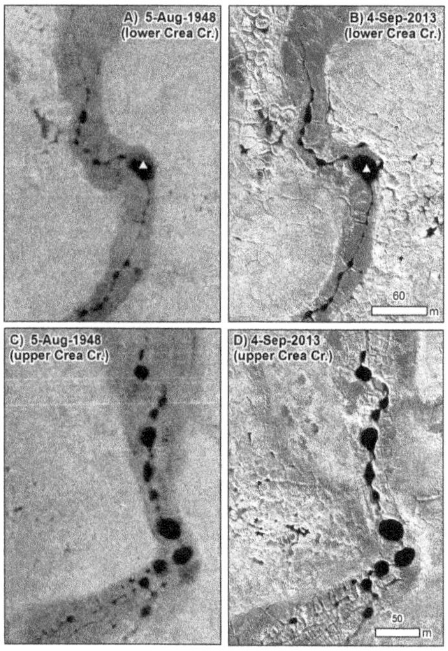

Figure 9. Comparison of two segments of the Crea Creek channel in 1948 **(a, c)** and 2013 **(b, d)** showing that pools, the riparian gulch, and adjacent thaw pits can be clearly observed in each image. The location of a sediment core collected for ^{14}C dating is indicated with a yellow triangle (location of Crea Creek is shown in Fig. 3).

or migrating river (Grosse et al., 2013). Alluvial channels on the ACP are considered highly dynamic and often with very high rates of bank erosion due to interactions with permafrost such that major changes in channel course can occur over short time periods (Scott, 1978). Our observations of a stable course along Crea Creek over 64 a, along with an apparent lack of beaded channels that appear abandoned on the ACP, suggest long-term behavior more similar to bedrock channels (Wohl, 2000). However, Tarbeeva and Surkov (2013) suggest the beaded streams are transient features and become easily filled with sediment from headwater thermokarst and other hillslope erosive processes. We suggest that sediment delivery plays a role in how beaded streams transition to other fluvial forms, but his typically operates at lower positions in the watershed.

We also delineated the riparian zone or gulch of this beaded stream, indicated in plan view by higher moisture and the contrast between upland tussock tundra and vegetation composed of willows, tall sedges, and dwarf birch, to see if other changes beyond the main channels were evident (Table 2). Such changes could correspond to progressive subsidence of ice-rich permafrost by thermokarst degradation or shrub expansion as has been noted throughout many areas of the Arctic (Sturm et al., 2001). Consistent with what can be observed in the shorter reaches in Fig. 9, the overall change in riparian gulch width was slight, a 9 % increase (Table 2). Analysis of repeat photography in this same area has shown a recent increase in degrading ice-wedge polygons to form thaw pits (Jorgenson et al., 2006). We also recorded and tracked thaw pits (ice-wedge junctions with ponded water) between the two images within a 100 m zone on either side of the channel, but outside of the riparian gulch. This showed a somewhat similar pattern as that found when tracking pools in the channel of Crea Creek. In total, we found 120 individual thaw pits or 1 pit per 2500 m^2, typically in clusters associated with high-centered polygons. In 1948 we found 74 thaw pits, 55 of which were not observed in 2013, and in 2013 we found 66 thaw pits, 47 of which were not observed in 1948 (Table 2). This suggests that thaw pits may progress through a form of succession in which they degrade, collect water, paludify and/or partly drain or dry, such that detection is obscured after several decades. This is a similar sequence to that demonstrated for denser networks of thaw pits in polygonized tundra in adjacent upland areas of the Fish Creek watershed (Jorgenson et al., 2006). We suggest that beaded channels may evolve in a similar manner with most pools gradually expanding and some contracting with changing vegetation. Such behavior seems particularly apparent in viewing coalesced beads of some channels (Fig. 4c). Yet our impression based on this photographic comparison and qualitative observation of other channels with repeat photography is that channel courses and networks appear to behave more like bedrock channels that are set in place and potentially very old.

Analyzing the stratigraphy and geochronology of sediments in a large pool of Crea Creek may attest to the timing of stream channel formation and the depositional environment since initiation. A fibrous, organic-rich layer with abundant terrestrial plant material separated the transition from organic-poor, medium-grained sand to organic-rich, silty sediment that is the uppermost unit – we interpreted this layer as basal sediments that were dated to 9.0 (±40), and 13.6 (±215) ka cal BP (before present) (Fig. 10). The terrestrial macrofossils (shrub twigs) in this fibrous unit and the two dates that span 4 ka suggest this layer may have been a terrestrial soil that persisted for millennia on top of eolian or alluvial sand deposits, but predated the initiation of the beaded stream pool. Alternatively, this layer may represent the depositional environment of an early stage of the beaded stream pool where terrestrial vegetation was overhanging and being deposited, and adjacent soils were being eroded by ice-wedge degradation and supplying a range of reworked material with different ^{14}C ages to be deposited onto this fibrous layer. Regardless, we interpret the 9.0 ka moss macrofossil sampled from the upper portion of the fibrous layer to be a conservative upper limit age on the initiation of the beaded stream pool. At this time, we do not know whether the lower limit of this age estimate is near the 9.0 ka time period, or represents the late Holocene. The large age gap from 9.0 ka at 42 cm to ~0.7 ka at 22 cm suggests that either a water-level lowering event caused a hiatus of sedimentation through much of the Holocene, or that high-flow events

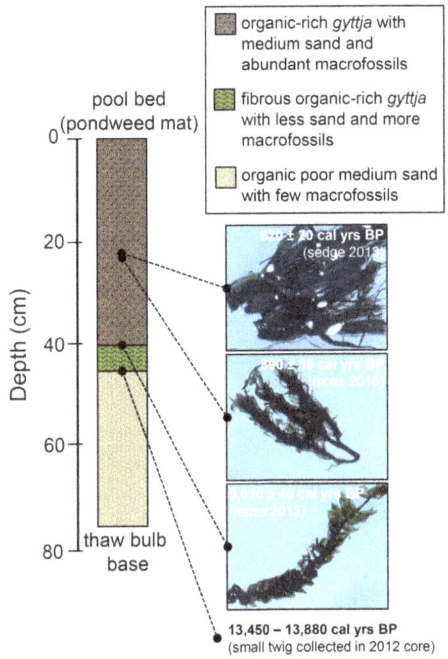

Figure 10. Diagram of generalized sediment core stratigraphy from a large pool in Crea Creek (indicated in Fig. 9) collected in both 2012 and 2013 showing the location of macrofossil fragments collected for radiocarbon dating. The sharp transition from organic-rich *gyttja* to medium sand is interpreted to be the base of the pool at its time of formation.

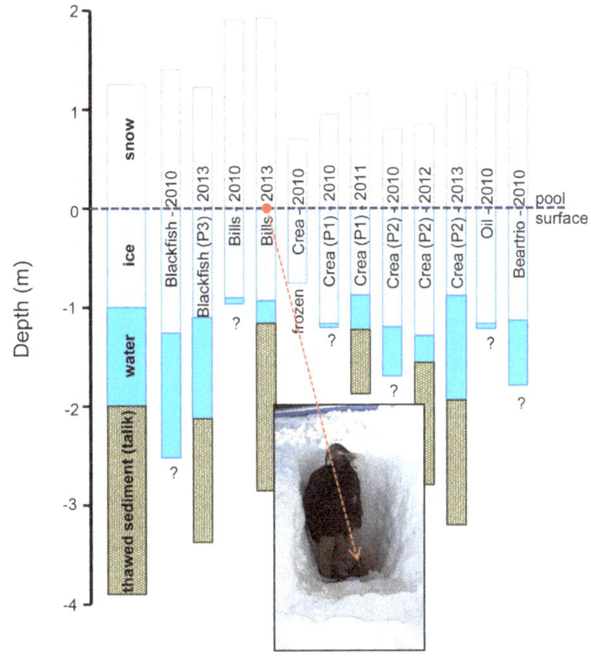

Figure 11. Late winter profiles (March or April) of several pools (beads) surveyed in multiple beaded streams from 2010 to 2013 ("?" indicated that no measurement of thawed sediment depth was attempted). An example photograph from one pool surveyed in 2013 shows a 1.9 m tall person (G. Grosse) standing on the frozen pool surface for scale.

or other processes eroded the sediment deposits representing most of the Holocene (Fig. 10). However, there was no preserved wetland or terrestrial soil layer interrupting the *gyttja* unit, which would have accompanied a water-lowering event. The Crea beaded stream pool we examined appears to have had episodic sedimentation during the Holocene that is periodically eroded by either high-flow events or ice scouring.

The stratigraphy and ^{14}C dates from a core in a deep pool in Blackfish Creek also suggest unconformities in sedimentation of beaded stream pools. The Blackfish pool had sandy organic-rich *gyttja* with several 3–6 cm bands of coarse sand that graded upward to fine sand. These suggested upstream scouring events that mobilized and transported high- and coarse-sediment loads episodically, potentially from the catastrophic drainage of upstream lakes. A number of DTLBs occur upstream of this site and their drainage dates are currently unknown, but may correspond to these events. The basal age of this unit from a sedge fragment yielded a date of 590 (±30) a BP, considerably younger than we found at Crea Creek (Fig. 10). A paired sedge and willow macrofossil extracted from above a coarse sand horizon at 20–30 cm indicated ages of 1430 (±25) a BP and 125 (±25) a BP. Our interpretation of this core and analyzed ages is that the basal material was either not reached or had been remobilized and that a number of very high-flow events in this stream's recent history had deposited upstream material of varying ages. These flow events may have partially eroded some of the late-Holocene record and/or deposited reworked macrofossils, which yielded less certain ^{14}C ages. The depositional environments of beaded streams seem discontinuous and difficult to interpret because of unconformities and reworked plant macrofossils. In the right situation, however, pool sediments may record upstream watershed events such as lake drainage, as we think is preserved in the Blackfish Creek core. At this time, the typical lifespan of the beaded streams we studied remains uncertain, but our best estimate places the Crea Creek channel's formation near the Pleistocene–Holocene transition. The Blackfish Creek core was much more complicated and provided no apparent clues to the age of this beaded channel.

3.4 Physical processes affecting morphology and habitat

3.4.1 Winter processes

Because winter is the dominant season in the Arctic and most beaded streams are ice-covered and likely stop flowing from October to late May or early June, understanding their state during this period is of great interest. An important characteristic of beaded stream channels on the ACP is that their often deep gulches, 0.5–2.0 m, rapidly fill with blowing snow early in the winter, effectively leveling the snow-surface topography with the surrounding tundra. This deep snow insulates ice on pool surfaces, reducing its rate of thickening, and impacting soil active-layer dynamics as well. Measured snow depths above beaded streams averaged 122 cm and ranged from 70 cm on a small pool in Crea Creek to 192 cm above a pool in Bill's Creek (both in the lower Fish Creek watershed) (Fig. 11). In contrast, surrounding tundra snowpack rarely exceeds 40 cm depth by late winter. Not only does this thick snowpack insulate ice and soil, but it also persists much longer in the spring and contributes a much larger portion of snow-water per unit area directly to runoff (Arp et al., 2010). From 12 beads we surveyed from 2010 to 2013, only one was found to be entirely frozen to the bed by March or April (Fig. 11). A more detailed and extensive survey of water below ice was conducted in March and April of 2013 using ground-penetrating radar (GPR) and high-resolution synthetic aperture radar (TerraSAR-X) in this area and found the majority of pools had liquid water below ice (Jones et al., 2013). The average ice thickness of pools surveyed was 106 cm and ranged from 89 to 129 cm (Fig. 9). For comparison, lake ice thickness in this same region and years ranged from 118 cm in 2011 to 171 cm in 2013 (Arp et al., 2012a; Jones et al., 2013). The average depth of water we found below the ice was 44 cm and ranged from 4 up to 106 cm (Fig. 11). This water was typically under pressure from ice expansion and the weight of snow, such that upon drilling through the ice, water typically floods the frozen pool surface. On at least two occasions live fish (Alaska blackfish, *Dallia pectoralis*) were pushed out of the drill hole to the surface by flowing water during these surveys. Monitored dissolved oxygen levels in one bead showed a rapid drop to hypoxic conditions by mid-January and measurements in March typically showed levels below 5 % of saturation or $< 1\ \mathrm{mg\,L^{-1}}$. Alaska blackfish, however, are known to tolerate such conditions (Scott and Crossman, 1973; Crawford, 1974), providing evidence that some beaded stream pools can function as overwintering habitat for select Arctic fish species. While we suspect that these stream pools are not preferred overwintering locations for most fishes, these relatively warm unfrozen sediments may be important habitat for invertebrate and microbial communities.

Despite the relatively small diameter of pools, thawed sediment underlie most of them and measured depths averaged 120 cm and were up to 170 cm in one pool with sand-gravel sediment (Fig. 11). Similar talik depths are reported for pools or broadenings in beaded channels in Russia (Tarbeeva and Surkov, 2013). This suggests that beaded stream channels further disrupt the ground thermal regimes of otherwise continuous permafrost landscapes at a scale relative to their size, whereas large river channels and lakes with floating ice result in taliks reaching tens of meters or more in depth (Brewer, 1958; Lachenbruch et al., 1962). Since 2009, we have been monitoring bed temperatures in a set of pools within beaded stream systems in the lower Fish Creek watershed. Typically winter bed temperatures rapidly approach the zero-degree curtain and average winter temperatures (November–April) consistently average 0 °C (±0.1). Similarly, mean annual bed temperatures (MABTs) fall within a narrow range averaging 2.9 °C and varying interannually almost entirely according to summer temperatures (Fig. 12a). Such MABTs above freezing also suggest the presence of a talik (Burn, 2002; Ensom et al., 2012), as we confirmed with field measurements. The presence of year-round unfrozen sediment and some liquid water in pools may be an essential factor supporting microbial- and invertebrate-based food webs, which then feed summer productivity and the use of beaded streams as important foraging habitat. Additionally, perennially thawed sediment also likely enhances the survival and productivity of macrophytes that provide additional habitat and forage.

3.4.2 Summer processes

Much of the variation in the MABT of pools is determined by whether pools become thermally stratified during the summer. Monitoring of surface temperatures relative to the pool beds and temperature in the channel runs suggests a wide range of mixing behaviors and stratification regimes among pools both between different stream systems and from pool to pool in a single stream. For example in three beaded streams monitored from 2009 to 2012, a 1.3 m pool never became stratified, another 1.4 m pool was stratified by 10 % or more (i.e., surface temperature / bed temperature > 1.1) for 13 days per summer on average, and a 2.1 m pool had a stratification ratio of 1.2 and was stratified for over a month on average per year (Fig. 12b). This generally suggests that deeper pools stratify to a greater degree and for longer periods. To assess interpool variability, we instrumented an additional three pools in Crea and Blackfish creeks from June 2013 through August 2013 with surface and bed thermistors. In Crea Creek with pools depths of 1.6, 1.7, and 2.0 m, the corresponding average stratification ratios (and durations with ratios > 1.1) were 1.05 (5 days duration), 1.09 (23 days), and 1.03 (4 days), respectively (Fig. 12b). In Blackfish Creek with deeper and coalesced pools, instrumented pools had 1.5, 2.2, and 2.6 m depths and the corresponding stratification ratios and durations were 1.04 (5 days), 1.16 (24 days), and 1.10 (19 days). Thus, there is, as expected, some relationship between pool depth and stratification, but this is generally weak and suggests other factors control how water mixes among different pools. A single densely instrumented pool in Imnavait Creek was shown to stratify in a complex and dynamic manner (Merck and Neilson, 2012), similar to more extensive work completed there originally (Oswood et al., 1989). The velocity of upstream runs and morphology of pools at run inflows is certainly one factor. A steeper run upstream of Bill's Creek was likely the cause of continuous mixing during all flows, ambient air temperatures, and wind regimes, which produced higher MABTs (Fig. 12a) and possibly the deepest talik we measured (Fig. 11).

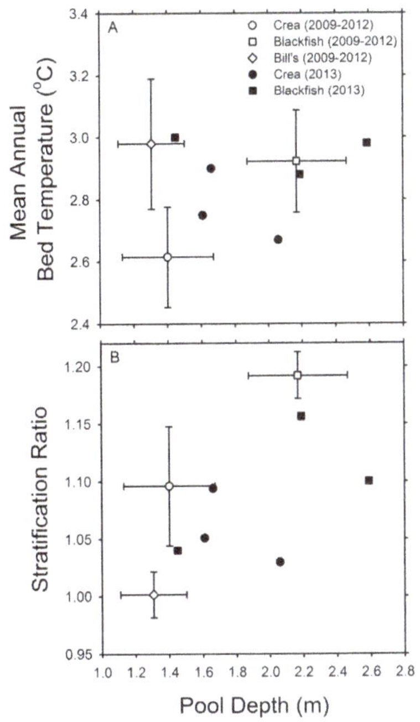

Figure 12. Thermal regime characteristics of single pools at three beaded streams averaged over 4 a (error bars are standard deviations). In 2013, three additional pools within two of these beaded streams were monitored to assess within-stream variability of thermal characteristics. Thermal regimes were characterized by mean annual temperatures at pool beds **(a)** and stratification ratios as the average ratio between the pool surface and bed during the period from July to mid-August in each year **(b)**. Pool depths are averaged during the same period that temperature was summarized in each plot.

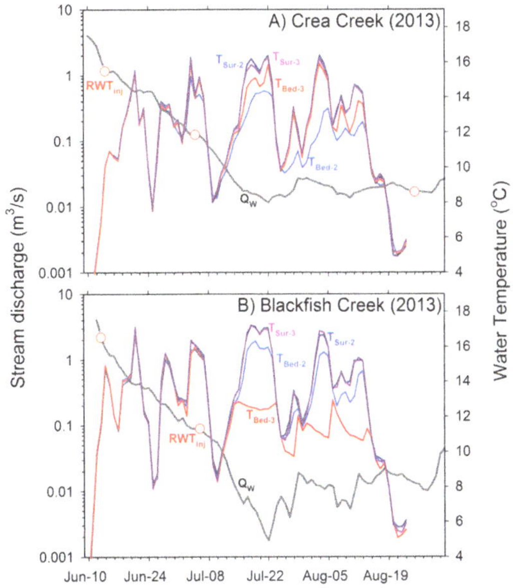

Figure 13. Streamflow hydrographs and temperature regimes for two beaded streams (Crea **(a)** and Blackfish **(b)** creeks) with contrasting channel and watershed morphology. Bed and surface temperatures were monitored in multiple pools within each reach to document the timing, magnitude, and variation in stratification in relation to streamflow (streamflow is indicated by Q_W, temperatures are indicated at pool beds by T_{bed} and pool surface by T_{sur}, and timing of water tracer injection studies are indicated with red circles by RWT_{inj}; all data are presented as mean daily values from hourly measurements).

The extent and structure of emergent aquatic macrophytes in pools likely also plays a role, where some shallow beads have very dense macrophyte beds (*Potamogeton* spp., *Arctophila fulva*, and *Hippuris vulgaris* are the most common plants) that likely create a rough and thick boundary layer enhancing stratification. Adjacent pools of seemingly similar depth and surface area are often devoid of vegetation, creating greater habitat heterogeneity within beaded stream systems. Variation in water color due to dissolved organic carbon may play some role; however, rarely do beaded streams in this part of the ACP have highly stained water from organic acids as has been observed in other beaded stream systems at foothill locations (Merck and Neilson, 2012; Oswood et al., 1989).

Ecologically, the important point in terms of fish habitat is that within a single beaded stream, varying degrees of mixing and thermal stratification from pool to pool likely create a range of temperature zones that can be utilized to either avoid thermal stress or optimize energetics for foraging and other activities. For example, some salmonids behaviorally thermoregulate by moving to warmer areas after foraging bouts in cooler water in order to accelerate metabolism and assimilate more quickly (Armstrong et al., 2013). Stratification within a single bead and heterogeneity in thermal characteristics of nearby beads within a network may provide similar opportunities to behaviorally optimize growth and foraging efficiency during summer. This thermal variability may also play a key role in the distribution of fish prey items, including the forage fish ninespine stickleback (*Pungitius pungitius*) as well as invertebrate and plankton communities (McFarland, 2012).

Similar to the development of stratification in Arctic lakes, stream pools tend to stratify starting in early July following snowmelt runoff and associated cold temperatures and turbulent mixing. An episode of intense summer warming leading to stratification was clearly observed in pools at Crea and Blackfish creeks starting on 9 July 2013 when the surface water temperature rose rapidly from 8 to 16 °C over several days while beds warmed more slowly, albeit to differing degrees (Fig. 13). In Crea Creek, the mean daily temperature difference between the pool surface and bed was as high as 2.5 °C in one pool and only 0.9 °C in the other (Fig. 13a). For the same warming event in Blackfish Creek, levels of stratification were 1.1 °C in one pool and 4.7 °C in the other (Fig. 13b). Another warming event in late July caused even higher stratification, up to 5 °C, in pools of both streams.

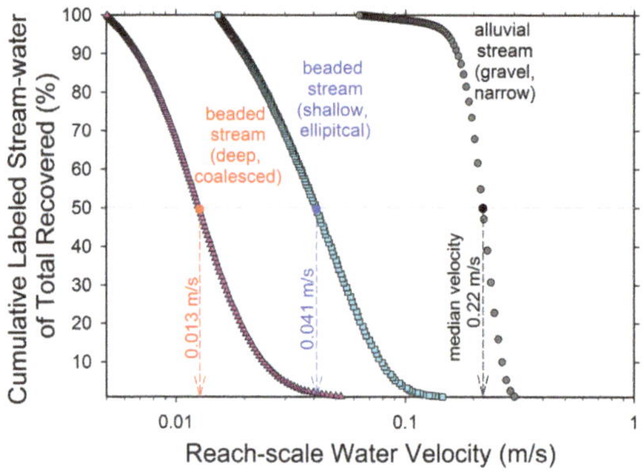

Figure 14. Examples of reach-scale water velocity distributions (reach length/travel time) measured using hydrologic tracer tests (rhodamine WT pulse additions) shown as cumulative tracer recovered downstream. Results from two beaded streams, Crea Creek (blue squares) and Blackfish Creek (red triangles) are compared to an alluvial stream (black circles) in a mountain meadow (Arp unpublished data, stream described in Arp et al., 2007); all three streams had similar discharges ranging from 85 to 140 $L\,s^{-1}$ during tracer tests and slopes ranging from 0.1 to 0.2 %, but with otherwise differing morphologies (experimental data and inverse modeling results shown in Table 3).

In beaded streams on the ACP, we have observed that peak flows predictably occur only 1–2 days after streams begin to flow initially, which is first on top of the ice and often partly beneath the rapidly melting snowpack in stream gulches. Over 5 a of gauging on five separate beaded streams, the timing of peak flows ranged from 1 to 10 June with peak hourly discharges of 1–10 $m^3\,s^{-1}$, which typically exceeds summer flows by 2 orders of magnitude or more. This fast consistent response is similar to that observed for larger river systems of the ACP (Arp et al., 2012b; Bowling et al., 2003), which are fed predominantly by beaded streams and their source-water lakes. A related characteristic is that water temperatures are very near 0 °C at flow initiation and rise very rapidly directly following peak discharge, often warming to 10 °C or more over a 2–3 day period (Fig. 13). These rapid changes in flow and temperature regimes may provide important cues to fish migrating along larger river courses fed by beaded streams. Arctic graylings (*Thymallus arcticus*) are known to seek habitats that warm most rapidly in the spring to spawn, and the quickly rising temperatures of beaded streams may contribute to their importance as spawning habitats (Heim, 2014). In fact, we often see individual fish migrating up beaded channels with water flowing over bedfast ice just prior to peak flows, when their dark bodies can be easily observed crossing the white ice surface. Tracking studies of Arctic grayling tagged in Crea Creek, show a rapid pulse of upstream migration into the system during and after peak flow (Heim, 2014). This early upstream migration may represent an adaption to maximize time spent in productive spawning habitats at the earliest possible time in order to provide a longer period of growth for offspring.

More broadly, the period of peak flow across this hydrologic landscape represents a period of high connectivity among aquatic habitats, where fish can disperse from relatively limited deepwater overwintering habitats and move into shallow, seasonally flowing habitats like beaded streams. Again, in late August through September, changes in flow and temperature may become important environmental cues that fish use to time migratory movements out of beaded streams (Heim, 2014). Migration out of Crea Creek in the fall was strongly correlated to decreases in stream temperature, as the channel connection to the Ublutuoch River became restricted due to ice formation. Low flows and colder temperatures increase the risks of utilizing Crea Creek (Arctic grayling were not found to overwinter within the drainage), yet persistence of fish within the drainage through September may be advantageous in terms of growth and acquisition of energy reserves prior to the onset of winter (Heim, 2014).

With respect to the basic physics of flow through stream systems characterized by multiple evenly spaced pools (storage zones), the attenuation of flows seems intuitive. This has implications for streamflow dynamics, movement and transformations of solutes (carbon, nutrients, and contaminants), the transport of particles including mineral and organic sediment, plankton (both semimobile and drift), and the movement of fish. Because most beaded streams are set within a permafrost framework without interactions with groundwater systems, the development of hyporheic flow through bed material or banks is unlikely. Storage processes have been investigated in Imnavait Creek and adjacent beaded streams around Toolik Lake in Alaska where the glaciated setting and corresponding porous substrates, and known spring systems, may allow hyporheic storage to play a significant role in beaded stream hydrology (Merck et al., 2012; Zarnetske et al., 2007). Still, we suggest that the characteristic large size and frequency of pools of beaded streams strongly dominates transient storage, even when groundwater systems are present and allowing for hyporheic exchange, which is probably rare in continuous permafrost zones of the ACP where surface-water interactions with groundwater are absent.

The distribution of water velocity at the reach scale in a beaded stream with large, deep and coalesced pools (Blackfish Creek) compared to a stream with shallower elliptical pools (Crea Creek) using tracer tests highlights how such morphology functions in water storage and residence time (Fig. 14). For example, the much more rapid velocities observed in an alluvial channel with otherwise similar discharge and slope underscores this impact on dense, evenly spaced pools on the hydrologic functioning of beaded channels. A similar range of reach-scale velocities are reported when comparing beaded channels to other channel types in Arctic drainage networks (Tarbeev and Surkov, 2013, Zarnetske et al., 2007).

Table 3. Results from reach-scale tracer injections for Crea (325 m, shallow elliptical beads) and Blackfish (232 m, deep coalesced beads) creeks during the summer of 2013 (RWT is rhodamine WT, Q is stream discharge, A is the advective cross-sectional area, U is advective zone velocity, D is the dispersion coefficient, A_S is the storage zone cross-sectional area, A_S/A is the relative storage zone area, α is the storage zone exchange coefficient, A_R and U_R are the cross-sectional area and velocity, respectively, of a single channel run). Comparisons of these results are made to two other RWT tracer studies of similar sized streams with beaded and other channel morphologies.

	Experiment data		Total channel hydraulics				Channel storage zone			Channel run (single)	
Site	Date	Solute added (RWT, g)	Q ($m^3 s^{-1}$)	A (m^2)	U ($m s^{-1}$)	D ($m^2 s^{-1}$)	A_S (m^2)	A_S/A	α (s^{-1})	A_R (m^2)	U_R ($m s^{-1}$)
	beaded streams in Fish Creek watershed in 2013 (this study)										
Crea	14-Jun	70.4	1.17	5.29	0.22	3.33	5.32	1.01	1.9E-03	2.48	0.54
Crea	5-Jul	49.9	0.13	1.89	0.04	0.88	2.71	1.43	5.9E-04	0.43	0.39
Crea	25-Aug	19.9	0.03	1.95	0.01	0.38	2.55	1.31	1.2E-03	0.10	0.33
Blackfish	13-Jun	92.4	1.73	9.81	0.18	1.90	9.08	0.93	3.2E-03	2.90	0.70
Blackfish	6-Jul	41.4	0.09	7.00	0.01	0.45	6.60	0.94	1.5E-03	0.52	0.33
Blackfish	24-Aug	19.1	0.03	–	–	–	–	–	–	0.36	0.15
	multiple stream types near Toolik Lake in 2004 (Zarnetske et al., 2007)										
Lake inlet	17-Jun	–	0.26	–	0.16	1.48	–	–	2.0E-4	–	–
Lake outlet	18-Jun	–	0.09	–	0.07	1.71	–	–	5.0E-4	–	–
Beaded	25-Jun	–	0.05	–	0.02	1.75	–	–	3.0E-4	–	–
Beaded	21-Jun	–	0.44	–	0.09	1.94	–	–	6.0E-4	–	–
	multiple stream types in a mountain meadow in 2004 (Arp unpublished data, streams described in Arp et al., 2007)										
Alluvial	11-Aug	–	0.14	–	0.22	–	–	0.69	1.6E-4	–	–
Lake outlet	10-Aug	–	0.17	–	0.06	–	–	0.23	6.7E-4	–	–

Residence times of water in these two beaded channels increase predictably with decreasing flows and relatively higher storage areas (Table 3). At the start of peak-flow recession, over 10 % of the water in both channels was still moving at velocities lower than 0.1 m s^{-1}. During summer flows, the fastest reach-scale velocities did not exceed 0.2 m s^{-1} in Crea Creak and 0.05 m s^{-1} in Blackfish Creek. Even though individual run velocities often exceed 0.5 m s^{-1} or greater, the water in the channel exchanges with storage zones (pools) sufficiently to slow the total movement of water by up to an order of magnitude or much more. Such slow transport rates of water in beaded stream systems may have important implications for maintaining in-stream flow during dry summers when evapotranspiration far exceeds rainfall on daily to weekly timescales. The major source of water to these channels are upstream lakes (Arp et al., 2012b; Bowling et al., 2003), and the evenly spaced storage-rich nature of these streams may function to maintain more constant flows and reduce evaporative losses during summer drought periods.

The summer of 2013 when these experiments were conducted was very wet and rainy compared to previous years when we have monitored discharge in these streams. Still, in 5 a of monitoring, starting in the summer of 2008, we have not yet observed interruptions in flow during summer drought periods in the five gauged streams. At least some alluvial streams in the Arctic foothills of Alaska have experienced prolonged periods of no flow over certain reaches during drought conditions when only minimal flow through interstitial gravels disrupt migration of the Arctic grayling (Betts and Kane, 2011). In some instances, individual Arctic graylings have been observed traveling over 160 km within a year visiting different key habitats within a "migratory circuit" (West et al., 1992). Thus, connectivity among spatially separated habitats is critical to this life history strategy, and beaded streams may function in maintaining hydrologic connectivity and fish passage between alluvial rivers, tundra lakes and ponds. Extreme drought conditions occurred on the ACP and foothills during the summer of 2007 and the hydrologic response has been well documented in rivers (Betts and Kane, 2011; Arp et al., 2012b), thermokarst lakes (Jones et al., 2009a), and upland tundra (Jones et al., 2009b) in this region. Whether beaded streams in this area maintained hydrologic connectivity between river and lake systems through this dry summer was undocumented and warrants reconstruction through hindcast modeling.

The other key function that the hydraulics of beaded streams provides is productive foraging habitat for Arctic fishes. This stems from the observation that larger foraging fishes (e.g., Arctic grayling) spend much of their time holding in channel runs downstream of pools, where they efficiently ambush drifting zooplankton, invertebrates, and ninespine stickleback (McFarland, 2012). The rapid shift in velocities from pools to runs may function as a key delivery system of forage that either resides primarily in beaded stream pools (i.e., ninespine stickleback and aquatic macroinvertebrates) or comes downstream as drift from lakes

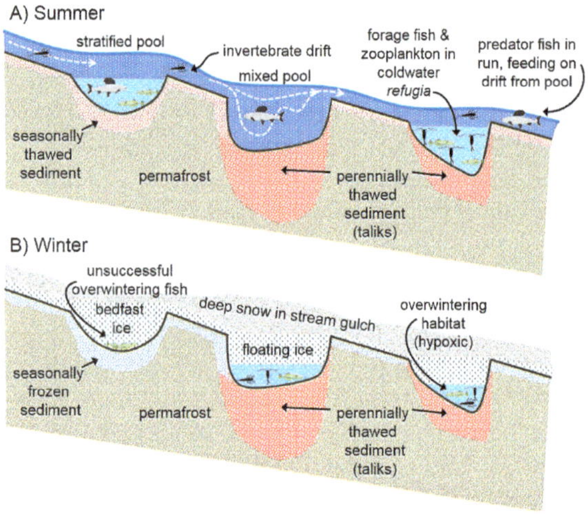

Figure 15. Conceptual diagram showing morphology, physical regimes, habitats, and organisms of a hypothetical pool-run system in the summer **(a)** and winter **(b)** based on observations and monitoring studies in multiple beaded stream systems during these time periods over many years.

(i.e., zooplankton) or laterally from riparian vegetation (i.e., terrestrial invertebrates). Such a setting may in part be the same reason why lake inlets and outlets are such productive ecosystems (Jones, 2010). The difference here is that along the course of beaded streams, this lake outlet delivery system is replicated multiple times over a short distance (i.e., 5 times per 100 m on average; Fig. 6). Approximately half of the Fish Creek drainage network is composed of beaded streams, the equivalent of 1200 km of stream length (Arp et al., 2012b). If we assume a pool density of five per 100 m, this gives us an estimated 60 000 pools (beads) throughout this watershed. Recently, the development of a Fish Creek watershed classification of lakes >1 ha shows 4362 lakes, of which 45 % have perennial stream outlets and another 30 % have at least ephemeral outlets (B. M. Jones, unpublished data). In terms of potential fish habitat for summer foraging, this comparison suggests that pools in beaded streams increase the number of potential fish habitat zones for ambush foraging by 18-fold across the landscape over lake inlets and outlets alone.

4 Conclusions

This body of research on beaded streams in continuous permafrost landscapes documents a wide and varied distribution across the circum-Arctic in relation to ground-ice content in the upper permafrost, topography, and elevation. On the inner coastal plain of northern Alaska, our surveys indicate that beaded streams compose the majority of drainage networks and most channels initiate from and are fed by lakes. At least in northern Alaska, lakes supply water for new development in the form of ice roads and other industrial and municipal uses. Knowing how such practices affect downstream ecosystems warrants investigation. Channels with beaded morphology are maintained downstream, eventually forming alluvial channels in relation to varied water and sediment supply. This suggests that new land disturbances, such as road construction or thermokarst processes that can alter these watershed fluxes, will factor into future drainage network changes. It also appears that beaded stream channels are relatively stable over time and potentially very old, such that any observations of rapid channel change may be indicative of more extreme forcing agents, either anthropogenic or climate driven. Given these concerns and the high density of beaded stream systems in many Arctic landscapes, expanded research into the role of these ecosystems in permafrost, hydrological, and biological processes will be essential.

The coupled biophysical processes of beaded stream systems that provide key ecosystem functionality are described conceptually in Fig. 15. We found high spatial and temporal thermal variability among pools, which likely play an important role in permafrost thaw and cold-water habitats (Fig. 15a). Beaded morphology appears to also play an important role in summer feeding habitats and hydrologic connectivity for migrating fish, the quality and availability of which is critical during short Arctic summers. During long Arctic winters, beaded stream gulches fill with deep snow that effectively insulates ice and permafrost and plays a role in creating taliks and providing overwintering habitats for certain fish and invertebrate communities (Fig. 15b). This conceptual understanding of beaded stream systems helps summarize seasonal and reach-scale ecosystem functions of interest to physical and biological scientists including managers concerned with changing human uses of Arctic lands and waters.

Acknowledgements. This research was supported primarily by the Bureau of Land Management's Arctic Field Office and the Arctic Landscape Conservation Cooperative. Additional funding was provided by Alaska EPSCoR Northern Test Case (OIA-1208927), the Circum-Arctic Lake Observation Network (ARC-1107481), and the National Fish and Wildlife Foundation. We thank F. Urban, R. Kemnitz, C. Couvillion, M. Lilly, J. Adams, J. Derry, H. Toniolo, J. Webster, V. Alexeev, and J. McFarland along with numerous helicopter pilots and ConocoPhilips-Alaska, Inc. (Alpine Facility) for assistance with fieldwork and logistics. This research benefited early on from enlightened conversations with Sveta Stuefer and Matthew Sturm. Two anonymous reviewers provided thoughtful and welcomed comments that improved this manuscript. Any use of trade, product, or firm names is for descriptive purposes only and does not imply endorsement by the US Government.

Edited by: J. Vonk

References

Armstrong, J. B., Schindler, D. E., Ruff, C. P., Brooks, G. T., Bentley, K. E., and Torgersen, C. E.: Diel horizontal migration in streams: Juvenile fish exploit spatial heterogeneity in thermal and trophic resources, Ecology, 94, 2066–2075, doi:10.1890/12-1200.1, 2013.

Arp, C. D., Schmidt, J. C., Baker, M. A., and Myers, A. K.: Stream geomorphology in a mountain lake district: hydraulic geometry, sediment sources and sinks, and downstream lake effects, Earth Surf. Proc. Land., 32, 525–543, doi:10.1002/Esp.1421, 2007.

Arp, C., Jones, B., Beck, R., Whitman, M., Derry, J., Lilly, M., and Grosse, G.: Variation in snow-water equivalent (SWE) among tundra, lakes, and streams on the Alaskan Arctic Coastal Plain: implications for regional SWE estimates and ice-thickness, Annual Conference of the American Water Resources Association, Phillidelphia, PA, November 2, 28, 2010.

Arp, C. D., Jones, B. M., Lu, Z., and Whitman, M. S.: Shifting balance of lake ice regimes on the Arctic Coastal Plain of northern Alaska, Geophys. Res. Lett., 39, 1–5, 2012a.

Arp, C. D., Whitman, M. S., Jones, B. M., Kemnitz, R., Grosse, G., and Urban, F. E.: Drainage network structure and hydrologic behavior of three lake-rich watersheds on the Arctic Coastal Plain, Alaska, Arct. Antarct. Alp. Res., 44, 385–398, 2012b.

Best, H., McNamara, J. P., and Liberty, L.: Association of ice and river channel morphology determined using ground-penetrating radar in the Kuparuk River, Alaska, Arct. Anarct. Alpi. Res., 37, 157–162, 2005.

Betts, E. and Kane, D. L.: Understanding the mechanisms by which a perennial Arctic stream appears intermittent, American Geophysical Union Fall Meeting, San Francisco, 2011.

Bowling, L. C., Kane, D. L., Gieck, R. E., Hinzman, L. D., and Lettenmaier, D. P.: The role of surface storage in a low-gradient Arctic watershed, Water Resour. Res., 39, 1087, doi:10.1029/2002WR001466, 2003.

Brewer, M. C.: The thermal regime of an arctic lake, Transactions of the American Geophysical Union, 39, 278–284, 1958.

Brosten, T. R., Bradford, J. H., McNamara, J. P., Zarnetske, J. P., Gooseff, M. N., and Bowden, W. B.: Profiles of temporal thaw depths beneath two arctic stream types using ground-penetrating radar, Permafrost Periglac., 17, 341–355, 2006.

Brown, J.: An estimation of ground ice, coastal plain, northern Alaska, US Army Cold Regions Research and Engineering Laboratory Hanover, NH, 22, 1968.

Brown, J., Ferrians, O. J., Heginbottom, J. A., and Melnikov, E. S.: Circum-arctic map of permafrost and ground ice conditions, National Snow and Ice Data Center, Boulder, CO, 1998.

Burn, C. R.: Tundra lakes and permafrost, Richards Island, western Arctic coast, Canada, Can. J. Earth Sci., 39, 1281–1298, doi:10.1139/E02-035, 2002.

Carter, L. D.: A Pleistocene sand sea on the Alaskan Arctic coastal plain, Science, 211, 381–383, 1981.

Carter, L. D. and Galloway, J. P.: Engineering Geologic Map of the Harrison Bay Quadrangle, Alaska, U.S. Geological Survey, Menlo Park, CA, 2005.

Craig, P. C. and McCart, P. J.: Classification of stream types in Beaufort Sea drainages between Prudhoe Bay, Alaska and the MacKenzie Delta, N.W.T., Canada, Arct. Alp. Res., 7, 183–198, 1975.

Crawford, R. H.: Structure of an air-breathing organ and the swim bladder in the Alaska blackfish, *Dallia pectoralis* Bean, Canadian Journal of Zoology, 52, 1221–1225, 1974.

Davis, N.: Permafrost: a guide to frozen ground in transition, University of Alaska Press, Fairbanks, 351 pp., 2001.

Ensom, T. P., Burn, C. R., and Kokelj, S. V.: Lake- and channel-bottom temperatures in the Mackenzie Delta, Northwest Territories, Can. J. Earth Sci., 49, 963–978, doi:10.1139/E2012-001, 2012.

Frohn, R. C., Hinkel, K. M., and Eisner, W. R.: Satellite remote sensing classification of thaw lakes and drained thaw lake basins on the North Slope of Alaska, Remote Sens. Environ., 97, 116–126, doi:10.1016/j.rse.2005.04.022, 2005.

Grosse, G., Jones, B., and Arp, C.: Thermokarst Lakes, Drainage, and Drained Basins, in: Treatise on Geomorphology, edited by: Schroder, J., Giardino, R., and Harbor, J., Academic Press, San Diego, 1–29, 2013.

Heim, K.: Seasonal Movements of Arctic Grayling in a Small Stream on the Arctic Coastal Plain, Alaska, M. S., School of Fisheries and Ocean Sciences, University of Alaska Fairbanks, Fairbanks, 82 pp., 2014.

Hinkel, K. M., Frohn, R. C., Nelson, F. E., Eisner, W. R., and Beck, R. A.: Morphometric and spatial analysis of thaw lakes and drained thaw lake basins in the western Arctic Coastal Plain, Alaska, Permafrost Periglac., 16, 327–341, 2005.

Jones, B. M., Arp, C. D., Hinkel, K. M., Beck, R. A., Schmutz, J. A., and Winston, B.: Arctic lake physical processes and regimes with implications for winter water availability and management in the National Petroleum Reserve Alaska, Environ. Manage., 43, 1071–1084, doi:10.1007/s00267-008-9241-0, 2009a.

Jones, B. M., Kolden, C. A., Jandt, R., Abatzoglou, J. T., Urban, F., and Arp, C. D.: Fire Behavior, Weather, and Burn Severity of the 2007 Anaktuvuk River Tundra Fire, North Slope, Alaska, Arct. Antarct. Alp. Res., 41, 309–316, doi:10.1657/1938-4246-41.3.309, 2009b.

Jones, B. M., Gusmeroli, A., Arp, C. D., Strozzi, T., Grosse, G., Gaglioti, B. V., and Whitman, M. S.: Classification of freshwater ice conditions on the Alaskan Arctic Coastal Plain using ground penetrating radar and TerraSAR-X satellite data, Int. J. Remote Sens., 34, 8267–8279, doi:10.1080/2150704x.2013.834392, 2013.

Jones, N. E.: Incorporating lakes within the river discontinuum: longitudinal changes in ecological characteristics in stream-lake networks, Can. J. Fish. Aquat. Sci., 67, 1350–1362, Doi 10.1139/F10-069, 2010.

Jorgenson, M. T. and Shur, Y.: Evolution of lakes and basins in northern Alaska and discussion of the thaw lake cycle, J. Geophys. Res., 112, 1–12, doi:10.1029/2006JF000531, 2007.

Jorgenson, M. T., Shur, Y. L., and Pullman, E. R.: Abrupt increase in permafrost degradation in Arctic Alaska, Geophys. Res. Lett., 33, 1–4, 2006.

Jorgenson, M. T., Romonovsky, V., Yoshikawa, K., Kanevskiy, M., Shur, Y., Marchenko, S., Brown, J., and Jones, B.: Permafrost characteristics of Alaska – a new permafrost map of Alaska, Ninth International Conference on Permafrost, Fairbanks, AK, 2008.

Kanevskiy, M., Shur, Y., Jorgenson, M. T., Ping, C-L., Michaelson, G. J., Fortier, D., Stephani, E., Dillon, M., and Tumskoy, V.:

Ground ice in the upper permafrost of the Beaufort Sea coast of Alaska, Cold Reg. Sci. Technol., 85, 56–70, 2013.

Lachenbruch, A. H.: Contraction theory of ice-wedge polygons: a qualitative discussion, Permaforst International Conference, Lafayette, Indiana, 63–70, 1966.

Lachenbruch, A. H., Brewer, M. C., Greene, G. W., and Marshall, B. V.: Temperatures in permafrost, in: Temperature, its Measurement and Control in Science and Industry, 791–803, 1962.

Manley, W. F. and Kaufman, D. S.: Alaska PaleoGlacier Atlas, Institute of Arctic and Alpine Research (INSTAAR), University of Colorado, Boulder, CO, 2002.

McFarland, J., Wipfli, M. W., and Whitman, M. S.: Feeding ecology of Arctic grayling in a small beaded stream on the Arctic Coastal Plain, Alaska, Conference of the Alaska Chapter of the American Fisheries Society, Kodiak, AK, 128, 23 October, 2012.

McKnight, D. M., Gooseff, M. N., Vincent, W. F., and Peterson, B. J.: High-latitude rivers and streams, in: Polar Rivers and Lakes, edited by: Vincent, W. F. and Laybourn-Parry, J., Oxford University Press, Oxford, 83–102, 2008.

McNamara, J. P., Kane, D. L., and Hinzman, L. D.: An analysis of an arctic channel network using a digital elevation model, Geomorphology, 29, 339–353, 1999.

Merck, M. F. and Neilson, B. T.: Modelling in-pool temperature variability in a beaded arctic stream, Hydrol. Process., 26, 3921–3933, doi:10.1002/Hyp.8419, 2012.

Merck, M. F., Neilson, B. T., Cory, R. M., and Kling, G. W.: Variability of in-stream and riparian storage in a beaded arctic stream, Hydrol. Process., 26, 2938–2950, doi:10.1002/Hyp.8323, 2012.

Montgomery, D. R. and Dietrich, W. E.: Source areas, drainage density, and channel initiation, Water Resour. Res., 25, 1907–1918, 1989.

Montgomery, D. R. and Buffington, J. M.: Channel-reach morphology in mountain drainage basins, Geol. Soc. Am. Bull., 109, 596–611, 1997.

Oswood, M. W., Everett, K. R., and Schell, D. M.: Some physical and chemical characteristics of an arctic beaded stream, Holarctic Ecol., 12, 290–295, 1989.

Pewé, T. L.: Ice-wedges in Alaska – classification, distribution, and climatic significance, Permafrost International Conference, Lafayette, Indiana, 1966, 76–81, 1966.

Reimer, P. J., Bard, E., Bayliss, A., Beck, J. W., Blackwell, P. G., Ramsey, C. B., Buck, C. E., Cheng, H., Edwards, R. L., Friedrich, M., Grootes, P. M., Guilderson, T. P., Haflidason, H., Hajdas, I., Hatte, C., Heaton, T. J., Hoffmann, D. L., Hogg, A. G., Hughen, K. A., Kaiser, K. F., Kromer, B., Manning, S. W., Niu, M., Reimer, R. W., Richards, D. A., Scott, E. M., Southon, J. R., Staff, R. A., Turney, C. S. M., and van der Plicht, J.: Intcal13 and Marine13 Radiocarbon Age Calibration Curves 0–50000 Years Cal Bp, Radiocarbon, 55, 1869–1887, 2013.

Rosgen, D. L.: A classification of natural rivers, Catena, 22, 169–199, 1994.

Runkel, R. L.: Using OTIS to model solute transport in streams and rivers, USGS, Denver, CO, Fact Sheet FS-138-99, 1–4, 2000.

Scott, K. M.: Effects of permafrost on stream channel behavior in Arctic Alaska, USGS, Washingtion, D. C., Geological Survey Professional Paper 1068, 1–19, 1978.

Scott, W. B. and Crossman, E. J.: Freshwater fishes of Canada, Fisheries Research Board of Canada, 966 pp., 1973.

Sturm, M., Racine, C., and Tape, K.: Increasing shrub abundance in the Arctic, Nature, 411, 546–547, 2001.

Tarbeeva, A. M. and Surkov, V. V.: Beaded channels of small rivers in permafrost zones, Geography and Natural Resources, 34, 27–32, doi:10.1134/S1875372813030049, 2013.

Vasudevan, D., Fimmen, R. L., and Francisco, A. B.: Tracer-grade rhodamine WT: structure of constituent isomers and their sorption behavior, Environ. Sci. Technol., 35, 4089–4096, 2001.

West, R. L., Smith, M. W., Barber, W. E., Reynolds, J. B., and Hop, H.: Autumn Migration and Overwintering of Arctic Grayling in Coastal Streams of the Arctic National Wildlife Refuge, Alaska, T. Am. Fish. Soc., 121, 709–715, doi:10.1577/1548-8659(1992)121<0709:Amaooa>2.3.Co;2, 1992.

Whiting, P. J. and Bradley, J. B.: A process-based classification system for headwater streams, Earth Processes and Landforms, 18, 603–612, 1993.

Whitman, M. S., Arp, C. D., Jones, B., Morris, W., Grosse, G., Urban, F., and Kemnitz, R.: Developing a long-term aquatic monitoring network in a complex watershed of the Alaskan Arctic Coastal Plain, in: Proceedings of the Fourth Interagency Conference on Research in Watersheds: Observing, Studying, and Managing for Change, 2011–5169 ed., edited by: Medley, C. N., Patterson, G., and Parker, M. J., Scientific Investigations Report, USGS, Reston, 15–20, 2011.

Wohl, E.: Mountain Rivers, Water Resources Monograph 14, American Geophysical Union, Washington, D.C., 320 pp., 2000.

Woo, M.: Permafrost Hydrology, Springer-Verlag, New York, 563 pp., 2012.

Zarnetske, J. P., Gooseff, M. N., Brosten, T. R., Bradford, J. H., McNamara, J. P., and Bowden, W. B.: Transient storage as a function of geomorphology, discharge, and permafrost active layer conditions in Arctic tundra streams, Water Resour. Res., 43, 1–13, doi:10.1029/2005WR004816, 2007.

Zarnetske, J. P., Gooseff, M. N., Bowden, W. B., Greenwald, M. J., Brosten, T. R., Bradford, J. H., and McNamara, J. P.: Influence of morphology and permafrost dynamics on hyporheic exchange in arctic headwater streams under warming climate conditions, Geophys. Res. Lett., 35, 1–5, doi:10.1029/2007GL032049, 2008.

17

A laboratory experiment on the behaviour of soil-derived core and intact polar GDGTs in aquatic environments

F. Peterse[1,*], C. M. Moy[2], and T. I. Eglinton[1]

[1]ETH Zürich, Geological Institute, Sonneggstrasse 5, 8092 Zürich, Switzerland
[2]University of Otago, Geology Department, Dunedin, New Zealand
[*]now at: Utrecht University, Department of Earth Sciences, Utrecht, the Netherlands

Correspondence to: F. Peterse (f.peterse@uu.nl)

Abstract. We have performed incubation experiments in order to examine the behaviour of soil-derived branched glycerol dialkyl glycerol tetraether (brGDGT) membrane lipids upon entering an aquatic environment and to evaluate the processes that potentially take place during their fluvial transport from land to sea. We incubated a soil from the Rakaia River catchment on the South Island of New Zealand using Rakaia River water and ocean water collected near the river mouth as inocula for a period of up to 152 days. The concentrations, as well as the relative distribution of brGDGTs derived from intact polar ("living"; IPL) lipids and core ("fossil"; CL) lipids remained unaltered over the course of the experiment. Although the stability of the brGDGTs may be a consequence of the higher than natural soil : water ratio used in the laboratory experiment, the substantial increase (27–72 %) in the total pool of isoprenoid GDGTs (isoGDGTs) in all incubation setups, including the control using distilled water, indicates that entering an aquatic environment does influence the behaviour of soil-derived GDGTs. However, the availability of water appears to be more important than its properties. As a consequence of increasing isoGDGT concentrations, a decrease in Branched and Isoprenoid Tetraether (BIT) index values – a proxy for the relative input of fluvially discharged soil material into a marine system – became evident after an incubation period of 30 days, with a maximum final decrease of 0.88 to 0.74 in the experiment with river water. The relative distribution within the isoGDGT pool shows changes with time, suggesting that isoGDGT producers may either have different rates of membrane adaptation or production/degradation, or that preferential release from the soil matrix or a shift in source organism(s) may take place. While the apparent stability of soil brGDGTs during this incubation experiment reinforces their potential as tracers for land–sea transport of soil organic carbon and their use in paleoclimate reconstructions, the distributional differences between GDGTs in river water and nearby soil, as well as in river and ocean water, indicate that further research is needed to pinpoint the sources of GDGTs that are ultimately discharged to the oceans and are subsequently archived in continental margin sediments.

1 Introduction

The global carbon cycle encompasses a myriad of biogeochemical processes that influence our climate and link all carbon reservoirs on the Earth surface. Soils are considered to play a very active and fundamental role in this cycle, as their mobilization from land and subsequent deposition in marine sediments serves as a long-term sink of atmospheric CO_2. However, the exact magnitude and mechanisms of soil organic carbon (OC) transfer from terrestrial source to marine sink remain elusive (e.g. Cole et al., 2007; Weyhenmeyer et al., 2012), which partly stems from the lack of a suitable tracer of the soil OC pool. Our current insights in land–sea soil OC transport dynamics are primarily based on bulk properties of river suspended particulate matter (SPM). However, next to soil OC, river SPM also comprises carbon derived from aquatic production and "fossil" carbon from rock erosion, each of which influences the bulk properties of SPM (Blair et al., 2010).

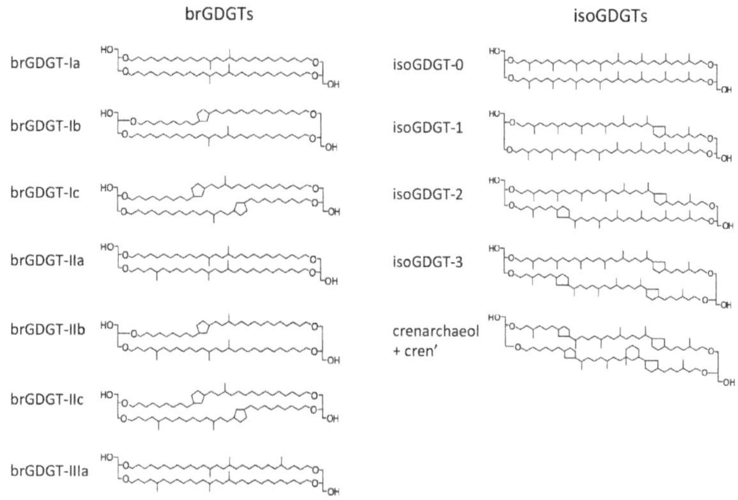

Figure 1. Molecular structures of the brGDGTs and isoGDGTs monitored during the incubation experiment.

Analytical advances in the field of organic biogeochemistry have yielded a growing number of powerful tools with the potential to exclusively target components of specific pools, including soil OC. In this context, branched glycerol dialkyl glycerol tetraethers (brGDGTs; Fig. 1) have been put forward as tracers for soil-derived OC in carbon transport studies (e.g. Hopmans et al., 2004; Kim et al., 2006). Since their discovery (Sinninghe Damsté et al., 2000; Schouten et al., 2000), brGDGTs have been found in soils and peats worldwide (Weijers et al., 2007b, 2009; Liu et al., 2010). The exact organism(s) that produce(s) these compounds have yet to be identified, but current evidence points towards an origin from heterotrophic soil bacteria (Pancost and Sinninghe Damsté, 2003; Oppermann et al., 2010; Weijers et al., 2010) from the phylum of *Acidobacteria* (Weijers et al., 2009; Peterse et al., 2010; Sinninghe Damsté et al., 2011). BrGDGTs have been used as a proxy for the relative input of fluvially transported soil OC in marine systems based on their abundance in coastal marine sediments relative to that of crenarchaeol (Fig. 1), an isoprenoid GDGT (isoGDGTs) produced by marine *Thaumarchaeota* (Sinninghe Damsté et al., 2002), parameterized as the Branched and Isoprenoid Tetraether (BIT) index (Hopmans et al., 2004). Moreover, subtle variations in the molecular distribution of the brGDGTs have been found to relate to mean annual air temperature and soil pH (Weijers et al., 2007b), indicating their potential for utility in paleoclimate studies. Indeed, using a combination of the Methylation of Branched Tetraethers (MBT) and Cyclization of Branched Tetraether (CBT) indices (the so-called "MBT-CBT" proxy; Weijers et al., 2007b, recently revised as MBT′-CBT by Peterse et al., 2012), down-core variations in brGDGT distribution in fluvially dominated continental margin sediments have been used to obtain an integrated climate history of the adjacent drainage basin (e.g. Weijers et al., 2007a).

Although GDGT-based proxies are finding increased use, our understanding of mobilization and transfer of soil OC and associated GDGT signals to fluvial environments, and the processes acting upon these components during transport are still poorly understood. There is emerging evidence, for example, for in situ production of brGDGTs in aquatic environments including high-latitude fjord systems (Peterse et al., 2009), open shelf sediments (Zhu et al., 2011), lakes (e.g. Sinninghe Damsté et al., 2009; Tierney and Russell, 2009) and, most recently, rivers (De Jonge et al., 2014; Kim et al., 2012; Zell et al., 2013b; Zhang et al., 2012). The presence of intact polar lipid (IPL) precursors of brGDGTs in SPM from the Amazon (Zell et al., 2013b), Pearl (Zhang et al., 2012), and Yenisei rivers (De Jonge et al., 2014) provides strong evidence for aquatic brGDGT production, as the IPL head groups are thought to be rapidly lost upon cell death (Harvey et al., 1986; White et al., 1979), although recent studies suggest that IPLs with ether-bound head groups may actually have very slow turnover rates in marine sediments (e.g. Logemann et al., 2011; Xie et al., 2013). Regardless, the core lipids (CLs) that are released after IPL degradation are considered to represent "dead", or fossil material – the fraction that is considered to be stored in sedimentary archives – and targeted for paleoclimate reconstruction. Hence, during fluvial transport from land to sea, the initial soil-derived brGDGT distribution entering a river may be modified by the addition of aquatic produced brGDGTs with a potentially different signature, but also by microbial degradation and transformation of the soil brGDGTs.

In order to explore the latter process, i.e. the behaviour of soil brGDGTs upon entering an aquatic system, we performed a laboratory experiment in which we incubated soil using river water or ocean water from near the mouth of the same river as microbial inocula. In the frame of a larger study on fluvial transport of soil organic carbon from land to sea on the South Island of New Zealand, we used soil and river water from the Rakaia River catchment for the experiment. In addition, the braided character of the Rakaia River may reduce the potential contribution of in situ produced brGDGTs due to the generally harsh and unstable conditions in such river systems. This should enable the monitoring in a controlled setting of what could also be an actual soil-derived brGDGT signal in the river. For that matter, concentration and distributional changes in IPL-derived and CL-brGDGTs were determined at different time intervals during the incubation experiment. Although isoGDGTs (Fig. 1) are primarily associated with marine archaea, they are also produced by soil *Thaumarchaeota* (Sinninghe Damsté et al., 2012), albeit in small amounts relative to the brGDGTs (Weijers et al., 2006b). In the marine realm, changes in isoGDGT distributions form the basis for the TEX_{86} index, which is used as a proxy for sea surface temperature (SST; Schouten et al., 2002; Kim et al., 2010). Large contributions from soil isoGDGTs to the total isoGDGTs pool in marine sediments may consequently complicate the interpretation of TEX_{86}-derived SST records from near continental margins (Weijers et al., 2006b). Therefore, all soil-derived GDGTs, i.e. both branched and isoprenoidal, were monitored during the experiment.

2 Materials and methods

2.1 Sample collection and incubation setup

The Rakaia River is one of the largest braided rivers on the South Island of New Zealand and has a mean annual discharge of 203 $m^3 s^{-1}$. The Rakaia is estimated to contribute 4.15 Mt of sediment a year, accounting for approximately 5 % of the total South Island sediment yield to the adjacent continental shelf (Hicks et al., 2011). The river originates in the Southern Alps and flows through the Canterbury Plains before reaching the Pacific Ocean approximately 150 km from its source. Fresh water from the river was sampled in jerry cans close to the town Rakaia (43°44′57.9″ S, 172°01′52.9″ E), about 20 km upstream from the river mouth and outside of the tidal influence. The ocean water used as inoculum was collected from the shoreline (43°54′15.1″ S, 172°11′41.9″ E), close to the river mouth (Fig. 2). Based on pH measurements, this water (pH 7.6) was strongly fluvially influenced (pH 7.5). Surface soil (0–10 cm; pH 5.8) was collected on the Canterbury Plains within the Rakaia River catchment (43°52′00.8″ S, 172°09′10.8″ E). All soil material and incubation water was collected in January 2012 and directly transported to the University of Otago, where the incubation experiment was set up the next day. SPM was collected by filtration of river and ocean water (100 L onto a 0.2 μm polyethersulfone membrane and 10 L onto a pre-combusted 0.7 μm GF/F, respectively) in the field, after which the filters were stored frozen.

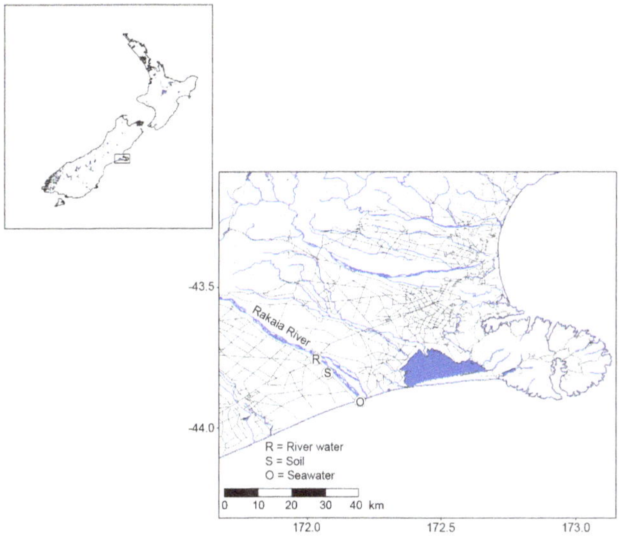

Figure 2. Overview map of the Rakaia River, South Island, New Zealand, indicating the sampling locations.

In the laboratory at the University of Otago, the soil was sieved over a 2 mm mesh to remove roots and homogenize the sample, after which two 10 g subsamples of the soil were directly frozen at −20 °C for subsequent determination of initial GDGT composition. For the incubations, twenty-eight 250 mL bottles were filled with 10 g soil subsamples, after which 100 mL of river water or ocean water, untreated to preserve the natural microbial community, was added to 12 bottles each. Distilled water was added to the remaining four bottles to serve as control setup. The incubation bottles were loosely plugged with cotton wool, creating quasi-aerobic conditions, and placed on a shaker table from January to June 2012. Based on IPL degradation rates observed in previous studies (Harvey et al., 1986; Logemann et al., 2011), two bottles with river water and two bottles with ocean water were taken from the shaker table and immediately placed into a freezer (−20 °C) at 1, 7, 14, 30, 91, and 152 days after the start of the experiment. Duplicate control samples were frozen after 91 and 152 days. At the end of the experiment all bottles were shipped on ice to ETH Zürich, Switzerland, where they were briefly thawed and the supernatant was pipetted off. The soils were then freeze dried, and stored frozen at −20 °C until subsequent sample work-up and analysis.

2.2 Total organic carbon and nitrogen analysis

The freeze-dried, homogenized soils were weighed in silver capsules and fumigated with 1 M HCl for 3 days and neutralized with NaOH for 2 days in a desiccator to remove carbonates prior to total organic carbon (TOC) and total nitrogen (TN) analysis. The measurements were performed on a vario MICRO elemental analyser at ETH Zürich.

2.3 GDGT extraction and analysis

Freeze-dried soil samples (3–4 g) were solvent-extracted three times using a Bligh and Dyer technique modified from Sturt et al. (2004). In short, a solvent mixture of MeOH : dichloromethane (DCM): phosphate buffer at pH 7.4 ($2:1:0.8$, $v/v/v$) was added to the soils and ultrasonically extracted for 10 min. The extract was collected each time after centrifuging at 1000 rpm for 5 min. DCM and phosphate buffer were added to the combined extracts to a volume ratio of $1:1:0.9$ to obtain phase separation. The DCM phase, containing the GDGTs, was collected after centrifuging, and the remaining solvent was rinsed twice with DCM. The combined DCM phases were dried under N_2 and passed over a silica column (deactivated with 1 weight % water) to separate CLs and IPLs according to Pitcher et al. (2009), with the exception that hexane:ethyl acetate $1:1$ (v/v) was used to elute the CLs. An aliquot of the IPL fraction was analysed directly for CLs in order to assess potential carry over. The remainder of IPL fraction was dissolved in 6N HCl in MeOH and heated at 100 °C for at least 3 h to release IPL-bound CLs. A known amount of C_{46} GDGT standard (Huguet et al., 2006) was added to all fractions prior to analysis.

The filters were freeze dried and extracted with a MARS Xpress microwave extraction system, using DCM : MeOH $9:1$ (v/v). After centrifugation (5 min at 400 rpm), the total lipid extract (TLE) was pipetted off and the residues were rinsed twice with DCM : MeOH $9:1$ (v/v). The combined extracts were dried under N_2 with a known amount of C_{46} GDGT standard, after which the TLEs were separated into an apolar and a polar (GDGT) fraction by passing them over a silica (1 % water deactivated) column using hexane : DCM $9:1$ (v/v) and DCM : MeOH $1:1$ (v/v), respectively.

All GDGT fractions were dissolved in hexane : isopropanol $99:1$ (v/v), filtered over a 0.45 μm PTFE filter, and analysed using high-performance liquid chromatography/atmospheric pressure chemical ionization – mass spectrometry (HPLC/APCI-MS) with an Agilent 1260 Infinity series LC/MS at ETH Zürich according to Schouten et al. (2007). Separation of the GDGTs was achieved with a Grace Prevail Cyano column (3 μm, 150 × 2.1 mm) after passing through a guard column of the same material (5 μm, 7.5 × 2.1mm) with hexane : isopropanol ($99:1$, v/v) as an eluent at a flow rate of 0.2 mL min^{-1}. The GDGT fractions eluted isocratically with 90 % A and 10 % B for 5 min, and then with a linear gradient to 18 % B for 34 min, where A = hexane and B = hexane : isopropanol $9:1$ (v/v). Selective ion monitoring of the $[M+H]^+$ was used to detect the different GDGTs. Although brGDGTs appear to have a higher response factor than isoGDGTs (Schouten et al., 2013a), quantification was done assuming similar response factors for all GDGTs and the internal standard. Hence, reported brGDGT concentrations are likely overestimated compared to those of the isoGDGTs. The amounts of all GDGTs are given as the average and the range of variation of the duplicate incubation samples. The amount of GDGTs in each sample has been corrected for carry over of CLs into the IPL fraction, which was on average <8 % for brGDGTs and <1 % for isoGDGTs.

2.4 GDGT-based index calculations

The CBT and MBT′ indices based on brGDGTs were calculated according to (Weijers et al., 2007b)

$$\mathrm{CBT} = -\log((\mathrm{Ib}+\mathrm{IIb})/(\mathrm{Ia}+\mathrm{IIa})) \tag{1}$$

and (Peterse et al., 2012)

$$\mathrm{MBT'} = (\mathrm{Ia}+\mathrm{Ib}+\mathrm{Ic})/(\mathrm{Ia}+\mathrm{Ib}+\mathrm{Ic}+\mathrm{IIa} + \mathrm{IIb}+\mathrm{IIc}+\mathrm{IIIa}). \tag{2}$$

The BIT index was calculated following Hopmans et al. (2004):

$$\mathrm{BIT} = (\mathrm{Ia}+\mathrm{IIa}+\mathrm{IIIa})/(\mathrm{cren}+\mathrm{Ia}+\mathrm{IIa}+\mathrm{IIIa}) \tag{3}$$

and for the TEX_{86} index based on isoGDGTs the equation of Schouten et al. (2002) was used:

$$\mathrm{TEX_{86}} = (\text{GDGT-2}+\text{GDGT-3}+\mathrm{cren'})/(\text{GDGT-1} + \text{GDGT-2}+\text{GDGT-3}+\mathrm{cren'}). \tag{4}$$

Roman numerals and GDGT names refer to the molecular structures in Fig. 1. Regular reruns of selected samples on the HPLC-MS at ETH show that the analytical error on the indices is <0.01.

3 Results and discussion

3.1 TOC and TN concentrations over time

The TOC concentration in the soil is 1.6–1.7 % at $t=0$, and varies in a range from 1.2–1.7 %, whereas the TN concentration is 0.15 % at $t=0$ and remains practically stable (0.12–0.15) over the course of the experiment (Fig. 3). Given the limited range of variation in TOC and TN concentrations, no direct influence of TOC or TN on the GDGTs in this experiment is assumed during the following discussion.

3.2 BrGDGT concentrations over time

The total brGDGT pool (IPLs + CLs) initially present in the soil is 239 ± 13 ng g^{-1}. In the different incubation experiments these concentrations varied between 187–325 ng g^{-1} (river water), 186–322 ng g^{-1} (ocean water), and 252–407 ng g^{-1} (distilled water control) (Fig. 4). Although the abundance of CL-brGDGTs seems to increase with time in the control setup, the trend is only weak and not significant ($r^2=0.34$, $p=0.122$) due to the large spread in concentration between the two replicate samples at $t=152$ days

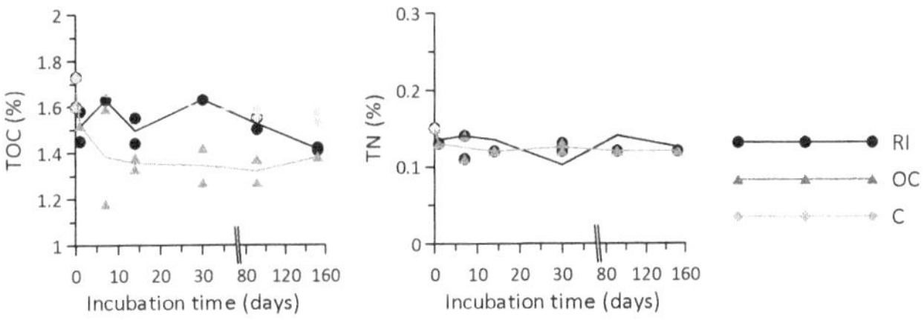

Figure 3. Total organic carbon (TOC) and total nitrogen (TN) concentrations in a sandy loam soil from the Rakaia River catchment, New Zealand, during incubation ($t = 152$ days) in river (RI), ocean (OC), and distilled (C) water.

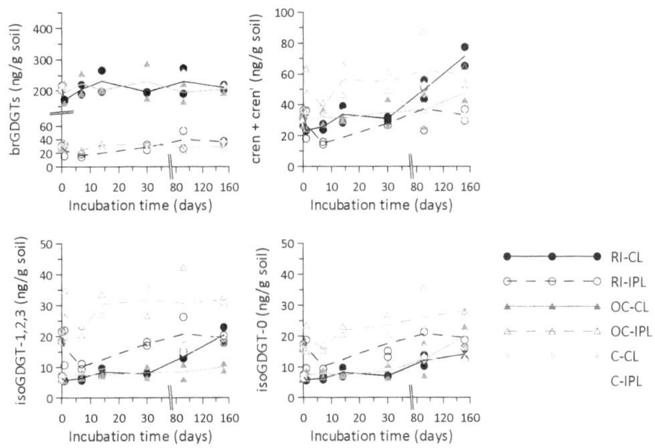

Figure 4. Concentrations of core lipid (CL) and intact polar lipid (IPL)-derived **(a)** brGDGTs, **(b)** crenarchaeol + cren′, **(c)** isoGDGT-1,2,3, and **(d)** isoGDGT-0 in a sandy loam soil from the Rakaia River catchment, New Zealand, during incubation ($t = 152$ days) in river (RI), ocean (OC), and distilled (C) water. Average concentrations are indicated by solid (CLs) and dashed (IPL-derived) lines.

(Fig. 4). Overall, the CL concentrations are within the same range between different experimental setups and remain essentially constant through time. Since the soils were incubated under natural light conditions, this suggests that soil-derived brGDGTs might not be sensitive to photodegradation during the incubation.

The contribution of IPLs to the total amount of brGDGTs in the soil at $t = 0$ is 12.4 ± 0.3 % (Fig. 5), which is in the same range as has previously been reported for temperate soils from the Netherlands, Scotland, and the UK (Peterse et al., 2010; Weijers et al., 2011). The contribution of IPL-derived brGDGTs to the total pool varies mostly within the same range in the river water (7–16 %), ocean water (9–15 %) and control (9–17 %) setups, and like the total amount of CL-brGDGTs, IPL-brGDGT concentrations show no strong trends or changes over time (river water: $r^2 = 0.34$, $p = 0.014$; ocean water: $r^2 = 0.00$, $p = 0.944$; distilled water: $r^2 = 0.20$, $p = 0.268$). This suggests that brGDGTs are either produced at the same rate that they are degraded, or that production and/or degradation of brGDGTs occurs at such low rates that these processes are not detected within the time frame of this experiment. The apparent stability of the IPL-derived, or "living" brGDGT signature (Fig. 4) during the experiment is surprising, given the perceived lability of IPLs in general and their susceptibility to loss of their head groups within a few days after cell lysis (Harvey et al., 1986; White et al., 1979). However, previous studies have indicated that the type of head group and the bond through which it is attached to a core lipid may influence the rate of degradation, offering a potential explanation for the absence of changes in IPL-brGDGT concentration in our experiment. For example, after a 96 h aerobic incubation of a mixture of IPLs with different head groups in beach sediment, 70 % of the phospholipids had degraded, compared to only 3 % of the glycolipids (Harvey et al., 1986). Although a primarily glycosidic head-group composition would be expected based on these former results, the majority of IPL-brGDGTs in soils, peat, and lake sediments are thus far identified as phospholipids (Weijers et al., 2011; Peterse et al., 2011; Tierney et al., 2012), implying that they should be sensitive to degradation, counter to our observations. That this may not be the case may potentially be explained by the findings of Logemann et al. (2011), who recently showed that the bond type within the CL, rather than the type of head group and its connection to that CL, determines IPL sensitivity to degradation. In their experiment, membrane lipids in which the side chains were ester bound, as is common for bacteria, started to rapidly degrade within the first 5 days of the incubation period, whereas the abundance of lipids with ether-bound chains remained invariant during the entire 97-day experiment. BrGDGTs are thought to be produced by bacteria, but they also possess archaeal traits, including their tetraether structure (Weijers et al., 2006a). As a consequence, the presence of at least four ether bonds in each brGDGT may thus not only protect the IPL, but also the CL-brGDGTs from degradation during our experiment, even when the IPLs possess a phospho head group.

Weijers et al. (2010) determined a turnover rate for the total pool of CL brGDGTs in a soil of about 2 decades, which implies that any changes in brGDGT concentration or distribution may indeed not yet be detectable after 152 days of incubation. On the other hand, a recent study by Huguet et al. (2013) indicated that the brGDGT signature in a French peat bog had completely adapted to the 2 °C maximal daytime temperature increase induced by the placement of open-top mini-greenhouses within a period of less than 26 months. Although these results suggest that brGDGTs are turned over at a substantially faster rate, significant changes were only observed in the latter phase of the warming experiment, and were primarily distributional, as their concentration remained the same (Huguet et al., 2013). The unaltered IPL-brGDGT concentrations in our experiment may thus be an indication that the time frame of our incubation has been

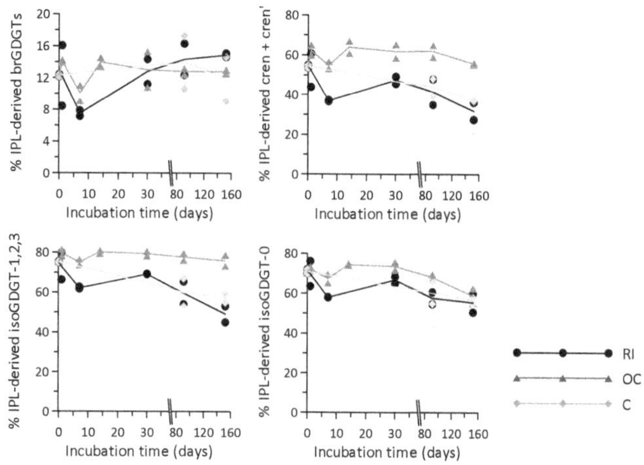

Figure 5. Percentages of the total pool of **(a)** brGDGTs, **(b)** crenarchaeol + cren′, **(c)** isoGDGTs-1,2,3, and **(d)** isoGDGT-0 present in "living", or intact polar lipid (IPL)-derived form. Average concentrations are indicated by solid lines.

too short to reveal any degradation processes based on IPL-brGDGT abundance. However, since the type of head group attached to the GDGTs has not been identified, our data do not allow us to fully exclude equal production and degradation rates as a possible explanation for the apparently stable IPL-brGDGT concentrations during the experiment.

While nutrient conditions in the river and ocean water added to the soil are presumably different, this does not seem to have had any effect on brGDGT concentrations. Moreover, given the soil : water ratio (10 g in 100 mL) used in this experiment, the amounts of nutrients that are released from the soil after the addition of water is likely substantially higher than the concentrations initially present in the river and ocean water. Nevertheless, the effect of nutrient availability on brGDGTs in soils has not yet received much attention, although recently it was shown that variations in the amount and distribution of brGDGTs in lake sediments did not relate to water column nutrient concentrations (Loomis et al., 2014).

3.3 IsoGDGT concentrations over time

Crenarchaeol is the most abundant isoGDGT in the New Zealand soil at the start of the incubation, although its total amount (CLs + IPLs; 63 ± 7 ng g^{-1}), as well as that of all other isoGDGTs (50 ± 5 ng g^{-1}), is low compared to the total pool of brGDGTs (239 ± 13 ng g^{-1}). In contrast, the average fraction of isoGDGTs initially present in the soil as IPL is 54 % for crenarchaeol to 76 % for isoGDGT-2, which is substantially higher than the proportion of brGDGTs occurring as IPLs (12 %) at the commencement of the experiment. Similar proportions of IPL-derived crenarchaeol have been found in soils from the Amazon (~ 50 %; Zell et al., 2013b). The total concentration (IPLs + CLs) of all isoGDGTs increases substantially during the experiment (Fig. 4), and that of isoGDGT-0 in the ocean water setup has even doubled by the end of the incubation (24 ± 2 to 48 ± 5 ng g^{-1}, $r^2 = 0.58$, $p = 0.000$).

The concentrations of the individual IPL-derived isoGDGTs vary between the different incubation setups, but are overall highest in ocean water (Fig. 4; Table in the Supplement). The difference in IPL-isoGDGT concentration between fresh and saline water incubations may potentially be a result of the addition of varying amounts of aquatic isoGDGTs to the soil-derived isoGDGT pool. However, 100 mL of ocean water only contained ~ 0.5 ng isoGDGTs, which is less than 1 % of their initial concentration in the soil. Alternatively, the isoGDGTs could be increasingly released from the soil matrix upon mixing with (saline) water, although the absence of a simultaneous increase in brGDGTs suggests that this is not the most likely scenario. Finally, since *Thaumarchaeota* play an important role in the nitrogen cycle in soils and the marine environment (Leininger et al., 2006; Wuchter et al., 2006; Zhang et al., 2010), nutrient availability may provide another explanation for the concentration differences between experiments. So far, crenarchaeol has been found in all (enrichment) cultures of ammonia-oxidizing *Thaumarchaeota*, and is thus considered as a biomarker lipid for this group (de la Torre et al., 2008; Pitcher et al., 2010, 2011; Sinninghe Damsté et al., 2012), although a recent study suggested that Marine Group II Euryarchaeota may also produce crenarchaeol (Lincoln et al., 2014). Nevertheless, the relatively high crenarchaeol concentration in the incubation setup using ocean water could then be a result of ammonium release due to the higher abundance of exchangeable cations (e.g. Na^+, Mg^+) in ocean water compared to fresh water. This is in agreement with the common method to determine available NH_4^+ in soils, which involves shaking soil in a KCl or Na_2SO_4 solution to extract ammonium from the soil matrix (e.g. Mehlich, 1953). The concentration profiles of the other isoGDGTs are comparable to that of crenarchaeol (Fig. 4), suggesting that they are most likely also primarily derived from soil *Thaumarchaeota*.

Despite the significant increase of the total pools of crenarchaeol and isoGDGTs 0–3 in the river ($r^2 = 0.74$, $p = 0.000$ for crenarchaeol; $r^2 = 0.56$, $p = 0.000$ for isoGDGTs), ocean water ($r^2 = 0.30$, $p = 0.014$ for crenarchaeol; $r^2 = 0.40$, $p = 0.003$ for isoGDGTs), as well as control setups ($r^2 = 0.52$, $p = 0.034$ for crenarchaeol; the increase of isoGDGTs is not significant $r^2 = 0.24$, $p = 0.215$), there is no statistically significant increase in the concentration of individual IPL-derived isoGDGTs during the course of the incubation experiment (e.g. max. $r^2 = 0.18$ for IPL-derived crenarchaeol in ocean water; Fig. 4). We assume therefore that the production rate of IPL-isoGDGTs is approximately constant during this time. The proportional decrease of IPL-isoGDGTs with time (Fig. 5) subsequently indicates that the increase in the overall isoGDGT pool is a

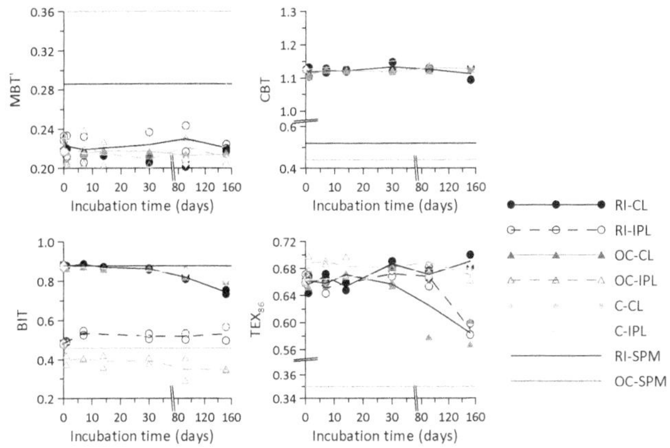

Figure 6. Distribution of core lipid (CL) and intact polar lipid (IPL)-derived GDGTs in a sandy loam soil from the Rakaia River catchment, New Zealand, as reflected by the **(a)** MBT′ index, **(b)** CBT index, **(c)** BIT index, and **(d)** TEX_{86} index during incubation ($t = 152$ days) in river (RI), ocean (OC), and distilled (C) water under quasi-aerobic conditions. Horizontal straight lines represent GDGT composition in river (black) and ocean water (grey) suspended particulate material (SPM). Average concentrations are indicated by solid (CLs) and dashed (IPL-derived) lines.

result of CL accumulation. This in turn confirms the general finding that the turnover of the IPL pool, and thus the release of CL-isoGDGTs, is faster than the degradation of these CLs. Regardless, the increase in isoGDGT concentrations in all incubation setups suggests that availability of water is likely more important that the type of water that is added to the soil.

3.4 GDGT distribution changes

To evaluate if and how soil GDGT signatures are modified during the incubations, we calculated brGDGT- and isoGDGT-based indices commonly used in paleoclimate and carbon cycle studies (i.e. the MBT′, CBT, BIT, TEX_{86}: Table in the Supplement).

Next to the absence of changes in brGDGT concentration, also the MBT′ and CBT indices for the ocean and river water experiments exhibit minimal variation over time, remaining within a range of 0.04 for the MBT′ index, and 0.02 for the CBT (Fig. 6). Surprisingly, the largest changes are observed in the control experiment using distilled water. The MBT′ index for the IPL-derived brGDGT fraction in the control setup increases from 0.23 ± 0.00 to 0.29 ± 0.04 ($r^2 = 0.57$, $p = 0.023$; Fig. 6), and is primarily caused by an increase in the concentration of brGDGT-Ia (from 23 ± 0.3 to 29 ± 3.9 %). A closer look at the relative distributions of each of the brGDGT types at $t = 0$ and $t = 152$ days reveals a subtle shift in the contribution of IPL-derived brGDGT-III (decreasing from 23 ± 1.3 to 19 ± 0.1 %) to brGDGT-II (increasing from 54 ± 0.4 to 58 ± 0.4 %) in river water (data not shown). Since both brGDGTs-II and III in the denominator of the MBT′ index (Eq. 2), and the fraction of IPL-derived brGDGT-I remains stable, this change is not reflected by the index values. Given that the absolute amounts of IPL-derived brGDGTs do not significantly change with time, it is hard to determine the exact processes that cause the distributional changes, as both in situ production and preferential degradation of specific brGDGTs could influence the total concentration. Since brGDGTs with cyclopentane moieties were only present below detection limit in most of the IPL-derived fractions, it was not possible to reliably calculate CBT index values for any of these fractions.

The BIT index is clearly influenced by the substantial increase in crenarchaeol during the experiment, and starts to decrease between $t = 30$ and $t = 91$ days in all water types (Fig. 6). Due to the large percentage of crenarchaeol that is present as IPL compared to that of the brGDGTs, the BIT index of the IPL-derived fraction is much (0.48) lower than that of the CL fraction (0.88) at $t = 0$. The lowest BIT index value for the IPL-derived fraction is 0.35 at $t = 152$ days in ocean water (Fig. 6), and stems from the increase in crenarchaeol (and isoGDGTs in general) in this experiment. The trend of a lower BIT value for the IPL versus the CL fraction has also been found for river SPM from the Yenisei (De Jonge et al., 2014) and Amazon (Zell et al., 2013b), where the difference was explained by a lower degradation rate of soil-derived crenarchaeol IPLs compared to brGDGTs and/or a contribution of in situ produced crenarchaeol IPLs in the river (Zell et al., 2013b). Our incubation results suggest that the degradation of IPLs from soil-derived crenarchaeol is faster than that of brGDGTs, and that the difference in BIT index between the IPL-derived and CL fractions is thus mainly caused by the higher production rate of crenarchaeol. This is in agreement with the recent finding that the BIT index of river SPM primarily tracks the seasonal aquatic production of crenarchaeol rather than that of soil input (Yang et al., 2013; Zell et al., 2013a). Our results furthermore indicate that care should be taken with using the BIT index to constrain the input of fluvially transported soil OC into a marine system. However, the stability of the soil brGDGT pool supports earlier observations that tracing the absolute amount (as opposed to the relative abundance) of brGDGTs in rivers during land–sea transport may provide a more reliable tracer of soil OC at our study site (see Fietz et al., 2011; Smith et al., 2012; Zell et al., 2013b). Moreover, the spread in GDGT response factors between laboratories resulting in a range of BIT values for the same sample provides another, independent argument to use absolute concentration measurements (Schouten et al., 2013a).

TEX_{86} index values start to change between 30 and 91 days of incubation for the IPL-fraction in river water and the CL-fraction in ocean water experiments (Fig. 6). We find no clear explanation for the diverging trends for the different water types or IPL versus CL fractions, but there are several factors that may contribute to the observed changes. For example, shifts in the initial isoGDGT community due to water contact, different membrane adaptation rates, or preferential release from the soil matrix may all have occurred. Alternatively, the relative distribution of isoGDGTs may differ per type of head group in both marine (Schouten et al., 2008; Pitcher et al., 2011; Lengger et al., 2012) and soil *Thaumarchaeota* (Sinninghe Damsté et al., 2012), so that the deviation in TEX_{86} values may also be explained by different turnover rates per head group. However, even though the total variation in TEX_{86} corresponds with a 4 °C SST change in the experiment, given the low absolute amounts of terrestrial isoGDGTs that are finally discharged to the ocean in this case, this will have a minor impact on TEX_{86} records from the marine environment where their abundances are much higher.

3.5 Experimental setup evaluation

Despite prior evidence for in situ production of brGDGTs in several large rivers (e.g. Kim et al., 2012; Zhang et al., 2012; Zell et al., 2013b; Yang et al., 2013; De Jonge et al., 2014), we have not observed this in our experiment. The absence of an in situ production signal in our data may be a consequence of the soil : water ratio of 1 : 10 that we used in our experimental setup, which is likely more concentrated than in most natural systems, so that the relatively large amount of soil-brGDGTs may have overprinted any evidence of aquatic brGDGT production during the experiment. An indication that in situ production in the Rakaia River may take place after all is reflected by the offset between brGDGTs in river SPM and that in the soil (Fig. 6). Although this soil-SPM offset is based on a single sample location in the entire river basin and brGDGT distributions in soils may show substantial variation within a small area (e.g. Weijers et al., 2007b; Naeher et al., 2014), it may also support the idea that brGDGTs in soils and river SPM may have different sources. Notably, also the brGDGT distributions in river and ocean SPM show an offset (Fig. 6), despite the ocean water sample location being close to the river mouth. Differences in biomarker distributions in SPM from the fresh–saline water transition have previously been attributed to hydrological sorting (e.g. Goñi et al., 1998), or, more specifically in the case of brGDGTs, additional marine production (e.g. Zhu et al., 2011; Zell et al., 2014). In our case, this offset may have also been introduced by the use of filters with different pore size to obtain the SPM from the river and ocean water (0.2 and 0.7 μm, respectively). This has resulted in a comparison of "free living" and "suspended" lipid fractions, which at least in marine water may have a different composition (Ingalls et al., 2012; Close et al., 2014).

Furthermore, 152 days of incubation appears not to have been long enough to determine the sensitivity of brGDGTs to degradation in aquatic environments based on changes in their distribution or abundance. The unexpected apparent stability of the IPL-derived brGDGTs suggests that the presence of IPL-brGDGTs in aquatic systems does not necessarily indicate that they are produced in situ, and rather support the earlier finding that IPLs containing ether lipids are relatively resistant against degradation, even on a longer timescale than has previously been shown (t = 97 days; Logemann et al., 2011). Alternatively, the water-saturated conditions may have inhibited both growth and degradation of brGDGTs to take place. However, without analysing the exact head-group composition of the brGDGTs in our experiment it is not possible to distinguish between these processes.

For this study, it was assumed that the soil-derived (IPL-)GDGTs would be degraded in aquatic environments, yet our data indicate that the concentration of isoGDGTs actually increased over time in all setups (Fig. 4). This suggests that isoGDGTs are either produced, or that they are (preferentially) released from the soil matrix during the experiment. In order to get a better control on the exact processes taking place, the control setup could have benefitted from the addition of a chemical agent to prevent any microbial activity, and thus to avoid additional GDGT-production during the experiment.

To summarize, follow-up experiments could benefit from a more natural soil : water ratio, so that a potential contribution of aquatic GDGTs can be detected by analysing the soils. The control setup should be treated to prevent any microbial activity. In addition, determining the abundance and distribution of GDGTs present in the incubation water could likely provide further information on water column processes, provided that the amount of water added to the soil is sufficient to obtain a large enough sample upon filtering for GDGT analysis. Finally, to better understand the apparent stability of IPL-brGDGTs in aquatic environment, the type of head group should be monitored during the incubation experiment, which should run longer than the 152 days of this study.

4 Conclusions

Laboratory incubation experiments involving admixture of soil and fresh/ocean water indicate that soil-derived brGDGTs appear to be surprisingly stable in aquatic environments in both IPL- and CL configurations. Our observations suggest that soil brGDGT signatures will likely be unaltered during fluvial transport from land to sea, although any influence of aquatic degradation and/or production processes may be obscured due to the higher-than-natural soil : water ratio used in our experimental setup. Indeed, distributional offsets between brGDGTs in river SPM and catchment soil, as well as between brGDGTs in river SPM and ocean SPM indicate that additional sources may contribute to the final distribution in which brGDGTs are delivered to the ocean. This means that the exact sources of brGDGTs in a river system need to be well constrained before these compounds can be used as reliable tracers for land–sea transport of soil organic carbon, as well as for paleoclimate reconstructions.

As opposed to brGDGTs, the concentration of isoGDGTs increased during all incubation experiments, altering their initial relative distribution. Although the changes in TEX_{86} index values correspond with a maximum change of about 4 °C in reconstructed SSTs in the ocean water setup, the low abundance of isoGDGTs in soils relative to in marine settings suggests that their land–sea transport would have minimal impact on TEX_{86} recorded in the marine sedimentary record. Nevertheless, after 30 days of incubation, the increase in isoGDGTs was sufficient to affect BIT index values. The observed increase in isoGDGT concentration in the control setup using distilled water indicates that the availability of water is more important for the behaviour of isoGDGTs than the properties of the aquatic system in which the soil is introduced. Our findings may have consequences for the interpretation of TEX_{86}-based SST records for coastal marine settings with a BIT index close to the cutoff value of 0.3 (Weijers et al., 2006b), as the actual input of soil-derived GDGTs upon delivery to the ocean may be larger than anticipated.

Acknowledgements. Three anonymous reviewers are thanked for their comments that have improved this paper. This study received funding from ETH Fellowship (FEL-36 11-1) and ESF-MOLTER Exchange Grant (no. 3695) awarded to F. Peterse. We thank John Williams (University of Otago) for sampling the incubation experiment, Daniel Montluçon and Negar Haghipour (ETH) for laboratory support, and Lukas Jonkers (Universidad Autònoma de Barcelona/Cardiff University) for help in the field.

Edited by: S. Pantoja

References

Blair, N. E., Leithold, E. L., Brackley, H., Trustrum, N., Page, M., and Childress, L.: Terrestrial sources and export of particulate organic carbon in the Waipaoa sedimentary system: Problems, progress and processes, Mar. Geol., 270, 108–118, 2010.

Close, H. G., Wakeham, S. G., and Pearson, A.: Lipid and ^{13}C signatures of submicron and suspended particulate organic matter in the Eastern Tropical North Pacific: Implications for the contribution of Bacteria, Deep-Sea Res. Pt. I, 85, 15–34, 2014.

Cole, J. J., Prairie, Y. T., Caraco, N. F., McDowell, W. H., Tranvik, L. J., Striegl, R. G., Duarte, C. M., Kortelainen, P., Downing, J. A., Middelburg, J. J., and Melack, J.: Plumbing the global carbon cycle: Integrating inland waters into the terrestrial carbon budget, Ecosystems, 10, 171–184, 2007.

De Jonge, C., Stadnitskaia, A., Hopmans, E. C., Cherkashov, G., Fedotov, A., and Sinninghe Damsté, J. S.: In-situ produced branched glycerol dialkyl glycerol tetraethers in suspended particulate matter from the Yenisei River, Eastern Siberia, Geochim. Cosmochim. Ac., 125, 476–491, 2014.

De la Torre, J. R., Walker, C. B., Ingalls, A. E., Könneke, M., and Stahl, D. A.: Cultivation of a thermophilic ammonia oxidizing archaeon synthesizing crenarchaeol, Environ. Microbiol., 10, 810–818, 2008.

Fietz, S., Martínez-García, A., Huguet, C., Rueda, G., and Rosell-Melé, A.: Constraints in the application of the Branched and Isoprenoid Tetraether index as a terrestrial input proxy, J. Geophys. Res., 116, C10032, doi:10.1029/2011JC007062, 2011.

Goñi, M. A., Ruttenberg, K. C., and Eglinton, T. I.: A reassessment of the sources and importance of land-derived organic matter in surface sediments from the Gulf of Mexico, Geochim. Cosmochim. Ac., 62, 3055–3075, 1998.

Harvey, H. R., Fallon, R. D., and Patton, J. S.: The effect of organic matter and oxygen on the degradation of bacterial membrane lipids in marine sediments, Geochim. Cosmochim. Ac., 50, 795–804, 1986.

Hicks, D. M., Shankar, U., Mckerchar, A. I., Basher, L., Lynn, I., Page, M., and Jessen, M.: Suspended sediment yields from New Zealand rivers, J. Hydrol., 50, 81–142, 2011.

Hopmans, E. C., Weijers, J. W. H., Schefuss, E., Herfort, L., Sinninghe Damsté, J. S., and Schouten, S.: A novel proxy for terrestrial organic matter in sediments based on branched and isoprenoid tetraether lipids, Earth Planet. Sci. Lett., 224, 107–116, 2004.

Huguet, A., Fosse, C., Laggoun-Défarge, F., Delarue, F., and Derenne, S.: Effects of a short-term experimental microclimate warming on the abundance and distribution of branched GDGTs in a French peatland, Geochim. Cosmochim. Ac., 105, 294–315, 2013.

Huguet, C., Hopmans, E. C., Febo-Ayala, W., Thompson, D. H., Sinninghe Damsté, J. S., and Schouten, S.: An improved method to determine the absolute abundance of glycerol dibiphytanyl glycerol tetraether lipids, Org. Geochem., 37, 1036–1041, 2006.

Ingalls, A. E., Huguet, C., and Truxal, L. T.: Lipids of archaeal glycerol dialkyl glycerol tetraethers among size-fractioned particulate organic matter in Hood Canal, Puget Sound, Appl. Environ. Microbiol., 78, 1480–149, 2012.

Kim, J.-H., Schouten, S., Buscail, R., Ludwig, W., Bonnin, J., Sinninghe Damsté, J. S., and Bourrin, F.: Origin and distribution of terrestrial organic matter in the NW Mediterranean (Gulf of Lions): Exploring the newly developed BIT index, Geochem. Geophys. Geosys., 7, Q11017, doi:10.1029/2006GC001306, 2006.

Kim, J.-H., van der Meer, J., Schouten, S., helmke, P., Willmott, V., Sangiorgi, F., Koç, N., Hopmans, E. C., and Sinninghe Damsté, J. S.: New indices and calibrations derived from the distribution of crenarchaeol isoprenoid tetraether lipids: Implications for past sea surface temperature reconstructions, Geochim. Cosmochim. Ac., 74, 4639–4654, 2010.

Kim, J.-H., Zell, C., Moreira-Turcq, P., Pérez, M.A.P., Abril, G., Mortillaro, J.-M., Weijers, J. W. H., Meziane, T., and Sinninghe Damsté, J. S.: Tracing soil organic carbon in the lower Amazon River and its tributaries using GDGT distributions and bulk organic matter properties, Geochim. Cosmochim. Ac., 90, 163–180, 2012.

Leininger, S., Urich, T., Schloter, M., Schwark, L., Qi, J., Nicol, G. W., Prosser, J. I., Schuster, S. C., and Schleper, C.: Archaea predominate among ammonia-oxidizing prokaryotes in soils, Nature, 442, 806–809, 2006.

Lengger, S. K., Hopmans, E. C., Reichart, G.-J., Nierop, K. G. J., Sinninghe Damsté, J. S., and Schouten, S.: Distribution of core and intact polar glycerol dibiphytanyl glycerol tetraether lipids in the Arabian Sea Oxygen Minimum Zone, II: evidence for selective preservation and degradation in sediments and consequences for the TEX_{86}, Geochim. Cosmochim. Ac., 98, 244–258, 2012.

Lincoln, S. A., Wai, B., Eppley, J. M., Church, M. J., Summons, R. E., and DeLong, E. F.: Planktonic Euryarchaeota are a significant source of archaeal tetraether lipids in the ocean, P. Natl. Acad. Sci. USA, 111, 9858–9863, 2014.

Liu, X.-L., Leider, A., Gillespie, A., Gröger, J., Versteegh, G. J. M., and Hinrichs, K.-U.: Identification of polar lipid precursors if the ubiquitous branched GDGT orphan lipids in a peat bog in Northern Germany, Org. Geochem., 41, 653–660, 2010.

Logemann, J., Graue, J., Köster, J., Engelen, B., Rullkötter, J., and Cypionka, H.: A laboratory experiment of intact polar lipid degradation in sandy sediments, Biogeosciences, 8, 2547–2560, doi:10.5194/bg-8-2547-2011, 2011.

Loomis, S. E., Russell, J. M., Eggermont, H., Verschuren, D., and Sinninghe Damsté, J. S.: Effects of temperature, pH, and nutrient concentration on branched GDGT distributions in East African lakes: implications for paleoenvironmental reconstruction, Org. Geochem., 66, 25–37, 2014.

Mehlich, A.: Determination of P, Ca, Mg, K, Na, and NH_4, Pub. No. 1–53, NC Soil Testing Division, Raleigh, NC, 1953.

Naeher, S., Peterse, F., Smittenberg, R. H., Niemann, H., Zigah, P. K., and Schubert, C. J.: Sources of glycerol dialkyl glycerol tetraethers (GDGTs) in catchment soils, water column and sediments of Lake Rotsee (Switzerland) – Implications for the application of GDGT-based proxies for lakes, Org. Geochem., 66, 164–173, 2014.

Oppermann, B. I., Michaelis, W., Blumenberg, M., Frerichs, J., Schulz, H. M., Schippers, A., Beaubien, S. E., and Krüger, M.: Soil microbial community changes as a result of long-term exposure to a natural CO_2 vent, Geochim. Cosmochim. Ac., 74, 2697–2716, 2010.

Pancost, R. D. and Sinninghe Damsté, J. S.: Carbon isotopic compositions of prokaryotic lipids as tracers of carbon cycling in diverse settings, Chem. Geol., 195, 29–58, 2003.

Peterse, F., Kim, J.-H., Schouten, S., Klitgaard Kristensen, D., Koç, N., and Sinninghe Damsté, J. S.: Constraints on the application of the MBT/CBT palaeothermometer at high latitude environments (Svalbard, Norway), Org. Geochem., 40, 692–699, 2009.

Peterse, F., Nicol, G. W., Schouten, S., and Sinninghe Damsté, J. S.: Influence of soil pH on the abundance and distribution of core and intact polar lipid-derived branched GDGTs in soil, Org. Geochem., 41, 1171–1175, 2010.

Peterse, F., Hopmans, E. C., Schouten, S., Mets, A., Rijpstra, W. I. C., and Sinninghe Damsté, J. S.: Identification and distribution of intact polar branched tetraether lipids in peat and soil, Org. Geochem., 42, 1007–1015, 2011.

Peterse, F., van der Meer, J., Schouten, S., Weijers, J. W. H., Fierer, N., Jackson, R. B., Kim, J.-H., and Sinninghe Damsté, J. S.: Revised calibration of the MBT-CBT paleotemperature proxy based on branched tetraether membrane lipids in surface soils, Geochim. Cosmochim. Ac., 96, 215–229, 2012.

Pitcher, A., Hopmans, E. C., Schouten, S., and Sinninghe Damsté, J. S.: Separation of core and intact polar archaeal tetraether lipids using silica columns: Insights into living and fossil biomass contributions, Org. Geochem., 40, 12–19, 2009.

Pitcher, A., Rychlik, N., Hopmans, E. C., Spieck, E., Rijpstra, W. I. C., Ossebaar, J., Schouten, S., Wagner, M., and Sinninghe Damsté, J. S.: Crenarchaeol and its regioisomer dominate the membrane lipids of "Candidatus Nitrososphaera gargensis", a thermophilic Group I, 1b Crenarchaeote, ISME J., 4, 542–552, 2010.

Pitcher, A., Hopmans, E. C., Mosier, A. C., Francis, C. A., Reese, S. K., Schouten, S., and Sinninghe Damsté, J. S.: Distribution of core and intact polar tetraether lipids in enrichment cultures of Thaumarchaeota from marine sediments, Appl. Environ. Microbiol., 77, 3468–3477, 2011.

Schouten, S., Hopmans, E. C., Pancost, R. D., and Sinninghe Damsté, J. S.: Widespread occurrence of structurally diverse tetraether membrane lipids: Evidence for the ubiquitous presence of low-temperature relatives of hyperthermophiles, P. Natl. Acad. Sci. USA, 97, 14421–14426, 2000.

Schouten, S., Hopmans, E. C., Schefuss, E., and Sinninghe Damsté, J. S.: Distributional variations in marine crenarchaeotal membrane lipids: a new tool for reconstructing ancient sea water temperatures?, Earth Planet. Sci. Lett., 204, 265–274, 2002.

Schouten, S., Huguet, C., Hopmans, E. C., Kienhuis, M. V. M., and Sinninghe Damsté, J. S.: Analytical methodology for TEX86 paleothermometry by high-performance liquid chromatography/atmospheric pressure chemical ionization-mass spectrometry, Anal. Chem., 79, 2940–2944, 2007.

Schouten, S., Hopmans, e. C., Baas, M., Boumann, H., Standfest, S., Könneke, M., Stahl, D. A., and Sinninghe Damsté, J. S.: Intact membrane lipids of "Candidatus Nitrosopumilus maritimus", a cultivated representative of the cosmopolitan mesophilic group I crenarchaeota, Appl. Environ. Microbiol., 74, 2433–2440, 2008.

Schouten, S., Hopmans, E. C., Rosell-Melé, A., Pearson, A., Adam, P., Bauersachs, T., Bard, E., Bernasconi, S. M., Bianchi, T. S., Brocks, J. J., Carlson, L. T., Castañeda, I. S., Derenne, S., Selver, A. D., Dutta, K., Eglinton, T., Fosse, C., Galy, V., Grice, K., Hinrichs, K. U., Huang, Y., Huguet, A., Huguet, C., Hurley, S., Ingalls, A., Jia, G., Keely, B., Knappy, C., Kondo, M., Krishnan, S., Lincoln, S., Lipp, J., Mangelsdorf, K., Martínez-García, A., Ménot, G., Mets, A., Mollenhauer, G., Ohkouchi, N., Ossebaar, J., Pagani, M., Pancost, R. D., Pearson, E.J., Peterse, F., Reichart, G.-J., Schaeffer, P., Schmitt, G., Schwark, L., Shah, S. R., Smith, R. W., Smittenberg, R. H., Summons, R. E., Takano, Y., Talbot, H. M., Taylor, K. W. R., Tarozo, R., Uchida, M., Dongen, B. E. van, Van Mooy, B. A. S., Wang, J., Warren, C., Weijers, J. W. H., Werne, J. P., Woltering, M., Xie, S., Yamamoto, M., Zhang, C. L., Zhang, Y., Zhao, M., and Sinninghe Damsté, J. S.: An interlaboratory study of TEX_{86} and BIT analysis of sediments, extracts, and standard mixtures, Geochem. Geophys. Geosys., 14, 5263–5285, 2013a.

Sinninghe Damsté, J. S., Hopmans, E. C., Pancost, R. D., Schouten, S., and Geenevasen, J. A. J.: Newly discovered non-isoprenoid glycerol dialkyl glycerol tetraether lipids in sediments, Chem. Commun., 17, 1683–1684, 2000.

Sinninghe Damsté, J. S., Rijpstra, W. I. C., Hopmans, E. C., Prahl, F. G., Wakeham, S. G., and Schouten, S.: Distribution of membrane lipids of planktonic Crenarchaeota in the Arabian Sea, Appl, Environ. Microbiol., 68, 2997–3002, 2002.

Sinninghe Damsté, J. S., Ossebaar, J., Abbas, B., Schouten, S., and Verschuren, D.: Fluxes and distribution of tetraether lipids in an equatorial African lake: Constraints on the application of the TEX(86) palaeothermometer and BIT index in lacustrine settings, Geochim. Cosmochim. Ac., 73, 4232–4249, 2009.

Sinninghe Damsté, J. S., Rijpstra, W. I. C., Hopmans, E. C., Weijers, J. W. H., Foesel, B. U., Overmann, J., and Dedysh, S. N.: 13, 16-Dimethyl Octacosanedioic Acid (iso-Diabolic Acid), a Common Membrane-Spanning Lipid of Acidobacteria Subdivisions 1 and 3, Appl. Environ. Microbiol., 77, 4147–4154, 2011.

Sinninghe Damsté, J. S., Rijpstra, W. I. C., Hopmans, E. C., Jung, M. Y., Kim, J. G., Rhee, S. K., Stieglmeier, M., and Schleper, C.: Intact polar and core glycerol dibiphytanyl glycerol tetraether lipids of group I.1a and I.1b Thaumarchaeota in soil, Appl. Environ. Microbiol., 78, 6866–6874, 2012.

Smith, R. W., Bianchi, T. S., and Li, X.: A re-evaluation of the use of branched GDGTs as terrestrial biomarkers: Implications for the BIT index, Geochim. Cosmochim. Ac., 80, 14–29, 2012.

Sturt, H. F., Summons, R. E., Smith, K., Elvert, M., and Hinrichs, K.-U.: Intact polar membrane lipids in prokaryotes and sediments deciphered by high-performance liquid chromatography/electrospray ionization multistage mass spectrometry – new biomarkers for biogeochemistry and microbial ecology, Rapid Commun. Mass Sp., 18, 617–628, 2004.

Tierney, J. E. and Russell, J. M.: Distributions of branched GDGTs in a tropical lake system: Implications for lacustrine application of the MBT/CBT paleoproxy, Org. Geochem., 40, 1032–1036, 2009.

Tierney, J. E., Schouten, S., Pitcher, A., Hopmans, E. C., and Sinninghe Damsté, J. S.: Core and intact polar glycerol dialkyl glycerol tetraethers (GDGTs) in Sand Pond, Warwick, Rhode Island (USA): Insights into the origin of lacustrine GDGTs, Geochim. Cosmochim. Ac., 77, 561–581, 2012.

Weijers, J. W. H., Schouten, S., Hopmans, E. C., Geenevasen, J. A. J., David, O. R. P., Coleman, J. M., Pancost, R. D., and Sinninghe Damsté, J. S.: Membrane lipids of mesophilic anaerobic bacteria thriving in peats have typical archaeal traits, Environ. Microbiol., 8, 648–657, 2006a.

Weijers, J. W. H., Schouten, S., Spaargaren, O. C., and Sinninghe Damsté, J. S.: Occurrence and distribution of tetraether membrane lipids in soils: Implications for the use of the TEX(86) proxy and the BIT index, Org. Geochem., 37, 1680–1693, 2006b.

Weijers, J. W. H., Schefuss, E., Schouten, S., and Sinninghe Damsté, J. S.: Coupled thermal and hydrological evolution of tropical Africa over the last deglaciation, Science, 315, 1701–1704, 2007a.

Weijers, J. W. H., Schouten, S., van den Donker, J. C., Hopmans, E. C., and Sinninghe Damsté, J. S.: Environmental controls on bacterial tetraether membrane lipid distribution in soils, Geochim. Cosmochim. Ac., 71, 703–713, 2007b.

Weijers, J. W. H., Panoto, E., van Bleijswijk, J., Schouten, S., Rijpstra, W. I. C., Balk, M., Stams, A. J. M., and Sinninghe Damsté, J. S.: Constraints on the biological source(s) of the orphan branched tetraether membrane lipids, Geomicrob. J., 26, 402–414, 2009.

Weijers, J. W. H., Wiesenberg, G. L. B., Bol, R., Hopmans, E. C., and Pancost, R. D.: Carbon isotopic composition of branched tetraether membrane lipids in soils suggest a rapid turnover and a heterotrophic life style of their source organism(s), Biogeosciences, 7, 2959–2973, doi:10.5194/bg-7-2959-2010, 2010.

Weijers, J. W. H., Bernhardt, B., Peterse, F., Werne, J. P., Dungait, J. A. J., Schouten, S., and Sinninghe Damsté, J. S.: Absence of seasonal patterns in MBT-CBT indices in mid-latitude soils, Geochim. Cosmochim. Ac., 75, 3179–3190, 2011.

Weyhenmeyer, G. A., Fröberg, M., Karltun, E., Khalili, M., Kothawala, D., Temnerud J., and Tranvik, L. J.: Selective decay of terrestrial organic carbon during transport from land to sea, Global Change Biol., 18, 349–355, 2012.

White, D. C., Davis, W. M., Nickels, J. S., King, J. D., and Bobbie, R. J.: Determination of the sedimentary microbial biomass by extractible lipid phosphate, Oecologia, 40, 51–62, 1979.

Wuchter, C., Abbas, B., Coolen, M. J. L., Herfort, L., Timmers, P., Strous, M., van Bleijswijk, J., Teira, E., Herndl, G. J., Middelburg, J. J., Schouten S., and Sinninghe Damsté, J. S.: Archaeal nitrification in the ocean, P. Natl. Acad. Sci. USA, 103, 12317–12322, 2006.

Xie, S., Lipp, J. S., Wegener, G., Ferdelman, T. G., and Hinrichs, K.-U.: Turnover of microbial lipids in the deep biosphere and growth of benthic archaeal populations, P. Natl. Acad. Sci. USA, 110, 6010–1014, 2013.

Yang, G., Zhang, C. L., Xie, S., Chen, Z., Gao, M., Ge, Z., and Yang, Z.: Microbial glycerol dialkyl glycerol tetraethers from river water and soil near the Three Gorges Dam on the Yangtze River, Org. Geochem., 56, 40–50, 2013.

Zell, C., Kim, J.-H., Abril, G., Lima Sobrinho, R., Dorhout, D., Moreira-Turcq, P., and Sinninghe Damsté, J. S.: Impact of seasonal hydrological variation on the distributions of tetraether lipids along the Amazon River in the central Amazon basin: Implications for the MBT/CBT paleothermometer and the BIT index, Frontiers in Terrestrial Microbiology, 4, 228, doi:10.3389/fmicb.2013.00228, 2013a.

Zell, C., Kim, J.-H., Moreira-Turcq, P., Abril, G., Hopmans, E. C., Bonnet, M.-P., Lima Sobrinho, R., and Sinninghe Damsté, J. S.: Disentangling the origins of branched tetraether lipids and crenarchaeol in the lower Amazon River: Implications for GDGT-based proxies, Limnol. Oceanogr., 58, 343–353, 2013b.

Zell, C., Kim, J.-H., Hollander, D., Lorenzoni, L., Baker, P., Guizan Silva, C., and Nittrouer, C.: Sources and distributions of branched and isoprenoid tetraether lipids on the Amazon shelf and fan: Implications for the use of GDGT-based proxies in marine sediments, Geochim. Gosmochim. Ac., 139, 293–312, 2014.

Zhang, C. L., Wang, J., Wei, Y., Zhu, C., Huang, L., and Dong, H.: Production of branched tetraether lipids in the lower Pearl River and estuary: Effects of extraction methods and impact on bGDGT proxies, Frontiers in Terrestrial Microbiology, 2, 274, doi:10.3389/fmicb.2011.00274, 2012.

Zhang, L.-M., Offre, P. R., he, J.-Z., Verhamme, D. T., Nicol, G. W., and Prosser, J. I.: Autotrophic ammonia oxidation by soil Thaumarchaeota, P. Natl. Acad. Sci. USA, 107, 17240–17245, 2010.

Zhu, C., Weijers, J. W. H., Wagner, T., Pan, J.-M., Chen, J.-F., and Pancost, R. D.: Sources and distributions of tetraether lipids in surface sediments across a large river-dominated continental margin, Org. Geochem., 42, 376–386, 2011.

18

New highlights of phytolith structure and occluded carbon location: 3-D X-ray microscopy and NanoSIMS results

A. Alexandre[1], I. Basile-Doelsch[1], T. Delhaye[2], D. Borshneck[1], J. C. Mazur[1], P. Reyerson[3], and G. M. Santos[4]

[1]Centre Européen de Recherche et d'Enseignement des Géosciences de l'Environnement (UMR 7330), CNRS, Aix-Marseille Université, Europôle méditerranéen de l'Arbois BP 80, 13545 Aix en Provence CEDEX 04, France
[2]Plateforme NanoSIMS, OSUR, Université de Rennes 1, Campus de Beaulieu, 35042 Rennes CEDEX, France
[3]Department of Geography, University of Wisconsin-Madison, 550 North Park Street, Madison, WI 53706, USA
[4]Department of Earth System Science, University of California, Irvine, B321 Croul Hall, Irvine, CA 92697-3100, USA

Correspondence to: A. Alexandre (alexandre@cerege.fr)

Abstract. Phytoliths contain occluded organic compounds called phytC. Recently, phytC content, nature, origin, paleoenvironmental meaning and impact in the global C cycle have been the subject of increasing debate. Inconsistencies were fed by the scarcity of in situ characterizations of phytC in phytoliths. Here we reconstructed at high spatial resolution the 3-D structure of harvested grass short cell (GSC) phytoliths using 3-D X-ray microscopy. While this technique has been widely used for 3-D reconstruction of biological systems it has never been applied in high-resolution mode to silica particles. Simultaneously, we investigated the location of phytC using nanoscale secondary ion mass spectrometry (NanoSIMS). Our data evidenced that the silica structure contains micrometric internal cavities. These internal cavities were sometimes observed isolated from the outside. Their opening may be an original feature or may result from a beginning of dissolution of silica during the chemical extraction procedure, mimicking the progressive dissolution process that can happen in natural environments. The phytC that may originally occupy the cavities is thus susceptible to rapid oxidation. It was not detected by the NanoSIMS technique. However, another pool of phytC, continuously distributed in and protected by the silica structure, was observed. Its N / C ratio (0.27) is in agreement with the presence of amino acids. These findings constitute a basis to further characterize the origin, occlusion process, nature and accessibility of phytC, as a prerequisite for assessing its significance in the global C cycle.

1 Introduction

When absorbing nutrients in the soil, plants roots also uptake a significant amount of silicon (Si). The Si fluxes recycled by plants are substantial; for example, Si take up by tropical forests or grasslands can reach 2 to 10 times Si fluxes generated from the dissolution of soil silicates that are exported to stream water (e.g., Blecker et al., 2006; Struyf and Conley, 2009; Cornelis et al., 2011; Alexandre et al., 2011). Inside the plant, Si is transported in the sap and deposited inside the cells, in the cell walls and in extracellular spaces of stems and leaves as micrometric hydrous amorphous silica particles called phytoliths. Upon plant decay, part of the phytolith production can be incorporated into soils or sediments and preserved for as long as millions of years (Alexandre et al., 2011; Miller et al., 2012; Strömberg et al., 2013). These fossil phytolith assemblages can be used for reconstructing past vegetation and climate conditions via their morphological and geochemical signatures (Piperno, 2006; Alexandre et al., 2012). Phytoliths occlude small amounts of organic compounds, first evidenced by the production of carbon (C) and nitrogen (N) during dry ashing (Jones and Beavers, 1963). Later on, scanning transmission electron microscopy (STEM) and energy dispersive X-ray (EDX) analyses of phytoliths in the plant tissues confirmed that the occluded organic compounds contained C, N and phosphorus (P) (Laue et al., 2007). By extension, these occluded compounds are here called phytC. PhytC, which is assumed to be protected from natural oxidation by the siliceous structure, has been the subject of increasing attention and debate.

Based on the assumption that phytC originated from the photosynthesis of atmospheric CO_2 in the host plant, several studies used phytC^{14}C and δ^{13}C signatures respectively as a dating tool (Piperno and Becker, 1996; Piperno and Stothert, 2003; McMichael et al., 2012) and a paleoenvironmental proxy (Kelly et al., 1991; Smith and White, 2004; Carter, 2009; Webb and Longstaffe, 2010; McInerney et al., 2011). However, very recently, ^{14}C-AMS measurements of phytC samples from modern grasses yielded ages of several thousand years, which suggested that phytoliths may incorporate a substantial amount of old carbon, potentially from the soil (Santos et al., 2010, 2012). Amino acids from soils have been shown to be taken up by plants, and transported in small proportions to roots, stems and shoots (Paungfoo-Lonhienne et al., 2008; Whiteside et al., 2009, 2012; Gao et al., 2010; Warren, 2012). Thus it is not inconsistent to assume that C and N derived from these soil amino acids have been trapped in phytoliths. Although the hypothesis still needs to be verified, it raises the question of the molecular nature of phytC. Several techniques such as high-performance liquid chromatography (HPLC), gas chromatography mass spectrometry (GC–MS), protein staining, micro-Raman analysis and X-ray photoelectron spectroscopy (XPS) have been used to characterize phytC and led to contradictory results, especially regarding the presence or not of amino acids (Harrison, 1996; Pironon et al., 2001; Smith and Anderson, 2001; Elbaum et al., 2009; Watling et al., 2011). The problem is that these methods were applied on phytolith concentrates that were not proven to be completely devoid of extraneous organic remains. Chemical extractions leading to high-purity phytolith concentrates are indeed difficult to implement. Although the absence of organic particles can be checked by scanning electron microscopy (SEM) coupled with EDX (Corbineau et al., 2013), the presence of extraneous organic remains on the phytolith surface cannot be accurately detected.

Differences in the efficiency of phytolith extraction protocols may also explain the inconsistencies in phytC quantification. Accurately quantifying the phytC is important for the assessment of its significance in the terrestrial C cycle. Multiple studies have recently claimed that phytC may play a role in atmospheric CO_2 sequestration and climate change mitigation (Parr and Sullivan, 2005; Parr et al., 2010; Song et al., 2014; Huang et al., 2014; Li et al., 2014; Zuo et al., 2014), although the fluxes of phytC from vegetation to soils and the residence time of phytC in soils are still largely unknown. PhytC content as high as 20 % dry weight was obtained when using a phytolith extraction method based on microwave digestion (Parr and Sullivan, 2014). This value was more than 20 to 200 times higher than the values obtained using a chemical method verified to be 100 % efficient for removing extraneous organic particles (from 0.1 to 1 % dry weight; Smith and White, 2004). The difference was somewhat justified by partial dissolution of phytC when using aggressive protocols. The assumption that phytC may be located at different sites in the silica structure, with different accessibility to oxidation, has been put forward (Parr and Sullivan, 2014). This assumption supplemented a previous one, widely found in the literature, that micrometric opaque areas observed by natural light (NL) microscopy on some phytoliths were holes containing the phytC (Prychid et al., 2003; Piperno, 2006; Carter et al., 2009; Song et al., 2012; Parr and Sullivan, 2014). No measurements were however performed to support any of these hypotheses.

Finally, the debates on content, location, nature, origin and paleoenvironmental meaning of phytC have been fed by the scarcity of in situ characterizations of phytC in phytoliths, despite few seminal works (Harrison, 1996; Laue et al., 2007). Here we reconstructed, at high spatial resolution, the 3-D structure of grass phytoliths using 3-D X-ray microscopy. Simultaneously, we characterized the location of phytC using nanoscale secondary ion mass spectrometry (NanoSIMS).

2 Material and methods

Grasses are among the main producers of phytoliths. The leaves of *Triticum durum* wheat (TD-F-L), were harvested in 2012 at the Genomics Research Centre in Fiorenzuola d'Arda (Italy). Hundreds of grams were made available to us for phytC investigation. Phytoliths were extracted from 50 g of dry leaves using a wet chemical protocol recently set up for geochemical analysis of phytC. The protocol was described in detail in Corbineau et al. (2013). The organic matter was oxidized with H_2SO_4, H_2O_2, HNO_3 and $KClO_3$, and potential remains on the phytolith surface were dissolved using KOH (pH of 11). The absence of residual extraneous organic particles was checked using SEM-EDX (Corbineau et al., 2013). Dominant phytolith types were recognized according to Madella et al. (2005) using NL microscopy at 600× and 1000× magnifications. As expected, the grass short cell group (GSC) and the bulliform cell group dominated the assemblage. These groups, which form in all grass epidermis, also dominate phytolith assemblages produced by grasslands and recovered from soils (e.g., Alexandre et al., 2011). Several NL microscopy and SEM pictures, illustrating the composition of the TD-F-L phytolith assemblage were taken. For the purpose of morphological comparison, pictures of fossil GSC and bulliform phytoliths from available soil assemblages described in previous papers, were additionally taken.

The 3-D structure of the GSC phytoliths was reconstructed by X-ray imaging at the micro-scale, using a 3-D X-ray microscope Zeiss Ultra XRM L 200. A few phytoliths, randomly selected from the bulk sample, were deposited on the inner surface of a bevel-cut Kapton tube of 50 μm internal diameter. Five individual GSC phytoliths were recognized by optical microscopy at 200× magnification and their position located for 3-D visualization. The principle of the 3-D X-ray microscopy technique is based on focusing the X-ray beam on a rotating sample using an optical lens; then transmitted X-rays are diffracted by a Fresnel zone plate on a scintillator in front of an optical device to produce a 200× magnified image of the phytolith captured by a charge-coupled device (CCD) image sensor. Using a 1K × 1K detector, it leads to a voxel size of 63 nm. The X-ray beam path is continuously flushed with helium to minimize the absorption of X-rays by air, the sample and the optics excepted. While this technique has been widely used for 3-D reconstruction of biological systems it has never been applied in high-resolution mode to silica particles. Analysis of the phytoliths proceeded at 150 nm resolution for a 65 μm field of view, in conventional absorption contrast imaging mode at 8 keV (copper rotating anode; power set at 40 kV and 30 mA). Using this mode, the contrast was generated both from the different X-ray attenuation coefficients of the chemical elements composing the sample and from the density. A total of 901 X-ray projections were recorded between −90 and +90° at an angle step of 0.2° and an exposure time of 80 s per view. After 20 h of analysis, reconstruction of the phytolith volume was performed using XMReconstructor (Zeiss Xradia software). The resulting stack of 2-D grayscale slices was then exported to Avizo Fire (FEI group) for further image processing.

NanoSIMS analyses were performed on cross sections of TD-F-L phytoliths embedded in epoxy resin. One milligram of phytoliths was deposited on polytetrafluoroethylene (PTFE) filters (9 mm i.d.) stuck onto double face tape. Polypropylene (PP) tubes (10 mm i.d. and 15 mm long) were placed on the tape, encircling the phytoliths. Epoxy resin (Araldite 100/Hardener 16) was slipped into the tubes up to 3 mm height and left to dry for 3 h at 40 °C. Resin of 7 mm height was added and left to dry for 48 h at 40 °C. Those two steps prevented the resin from leaking from the base of the tube. Embedded samples were taken off the tubes and polished with diamond paste up to 0.1 μm, until the PTFE filter was completely removed and cross sections of phytoliths were visible in NL microscopy. Samples were sawn into 4 mm thick blocks. Dozens of GSC phytoliths cross sections to be analyzed with the NanoSIMS were located by SEM. The NanoSIMS technique is based upon the sputtering of a few atomic layers from the surface of a sample induced by a primary ion bombardment. The primary ion impact triggers a cascade of atomic collision. Atoms and atomic clusters are ejected. During the ejection process, some atoms and clusters are spontaneously ionized. These secondary ions are characteristic of the composition of the analyzed area. They are separated according to their mass and an image of the intensity of the secondary ion beam is made for a selected mass (http://www.cameca.com/instruments-for-research/sims.aspx. Over the past few years, the NanoSIMS technique has increasingly been used in geosciences to investigate the elemental and isotopic composition of organic and inorganic materials (Herrmann et al., 2007; Hatton et al., 2012; Mueller et al., 2012, 2013). The NanoSIMS technique however has only scarcely been used for measuring secondary ion emission from amorphous silica. One study showed NanoSIMS images of a thin section of a giant siliceous sponge spicule (several millimeters in diameter). A micrometric proteinaceous scaffold, which averaged 2 % C dry weight, could be detected in the siliceous structure (Müller et al., 2010). The NanoSIMS technique was also used for identifying silicification sites in rice roots (Moore et al., 2011). Here, we analyzed the intensities of $[^{28}Si]^-$, $[^{16}O]^-$, $[^{24}C_2]^-$ and $[^{26}CN]^-$ ions produced by selected areas of the GSC phytoliths polished cross sections using a Cameca NanoSIMS 50. The section was coated with 25 nm gold and introduced in the NanoSIMS. A $[Cs]^+$ primary ion probe with 16 kV primary ion impact energy and a 8 kV secondary ion extraction voltage was used. The best adjustment for obtaining secondary ion images of $[^{28}Si]^-$, $[^{16}O]^-$, $[^{12}C]^-$ and $[^{26}CN]^-$ was the following: the selected phytolith surface were first pre-sputtered with a de-focused primary beam on a 60 μm × 60 μm area for 3 min. Then 256 × 256 pixel images were made using a 2.2 pA primary ion current (primary diaphragm diameter of 300 μm) and a counting time of 10 ms per pixel for areas of 30 μm × 30 μm. Analyses with longer counting time or larger primary diaphragm/higher primary beam intensity were also tested. Secondary ion images of $[^{28}Si]$, $[^{16}O]^-$, $[^{12}C]^-$ and $[^{26}CN]^-$ were processed using the ImageJ software (http://imagej.nih.gov/ij). Colors were assigned to different signal intensities, increasing from black to red. Images of the $[^{26}CN]^- / [^{12}C]^-$ ratio were also created. Line scans were drawn across the analyzed surfaces and ion intensity vs. distance along the line were plotted.

For comparison with the NanoSIMS results, the C and N contents of the bulk TD-F-L phytolith sample were measured by chemiluminescence after combustion at 1350 °C (for C) and 1000 °C (for N). The C and N contents of the epoxy resin were measured with an elemental analyzer (EA) after combustion at 1350 °C.

3 Results

Three morphological categories of phytoliths, commonly found in grasses, constituted the bulk sample. SEM pictures of phytoliths placed on the aluminum mount illustrate these categories in Fig. 1. SEM pictures of cross sections of the same categories are shown in Fig. 2. For each category, the mode of silica deposition is specified below when it has been previously evidenced in SEM, TEM, fluorescence microscopy or NanoSIMS images of plant cross sections (Sangster and Parry, 1969; Sowers and Thurston, 1979; Harrison, 1996; Currie and Perry, 2007; Law and Exley, 2011; Moore et al., 2011). The first phytolith category is constituted by thin fragments of multi-cellular silica sheets, several tenths of a micrometer long and wide but less than a few micrometers thick (Figs. 1a, b, 2a, b). These silica "skeletons" (Sang-

Figure 1. SEM images of TD-F-L wheat phytolith assemblage, deposited on an aluminum mount. Three categories are illustrated: (1) silica sheets (**a, b**), (2) stellate type from intercellular space (**c**) and (3) GSC phytoliths including rondel (**d, e**) and polylobate types (**f**).

ster and Parry, 1969; Law and Exley, 2011) were shown to result from the silicification of the middle lamella of the cells walls in grass epidermis and mesophyll, possibly as an early step of silicification (Laue et al., 2007; Law and Exley, 2012). Although abundant in plants (Piperno, 2006), the multi-cellular silica sheets are rapidly subjected to fragmentation and dissolution and are scarcely preserved in soils and sediments (Alexandre et al., 1994). The second phytolith category is constituted by stellate silica particles, of 10–15 μm width (Fig. 1c) that were shown to form in the intercellular spaces of the grass epidermis (Lins et al., 2002). This mode of silica precipitation was described as centripetal, starting as a narrow band lining the cavity, then partially or completely infilling the intercellular space (Sangster and Parry, 1981). The third category dominated the wheat phytolith assemblage. It is made of monocellular phytolith types of 10–50 μm of length, width and thickness. Most of them are GSC phytoliths and belong to the rondel (Fig. 1d, e) and polylobate (Fig. 1f) types. The formation of the monocellular phytolith type was also shown to be centripetal, starting in or against the cell walls and progressively infilling the lumen (e.g., Zhang et al., 2013). The processes that lead to complete silicification of the cells and to organic compounds occlusion are still unknown. Cellulose fibrils from the cell wall may regulate the silica formation (Laue et al., 2007).

SEM observations of cross sections of dozens of GSC phytoliths evidenced one or two internal cavities a few micrometers in diameter in the silica structure (Fig. 2c, d, e). They were similar in shape and size to the low electron density round area visible on one of the TEM image of phytoliths shown by Laue et al. (2007) (Fig. 2b of Laue et al., 2007). However, SEM observation of the GSC phytoliths just placed on the aluminum mount did not evidence any holes on the phytolith surface. When observed with NL microscopy (Fig. 3) the internal cavities were recognizable as opaque areas.

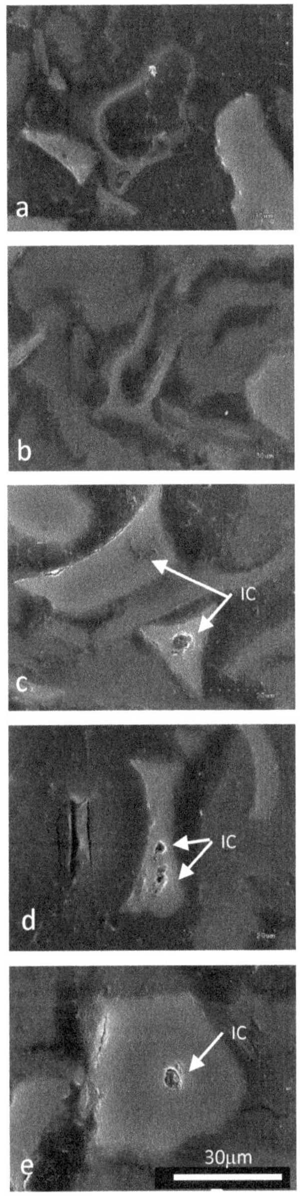

Figure 2. SEM images of the thin section of the TD-F-L wheat phytolith types including silica sheets (**a, b**), and GSC phytoliths of the rondel type (**c, d, e**). GSC types show micrometric internal cavities (IC).

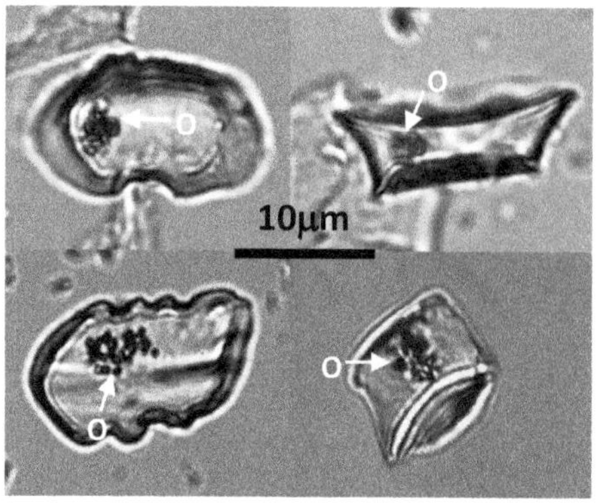

Figure 3. NL microscopy images of GSC phytolith types from the wheat TD-F-L sample showing opaque areas (O).

Two examples of reconstructed 3-D X-ray microscopy volumes are presented in Figs. 4 and 5. The observed patterns were common to the five analyzed GSC particles. The siliceous structure appeared porous at the sub-micrometer scale (Figs. 4a and 5a). Inside the structure, areas of a few micrometers in diameter, with significantly lower X-ray absorption than the surrounding, were observed (Fig. 4a). 2-D planes of the reconstructed volumes evidenced that these heterogeneities were the cavities several micrometers wide previously identified on the cross sections by SEM. The cavities were interconnected (Figs. 4b, 5b). Some particles showed cavities isolated from the phytolith surface by a few micrometer thick silica wall (Fig. 4b). Other particles showed cavities connected to the phytolith surface by small holes of 0.1 μm diameter only (Fig. 5b). These cavities appeared to be filled with air (no X-ray absorption), although the high contrast in X-ray absorbance between silica and air may have masked the presence of organic compounds.

The NanoSIMS results, common to the dozens of analyzed phytolith thin sections, are illustrated in Figs. 6–8. Adjustments were done to find the pre-sputtering duration (3 min), the primary ion beam intensity (L1 = 2 kV), the primary diaphragm diameter (750 μm) and the duration of analyses (11 min) appropriate for obtaining sufficient total ion current (TIC) and avoid charging effects (Figs. 6a, 7a). When the primary ion beam intensity was increased to L1 = 4 kV (Fig. 8a), when the primary diaphragm diameter was decreased to 300 μm (Fig. 8b), or when a succession of analyses resulted in increasing the duration of sputtering (Fig. 8c), a zone devoid of secondary ion signal appeared at the center of the silica surface. This was probably due to charging (Mueller et al., 2012) and/or to topographic heterogeneity (Winterholler et al., 2008). As silica was more resistant to polishing than the epoxy, silica surfaces were often convex (Fig. 8). The tests conducted here emphasized the importance of looking for the most efficient adjustment (i.e., avoiding charging and topographic effects) before performing NanoSIMS analyses on silica surfaces.

$[^{28}Si]^-$, $[^{16}O]^-$, $[^{12}C]^-$ and $[^{26}CN]^-$ images clearly individualized phytoliths from the epoxy resin. The $[^{28}Si]^-$ and $[^{16}O]^-$ images and scan lines showed that phytoliths were made of a continuous silica structure (Figs. 6 and 7) sometimes interrupted by central micrometric areas devoid of silica (Fig. 7). This is again in concordance with the central cavities identified in SEM and 3-D X-ray imaging. Carbon was present in the cavities and in the silica structure itself. However when values of $[^{12}C]^-$ intensity were similar in the cavities and in the epoxy resin, they were 10 to 20 times lower in the silica structure than in the epoxy resin (Figs. 6 and 7). N was also present in the silica structure and $[^{26}CN]^-$ intensity was 3 to 4 times lower in the silica structure than in the cavities or the epoxy (Figs. 6 and 7). Interestingly, the $[^{26}CN]^- / [^{12}C]^-$ ratio ranged between 20 and 30 in the silica structure and between 5 and 10 in the cavities and the epoxy. The silica structure was thus enriched in N by a factor of 4 to 8 relative to the surrounding epoxy. These features were reproducible from a particle to another. Bulk C and N contents in phytoliths, measured by chemiluminescence and EA (cf. Sect. 2), were 0.4 and 0.1 % dry weight for phytoliths, and 68.8 and 2.8 % dry weight for the epoxy resin, respectively. The N / C ratio was 0.27 for the phytoliths and 0.04 for the epoxy resin. The bulk phytolith sample was thus enriched in N relative to the epoxy resin by a factor of 6.8, in agreement with N enrichments calculated from the NanoSIMS data. This consistency strengthened the accuracy of the $[^{12}C]^-$ and $[^{26}CN]^-$ relative intensities measured with the NanoSIMS. Finally, $[^{26}CN]^- / [^{12}C]^-$ NanoSIMS images clearly showed that organic compounds, with N content significantly higher than in the resin, were continuously distributed (at the sub-micrometer scale) in the silica structure. On the contrary, cavities appeared to be filled with the epoxy resin.

4 Discussion

4.1 PhytC locations in the silica structure of GSC phytoliths

SEM, 3-D X-ray microscopy and NanoSIMS images showed that the silica structure of GSC phytoliths was homogeneous at the micrometric scale and systematically contained central micrometric interconnected cavities. The fact that some particles contained cavities isolated from the outside suggests that the opening to the outside can be either original or result from dissolution after the phytolith formation. Phytoliths often contain up to a few percent by dry weight of aluminium (Al) by dry weight (Bartoli and Wilding, 1980; Carnelli et al., 2004) co-precipitating with silica (Hodson and Sangster, 1993). As Al dissolves in strong acids and in strong bases, the phytolith chemical extraction procedure that included HNO_3 and H_2SO_4 steps may have initiated phytolith surficial dissolution and opened the few micrometer thick sil-

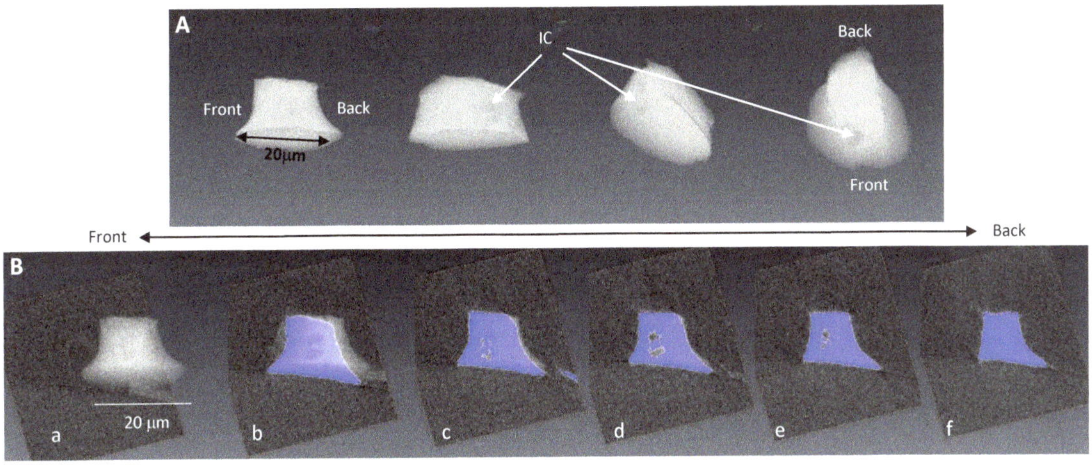

Figure 4. 3-D X-ray microscopy of a GSC phytolith from wheat (TD-F-L): **(a)** four views of the reconstructed volumes; internal cavities (IC) are distinguishable. **(b)** 2-D X-ray slices superimposed on the phytolith volume rendering, showing from front to back the internal cavity (IC). No connection to the surface was evidenced. The blue area corresponds to the thresholding of the phytolith grayscale values.

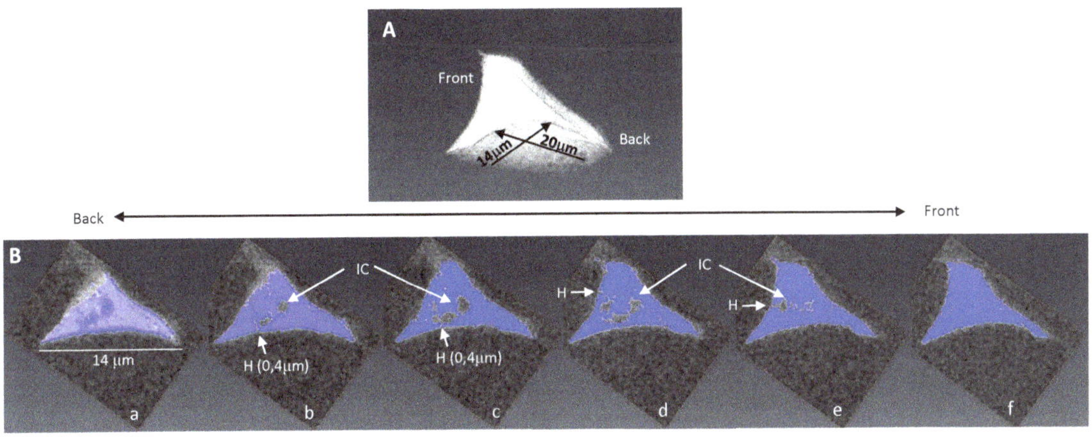

Figure 5. 3-D X-ray microscopy of a GSC phytolith from wheat (TD-F-L): **(a)** reconstructed volume. **(b)** 2-D X-ray images from back to front of the phytolith showing the internal cavity (IC) and its connection to the surface, forming holes (H). The blue area corresponds to the thresholding of the phytolith grayscale values.

ica wall between the cavities and the phytolith surface. The procedure also included a final alkaline step (KOH solution at pH 11) that may also have increased the dissolution features on the silica surfaces. As phytoliths were directly extracted from the plant, the surficial dissolution was revealed here at its beginning. It is expected to reach higher degrees over time in natural environments where multiple dissolution factors come into play (Iler, 1979). Large dissolution features were indeed often observed on fossil phytoliths and were quantified to assess the degree of weathering of soil phytolith assemblages (Alexandre et al., 1999; Oleschko et al., 2004). To illustrate this point, SEM and NL microscopy images of the entirety and cross sections of fossil monocellular phytoliths collected from soils are shown in Fig. 9. The phytolith types are characteristic of grass epidermis (GSC types and Cuneiform bulliform types; Madella et al., 2005) (Fig. 9a, b) and wood parenchyma (globular granulate type; Madella et al., 2005) (Fig. 9c). The dissolution of silica has made central depressions of several micrometers wide. The particles appear empty inside, which is consistent with dissolution starting from the silica walls located between the cavities and the phytolith surface, then slightly opening, or increasing the opening of the cavities to the outside, and then enlarging the cavities into dissolution depressions. Such dissolution depressions are not limited to GSC phytoliths. They were observed on many types of monocellular phytoliths from grasses and non-grasses extracted from soils and sediments as illustrated in Fig. 9a5 (Acicular type), 9b2 and 9b3 (globular granulate). This implies that the inner part of all these phytolith types was constituted by silica less dense than the outer part, either due to phytC occlusion or to a lack of

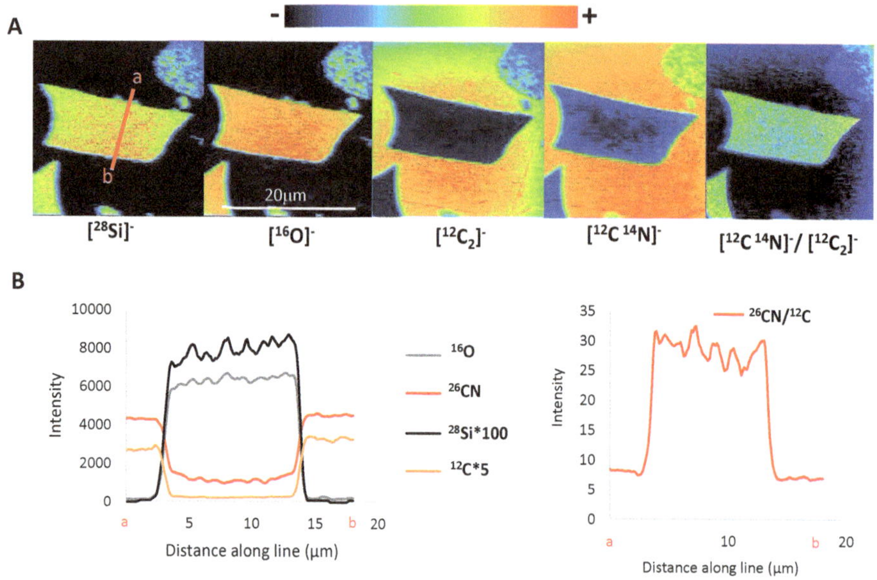

Figure 6. NanoSIMS images and intensities of a first typical GSC phytolith (rondel type) from TD-F-L (wheat) embedded in epoxy resin (polished section). Pre-sputtering: L1 = 2 kV de-focused (60 × 60 mm) $[Cs]^+$ primary beam, for 3 min **(a)** $[^{28}Si]^-$, $[^{16}O]^-$, $[^{12}C_2]^-$, $[^{12}C\,^{14}N]^-$ and $[^{12}C^{14}N]^- / [^{12}C_2]^-$ images obtained with a $[Cs]^+$ primary beam with L1 = 2 kV, D1-1 primary diaphragm (750 μm), for 11 min; **(b)** secondary ion intensities along line scans (red line in **a**).

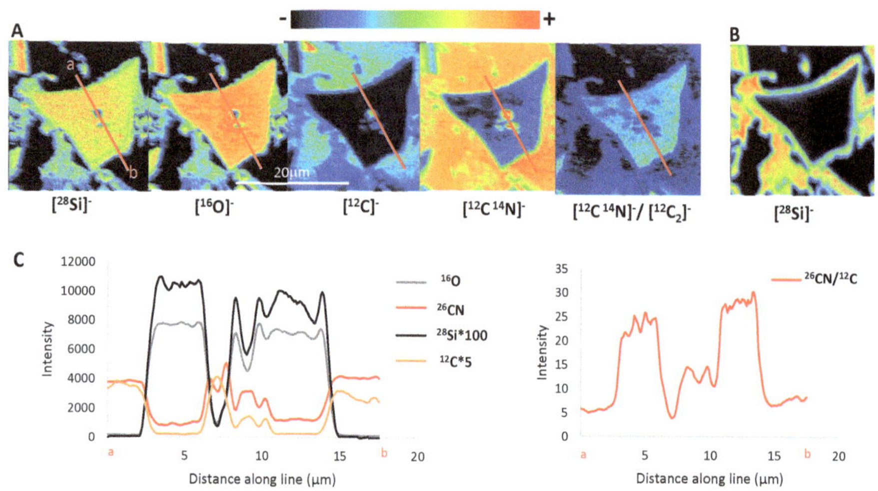

Figure 7. NanoSIMS images and intensities of a second typical GSC phytolith (rondel type) from TD-F-L (wheat) embedded in epoxy resin (polished section). Pre-sputtering: L1 = 2 kV de-focused (60 × 60 mm) $[Cs]^+$ primary beam, for 3 min **(a)** $[^{28}Si]^-$, $[^{16}O]^-$, $[^{12}C_2]^-$, $[^{12}C^{14}N]^-$ and $[^{12}C^{14}N]^- / [^{12}C_2]^-$ images obtained with a $[Cs]^+$ primary beam with L1 = 2 kV, D1-1 primary diaphragm (750 μm), for 11 min; **(b)** $[^{28}Si]^-$ image obtained with a $[Cs]^+$ primary beam increased with L1 = 4 kV, D1-1 primary diaphragm, for 11 min; **(c)** secondary ion intensities along line scans (red line in Fig. 5a).

dissolved Si available for precipitation during the phytolith formation.

Inside the internal cavities, no original organic compounds could be detected by NanoSIMS. If initially present, they may have been squeezed out and replaced by the epoxy resin during the polishing step. On the contrary, the $[^{26}CN]^- / [^{12}C]^-$ images clearly evidenced the presence of organic compounds rich in N continuously distributed in the silica structure and clearly differentiated from the epoxy resin. The absolute composition of $[^{26}CN]^-$ and $[^{12}C]^-$ was not calculated. This would have required including standard materials with known composition in the analyzed section. However, the consistency of N enrichment of the organic compound in the silica structure (measured by NanoSIMS) with N enrichment of the bulk phytC (measured by chemiluminescence/EA) supports the claim that the organic com-

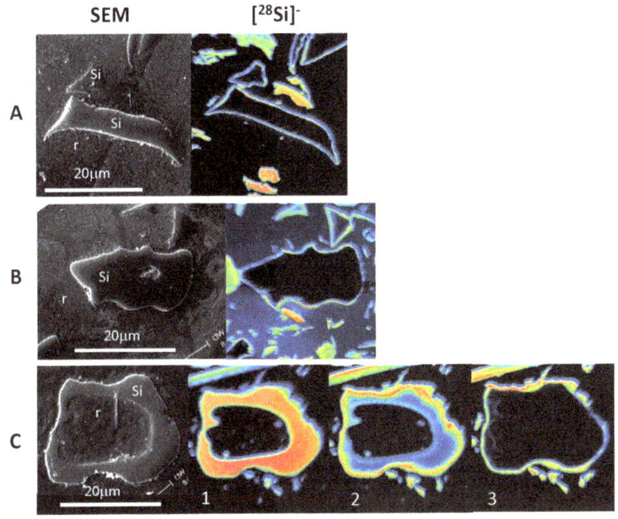

Figure 8. SEM images of the polished section showing convex silica surfaces (Si) in the epoxy resin (r). Associated NanoSIMS $[^{28}Si]^-$ images showing central areas devoid of secondary ion signal. **(a)** $[Cs]^+$ primary beam with L1 = 4 kV, D 1-1 primary diaphragm (750 mm), 11 min; **(b)** $[Cs]^+$ primary beam with L1 = 2 kV, D 1–2 primary diaphragm (300 μm), 11 min; **(c)** $[Cs]^+$ primary beam with L1 = 2 kV, D1-1 primary diaphragm (750 μm), 3 min analyses for 1, 2 and 3 successively.

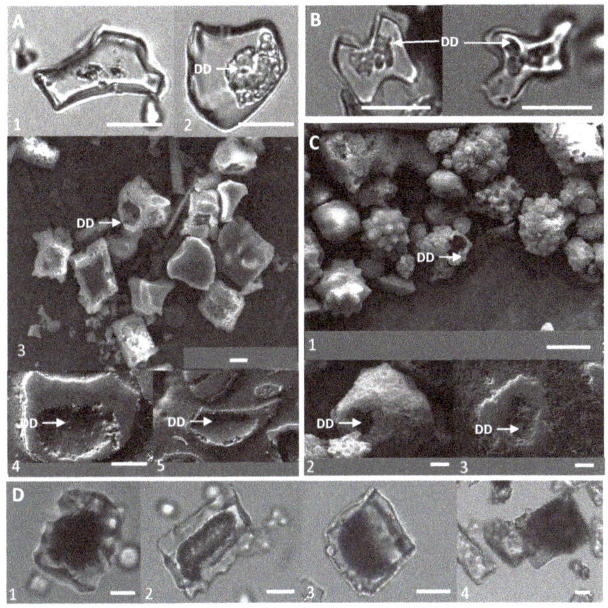

Figure 9. NL microscopy and SEM images of dissolution depressions (DD) affecting fossil phytoliths from soils. **(a)** Grass epidermis monocellular phytoliths (cuneiform bulliform types and acicular type) from Mascareignite (MSG 70, La Réunion, France) (Crespin et al., 2008); NL microscopy phytolith surface (1, 2), SEM phytolith volume (3) and polished section (4, 5). **(b)** Grass epidermis monocellular phytoliths from a ferrugineous soil (Salitre, Brazil) (Alexandre et al., 1999); NL microscopy phytolith. **(c)** Phytoliths from palms and trees from a ferallitic soil (Dimonika, RDA) (Alexandre et al., 1997); SEM globular granulate type volumes (1, 2) and polished section (3). **(d)** Opaque areas observed with NL microscopy on bulliform cell phytoliths from MSG 70 (1, 2) and Salitre (3, 4). Scale bars: 10 μm.

pound measured by NanoSIMS is phytC. Finally, although our data cannot be used to conclude the presence or absence of any phytC in the internal cavities, they demonstrate that the phytC is, in whole or in part, continuously distributed in the silica structure.

4.2 Implications regarding phytC occlusion and phytC accessibility

Evidence of the continuous distribution of phytC in the silica structure, at the sub-micrometric scale, suggests that it had been occluded since the early stage of silicification. SEM, environmental scanning electron microscope (ESEM) and TEM-EDX analyses showed that silica first precipitates in the inner cell wall, probably triggered by the presence of callose or lignin (Laue et al., 2007; Law and Exeley, 2011; Zhang et al., 2013). Silica nanospheres are then organized in a variety of structural motifs such as sheet-like, globular and fibrillar bundles that, from the cell wall, infill the cell lumen in a centripetal way (e.g., Kaufman et al., 1981; Sangster and Parry, 1981; Perry et al., 1987; Laue et al., 2007; Zhang et al., 2013), until most of the cell becomes silicified (Motomura, 2004; Laue et al., 2007). As previously noted, an organic template may participate to the silica formation Harrison, 1996; Laue et al., 2007). This organic template, progressively trapped in the silica structure, may constitute the phytC evidenced by NanoSIMS in the phytoliths. Its N / C value (0.27) is in the range of N / C values characteristic of amino acids. Amino acids may originate either from the cell itself or from the extra-cellular space. Different families of transporters have been identified for their import into plant cells (Tegeder, 2012). At the same time, amino acids entering the cell simultaneously to silica thanks to an invagination/vesicle formation mechanism previously evidenced (Neumann and De Figueiredo, 2002) may occur.

At the end of the cell silicification, residual cell organic compounds that were not already occluded may gather in any remaining space and delimitate the micrometric central cavities. This second pool of phytC should be rapidly oxidized when phytoliths start to dissolve after their deposition in litter, soil or sediment (Fig. 9). This suggests that this phytC pool participates to a limited extent in long-term atmospheric CO_2 sequestration. These considerations rise the need to further estimate the respective contributions to C contents measured from bulk phytolith concentrates of (i) phytC in the silica structure, (ii) phytC in the central cavities and (iii) extraneous C that may remain on porous phytolith surfaces. This is a prerequisite for reliable assessments of the significance of phytC in atmospheric CO_2 sequestration. For that purpose, phytC contents measured from phytolith assemblages characterized by 3-D X-ray microscopy as dominated by phytoliths with closed internal cavities or by phytoliths with open cavities should be compared.

4.3 Reassessment of NL microscopy observations

Several studies have speculated that opaque areas observed by NL microscopy in fossil phytoliths from soils and sediments were burnt organic remains indicative of past fire occurrence (Kealhofer and Penny, 1998; Elbaum et al., 2003; Parr, 2006; Piperno, 2006). However, when observed by NL microscopy, the empty dissolution depressions evidenced by SEM on monocellular phytoliths from soils (Fig. 9a) also appeared as opaque areas, especially when they were oriented downwards (Fig. 9c). This is probably due to trapped air in the dissolution depressions that caused an optical artifact at the place where the air met the mounting medium. This feature implies that opaque areas in fossil phytoliths should not be considered as unequivocal evidence of burnt organic compounds. Similarly, internal cavities may also appear as opaque spots due to the occurrence of trapped air, independent of the presence of organic compounds.

5 Conclusions

3-D X-ray microscopy reconstructions of GSC phytoliths from harvested grasses, and SEM observations of their cross sections, showed that the silica structure contains micrometric internal cavities. These cavities were sometimes observed isolated from the outside. Their opening may be an original feature or may result from the silica dissolution during the chemical extraction procedure, mimicking the beginning of dissolution process that may happen in natural environments. The phytC that may originally occupy those cavities is thus susceptible to rapid oxidation. It was not detected by the NanoSIMS technique. On the contrary, another pool of phytC, continuously distributed in and protected by the silica structure was evidenced by NanoSIMS. Its N / C ratio (0.27) is in agreement with the presence of amino acids. These findings constitute a basis to further characterize the origin, occlusion process, nature and accessibility of phytC, necessary for assessing its significance in the global C cycle.

Acknowledgements. The authors gratefully acknowledge support from LABEX OT-Med (France) (2013–2014 grant to A. Alexandre), from the U.S. National Science Foundation (DEB-1144888 to GMS) and from the "Investissements d'Avenir" French Government program of the French National Research Agency that funded the French X-ray CT platform Nano-ID (ANR-10-EQPX-39-01). NanoSIMS analyses were conducted at the Plateforme NanoSIMS, OSUR, Université de Rennes 1, France. We express special thanks to Franz Badeck and Luigi Cattivelli (CRA – Genomic Research Centre, Fiorenzuola d'Arda, Italy) for providing the wheat sample. The authors would like to thank two anonymous reviewers and Jack Middelburg, Biogeosciences associate editor, for their valuable comments and suggestions for improving the manuscript.

Edited by: J. Middelburg

References

Alexandre, A., Colin F., and Meunier J.-D., Phytoliths as indicators of the biogeochemical turnover of silicon in equatorial forest, Compte-rendus de l'Académie des Sciences, Paris, 319, 453–458, 1994.

Alexandre, A., Meunier, J. D., Colin, F., and Koud, J. M., Plant Impact on the Biogeochemical Cycle of Silicon and Related Weathering Processes, Geochim. Cosmochim. Ac., 61, 677–682, 1997.

Alexandre, A., Meunier, J. D., Mariotti, A., and Soubies, F.: Late Holocene phytolith and carbon isotope record from a latosol at Salitre, south-central Brazil, Quaternary Res., 51, 187–194, 1999.

Alexandre, A., Bouvet, M., and Abbadie, L.: The role of savannas in the terrestrial Si cycle: a case-study from Lamto, Ivory Coast, Global Planet. Change, 78, 162–169, 2011.

Alexandre, A., Crespin, J., Sylvestre, F., Sonzogni, C., and Hilbert, D. W.: The oxygen isotopic composition of phytolith assemblages from tropical rainforest soil tops (Queensland, Australia): validation of a new paleoenvironmental tool, Clim. Past, 8, 307–324, doi:10.5194/cp-8-307-2012, 2012.

Bartoli, F. and L. P. Wilding. Dissolution of Biogenic Opal as a Function of Its Physical and Chemical Properties, Soil Sci. Soc. Am. J., 44, 873–878, 1980.

Blecker, S. W., McCulley, R. L., Chadwick, O. A., and Kelly, E. F.: Biologic cycling of silica across a grassland bioclimosequence, Global Biogeochem. Cy., 20, GB3023, doi:10.1029/2006gb002690, 2006.

Carnelli, A. L., Theurillat, J. P., and Madella, A.: Phytolith types and type-frequencies in subalpine-alpine plant species of the European Alps, Rev. Palaeobot. Palyno., 129, 39–65, 2004.

Carter, J. A.: Atmospheric carbon isotope signatures in phytolith-occluded carbon, Quatern. Int., 193, 20–29, 2009.

Corbineau, R., Reyerson, P. E., Alexandre, A., and Santos, G. M.: Towards producing pure phytolith concentrates from plants that are suitable for carbon isotopic analysis, Rev. Palaeobot. Palyno., 197, 179–185, 2013.

Cornelis, J.-T., Delvaux, B., Georg, R. B., Lucas, Y., Ranger, J., and Opfergelt, S.: Tracing the origin of dissolved silicon transferred from various soil-plant systems towards rivers: a review, Biogeosciences, 8, 89–112, doi:10.5194/bg-8-89-2011, 2011.

Currie, H. A. and Perry, C. C.: Silica in plants: biological, biochemical and chemical studies, Ann. Bot., 100, 1383–1389, 2007.

Elbaum, R., Weiner, S., Albert, R. M., and Elbaum, M.: Detection of burning of plant materials in the archaeological record by changes in the refractive indices of siliceous phytoliths, J. Archaeol. Sci., 30, 217–226, 2003.

Elbaum, R., Melamed-Bessudo, C., Tuross, N., Levy, A. A., and Weiner, S.: New methods to isolate organic materials from silicified phytoliths reveal fragmented glycoproteins but no DNA, Quatern. Int., 193, 11–19, 2009.

Gao, Y., Cheng, Z., Ling, W., and Huang, J.: Arbuscular mycorrhizal fungal hyphae contribute to the uptake of polycyclic aromatic hydrocarbons by plant roots, Bioresource Technol., 101, 6895–6901, 2010.

Harrison, C. C.: Evidence for intramineral macromolecules containing protein from plant silicas, Phytochemistry, 41, 37–42, 1996.

Hatton, P.-J., Remusat, L., Zeller, B., and Derrien, D.: A multiscale approach to determine accurate elemental and isotopic ratios by nano-scale secondary ion mass spectrometry imaging, Rapid Commun. Mass Sp., 26, 1363–1371, 2012.

Herrmann, A. M., Clode, P. L., Fletcher, I. R., Nunan, N., Stockdale, E. A., O'Donnell, A. G., and Murphy, D. V.: A novel method for the study of the biophysical interface in soils using nano-scale secondary ion mass spectrometry, Rapid Commun. Mass Sp., 21, 29–34, 2007.

Hodson, M. J. and Sangster, A. G.: The interaction between silicon and aluminium in *Sorghum bicolor* (L.) Moench: growth analysis and X-ray microanalysis, Ann. Bot., 72, 389–400, 1993.

Huang, Z., Li, Y., Jiang, P., Chang, S. X., Song, Z., Liu, J., and Zhou, G.: Long-term intensive management increased carbon occluded in phytolith (PhytOC) in bamboo forest soils, Scientific Report, 4, 3602, doi:10.1038/srep03602, 2014.

Iler, R.: The Chemistry of Silica. Solubility, Polymerization, Colloid and Surface Properties, and Biochemistry, Chichester: John Wiley and Sons, 1979.

Instruments-for-research: SIMS: http://www.cameca.com/instruments-for-research/sims.aspx, 2014.

Jones, R. L. and Beavers, A. H. Some mineralogical and chemical properties of plant opal. Soil Sci., 96, 375–379. 1963.

Kaufman, P. B., Dayanandan, P., Takeoka, Y., Bigelow, W. C., Jones, J. D., and Iler, R.: Silica in shoots of higher plants, in: Silicon and Siliceous structures in biological systems, edited by: Simpson, J. L. and Volcani, B. E., Springer Verlag, NY, 409–449, 1981.

Kealhofer, L. and Penny, D.: A combined pollen and phytolith record for fourteen thousand years of vegetation change in northeastern Thailand, Rev. Palaeobot. Palyno., 103, 83–93, 1998.

Kelly, E. F., Amundson, R. G., Marino, B. D., and Deniro, M. J.: Stable isotope ratios of carbon in phytoliths as a quantitative method of monitoring vegetation and climate change, Quaternary. Res., 35, 222–233, 1991.

Laue, M., Hause, G., Dietrich, D., and Wielage, B.: Ultrastructure and microanalysis of silica bodies in *Dactylis glomerata* L., Microchim. Acta, 156, 103–107, 2007.

Law, C. and Exley, C.: New insight into silica deposition in horsetail (*Equisetum arvense*), BMC Plant Biol., 11, 112, doi:10.1038/srep03602, 2011.

Li, B., Song, Z., Li, Z., Wang, H., Gui, R., and Song, R.: Phylogenetic variation of phytolith carbon sequestration in bamboos, Scientific Report, 4, 4710, doi:10.1038/srep04710, 2014.

Lins, U., Barros, C. F., da Cunha, M., and Miguens, F. C.: Structure, morphology, and composition of silicon biocomposites in the palm tree *Syagrus coronata* (Mart.) Becc., Protoplasma, 220, 89–96, 2002.

Madella, M., Alexandre, A., Ball, T., and Grp, I. W.: International code for phytolith nomenclature 1.0, Ann. Bot., 96, 253–260, 2005.

McInerney, F. A., Strömberg, C. A. E., and White, J. W. C.: The Neogene transition from C3 to C4 grasslands in North America: stable carbon isotope ratios of fossil phytoliths, Paleobiology, 37, 23–49, 2011.

McMichael, C. H., Bush, M. B., Silman, M. R., Piperno, D. R., Raczka, M., Lobato, L. C., Zimmerman, M., Hagen, S., and Palace, M.: Historical fire and bamboo dynamics in western Amazonia, J. Biogeogr., 40, 299–309, doi:10.1111/jbi.12002, 2012.

Miller, L. A., Smith, S. Y., Sheldon, N. D., and Strömberg, C. A. E.: Eocene vegetation and ecosystem fluctuations inferred from a high-resolution phytolith record, Geol. Soc. Am. Bull., B30548.1, doi:10.1130/B30548.1, 2012.

Moore, K. L., Schröder, M., Wu, Z., Martin, B. G. H., Hawes, C. R., Mcgrath, S. P., Hawkesford, M. J., Ma, J. F., Zhao, F.-J., and Grovenor, C. R. M.: High-Resolution Secondary Ion Mass Spectrometry Reveals the Contrasting Subcellular Distribution of Arsenic and Silicon in Rice Roots, Plant Physiol., 156, 913–24, 2011.

Motomura, H.: Silica deposition in relation to ageing of leaf tissues in *Sasa veitchii* (Carriere) Rehder (Poaceae: Bambusoideae), Ann. Bot., 93, 235–248, 2004.

Mueller, C. W., Kölbl, A., Hoeschen, C., Hillion, F., Heister, K., Herrmann, A. M., and Kögel- Knabner, I.: Submicron scale imaging of soil organic matter dynamics using NanoSIMS – from single particles to intact aggregates, Org. Geochem., 42, 1476–1488, 2012.

Mueller, C. W., Weber, P. K., Kilburn, M. R., Hoeschen, C., Kleber, M., and Pett-Ridge, J.: Advances in the Analysis of Biogeochemical Interfaces: NanoSIMS to Investigate Soil Microenvironments, Adv. Agron., 121, 1–46, doi:10.1016/B978-0-12-407685-3.00001-3, 2013.

Müller, W. E. G., Wang, X., Sinha, B., Wiens, M., Schröder, H.-C., and Jochum, K. P.: NanoSIMS: insights into the organization of the proteinaceous scaffold within Hexactinellid sponge spicules, Chembiochem, 11, 1077–1082, 2010.

Neumann, D. and De Figueiredo, C.: A novel mechanism of silicon uptake, Protoplasma, 220, 59–67, 2002.

Oleschko, K., Parrot, J.-F., Ronquillo, G., Shoba, S., Stoops, G., and Marcelino, V.: Weathering: toward a fractal quantifying, Math. Geol., 36, 607–627, 2004.

Parr, J. F.: Effect of fire on phytolith coloration, Geoarchaeology, 21, 171–185, 2006.

Parr, J. F. and Sullivan, L. A.: Soil carbon sequestration in phytoliths, Soil Biol. Biochem., 37, 117–124, 2005.

Parr, J. F. and Sullivan, L. A.: Comparison of two methods for the isolation of phytolith occluded carbon from plant material, Plant Soil, 374, 45–53, 2014.

Parr, J., Sullivan, L., Chen, B., Ye, G., and Zheng, W.: Carbon biosequestration within the phytoliths of economic bamboo species, Glob. Change Biol., 16, 2661–2667, 2010.

Paungfoo-Lonhienne, C., Lonhienne, T. G. A., Rentsch, D., Robinson, N., Christie, M., Webb, R. I., Gamage, H. K., Carroll, B. J., Schenk, P. M., and Schmidt, S.: Plants can use protein as a nitrogen source without assistance from other organisms, P. Natl. Acad. Sci. USA, 105, 4524–4529, 2008.

Perry, C. C., Williams R. J. P., and S.C. Fry. Cell Wall Biosynthesis during Silicification of Grass Hairs, J. Plant Physiol., 126, 437–48, 1987.

Piperno, D. R. and Becker, P.: Vegetational history of a site in the central Amazon Basin derived from phytolith and charcoal records from natural soils, Quaternary Res., 45, 202–209, 1996.

Piperno, D. R. and Stothert, K. E.: Phytolith evidence for early Holocene Cucurbita domestication in southwest Ecuador, Science, 299, 1054–1057, 2003.

Piperno, D. R. Phytoliths. A Comprehensive Guide for Archaeologists and Paleoecologists. Lanham, New York, Toronto, Oxford: AltaMira Press (Rowman & Littlefield), 2006.

Pironon, J., Meunier, J. D., Alexandre, A., Mathieu, R., Mansuy, L., Grosjean, A., and Jarde, E.: Individual characterization of phytoliths: experimental approach and consequences on paleoenvironment understanding, A. A. Balkema, Lisse, 2001.

Prychid, C. J., Rudall, P. J., and Gregory, M.: Systematics and biology of silica bodies in monocotyledons, Bot. Rev., 69, 377–440, 2003.

Sangster, A. G. and Parry, D. W.: Some factors in relation to bulliform cell silicification in the grass leaf, Ann. Bot., 33, 315–323, 1969.

Sangster A. G. and Parry D. W.: Ultrastructure of silica deposits in higher plants. In Silicon and siliceous structures in biological systems, ed. Simpson and Volcani, 383–407, New York, Heidelberg and Berlin, Springer-Verlag, 1981.

Santos, G. M., Alexandre, A., Coe, H. H. G., Reyerson, P. E., Southon, J. R., and De Carvalho, C. N.: The phytolith C-14 puzzle: a tale of background determinations and accuracy tests, Radiocarbon, 52, 113–128, 2010.

Santos, G. M., Alexandre, A., Southon, J. R., Treseder, K. K., Corbineau, R., and Reyerson, P. E.: Possible source of ancient carbon in phytolith concentrates from harvested grasses, Biogeosciences, 9, 1873–1884, doi:10.5194/bg-9-1873-2012, 2012.

Smith, F. A. and Anderson, K. B.: Characterization of organic compounds in phytoliths: improving the resolving power of phytolith delta C-13 as a tool for paleoecological reconstruction of C3 and C4 grasses, in: Phytoliths – Applications in Earth Science and Human History, edited by: Meunier, J. D., and Colin, CRC Press, 317–327, 2001.

Smith, F. A. and White, J. W. C.: Modern calibration of phytolith carbon isotope signatures for C3/C4 paleograssland reconstruction, Palaeogeogr. Palaeocl., 207, 277–304, 2004.

Song, Z., Wang, H., Strong, P. J., Li, Z., and Jiang, P.: Plant impact on the coupled terrestrial biogeochemical cycles of silicon and carbon: implications for biogeochemical carbon sequestration, Earth-Sci. Rev., 115, 319–331, doi:10.1016/j.earscirev.2012.09.006, 2012.

Song, Z., Wang, H., Strong, P. J., and Guo, F.: Phytolith carbon sequestration in China's croplands, Eur. J. Agron., 53, 10–15, 2014.

Sowers, A. E. and Thurston, E. L.: Ultrastructural evidence for uptake of silicon-containing silicic acid analogs by *Urtica pilulifera* and incorporation into cell wall silica, Protoplasma, 101, 11–22, 1979.

Strömberg, C. A. E., Dunn, R. E., Madden, R. H., Kohn, M. J., and Carlini, A. A.: Decoupling the spread of grasslands from the evolution of grazer-type herbivores in South America, Nat. Commun., 4, 1478, doi:10.1038/ncomms2508, 2013.

Struyf, E. and D. J. Conley. Silica: An Essential Nutrient in Wetland Biogeochemistry, Front. Ecol. Environ., 7, 88–94. 2009.

Tegeder, M.: Transporters for amino acids in plant cells: some functions and many unknowns, Curr. Opin. Plant Biol., 15, 315–321, 2012.

Warren, C. R.: Post-uptake metabolism affects quantification of amino acid uptake, New Phytol., 193, 522–531, 2012.

Watling, K. M., Parr, J. F., Rintoul, L., Brown, C. L., and Sullivan, L. A.: Raman, infrared and XPS study of bamboo phytoliths after chemical digestion, Spectrochim. Acta A, 80, 106–111, 2011.

Webb, E. A. and Longstaffe, F. J.: Limitations on the climatic and ecological signals provided by the delta C-13 values of phytoliths from a C4 North American prairie grass, Geochim. Cosmochim. Ac., 74, 3041–3050, 2010.

Whiteside, M. D., Treseder, K. K., and Atsatt, P. R.: The brighter side of soils: quantum dots track organic nitrogen through fungi and plants, Ecology, 90, 100–108, 2009.

Whiteside, M. D., Garcia, M. O., and Treseder, K. K.: Amino acid uptake in arbuscular mycorrhizal plants, PLoS ONE, 7, e47643, doi:10.1371/journal.pone.0047643, 2012.

Winterholler, B., Hoppe, P., Foley, S., and Andreae, M. O.: Sulfur isotope ratio measurements of individual sulfate particles by NanoSIMS, Int. J. Mass Spectrom., 272, 63–77, 2008.

Zhang, C., Wang, L., Zhang, W., and Zhang, F.: Do lignification and silicification of the cell wall precede silicon deposition in the silica cell of the rice (*Oryza sativa* L.) leaf epidermis?, Plant Soil, 372, 137–149, 2013.

Zuo, X., Lu, H., and Gu, Z.: Distribution of soil phytolith-occluded carbon in the Chinese Loess Plateau and its implications for silica-carbon cycles, Plant Soil, 374, 223–232, 2014.

Permissions

 Every chapter published in this book has been scrutinized by our experts. Their significance has been extensively debated. The topics covered herein carry significant findings which will fuel the growth of the discipline. They may even be implemented as practical applications or may be referred to as a beginning point for another development.

The contributors of this book come from diverse backgrounds, making this book a truly international effort. This book will bring forth new frontiers with its revolutionizing research information and detailed analysis of the nascent developments around the world.

We would like to thank all the contributing authors for lending their expertise to make the book truly unique. They have played a crucial role in the development of this book. Without their invaluable contributions this book wouldn't have been possible. They have made vital efforts to compile up to date information on the varied aspects of this subject to make this book a valuable addition to the collection of many professionals and students.

This book was conceptualized with the vision of imparting up-to-date information and advanced data in this field. To ensure the same, a matchless editorial board was set up. Every individual on the board went through rigorous rounds of assessment to prove their worth. After which they invested a large part of their time researching and compiling the most relevant data for our readers.

The editorial board has been involved in producing this book since its inception. They have spent rigorous hours researching and exploring the diverse topics which have resulted in the successful publishing of this book. They have passed on their knowledge of decades through this book. To expedite this challenging task, the publisher supported the team at every step. A small team of assistant editors was also appointed to further simplify the editing procedure and attain best results for the readers.

Apart from the editorial board, the designing team has also invested a significant amount of their time in understanding the subject and creating the most relevant covers. They scrutinized every image to scout for the most suitable representation of the subject and create an appropriate cover for the book.

The publishing team has been an ardent support to the editorial, designing and production team. Their endless efforts to recruit the best for this project, has resulted in the accomplishment of this book. They are a veteran in the field of academics and their pool of knowledge is as vast as their experience in printing. Their expertise and guidance has proved useful at every step. Their uncompromising quality standards have made this book an exceptional effort. Their encouragement from time to time has been an inspiration for everyone.

The publisher and the editorial board hope that this book will prove to be a valuable piece of knowledge for researchers, students, practitioners and scholars across the globe.

List of Contributors

J. Zhang
Institute of Desertification Studies, Chinese Academy of Forestry, Beijing, China
Research Institute of Forestry, Chinese Academy of Forestry, Beijing, China

L. Gu
Environmental Sciences Division and Climate Change Science Institute, Oak Ridge National Laboratory, Oak Ridge, TN 37831, USA

F. Bao
Institute of Desertification Studies, Chinese Academy of Forestry, Beijing, China

Y. Cao
Institute of Desertification Studies, Chinese Academy of Forestry, Beijing, China

Y. Hao
The Experimental Center of Desert Forestry of the Chinese Academy of Forestry, Dengkou, Inner Mongolia Autonomous Region, China

J. He
Institute of Desertification Studies, Chinese Academy of Forestry, Beijing, China

J. Li
Institute of Desertification Studies, Chinese Academy of Forestry, Beijing, China

Y. Li
Institute of Desertification Studies, Chinese Academy of Forestry, Beijing, China

Y. Ren
Institute of Desertification Studies, Chinese Academy of Forestry, Beijing, China

F. Wang
Institute of Desertification Studies, Chinese Academy of Forestry, Beijing, China

R. Wu
Institute of Desertification Studies, Chinese Academy of Forestry, Beijing, China

B. Yao
Institute of Desertification Studies, Chinese Academy of Forestry, Beijing, China

Y. Zhao
The Experimental Center of Desert Forestry of the Chinese Academy of Forestry, Dengkou, Inner Mongolia Autonomous Region, China

G. Lin
Center for Earth System Science, Tsinghua University, Beijing, China

B. Wu
Institute of Desertification Studies, Chinese Academy of Forestry, Beijing, China

Q. Lu
Institute of Desertification Studies, Chinese Academy of Forestry, Beijing, China

P. Meng
Headquarters, Chinese Academy of Forestry, Beijing, China

S. K. von Bueren
Institute of Agriculture & Environment, Massey University, Palmerston North, New Zealand

A. Burkart
Institute of Bio- and Geosciences, IBG-2: Plant Sciences, Forschungszentrum Jülich GmbH, Jülich, Germany

A. Hueni
Remote Sensing Laboratories, University of Zurich, Zurich, Switzerland

U. Rascher
Institute of Bio- and Geosciences, IBG-2: Plant Sciences, Forschungszentrum Jülich GmbH, Jülich, Germany

M. P. Tuohy
Institute of Agriculture & Environment, Massey University, Palmerston North, New Zealand

I. J. Yule
Institute of Agriculture & Environment, Massey University, Palmerston North, New Zealand

J. M. Chen
International Institute of Earth System Science, Nanjing University, Nanjing, Jiangsu, China
Department of Geography and Program in Planning, University of Toronto, Toronto, Ontario, M5S 3G3, Canada

J. W. Fung
Department of Geography and Program in Planning, University of Toronto, Toronto, Ontario, M5S 3G3, Canada

G. Mo
Department of Geography and Program in Planning, University of Toronto, Toronto, Ontario, M5S 3G3, Canada

F. Deng
Department of Physics, University of Toronto, Toronto, Ontario, M5S 3G3, Canada

T. O. West
Joint Global Change Research Institute, Pacific Northwest National Laboratory, College Park, Maryland, USA

C. Metzger
Royal Institute of Technology, Department of Land and Water Resources Engineering, Stockholm, Sweden
Technische Universität München, Chair of Restoration Ecology, Freising, Germany

P.-E. Jansson
Royal Institute of Technology, Department of Land and Water Resources Engineering, Stockholm, Sweden

A. Lohila
Finnish Meteorological Institute, Atmospheric Composition Research, Helsinki, Finland

M. Aurela
Finnish Meteorological Institute, Atmospheric Composition Research, Helsinki, Finland

T. Eickenscheidt
University of Applied Sciences Weihenstephan-Triesdorf, Chair of Vegetation Ecology, Freising, Germany

L. Belelli-Marchesini
VU University, Department of Earth Sciences, Amsterdam, the Netherlands

K. J. Dinsmore
Centre for Ecology and Hydrology, Penicuik, Midlothian, UK

J. Drewer
Centre for Ecology and Hydrology, Penicuik, Midlothian, UK

J. van Huissteden
VU University, Department of Earth Sciences, Amsterdam, the Netherlands

M. Drösler
University of Applied Sciences Weihenstephan-Triesdorf, Chair of Vegetation Ecology, Freising, Germany

S. Karki
Aarhus University, Department of Agroecology, Blichers Allé 20, 8830 Tjele, Denmark

L. Elsgaard
Aarhus University, Department of Agroecology, Blichers Allé 20, 8830 Tjele, Denmark

P. E. Lærke
Aarhus University, Department of Agroecology, Blichers Allé 20, 8830 Tjele, Denmark

Y. Zhang
Institute for Space Sciences, Freie Universität Berlin, 12165 Berlin, Germany

M. Voigt
Institute for Space Sciences, Freie Universität Berlin, 12165 Berlin, Germany

H. Liu
Changjiang River Scientific Research institute, Changjiang Water Resources Commission, 430010 Wuhan, China

B. Bond-Lamberty
Pacific Northwest National Laboratory, Joint Global Change Research Institute at the University of Maryland–College Park, 5825 University Research Court, Suite 3500, College Park, Maryland, MA 20740, USA

J. P. Fisk
Department of Geographical Sciences, 1150 LeFrak, University of Maryland, College Park, Maryland, MA 20742, USA

J. A. Holm
Climate Sciences Department, Lawrence Berkeley National Laboratory, 1 Cyclotron Rd., MS 74-0171, Berkeley, CA 94720, USA

V. Bailey
Pacific Northwest National Laboratory, 902 Battelle Boulevard, Richland, WA 99352, USA

G. Bohrer
Department of Civil, Environmental and Geodetic Engineering, The Ohio State University, 470 Hitchcock Hall, 2070 Neil Avenue, Columbus, Ohio, OH 43210, USA

C. M. Gough
Virginia Commonwealth University, Department of Biology, P.O. Box 842012, 1000 West Cary Street, Richmond, VA 23284-2012, USA

X. H. Ye
State Key Laboratory of Vegetation and Environmental Change, Institute of Botany, Chinese Academy of Sciences, Beijing 100093, China

X. Pan
State Key Laboratory of Vegetation and Environmental Change, Institute of Botany, Chinese Academy of Sciences, Beijing 100093, China
College of Life Sciences, University of Chinese Academy of Sciences, Beijing 100049, China

W. K. Cornwell
Systems Ecology, Department of Ecological Science, VU University, Amsterdam, De Boelelaan 1085, 1081 HV Amsterdam, the Netherlands
School of Biological, Earth and Environmental Sciences, University of New South Wales, Sydney, Australia

S. Q. Gao
State Key Laboratory of Vegetation and Environmental Change, Institute of Botany, Chinese Academy of Sciences, Beijing 100093, China

M. Dong
State Key Laboratory of Vegetation and Environmental Change, Institute of Botany, Chinese Academy of Sciences, Beijing 100093, China
Key Laboratory of Hangzhou City for Ecosystem Protection and Restoration, College of Life and Environmental Sciences, Hangzhou Normal University, Hangzhou 310036, China

J. H. C. Cornelissen
Systems Ecology, Department of Ecological Science, VU University, Amsterdam, De Boelelaan 1085, 1081 HV Amsterdam, the Netherlands

L. Tremblay
Department of Chemistry and Biochemistry, Université de Moncton, Moncton, NB, E1A 3E9, Canada

J. Caparros
CNRS, UMR 7621, Laboratoire d'Océanographie Microbienne, Observatoire Océanologique, 66650 Banyuls/mer, France
Sorbonne Universités, UPMC Univ Paris 06, UMR 7621, Laboratoire d'Océanographie Microbienne, Observatoire Océanologique, 66650 Banyuls/mer, France

K. Leblanc
Aix Marseille Université, CNRS, Université de Toulon, IRD, Mediterranean Institute of Oceanography (MIO), UM 110, 13288, Marseille, France

I. Obernosterer
CNRS, UMR 7621, Laboratoire d'Océanographie Microbienne, Observatoire Océanologique, 66650 Banyuls/mer, France
Sorbonne Universités, UPMC Univ Paris 06, UMR 7621, Laboratoire d'Océanographie Microbienne, Observatoire Océanologique, 66650 Banyuls/mer, France

D. Bolshiyanov
State Research Center, Arctic and Antarctic Research Institute, St.Petersburg, Russia St.Petersburg State University, St.Petersburg, Russia

A. Makarov
State Research Center, Arctic and Antarctic Research Institute, St.Petersburg, Russia St.Petersburg State University, St.Petersburg, Russia

L. Savelieva
St.Petersburg State University, St.Petersburg, Russia

C. Wu
Institut für Energie und Klimaforschung, IEK-8, Forschungszentrum Jülich, 52425 Jülich, Germany

I. Pullinen
Institut für Energie und Klimaforschung, IEK-8, Forschungszentrum Jülich, 52425 Jülich, Germany

S. Andres
Institut für Energie und Klimaforschung, IEK-8, Forschungszentrum Jülich, 52425 Jülich, Germany

G. Carriero
Istituto Protezione Sostenibile Piante, Consiglio Nazionale Ricerche (CNR), Via Madonna del Piano10, 50019 Sesto Fiorentino, Florence, Italy

S. Fares
Consiglio per la Ricerca e la sperimentazione in Agricoltura (CRA), Research Centre for the Soil-Plant System, Rome, Italy

H. Goldbach
Department of plant nutrition (INRES), Universität Bonn, Karlrobert Kreiten Str. 13, 53115 Bonn, Germany

L. Hacker
Institut für Energie und Klimaforschung, IEK-8, Forschungszentrum Jülich, 52425 Jülich, Germany

T. Kasal
Department of plant nutrition (INRES), Universität Bonn, Karlrobert Kreiten Str. 13, 53115 Bonn, Germany

A. Kiendler-Scharr
Institut für Energie und Klimaforschung, IEK-8, Forschungszentrum Jülich, 52425 Jülich, Germany

E. Kleist
Institut für Bio- und Geowissenschaften, IBG-2, Forschungszentrum Jülich, 52425 Jülich, Germany

E. Paoletti
Istituto Protezione Sostenibile Piante, Consiglio Nazionale Ricerche (CNR), Via Madonna del Piano10, 50019 Sesto Fiorentino, Florence, Italy

A. Wahner
Institut für Energie und Klimaforschung, IEK-8, Forschungszentrum Jülich, 52425 Jülich, Germany

J. Wildt
Institut für Bio- und Geowissenschaften, IBG-2, Forschungszentrum Jülich, 52425 Jülich, Germany

Th. F. Mentel
Institut für Energie und Klimaforschung, IEK-8, Forschungszentrum Jülich, 52425 Jülich, Germany

P. E. Kauppi
Department of Environmental Sciences, P.O. Box 65, 0014, University of Helsinki, Helsinki, Finland

R. A. Birdsey
USDA Forest Service, Northern Research Station, Newtown Square, PA, USA

Y. Pan
USDA Forest Service, Northern Research Station, Newtown Square, PA, USA

A. Ihalainen
Natural Resources Institute Finland (Luke), P.O. Box 18, 01301 Vantaa, Finland

P. Nöjd
Natural Resources Institute Finland (Luke), P.O. Box 18, 01301 Vantaa, Finland

A. Lehtonen
Natural Resources Institute Finland (Luke), P.O. Box 18, 01301 Vantaa, Finland

C. Xiao
State Key Laboratory of Vegetation and Environmental Change, Institute of Botany, the Chinese Academy of Sciences, Beijing 100093, China

I. A. Janssens
Department of Biology, Research Group of Plant and Vegetation Ecology, University of Antwerp, Universiteitsplein 1, 2610 Wilrijk, Belgium

Y. Zhou
State Key Laboratory of Vegetation and Environmental Change, Institute of Botany, the Chinese Academy of Sciences, Beijing 100093, China

J. Su
State Key Laboratory of Vegetation and Environmental Change, Institute of Botany, the Chinese Academy of Sciences, Beijing 100093, China
University of Chinese Academy of Sciences, Beijing, 100049, China

Y. Liang
State Key Laboratory of Vegetation and Environmental Change, Institute of Botany, the Chinese Academy of Sciences, Beijing 100093, China
University of Chinese Academy of Sciences, Beijing, 100049, China

B. Guenet
CNRS-CEA-UVSQ, Laboratoire des Sciences du Climat et de l'Environnement, UMR8212, 91191 Gif-sur-Yvette, France

S. Blain
Sorbonne Universités, UPMC Univ Paris 06, UMR7621, Laboratoire d'Océanographie Microbienne, Observatoire Océanologique, 66650 Banyuls/mer, France
CNRS, UMR7621, Laboratoire d'Océanographie Microbienne, Observatoire Océanologique, 66650 Banyuls/mer, France

J. Capparos
Sorbonne Universités, UPMC Univ Paris 06, UMR7621, Laboratoire d'Océanographie Microbienne, Observatoire Océanologique, 66650 Banyuls/mer, France
CNRS, UMR7621, Laboratoire d'Océanographie Microbienne, Observatoire Océanologique, 66650 Banyuls/mer, France

A. Guéneuguès
Sorbonne Universités, UPMC Univ Paris 06, UMR7621, Laboratoire d'Océanographie Microbienne, Observatoire Océanologique, 66650 Banyuls/mer, France
CNRS, UMR7621, Laboratoire d'Océanographie Microbienne, Observatoire Océanologique, 66650 Banyuls/mer, France

I. Obernosterer
Sorbonne Universités, UPMC Univ Paris 06, UMR7621, Laboratoire d'Océanographie Microbienne, Observatoire Océanologique, 66650 Banyuls/mer, France
CNRS, UMR7621, Laboratoire d'Océanographie Microbienne, Observatoire Océanologique, 66650 Banyuls/mer, France

L. Oriol
Sorbonne Universités, UPMC Univ Paris 06, UMR7621, Laboratoire d'Océanographie Microbienne, Observatoire Océanologique, 66650 Banyuls/mer, France
CNRS, UMR7621, Laboratoire d'Océanographie Microbienne, Observatoire Océanologique, 66650 Banyuls/mer, France

P. J. Rayner
School of Earth Sciences, University of Melbourne, Melbourne, Australia

A. Stavert
CSIRO Oceans and Atmosphere Flagship, Melbourne, Australia

M. Scholze
Department of Physical Geography and Ecosystem Science, Lund University, Sweden

A. Ahlström
Department of Physical Geography and Ecosystem Science, Lund University, Sweden

C. E. Allison
CSIRO Oceans and Atmosphere Flagship, Melbourne, Australia

R. M. Law
CSIRO Oceans and Atmosphere Flagship, Melbourne, Australia

C. D. Arp
Water and Environmental Research Center, University of Alaska Fairbanks, Fairbanks, AK 99775, USA

M. S. Whitman
Arctic Field Office, Bureau of Land Management, Fairbanks, AK 99709, USA

B. M. Jones
Alaska Science Center, U.S. Geological Survey, Anchorage, AK 99508, USA

G. Grosse
Alfred Wegener Institute, Helmholtz Centre for Polar and Marine Research, Potsdam, Germany

B. V. Gaglioti
Water and Environmental Research Center, University of Alaska Fairbanks, Fairbanks, AK 99775, USA
Alaska Science Center, U.S. Geological Survey, Anchorage, AK 99508, USA

K. C. Heim
School of Fisheries and Ocean Sciences, University of Alaska Fairbanks, Fairbanks, AK 99775, USA

F. Peterse
ETH Zürich, Geological Institute, Sonneggstrasse 5, 8092 Zürich, Switzerland

C. M. Moy
University of Otago, Geology Department, Dunedin, New Zealand

T. I. Eglinton
ETH Zürich, Geological Institute, Sonneggstrasse 5, 8092 Zürich, Switzerland

A. Alexandre
Centre Européen de Recherche et d'Enseignement des Géosciences de l'Environnement (UMR 7330), CNRS, Aix-Marseille Université, Europôle méditerranéen de l'Arbois BP 80, 13545 Aix en Provence CEDEX 04, France

I. Basile-Doelsch
Centre Européen de Recherche et d'Enseignement des Géosciences de l'Environnement (UMR 7330), CNRS, Aix-Marseille Université, Europôle méditerranéen de l'Arbois BP 80, 13545 Aix en Provence CEDEX 04, France

T. Delhaye
Plateforme NanoSIMS, OSUR, Université de Rennes 1, Campus de Beaulieu, 35042 Rennes CEDEX, France

D. Borshneck
Centre Européen de Recherche et d'Enseignement des Géosciences de l'Environnement (UMR 7330), CNRS, Aix-Marseille Université, Europôle méditerranéen de l'Arbois BP 80, 13545 Aix en Provence CEDEX 04, France

J. C. Mazur
Centre Européen de Recherche et d'Enseignement des Géosciences de l'Environnement (UMR 7330), CNRS, Aix-Marseille Université, Europôle méditerranéen de l'Arbois BP 80, 13545 Aix en Provence CEDEX 04, France

P. Reyerson
Department of Geography, University of Wisconsin-Madison, 550 North Park Street, Madison, WI 53706, USA

G. M. Santos
Department of Earth System Science, University of California, Irvine, B321 Croul Hall, Irvine, CA 92697-3100, USA